# Remagnetization and Chemical Alteration of Sedimentary Rocks

## Geological Society books refereeing procedures

The Society makes every effort to ensure that the scientific and production quality of its books matches that of its journals. Since 1997, all book proposals have been refereed by specialist reviewers as well as by the Society's Books Editorial Committee. If the referees identify weaknesses in the proposal, these must be addressed before the proposal is accepted.

Once the book is accepted, the Society Book Editors ensure that the volume editors follow strict guidelines on refereeing and quality control. We insist that individual papers can only be accepted after satisfactory review by two independent referees. The questions on the review forms are similar to those for *Journal of the Geological Society*. The referees' forms and comments must be available to the Society's Book Editors on request.

Although many of the books result from meetings, the editors are expected to commission papers that were not presented at the meeting to ensure that the book provides a balanced coverage of the subject. Being accepted for presentation at the meeting does not guarantee inclusion in the book.

More information about submitting a proposal and producing a book for the Society can be found on its website: www.geolsoc.org.uk.

GEOLOGICAL SOCIETY SPECIAL PUBLICATION NO. 371

# Remagnetization and Chemical Alteration of Sedimentary Rocks

EDITED BY

R. D. ELMORE
University of Oklahoma, USA

A. R. MUXWORTHY
Imperial College, UK

M. M. ALDANA
Simon Bolivar University, Venezuela

and

M. MENA
University of Buenos Aires, Argentina

2012
Published by
The Geological Society
London

# THE GEOLOGICAL SOCIETY

The Geological Society of London (GSL) was founded in 1807. It is the oldest national geological society in the world and the largest in Europe. It was incorporated under Royal Charter in 1825 and is Registered Charity 210161.

The Society is the UK national learned and professional society for geology with a worldwide Fellowship (FGS) of over 10 000. The Society has the power to confer Chartered status on suitably qualified Fellows, and about 2000 of the Fellowship carry the title (CGeol). Chartered Geologists may also obtain the equivalent European title, European Geologist (EurGeol). One fifth of the Society's fellowship resides outside the UK. To find out more about the Society, log on to www.geolsoc.org.uk.

**The Geological Society Publishing House** (Bath, UK) produces the Society's international journals and books, and acts as European distributor for selected publications of the American Association of Petroleum Geologists (AAPG), the Indonesian Petroleum Association (IPA), the Geological Society of America (GSA), the Society for Sedimentary Geology (SEPM) and the Geologists' Association (GA). Joint marketing agreements ensure that GSL Fellows may purchase these societies' publications at a discount. The Society's online bookshop (accessible from www.geolsoc. org.uk) offers secure book purchasing with your credit or debit card.

To find out about joining the Society and benefiting from substantial discounts on publications of GSL and other societies worldwide, consult www.geolsoc.org.uk, or contact the Fellowship Department at: The Geological Society, Burlington House, Piccadilly, London W1J 0BG: Tel. +44 (0)20 7434 9944; Fax +44 (0)20 7439 8975; E-mail: enquiries@geolsoc.org.uk.

For information about the Society's meetings, consult *Events* on www.geolsoc.org.uk. To find out more about the Society's Corporate Affiliates Scheme, write to enquiries@geolsoc.org.uk.


Published by The Geological Society from:
The Geological Society Publishing House, Unit 7, Brassmill Enterprise Centre, Brassmill Lane, Bath BA1 3JN, UK

The Lyell Collection: www.lyellcollection.org
Online bookshop: www.geolsoc.org.uk/bookshop
Orders: Tel. +44 (0)1225 445046, Fax +44 (0)1225 442836

The publishers make no representation, express or implied, with regard to the accuracy of the information contained in this book and cannot accept any legal responsibility for any errors or omissions that may be made.





**British Library Cataloguing in Publication Data**

A catalogue record for this book is available from the British Library.
ISBN 978-1-86239-351-6
ISSN 0305-8719

**Distributors**
For details of international agents and distributors see:
www.geolsoc.org.uk/agentsdistributors

Typeset by Techset Composition Ltd, Salisbury, UK
Printed by MPG Books Ltd, Bodmin, UK

# Contents

# Remagnetization and chemical alteration of sedimentary rocks

R. D. ELMORE[1]*, A. R. MUXWORTHY[2] & M. ALDANA[3]

[1]*School of Geology and Geophysics, University of Oklahoma, Norman, OK 73019, USA*

[2]*Department of Earth Science and Engineering, Imperial College London, London, UK*

[3]*Earth Science Department, Simón Bolívar University, Baruta, Venezuela*

**Corresponding author (e-mail: delmore@ou.edu)*

**Abstract:** Chemical remagnetization is a very common phenomenon in sedimentary rocks and developing a greater understanding of the mechanisms has several benefits. Acquisition of a secondary magnetization is usually tangible evidence of a diagenetic event that can be dated by isolation of the chemical remanent magnetization and comparison of the pole position to the apparent polar wander path. This can be important because diagenetic investigations are frequently limited by the difficulty in constraining the time frames in which most past events have occurred. Remagnetization can commonly obscure a primary magnetization; developing a better understanding of remagnetization could improve our ability to uncover primary magnetizations. Many chemical remagnetization mechanisms have been proposed, including those associated with chemical alteration by a number of different fluids (orogenic, basinal and hydrocarbons), burial diagenetic processes (clay diagenesis and maturation of organic matter) or other processes. This paper summarizes our current knowledge of these chemical remagnetization mechanisms, with a focus on examples where there is a connection with chemical alteration.

Secondary magnetizations are common in sedimentary rocks and are widespread (e.g. McCabe & Elmore 1989). Since originally 'discovered' approximately 50 years ago, many palaeomagnetists have considered them a problem because they can obscure or remove primary magnetizations, which are commonly considered to be of greater interest. It is therefore important to better understand these secondary magnetizations in order to isolate primary magnetizations. Remagnetization or acquisition of a secondary magnetization can also be tangible evidence of a diagenetic event; remagnetizations can therefore be used to date diagenetic events. Diagenetic investigations can be limited by the difficulty in constraining the timeframes in which the events occurred. This dating approach is based on isolation of the chemical remanent magnetization (CRM) carried by diagenetic magnetic minerals and comparison of the pole position for the CRM to the appropriate apparent polar wander path (APWP). CRMs have also been used to constrain the timing of deformation (e.g. Stamatakos *et al.* 1996; Weil & Van der Voo 2002*a*). Magnetic susceptibility data and remagnetizations can also have applications in hydrocarbon exploration.

This Special Publication contains a selection of papers that focus on chemical remagnetization and magnetic changes associated with chemical alteration by hydrocarbons. In addition to several case studies, the book includes a paper on the history of remagnetization studies (Van der Voo & Torsvik 2012) as well as a number of review articles on various aspects of remagnetization. This introductory paper provides a general overview and reviews our current knowledge of chemical remagnetization mechanisms. We will not try to cover all papers on remagnetization, but will focus on examples where there is a connection with chemical alteration.

## Remagnetization mechanisms

Our understanding of remagnetization processes has improved significantly since the first hints of remagnetization in the 1950s and Creer's 'remagnetization hypothesis' (Creer 1968). Dating of diagenetic events using palaeomagnetism has also met with varying degrees of success (see below). Although we have significantly improved our understanding of chemical and other remagnetization mechanisms, the origins of some remagnetizations remain enigmatic.

Many secondary magnetizations are interpreted to be chemical or diagenetic in origin, although other remagnetization mechanisms (e.g. thermoviscous) are also important (e.g. Kent 1985; Hudson *et al.* 1989). Evidence cited in support of the chemical origin includes authigenic magnetic phases in the rocks (e.g. Elmore *et al.* 1985; Suk *et al.* 1990) and thermal maturities which are too low for a thermoviscous remagnetization based on blocking-temperature–relaxation-time relationships (e.g. Pullaiah *et al.* 1975; Jackson 1990; Dunlop *et al.*

*From*: ELMORE, R. D., MUXWORTHY, A. R., ALDANA, M. M. & MENA, M. (eds) 2012. *Remagnetization and Chemical Alteration of Sedimentary Rocks*. Geological Society, London, Special Publications, **371**, 1–21.
First published online September 18, 2012, http://dx.doi.org/10.1144/SP371.15

2                                            R. D. ELMORE *ET AL.*

2000). Rock magnetic data has also been cited as supporting a diagenetic origin for some magnetite (e.g. Jackson 1990; Jackson & Sun 1992). See Jackson & Swanson-Hysell (2012) for an update on this issue.

As described in the following sections, many chemical remagnetization mechanisms have been proposed. All involve alteration by fluids with a number of different driving mechanisms or sources for the fluids suggested, including tectonic processes, diagenetic reactions, and heat from igneous bodies. In this paper the chemical remagnetization mechanisms are divided into two general groups: alteration triggered by externally derived fluids and alteration associated with burial diagenetic processes. Tests of these remagnetization mechanisms are important in order to demonstrate a connection between a diagenetic event and a specific remagnetization, and thereby date a diagenetic event. For example, determining that a remagnetization is caused by fluids requires testing for a connection between the alteration caused by a particular fluid and a CRM. A basic approach is to conduct a presence/absence test where geochemical and petrographic data are compared with the distribution of a CRM (e.g. Elmore *et al.* 1985, 1994; Symons *et al.* 2005). A more powerful approach is to conduct a contact test around veins (Figs 1 & 2; Cochran & Elmore 1987; Elmore *et al.* 1993*a*) or conduits for fluid flow (Fig. 3; Elmore *et al.* 1998,

2010; Costanzo-Álvarez *et al.* 2000*a*; Evans *et al.* 2012). In this test, petrographic, geochemical, and palaeomagnetic data are collected and compared from in and around the fluid conduit (Fig. 2). If the geochemical evidence for alteration coincides with the distribution of a CRM, a strong case for a connection between the fluid which caused the alteration and the CRM is possible.

## Externally derived fluids

### Orogenic fluids

The action of orogenic fluids is a popular mechanism invoked for many CRMs (McCabe & Elmore 1989). For example, many Late Palaeozoic CRMs in North America are inferred to be related to alteration caused by fluids which migrated from the Appalachian and Ouachita fold-thrust belts (e.g. Oliver 1986, 1992). These fluids probably migrated along aquifers (e.g. Bethke & Marshak 1990) as a result of compression (e.g. squeegee model of Oliver 1992) or by gravitational flow of meteoric fluids from mountains (e.g. Bethke & Marshak 1990; Garven 1995).

Evidence supporting the hypothesized connection between orogenic fluids and remagnetization is a temporal association between the timing of many CRMs and orogenies as well as a spatial association between the CRMs and mountain belts

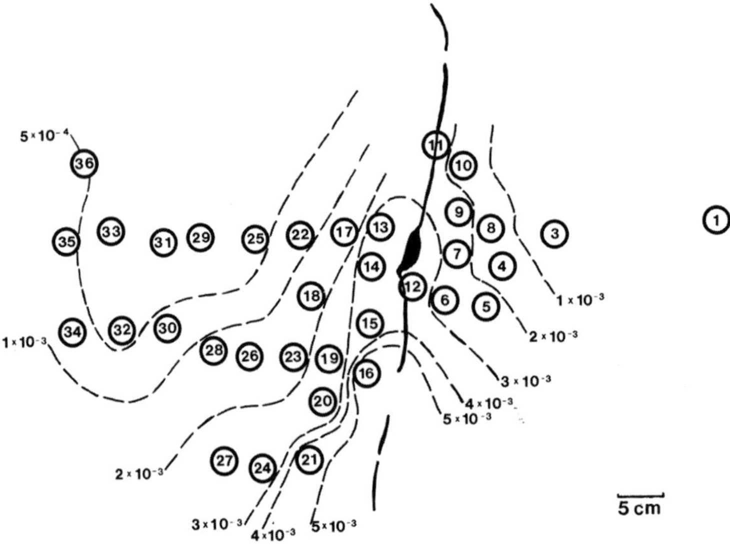

**Fig. 1.** Natural remanent magnetization (Am$^{-1}$) intensities decrease away from calcite-filled fractures in the Ordovician Kindblade Formation in southern Oklahoma. The highest intensities correspond to maximum amount of liesegang banding which is caused by haematite. Reddish specimens near the veins contain a post-tilting Permian CRM in haematite, whereas specimens away from veins contain an unstable and weak magnetization. These results suggest that the CRM was caused by the fluids which migrated away from the veins (after Cochran & Elmore 1987).

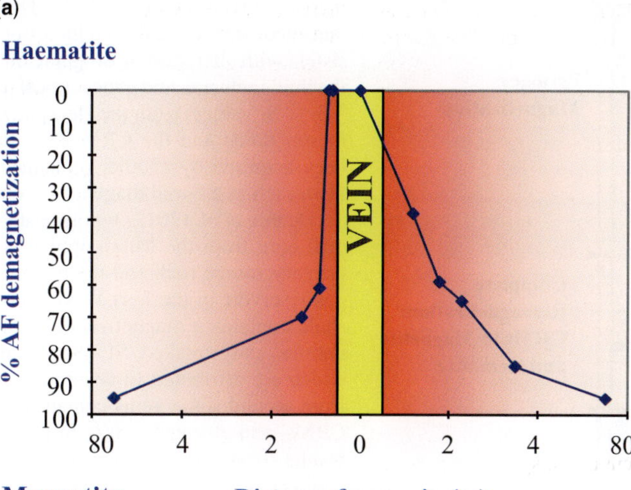


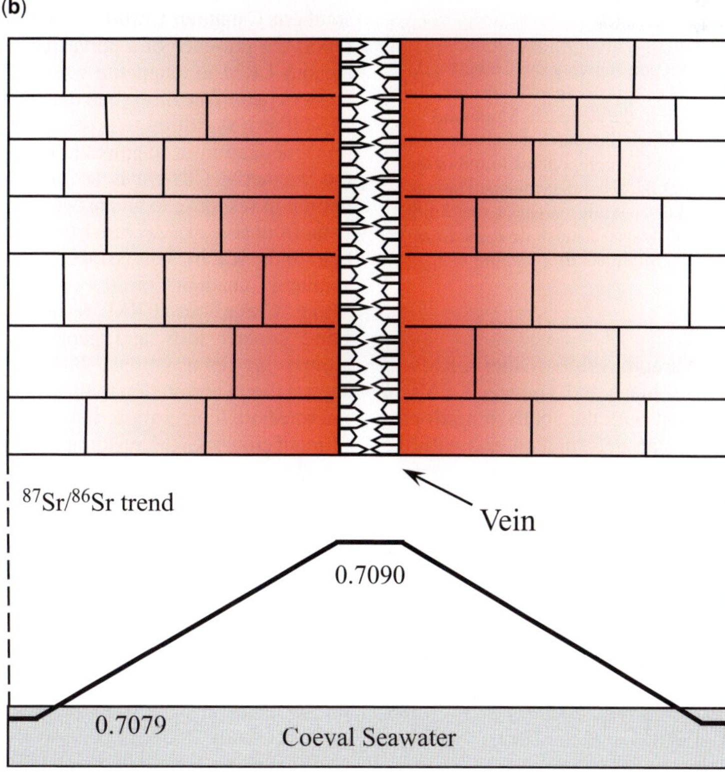


**Fig. 2.** (a) Vein in the Ordovician Viola Limestone (southern Oklahoma) showing distance from the vein v. percent alternating field (AF) decay. The rock near the vein contains a Permian CRM in haematite whereas the rock away from the vein contains a Pennsylvanian CRM in magnetite (Elmore *et al.* 1993*a*). The increase in AF decay away from the vein shows that the haematite CRM decreases with distance from the vein, whereas the magnetite CRM increases with distance from the vein. (b) Schematic diagram illustrating remagnetization and geochemical halos around a vein. The coincidence of the two halos suggests that they are related and that the CRM in haematite dates the migraton of fluid through the veins and the associated geochemical alteration.

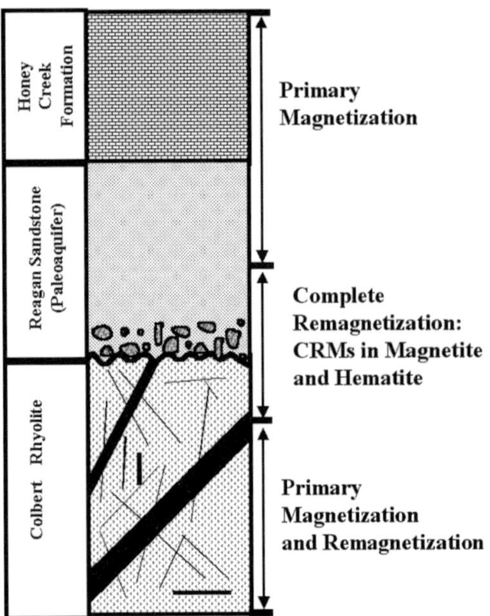

**Fig. 3.** Stratigraphic section showing the Colbert Rhyolite (with intruded dykes), Reagan Sandstone, and Honey Creek Limestone in the Arbuckle Mountains (southern Oklahoma), which illustrates the distribution of the primary and secondary components found in the units. The occurrence of the secondary magnetizations within and below the palaeoaquifer suggests that fluids that moved through the Reagan Sandstone caused the remagnetization events. Modified after Elmore *et al.* (1998).

(e.g. McCabe & Elmore 1989). Older CRMs in the southern Appalachians and younger CRMs in the north is consistent with the south to north progression of Alleghanian deformation and with the hypothesis that orogenic fluids caused remagnetization (e.g. Miller & Kent 1988). In the central Appalachians, post-folding remagnetizations near the hinterland, syntilting remagnetizations in the central part of the belt and pre-folding remagnetizations near the foreland were interpreted to be consistent with brine migration (Stamatakos *et al.* 1996). Migration of orogenic fluids has also been proposed as an explanation for some regional trends in magnetite authigenesis (e.g. Jackson *et al.* 1988; McCabe *et al.* 1989; Lu *et al.* 1991; Saffer & McCabe 1992).

Syntilting CRMs have been related to alteration caused by fluids activated during deformation (e.g. McCabe *et al.* 1983). In the Valley and Ridge province in West Virginia, the Devonian Helderberg Group (an aquitard) and the overlying Oriskany Formation (a palaeoaquifer) both contain similar syntilting Late Palaeozoic CRMs residing in

magnetite (Elmore *et al.* 2001). The Oriskany contains geochemical and fluid inclusion evidence consistent with alteration by orogenic fluids. In contrast, the fluid inclusion and geochemical data (e.g. coeval $^{87}$Sr/$^{86}$Sr values) from the Helderberg indicate only *in situ* fluids and the CRM is interpreted to have been acquired by another remagnetization mechanism such as a burial diagenetic process.

Dennie *et al.* (2012) report results from oriented drill core from the Mississippian Barnett Shale, a primary source rock and the major unconventional gas reservoir in the Fort Worth Basin, Texas. The shale contains a magnetization with shallow inclinations and streaked SE−S-directed declinations which are attributed to several CRMs. Specimens from around some veins contain Permian−Triassic CRMs, and elevated $^{87}$Sr/$^{86}$Sr ratios and S isotope results from vein minerals suggest the CRMs could be related to fluids sourced from the Ouachita thrust front that forms the eastern margin of the basin.

In a regional study of Palaeozoic rocks in the southern Canadian Cordillera, Enkin *et al.* (2000) report the presence of a pervasive pre-folding Cretaceous CRM in magnetite which was older in the western part of the thrust belt than in the eastern part, as well as a syntilting to post-tilting thermoviscous remagnetization in magnetite. They suggested that the magnetite CRM was acquired prior to deformation in response to an eastward-migrating diagenetic front (Fig. 4). Zechmeister *et al.* (2012) report that folds within Mississippian carbonates in the southern Canadian Cordillera contain a similar pretilting Cretaceous CRM residing in magnetite. Fluid contact tests and geochemical data from around pre-deformational bedding-parallel veins show a direct correlation to the magnetite CRM. An intermediate-temperature late syntilting to posttilting Tertiary CRM residing in pyrrhotite is also present. Based on the presence of thermal sulphate reduction (TSR) byproducts, the pyrrhotite CRM is interpreted to be the result of late-stage TSR of hydrocarbons that were exposed to warm fluids which migrated along faults and fractures. In contrast to examples where CRMs occur around fluid conduits, Evans *et al.* (2012) report that the Devonian Alamo Breccia (Nevada), a likely palaeoaquifer, was not a conduit for focused flow of remagnetizing fluids. Evans *et al.* (2012) suggest that externally derived fluids moved pervasively through the rocks, causing acquisition of a CRM in the Late Palaeozoic Era (Sonoma Orogeny?).

Many studies have also interpreted that CRMs were caused by orogenic fluids in areas other than North America. The role of orogenic fluids was evaluated for Late Palaeozoic CRMs in Europe (e.g. Weil & Van der Voo 2002*b*; Zegers *et al.* 2003). Font *et al.* (2012) summarize remagnetization in

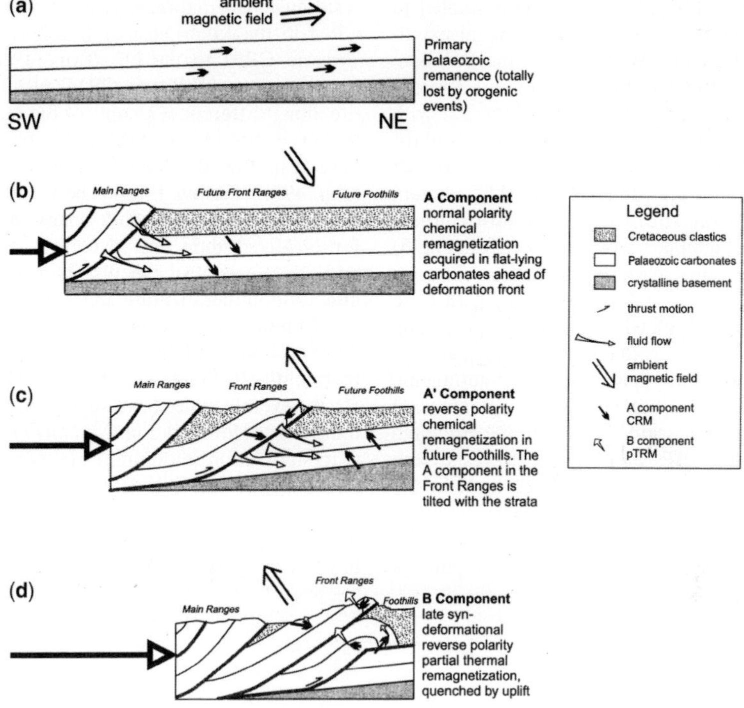



**Fig. 4.** Simplified diagram modelling magnetization acquisition in the southern Canadian Cordillera (after Enkin *et al.* 2000). (**a**) Shallow-inclination Palaeozoic remanence acquired in carbonates. (**b**) A component: Ahead of the eastward migrating deformation front, carbonates in the future Front Ranges acquire total chemical remagnetization with steep downwards direction. (**c**) Before the Front Ranges become entrained into the deformation front, the A component becomes fixed and rotates with the beds. A′ component: After a polarity reversal, a similar event remagnetizes the carbonates which form the Foothills. (**d**) Structures underneath the frontal thrusts of the Front Ranges uplift the Front Ranges and Foothills and, during one reverse-polarity chron, quench a partial thermal remanence acquired during burial. CRM: chemical remanent magnetization; pTRM: partial thermo-remanent magnetization.

South America, and report several remagnetization events related to major tectonic episodes.

Several palaeomagnetic-geochemical studies provide evidence that warm, saline fluids caused remagnetization in and around conduits for flow (Elmore *et al.* 1993*a*, 1998). The timing of these CRMs does not necessarily indicate that they are orogenic fluids. For example, the results of a study of the Viola Limestone in southern Oklahoma suggest that basinal fluids locally caused a Permian CRM in haematite within altered rock around veins which were the conduits for externally derived fluids (Fig. 2; Elmore *et al.* 1993*a*). The rock away from the veins was not altered by such fluids and contains a Pennsylvanian CRM in magnetite that was interpreted to be caused by a burial diagenetic mechanism. In another study in southern Oklahoma, the Upper Cambrian Reagan Sandstone (the basal palaeoaquifer in the Palaeozoic section) contains two Late Palaeozoic CRMs that can be related to fluid migration (Fig. 3; Elmore

*et al.* 1998); rocks above and below contain primary or early magnetizations, however. Other studies in the Arbuckle Mountains in Oklahoma document that basinal fluids migrated laterally through palaeoaquifers and vertically through faults/ fractures causing localized remagnetization and alteration in and around the fluid conduits during a time interval of *c.* 60 Ma in the Late Palaeozoic (Elmore 2001). The fact that some CRMs reside in haematite whereas others reside in magnetite indicates that remagnetization was caused by fluids with different chemistries.

Numerous studies have investigated CRMs in fault zones, relating them to igneous activity or to extensional tectonics (e.g. Torsvik *et al.* 1992; Preeden *et al.* 2009). A study of the Moine Thrust Zone (MTZ) in the Caledonides of Scotland suggests that four focused fluid-flow events occurred along the MTZ between the Devonian and Early Cenozoic (Blumstein *et al.* 2005). The localized CRMs coincide with post-Caledonian events such as the migration of hydrothermal fluids related to Devonian igneous activity; regional crustal extension of Scotland and NW Europe in the Permian; Proto-Atlantic rifting in the Triassic; and Tertiary intrusive activity. Red fault-related breccias with abundant authigenic haematite in the Cambro-Ordovician Durness Group in NW Scotland that are found in close proximity to the MTZ contain two CRMs (Elmore *et al.* 2010). The host Durness (grey dolomite) contains a Devonian CRM (Fig. 5) in magnetite. North–south veins contain a Triassic CRM whereas east–west veins contain a Jurassic CRM (Fig. 5), both of which reside in haematite. The two CRMs are interpreted as reflecting two separate fluid-flow events that precipitated authigenic haematite and caused brecciation during extension of the NE Atlantic margin. A study of the Highland Boundary Fault (HBF) in Scotland suggests there were multiple flow events along the fault in the Late Palaeozoic (Elmore *et al.* 2002). The Devonian Old Red Sandstone in the vicinity of the Great Glen Fault in Scotland contains two different components

residing in haematite: a post-tilting Carboniferous CRM in the Loch Ness area and a Cretaceous or Triassic CRM to the NE (Elmore *et al.* 2006a). The presence of different CRMs residing in haematite along different segments of the faults is similar to that reported from other studies of faults in Scotland (e.g. Van der Voo & Scotese 1981; Torsvik *et al.* 1983; Tarling 1985). These and other studies provide evidence that fault zones can be conduits for localized fluid-flow events at different times and thereby control the distribution of diagenetic alteration, mineralization and remagnetization.

A number of studies present evidence for connections between CRMs and basinal-type or hydrothermal fluids. For example, in the Western Canada Basin several studies have suggested that CRMs are related to dolomitization (Symons *et al.* 1999) and recrystallization in dolomites (Cioppa *et al.* 2000). Results from the NE Williston Basin (Szabo & Cioppa 2012) indicate the presence of a Jurassic remagnetization in magnetite that is interpreted to have been caused by basement fluids. These fluids

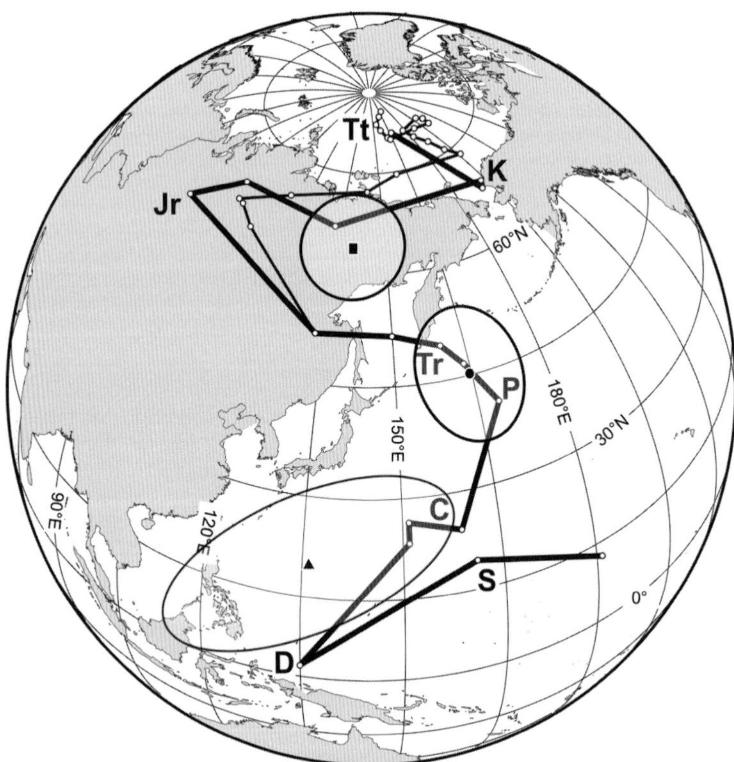


**Fig. 5.** Poles (with 95% error ellipses, dp and dm) from the red breccias and host dolomite in Scotland on the apparent polar wander paths from Van der Voo (1993) for the Phanerozoic (heavy line with circles representing the median age poles) and Besse & Courtillot (2002) for the last 200 Ma (thinner line with circles representing the 10 Ma mean poles). CRM1: triangle, Durness Group Dolomite; CRM2: circle, north–south veins; and CRM3: square, east–west veins (after Elmore *et al.* 2010).

moved along faults and fractures that formed as a result of the Hartney impact/volcanic structure and/or tectonic activity in the Superior Boundary Zone. Harlan *et al.* (1996) report a Late Cretaceous CRM in mafic dykes in Montana that is attributed to alteration by hydrothermal fluids.

## Mineralizing fluids

Many studies have used palaeomagnetism to date or constrain the timing of Mississippi Valley Type (MVT) deposits (e.g. Symons *et al.* 1996; Leach *et al.* 2001; Zegers *et al.* 2003). This approach is based on the hypothesis that the mineralizing fluids cause precipitation of magnetic phases (magnetite and pyrrhotite); see Symons *et al.* (1996) for a summary of the palaeomagnetic dating approach and for the results from a number of MVT deposits. Several studies have shown that the host rocks have distinct and older magnetizations than the mineralized rocks, which suggest that the ore magnetization is related to the mineralizing fluids (e.g. Lewchuk & Symons 1995; Symons *et al.* 2005). Many of the MVT deposits are interpreted to have formed from orogenic fluids that were triggered as a result of tectonic activity (Leach *et al.* 2001). The palaeomagnetic data from some MVT deposits have a streaked or oval distribution rather than a circular distribution (Lewchuk & Symons 1995). Although there are several possible explanations (e.g. overlapping components), one hypothesis is that the streaked pattern reflects apparent pole wander during the acquisition of the CRM (Lewchuk & Symons 1995).

## Weathering fluids

A number of studies present evidence that weathering fluids can cause remagnetization. Examples include Mesozoic limestones in Germany where goethite has been shown to carry a secondary remanence (Heller 1979; Johnson *et al.* 1984). Loucks & Elmore (1986) present evidence that surficial dedolomitization can cause a remagnetization in goethite. Meteoric/weathering fluids can cause a localized remagnetization in haematite around karst features (Nick & Elmore 1990).

Remagnetization by weathering is currently experiencing a surge in interest related to Late Carboniferous–Triassic remagnetizations, common in alteration zones below palaeoweathering surfaces in crystalline basement rocks in Europe (e.g. Edel & Schneider 1995). Many of these remagnetizations reside in authigenic haematite and are interpreted to be caused by weathering fluids (e.g. Ricordel *et al.* 2007; Franke *et al.* 2010; Fàbrega *et al.* 2012). A study of dolomite veins in the Precambrian–Early Palaeozoic Dalradian schist in Scotland reported a Carboniferous–Triassic CRM which was interpreted as being related to deep oxidizing weathering (Parnell *et al.* 2000). Late Palaeozoic remagnetizations residing in haematite are also reported from crystalline rocks in North America and have been related to weathering fluids (Hamilton *et al.* 2012) or to migration of brines along porous zones at the Precambrian–Carboniferous nonconformity (Geissman & Harlan 2002).

## Hydrocarbons

Numerous studies provide evidence for, or propose a relationship between, hydrocarbons and authigenic magnetite and/or pyrrhotite (e.g. Elmore *et al.* 1987, 1993*b*; McCabe *et al.* 1987; Reynolds *et al.* 1993; Cioppa & Symons 2000; Aldana *et al.* 2003; Costanzo-Álvarez *et al.* 2006). In some cases the magnetite or pyrrhotite carries a CRM (e.g. Benthien & Elmore 1987; Elmore & Crawford 1990; Elmore & Leach 1990; Gose & Kyle 1993; Katz *et al.* 1996; Lewchuk *et al.* 1998). Some studies have presented evidence that secondary magnetite could be responsible for magnetic enhancement in soil polluted by hydrocarbons (Rijal *et al.* 2010). Pyrrhotite can also form during thermochemical sulphate reduction of hydrocarbons (Pierce *et al.* 1998); this has been proposed as a mechanism for acquisition for a CRM in pyrrhotite (Manning & Elmore 2012; Zechmeister *et al.* 2012). In some hydrocarbon-impregnated units, magnetite is present but it does not carry a stable remanence; in some red beds, hydrocarbons can cause a net decrease in magnetization by dissolving haematite (e.g. Kilgore & Elmore 1989; Elmore & Leach 1990). Several studies also describe the chemical conditions in which magnetite can form in association with hydrocarbons (e.g. Machel 1995).

Based on this connection between magnetic anomalies and hydrocarbon micro-seepage, it has been proposed that petroleum reservoirs can be characterized based on the analysis of near-surface and secondary magnetic contrasts in high-resolution aeromagnetic surveys over oil fields. Donovan *et al.* (1979) identified pronounced noticeable aeromagnetic anomalies in the Cement oil field (Oklahoma). They associated these anomalies with the presence of authigenic magnetite produced by the chemical alteration of original Fe-oxides in a reducing environment produced by the underlying oil reservoir. Although aeromagnetic anomalies could also be due to other factors related to natural or anthropogenic processes (Gay 1992), a hydrocarbon-related origin is possible in geological settings dominated by underlying oil reservoirs. Donovan *et al.* (1984), Foote (1996) and Saunders *et al.* (1991) have argued that it is possible to use aeromagnetic studies to determine the presence of hydrocarbon reservoirs.

Studies utilizing bulk magnetic properties (e.g. magnetic susceptibility) measured in soils, sediments and drill cuttings can provide a better understanding and assessment of the origin of the magnetic anomalies (Aldana *et al.* 1999, 2003; Costanzo-Álvarez *et al.* 2000*b*, 2006; Díaz *et al.* 2000; González *et al.* 2002). It is crucial in these types of studies to analyse for and rule out contamination via anthropogenic factors. Some studies, based on the presence of spherical aggregates of magnetic minerals which points to an *in situ* formation, show that it is possible to discriminate between magnetic susceptibility anomalies related to hydrocarbon micro-seepage and those caused by lithological contrasts, that is sedimentation processes (Costanzo-Álvarez *et al.* 2000*b*).

Using magnetic properties of drill cuttings taken at shallow depth levels, Guzmán *et al.* (2011) gave a preliminary characterization of an exploratory area in the Maturin Sub-Basin, Venezuela. They mapped magnetic susceptibility and the S-ratio, as well as the concentration of free radicals in the extracted organic matter measured using Electron Paramagnetic Resonance (EPR) experiments (see Díaz *et al.* 2000), from drill cuttings. These maps show a region of anomalous values, probably associated either with the maximum accumulations of an unexplored deep reservoir and/or with the path of migrated hydrocarbons. They argue that this result could be employed for future exploration and production ventures in the region, in a similar way as seismic attributes are used in the oil industry to monitor the continuity of a particular sedimentary layer and/or features associated with hydrocarbon accumulations. Due to their low cost and non-invasive nature, magnetic methods can be used as an alternative tool to study the near-surface expression of hydrocarbon micro-seepage not only in oil fields but also in prospective areas (González *et al.* 2002).

The rock magnetic characterization of the stratigraphic well (Saltarín 1A) reports early and late diagenetic events that affected the Upper Cretaceous–Pliocene sequence of the distal Llanos foreland basin in Colombia (Costanzo-Álvarez *et al.* 2012). At shallow depths, anomalously high susceptibility values appear to be the result of partial replacement of pyrite framboids by magnetite. At deeper levels the magnetic anomaly observed is related to variation from oxidized palaeosols to alluvial plain sediment which accumulated in reducing conditions. This suggests a pedoclimatic control.

Emmerton *et al.* (2012) found an inverse correlation between the magnetic susceptibility and the extracted organic matter content for samples from the Wessex Basin in SW England. These results indicate a complex relationship between existing magnetic minerals within the sandstones and the alteration of these magnetic minerals due to the multiplex biological activity and biodegradation of the oil.

Magnetic properties from drill cuttings were compared with petrophysical properties from an oil well in the Golfo San Jorge Basin, Argentina (Mena & Walther 2012). Positive correlations were found between bulk magnetic susceptibility and relative hydrocarbon content, and between magnetic properties and neutron log porosity. A negative correlation exists between concentration indices for some magnetic species (magnetite and pyrrhotite) and resistivity. Pyrrhotite could be directly related to the presence or migration of hydrocarbons in porous units. The qualitative correlations between magnetic data and key petrophysical parameters suggest the potential utility of these techniques for subsurface exploration.

## Burial remagnetization mechanisms

Although the migration of externally derived fluids is a likely agent for many remagnetizations, widespread CRMs that occur in rocks that have not been altered by such fluids must be explained by other remagnetization mechanisms. Several burial diagenetic processes, such as clay diagenesis and maturation of organic matter, have received considerable attention as agents of remagnetization. These mechanisms can be thought of as 'cooking in their own juices'.

### Smectite to illite

A number of studies have proposed that the transformation from Fe-rich smectite to illite can release iron that can result in magnetite authigenesis (e.g. Lu *et al.* 1991; Hirt *et al.* 1993; Katz *et al.* 1998, 2000; Gill *et al.* 2002; Woods *et al.* 2002; Zegers *et al.* 2003; Blumstein *et al.* 2004; Moreau *et al.* 2005; Tohver *et al.* 2008). Many of these studies incorporated presence/absence tests (e.g. Katz *et al.* 2000; Gill *et al.* 2002; Woods *et al.* 2002). Weil & Van der Voo (2002*b*) noted the presence of microscale Fe oxides in a matrix of Fe-rich smectite and aluminous illite.

In a study of Mesozoic carbonates in the Vocontian Trough in SE France, the results support a hypothesized acquisition of a CRM residing in magnetite during burial diagenesis of smectite (Katz *et al.* 1998, 2000). Where smectite has altered to other clay minerals, conglomerate tests indicate that the magnetization is secondary and tilt tests indicate that the limestones contain a pre-tilting magnetization, interpreted as a CRM. Where significant smectite is still present, the CRM is absent/weakly developed (Fig. 6) and where the clays show no evidence for burial alteration, the units are

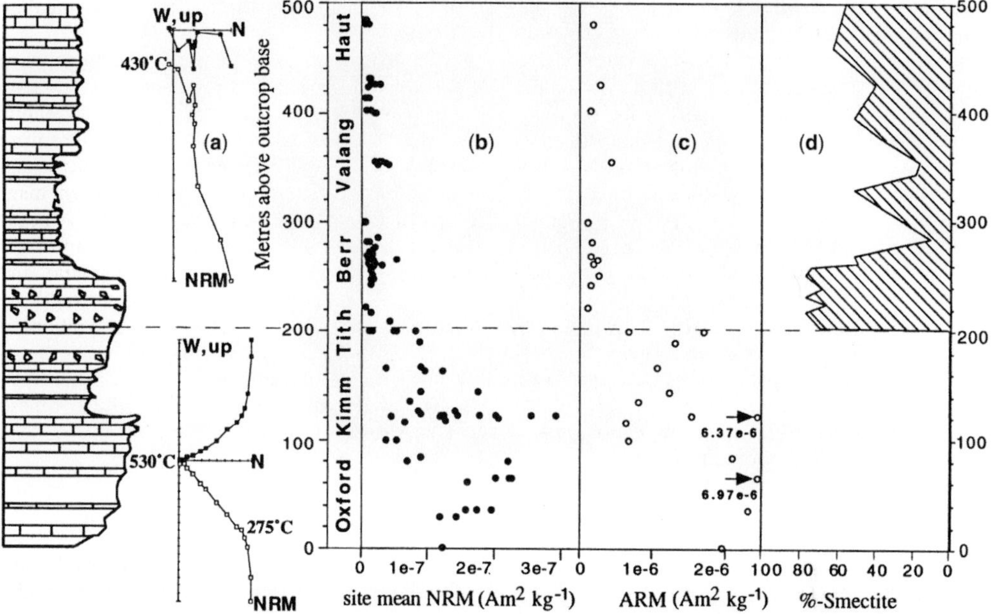



**Fig. 6.** Results from central part of Vocontian trough near Montclus as a function of stratigraphic position (left column; limestone, hachured; marl, blank; breccia, spotted). (**a**) Orthogonal projection diagrams for representative specimens (tilt-corrected) from younger units (top) and from older units (bottom). Most specimens from the older units contain a well-developed CRM. Closed symbols: horizontal projection; open symbols: vertical projection. (**b**) The natural remanent magentization (NRM) as a function of stratigraphic position. Intensities are higher in older units where the CRM is well developed. (**c**) Anhysteretic remanent magnetization (ARM) as a function of stratigraphic position is shown, indicating that older units contain more remanence-carrying magnetite. (**d**) The column on the right shows percent smectite of total clay fraction (after Katz *et al.* 1998).

characterized by a primary magnetization. The rocks with this CRM have $^{87}Sr/^{86}Sr$ values that are similar to coeval seawater; externally derived fluids are therefore not a likely agent of remagnetization.

While geochemical and petrographic studies provide important clues for establishing this relationship, the ultimate test of this hypothesis requires the application of independent dating methods to verify the palaeomagnetic ages. K–Ar dating of illite is one such approach that has worked in some cases (e.g. Elliot *et al.* 2006*a*; Tohver *et al.* 2008; Zwing *et al.* 2009), but in other cases it was not successful because of the presence of detrital illite (Elliot *et al.* 2006*b*).

Results of laboratory experiments provide evidence for a connection between authigenic magnetite and transformation of Fe-rich smectite to illite (Hirt *et al.* 1993). Other experimental studies (Cairanne *et al.* 2004; Moreau *et al.* 2005; Aubourg *et al.* 2008; Aubourg & Pozzi 2010) have presented evidence that low-temperature (95–250 °C) burial heating can cause acquisition of a CRM in magnetite. Aubourg *et al.* (2012) propose that burial remagnetization is continuous with greigite forming at or soon after deposition, magnetite at depths

greater than 2 km and pyrrhotite at depths greater than 6 km.

*Maturation of organic matter*

A number of studies provide evidence for a relationship between remagnetization and maturation of organic matter. For example, in a regional study of the Belden Formation, Colorado, Banerjee *et al.* (1997) reported that the timing of CRM acquisition is different across the basin and it agrees with the modelled time of maturation of organic matter for different localities. In addition, the oxygen isotopic data indicate that the diagenetic magnetite in the Belden formed from water having $\delta^{18}O$ near $0°/_{oo}$ or less, implying a meteoric or connate source rather than a highly evolved orogenic or basinal fluid (Ripperdan *et al.* 1998). The results of these studies are consistent with a study of a single fold in the Belden which indicates that a syntilting CRM which resides in authigenic magnetite that rims pyrite grains (Fig. 7) is not related to syndeformational orogenic fluids (Fruit *et al.* 1995). A study of the Mississippian Desert Limestone and Chainman Shale (Blumstein *et al.* 2004), source rocks in

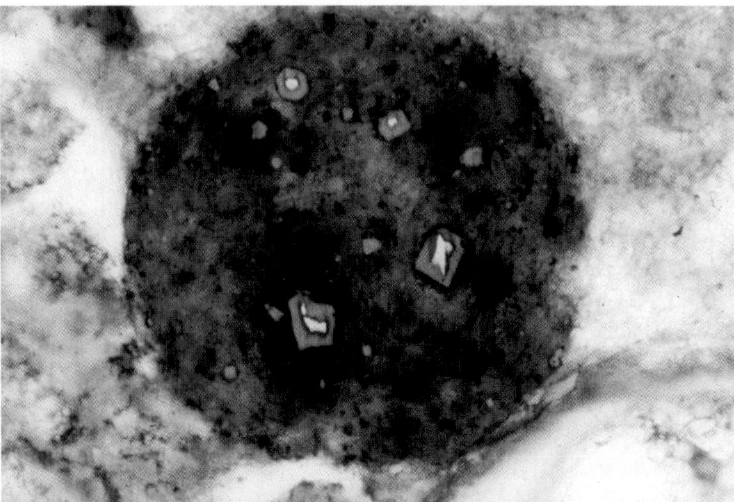

**Fig. 7.** Photomicrograph of a fecal pellet from the Belden Formation in reflected light. The pellets contain pyrite rimmed by magnetite. Pellet is 250 μm in diameter (after Fruit *et al.* 1995).

western Utah, found a Jurassic CRM that was interpreted to have formed as a result of the maturation of organic matter based on the overlap between the timing of CRM acquisition and the timing of oil generation from modelling studies (Fig. 8). The study also found that the timing of CRM acquisition did not overlap the timing of illitization based on kinetic modelling. In another study, the timing of a CRM in organic-rich beds in the Old Red Sandstone in Scotland agrees with independent estimates for the timing of thermal maturation (Plaster-Kirk *et al.* 1995). Other studies have also proposed a relationship between thermal maturity and remagnetization (e.g. Cioppa *et al.* 2002; Font *et al.* 2006).

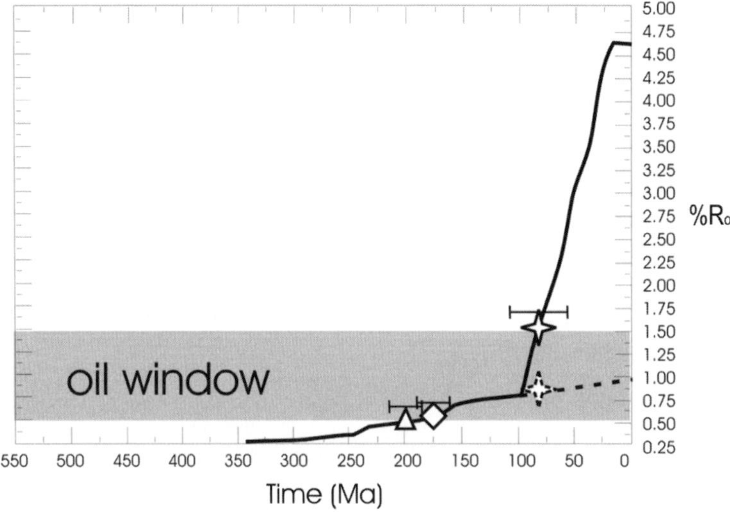


**Fig. 8.** Time versus vitrinite reflectance curve for the Mississippian Desert Limestone (with the oil window) and poles from the Desert Limestone/Chainman shale. The triangle (CR) and diamond (MHR) represent the Late Triassic–Early Jurassic component 1, and the star represents the Cretaceous–Early Tertiary component 2. Error bars for the poles were estimated from the APWP. The dashed line represents the estimated burial curve for the Chainman Shale in western Utah and the dashed star represents component 2 on the estimated curve, suggesting that component 2 could have been acquired within the centre of the oil window (modified after Blumstein *et al.* 2004).

Laboratory simulation experiments have also successfully produced magnetite by dissolution–reprecipitation of pyrite (Brothers *et al.* 1996). These experiments may simulate diagenesis at temperatures below 100 °C, one possible pathway for magnetite authigenesis. A study of organic-rich Jurassic sedimentary rocks adjacent to a Tertiary dyke in Scotland, considered as an analogue for burial heating, suggests that moderate burial depth might be sufficient to cause magneto-chemical changes (Katz *et al.* 1998).

It is important to note that the smectite to illite transformation and maturation of organic matter mechanisms for CRM acquisition are not necessarily mutually exclusive. Organic matter is intimately associated with clay minerals in black shales (Kennedy *et al.* 2002) and it is conceivable that the authigenesis of magnetic phases is facilitated by this interplay.

## Origin of magnetization in red beds

The reader is referred to Van der Voo & Torsvik (2012) for a summary of the history of the red bed debate. Some Mesozoic red beds contain a detrital remanent magnetization (DRM; e.g. Herrero-Bervera & Helsley 1983; Shive *et al.* 1984) and detrital specularite has been demonstrated as carrying a DRM in some Appalachian red beds (e.g. Kent & Opdyke 1985). Other red beds contain an early CRM that probably formed as a result of the breakdown of unstable Fe-bearing detrital grains such as hornblende and biotite (e.g. Butler 1992). Many other studies report Permian CRMs in Early and Middle Palaeozoic red bed units from the Appalachians that were deformed during the Alleghenian Orogeny (e.g. Kent & Opdyke 1985; Miller & Kent 1986; Miller & Kent 1988). It has been popular to relate the CRMs in some red beds to the actions of migrating orogenic fluids (e.g. McCabe & Elmore 1989). Many of these CRMs are found in rocks that do not contain evidence for such fluids (e.g. Cox *et al.* 2005) however, and their origins remain elusive.

## Origin of syntilting CRMs

Syntilting magnetizations are statistical observations that suggest that magnetizations were acquired during folding. Several possibilities exist for the origin of these magnetizations including: growth of new magnetic minerals during folding, physical rotation of magnetic minerals during folding (e.g. Kligfield *et al.* 1983; Hirt *et al.* 1986; van der Pluijm 1987; Kodama 1988), acquisition of a peizoremanent magnetization (PRM) (Xu & Merrill 1992; Borradaile & Jackson 1993; Borradaile 1997), and contamination by overlapping components (e.g.

Hudson *et al.* 1989). In terms of the growth of new magnetic grains during folding, a true syntilting CRM could be produced. Determining the time of growth during folding is difficult because syntilting results observed in incremental tilt tests do not give a unique result. Growth of new minerals is commonly invoked for many syntilting CRMs, although specific chemical mechanisms are rarely demonstrated. In terms of the other three mechanisms, they will produce a false syntilting result or will completely reset the magnetization.

Remanence-carrying Fe oxide grains may rotate during folding, which would alter the original magnetic direction. The rotated direction would not be related to the ambient field during folding. Based on modelling, Kodama (1988) proposed that shear strain during flexural flow folding could cause a prefolding magnetization to be rotated into a syntilting configuration. Other studies also indicate that strain could account for the syntilting characteristics in some clastic units (e.g. Stamatakos & Kodama 1991). On the other hand, Kodama (1988) determined that volume-loss strain (solution) could not account for the observed syntilting character in one carbonate unit.

Several studies have tested for alteration of a CRM by comparing the magnetic properties between limestones with strain indicators. For example, Lewchuk *et al.* (2003) and Elmore *et al.* (2006*b*) tested for a connection between strain and remagnetization by comparing the types and levels of strain with the magnetic properties between the generally coarse-grained thickly bedded Helderberg and the thinly bedded and finer-grained Tonoloway in West Virginia. Standard tilt tests, as well as optimal differential untilting, from two anticlines indicate that the Helderberg contains a pervasive late syntilting CRM. The Tonoloway however contains a well-defined early syntilting CRM1 as well as a CRM2 which is similar to the CRM in the Helderberg. The CRMs reside in magnetite and rock magnetic results from the two units are similar, although the Helderberg samples have a finer apparent grain size (perhaps as a result of magnetic hardening). Pressure solution strain is higher in the Helderberg and lower in the Tonoloway. Strain may have caused rotation of magnetic minerals but it is not clear that the differences in strain between the units were high enough to cause the rotations that are needed to explain the tilt test results. Another viable hypothesis is a strain-enhanced chemical process caused dissolution and precipitation of new magnetite in solution structures during folding (e.g. Evans & Elmore 2006).

Folds with different geometries and tilted thrust sheets in the Mississippian rocks in Montana (O'Brien *et al.* 2007) all have the same magnetic characteristics and are probably caused by the same

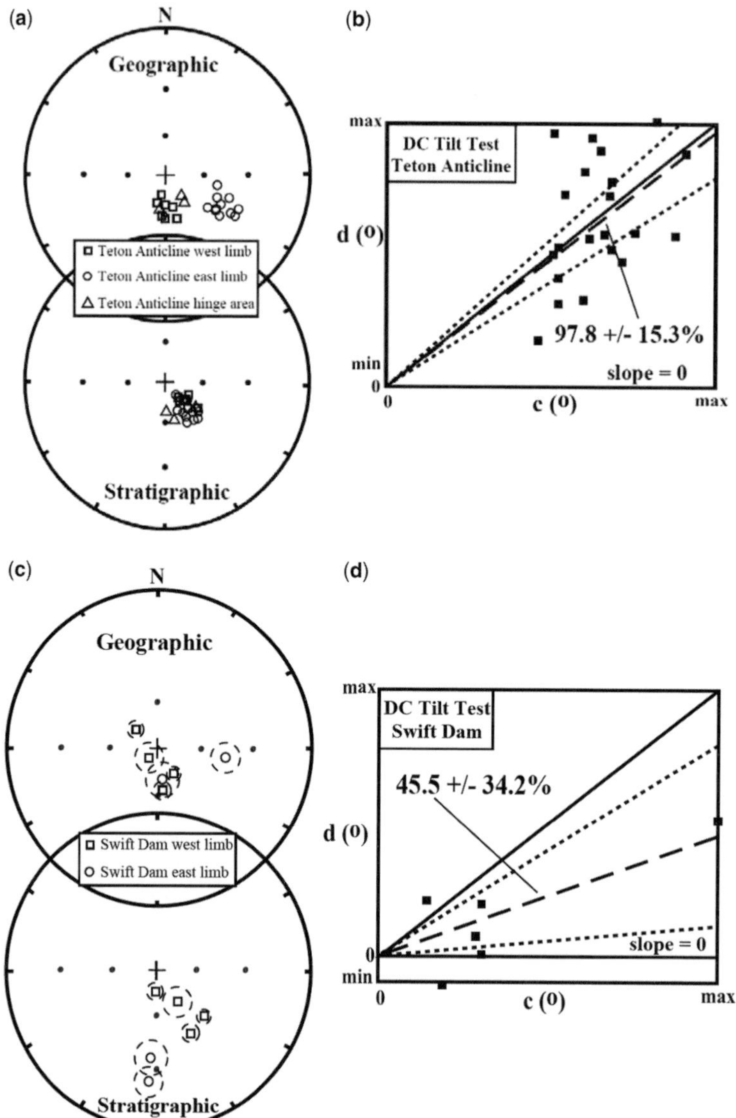

**Fig. 9.** Results from two folds with different geometries in Mississippian rocks in Montana (after O'Brien *et al.* 2007). (**a**) Equal-area projection for the site means of the Teton anticline (fault-bend fold) in geographic (0% untilting) and stratigraphic (100% untilting) coordinates. Circles represent the east limb, squares represent the west limb and triangles represent the hinge area of the fold. The mean $\alpha_{95}$ of the Teton anticline sites is 7.2 (standard deviation of 2.5, $n = 19$). Open circles are negative inclinations. (**b**) DC tilt test results (Enkin 2003). Optimal clustering is not significantly different than 100%, indicating a positive tilt test result. (**c**) Equal area projection for the site means of the Swift Dam fold (fault-propagation fold) in geographic (0% untilting) and stratigraphic (100% untilting) coordinates. Circles represent the east limb and squares represent the west limb on the fold. The dashed circles represent the $\alpha_{95}$ of the site means. (**d**) DC tilt test results. Optimal clustering indicates an indeterminate tilt test result in which there was a syntilting acquisition or an incomplete separation of pre- and post-tilting components.

remagnetization event. Tilt test results however suggest that the CRM is pre-tilting in both the thrust sheets and a fold with a fault-bend fold geometry and syntilting in folds with a fault-propagation fold geometry (Fig. 9) that probably experienced higher strains.

Experimental studies suggest that geologically reasonable differential stress levels can cause the acquisition of a PRM in carbonates (Borradaile 1994) as a result of magnetic domain migration in magnetite (Xu & Merrill 1992; Borradaile & Jackson 1993). A PRM should theoretically require more stress than most rocks have been subjected to during deformation, be partially reversible and have the greatest effect on the low-coercivity phases (e.g. Borradaile 1997).

In summary, the results from some studies suggest that pre-folding CRMs may have been altered into a syntilting configuration, perhaps as a result of strain. Additional studies are needed to further test this hypothesis, however. Although not widely applied in palaeomagnetic studies, optimal differential untilting (Enkin *et al.* 2000; Elmore *et al.* 2006*b*) or small circle intersection (SCI) analysis (Shipunov 1997; Waldhöer & Appel 2006) may help facilitate a better understanding of syntilting remagnetizations.

## Other remagnetization mechanisms

Numerous studies have documented that pyrrhotite can carry a CRM (e.g. Dekkers *et al.* 1989; Rochette *et al.* 1990; Jackson *et al.* 1993; Hirt *et al.* 1995; Xu *et al.* 1998; Weaver *et al.* 2002; Crouzet *et al.* 2003; Gillett & Karlin 2004; Font *et al.* 2006; Preeden *et al.* 2008; Manning & Elmore 2012). Although some of the remagnetizations are thermoviscous in origin (Kligfield & Channell 1981), many are interpreted to be chemical in origin with several remagnetization mechanisms proposed. These include pyrrhotite authigenesis caused by thermochemical sulphate reduction (Pierce *et al.* 1998; Zechmeister *et al.* 2012), migration of hydrocarbons (Machel & Burton 1991) or oxidation of pre-existing pyrite (Salmon *et al.* 1988). Chemical alteration by gas hydrates (Housen & Musgrave 1996; Larrasoaña *et al.* 2007) and pore fluids (Urbat *et al.* 2000) have also been suggested as mechanisms.

Pyrrhotite remagnetization in the Himalaya has been documented in numerous studies during the past 20 years and the reader is referred to Appel *et al.* (2012) for a history of such studies. Some of these remagnetizations are thermoremanences, but in cases where the peak metamorphic temperature ($T_{max}$) is less than the Curie temperature ($T_c \approx 325\ °C$) the magnetizations can be chemical, thermochemical or thermoremanent in origin (Appel *et al.* 2012).

Remagnetization by cooling during uplift (thermoviscous) is likely for some units that experienced moderate-to-elevated burial temperatures. A thermoviscous mechanism is the best explanation for some magnetizations in the Appalachians (e.g. Kent 1985). Careful attention to thermal histories is necessary when attempting to evaluate the possibility of a thermoviscous remagnetization.

Thermoviscous remagnetizations have also been used as geothermometers (e.g. Pullaiah *et al.* 1975; Middleton & Schmidt 1980; Crouzet *et al.* 1999; Dunlop *et al.* 2000).

Recent studies have also presented evidence that greigite ($Fe_3S_4$) can carry a diagenetic magnetization (e.g. Roberts & Weaver 2005; Rowan *et al.* 2009). Roberts & Weaver (2005) describe five mechanisms of CRM acquisition in greigite.

## Rock magnetic characterization of chemical remagnetization

In attempting to develop a better understanding of remagnetization processes and use palaeomagnetism to date diagenetic events, it is of obvious importance to identify the magnetic minerals that are carrying the remanence; rock magnetic studies are crucial in this endeavour. A general overview of this topic is covered in this paper, but the reader is also referred to other papers in this volume by Jackson & Swanson-Hysell (2012) and Dekkers (2012) as well as to previous publications (e.g. Dunlop & Özdemir 1997) for a more detailed discussion.

Although demagnetization characteristics provide some information, rock magnetic studies are needed to fully characterize magnetic mineralogy. Low-temperature demagnetization (LTD) by liquid nitrogen treatment prior to demagnetization has proven successful in removing magnetizations held in multi-domain (MD) magnetite (Dunlop & Argyle 1991). In addition to removing a modern magnetization in magnetite that could contaminate the characteristic magnetization, some specimens that were subjected to LTD also displayed more stable decay on the orthogonal plots than those that were not subject to the treatment (Manning & Elmore 2012).

Standard rock magnetic techniques include isothermal remanent magnetization (IRM) acquisition and thermal decay experiments (Lowrie 1990). As described in Dekkers (2012), end-member modelling of IRM acquisition curves has shown promise in determining the individual coercivity contributions. It is worth noting that these rock magnetic techniques provide information on all the magnetic phases in the rock, not just the minerals that carry the CRM.

The results of previous rock magnetic studies suggest that single-domain magnetite in remagnetized carbonates lacks shape anisotropy and that the magnetic properties are controlled by cubic magnetocrystalline anisotropy; this is consistent with the diagenetic origin for many secondary magnetizations (e.g. Jackson 1990; Jackson & Sun 1992). As pointed out by Jackson & Swanson-Hysell (2012) in this volume, recent studies have demonstrated that the magnetic signature in some units is not controlled by cubic magnetocrystalline anisotropy but by uniaxial anisotropy.

Low-temperature (LT) experiments to identify the 120 K Verwey transition in magnetite (Verwey 1939) and 34 K Besnus transition in pyrrhotite (Rochette *et al.* 2011) have proven useful in identifying magnetite and pyrrhotite in rocks (Dekkers *et al.* 1989; Rochette *et al.* 1990; Manning & Elmore 2012; Zechmeister *et al.* 2012). Hysteresis properties can also be used to identify the magnetic mineralogy and to characterize the type and size of magnetite (single-domain, pseudo-single-domain or multi-domain) in order to make inferences about the remagnetization processes. Many CRMs have wasp-waisted hysteresis loops, which are interpreted to reflect the presence of a range of magnetic grain sizes and/or magnetic minerals (e.g. Jackson 1990; Roberts *et al.* 1995). Determining the hysteresis parameters (including the coercivity of remanence $H_{cr}$, coercive force $H_c$, saturation remanence $M_{rs}$ and saturation magnetization $M_s$) can also be very useful (Tauxe 2005). For example, on a plot of coercivity ($H_{cr}/H_c$) and remanence ($M_{rs}/M_s$) ratios, CRMs commonly have high remanence values compared to the coercivity ratios (e.g. Jackson 1990; McCabe & Channell 1994). This property, along with wasp-waisted hysteresis loops, frequency-dependent susceptibility and high ratios of anhysteretic remanence to saturation remanence, is considered a rock magnetic fingerprint of remagnetization. The reader is referred to the paper by Jackson & Swanson-Hysell (2012) for a discussion of the current status of rock magnetism of remagnetized carbonates.


## Summary and unresolved issues

Considerable progress has been made on understanding remagnetization mechanisms during the last five decades. Chemical remanent magnetizations can be used to date a variety of different diagenetic events. We have also learned to identify remagnetization using both rock magnetic and palaeomagnetic data. Despite the progress, a number of important issues related to remagnetization remain to be resolved and warrant further study. Continued integration of magnetic data with diagenetic formation from geochemical and petrographic studies could help resolve many of the issues. Unresolved issues include the following.

(1)  Why are many remagnetizations of one polarity? Although palaeomagnetists are finding more dual-polarity CRMs, many CRMS are of one polarity and were acquired during the Late Palzeozoic–Cretaceous long-polarity intervals.

(2)  What are the origins of syntilting CRMs? Although many CRMs are syntilting in character, the suspicion exists that some CRMs are not truly syntilting but are instead a result of modification of a pre-folding CRM. Possible explanations include modification of a pre-folding component by contamination or strain, and more research is warranted on this issue.

(3)  What is the chemical origin of the widespread Late Palaeozoic CRMs in many Early and Middle Palaeozoic red bed units in the Appalachians? The timing contrasts with other red beds where the haematite is acquired relatively early. One popular model relates these CRMs to the migration of orogenic fluids (e.g. McCabe & Elmore 1989). Little geochemical or petrographic evidence which indicates that the rocks were altered by orogenic-type fluids has been presented.

(4)  The principal working hypothesis of the palaeomagnetic dating approach is that diagenetic processes can trigger the authigenesis of magnetic mineral phases and acquisition of a CRM. Geochemical and petrographic studies as well as field tests (vein or contact tests) provide important evidence for establishing these relationships. Additional tests of this hypothesis using independent dating methods to verify the palaeomagnetic ages would be useful.

(5)  There is empirical evidence that hydrocarbons can cause significant changes in the magnetic characteristics as well as remagnetization in a number of different rock types. The mechanisms are not understood, however. More research is needed to fully develop the use of magnetic studies in hydrocarbon exploration.

(6)  Although there is a significant body of literature on the role of bacteria in causing mineral authigenesis, this area warrants further research.

(7)  Although it is clear that LTD can remove low-coercivity unstable secondary components which contaminate a characteristic remanent magnetization, the procedure is not being universally applied. More research on which rocks should be subjected to LTD would be useful.

(8)  Although not necessarily directly related to acquisition of CRMs, the effect of inclination shallowing on pole positions used to determine APWPs is necessary if remagnetizations are to be used to date events. For example, Bilardello & Kodama (2010) presented evidence that a correction for

inclination shallowing could cause the Late Palaeozoic APWP to shift to a lower latitude by 4–10°.

(9) Many sedimentary rocks in the Appalachians and other areas contain intermediate-temperature components (unblocking temperatures of 300–350 °C) that are commonly interpreted as thermoviscous remanent magnetizations (TVRMs) in magnetite. Some studies suggest that some of these magnetizations could be CRMs in pyrrhotite. Additional rock magnetic studies could determine the origin of these remagnetizations.

(10) Additional palaeomagnetic and geochemical/petrographic studies of alteration below palaeoweathering surfaces can provide data on the extent of the event and age constraints on the surfaces which are commonly difficult to obtain. This can be useful information to evaluate the evolution (e.g. erosion rates) of ancient continents.

The authors thank R. Stephenson, S. Dulin, E. Manning, M. Zechmeister and an anomyous reviewer for useful reviews of a previous version of this manuscript. ARM is funded by the Royal Society.

# References

ALDANA, M., COSTANZO-ÁLVAREZ, V., VITIELLO, D., COMENARES, L. & GÓMEZ, G. 1999. Framboidal magnetic minerals and their possible association to hydrocarbons: La Victoria oil field (SW, Venezuela). *Geofísica Internacional*, **38**, 137–152.

ALDANA, M., COSTANZO-ÁLVAREZ, V. & DÍAZ, M. 2003. Magnetic and mineralogical studies to characterize oil reservoirs in Venezuela. *The Leading Edge*, **22**, 526–528.

APPEL, E., CROUZET, C. & SCHILL, E. 2012. Pyrrhotite remagnetizations in the Himalaya: a review. *In*: ELMORE, R. D., MUXWORTHY, A. R., ALDANA, M. & MENA, M. (eds) *Remagnetization and Chemical Alteration of Sedimentary Rocks*. Geological Society, London, Special Publications, **371**, first published online 22 August 2012, http://dx.doi.org/10.1144/SP371.1

AUBOURG, C. & POZZI, J.-P. 2010. Toward a new <250 °C pyrrhotite-magnetite geothermometer for claystones. *Earth and Planetary Science Letters*, **294**, 47–57.

AUBOURG, C., POZZI, J.-P., JANOTS, D. & SARAHOUI, L. 2008. Imprinting chemical remanent magnetization in claystones at 95 °C. *Earth and Planetary Science Letters*, **272**, 172–80.

AUBOURG, C., POZZI, J.-P., & KARS, M. 2012. Burial, claystones remagnetization and some consequences for magnetostratigraphy. *In*: ELMORE, R. D., MUXWORTHY, A. R., ALDANA, M. & MENA, M. (eds) *Remagnetization and Chemical Alteration of Sedimentary Rocks*. Geological Society, London, Special Publications, **371**, first published online 26 June 2012, http://dx.doi.org/10.1144/SP371.4

BANERJEE, S., ENGEL, M. & ELMORE, R. D. 1997. Chemical remagnetization and burial diagenesis of organic matter: testing the hypothesis in the Pennsylvanian Belden Formation, Colorado. *Journal Geophysical Research*, **102**, 24825–24842.

BENTHIEN, R. H. & ELMORE, R. D. 1987. Origin of magnetization in the Phosphoria Formation at Sheep Mountain, Wyoming: a possible relationship with hydrocarbons. *Geophysical Research Letters*, **14**, 323–326.

BESSE, J. & COURTILLOT, V. 2002. Apparent and true polar wander and the geometry of the geomagnetic field over the last 200 Myr. *Journal of Geophysical Research*, **107**, 2300, http://dx.doi.org/10.1029/2000JB000050

BETHKE, C. M. & MARSHAK, S. 1990. Brine migrations across North America – the plate tectonics of groundwater. *Annual Reviews Earth Planetary Sciences*, **18**, 287–315.

BILARDELLO, D. & KODAMA, K. P. 2010. A new inclination shallowing correction of the Mauch Chunk Formation of Pennsylvania, based on high-field AIR results: implications for the Carboniferous North American APW path and Pangea reconstructions. *Earth and Planetary Science Letters*, **299**, 218–227.

BLUMSTEIN, A. M., ELMORE, R. D., ENGEL, M. H., ELLIOT, C. & BASU, A. 2004. Paleomagnetic dating of burial diagenesis in Mississippian carbonates, Utah. *Journal of Geophysical Research*, **109**, B04101, http://dx.doi.org/10.1029/2003JB002698

BLUMSTEIN, R. D., ELMORE, R. D., ENGEL, M. H., PARNELL, J. & BARON, M. 2005. Date and origin of multiple fluid flow events along the Moine Thrust Zone, Scotland. *Journal Geological Society, London*, **162**, 1031–1045.

BORRADAILE, G. J. 1994. Remagnetisation of a rock analogue during experimental triaxial deformation. *Physics Earth Planetary Interiors*, **83**, 147–163.

BORRADAILE, G. J. 1997. Deformation and paleomagnetism. *Surveys in Geophysics*, **18**, 405–435.

BORRADAILE, G. J. & JACKSON, M. 1993. Changes in magnetic remanence during simulated deep sedimentary burial. *Physics Earth and Planetary Interiors*, **77**, 315–327.

BROTHERS, L. A., ENGEL, M. H. & ELMORE, R. D. 1996. A laboratory investigation of the late diagenetic conclusion of pyrite to magnetite by organically complexed ferric iron. *Chemical Geology*, **130**, 1–14.

BUTLER, R. F. 1992. *Paleomagnetism: Magnetic Domains to Geologic Terranes*. Blackwell Science Inc., Boston.

CAIRANNE, G., AUBOURG, C., POZZI, J.-P., MOREAU, M. G., DECAMPS, T. & MAROLLEAU, G. 2004. Experimental Chemical Remanent Magnetization in a natural claystone: a record of two magnetic polarities. *Geophysical Journal International*, **159**, 907–916, http://dx.doi.org/ 10.1111/j.1365-246X.2004.02439x

CIOPPA, M. T. & SYMONS, D. T. A. 2000. Timing hydrocarbon generation and migration: paleomagnetic and rock magnetic analysis of the Devonian Duvernay Formation, Alberta, Canada. *Journal of Geochemical Exploration*, **69–70**, 387–390.

CIOPPA, M. T., AL-AASM, I. S., SYMONS, D. T. A., LEWCHUK, M. T. & GILLEN, K. P. 2000. Correlating petrologic, geochemical, and magnetic to date diagenetic and fluid flow events in the Moose Reservoir, Alberta, Canada. *Sedimentary Geology*, **131**, 109–129.

CIOPPA, M. T., SYMONS, D. T. A. & FLORE, M. 2002. Initial analysis of the Jurassic 'Nordegg Member', Fernie Formation: using paleomagnetism to measure source rock thermal maturity. *Physics and Chemistry of the Earth*, **27**, 1161–1168.

COCHRAN, K. L. & ELMORE, R. D. 1987. Absolute dating of Liesegang bands. *Journal of Sedimentary Petrology*, **57**, 701–708.

COSTANZO-ÁLVAREZ, V., WILLIAMS, W., PILLOUD, A., MIRÓN VALDESPINO, O. & ALDANA, M. 2000*a*. Paleomagnetic results of remagnetized mid-cretaceous (Albian-Cenomanian) strata of northeastern Venezuela. *Geophysical Journal International*, **41**, 337–350.

COSTANZO-ÁLVAREZ, V., ALDANA, M., ARISTIGUIETA, O., MARCANO, M. & ACONCHA, E. 2000*b*. Study of magnetic contrast in the Guafita Oil Field (southwestern Venezuela). *Physics and Chemistry of the Earth*, **25**, 437–445.

COSTANZO-ÁLVAREZ, V., ALDANA, M., DÍAZ, M., BAYONA, G. & AYALA, C. 2006. Hydrocarbon-induced magnetic contrasts in some Venezuelan and Colombian oil wells. *Earth, Planets and Space*, **58**, 1401–1410.

COSTANZO-ÁLVAREZ, V., ALDANA, M., BAYONA, G., LÓPEZ-RODRÍGUEZ, D. & BLANCO, J. M. 2012. Rock magnetic characterization of early and late diagenesis in a stratigraphic well from the Llanos foreland basin (eastern Columbia). *In*: ELMORE, R. D., MUXWORTHY, A. R., ALDANA, M. & MENA, M. (eds) *Remagnetization and Chemical Alteration of Sedimentary Rocks*. Geological Society, London, Special Publications, **371**, first published online 20 September 2012, http://dx.doi.org/10.1144/SP371.13

COX, E., ELMORE, R. D. & EVANS, M . 2005. Paleomagnetism of devonian red beds in the Appalachian plateau and valley and ridge provinces. *Journal Geophyiscal Research*, **110**, B08102, http://dx.doi.org/10.1029/2005JB003640

CREER, K. M. 1968. Palaeozoic paleomagnetism. *Nature*, **219**, 246–250.

CROUZET, C., MENARD, G. & ROCHETTE, P. 1999. High-precision three-demensional paleothermometry from paleomagnetic data in an Alpine metamorphic unit. *Geology*, **27**, 503–506.

CROUZET, C., GAUTAM, P., SCHILL, E. & APPEL, E. 2003. Multicomponent magnetization in western Dolpo (Tethyan Himalaya, Nepal); tectonic implications. *Tectonophysics*, **377**, 179–196.

DEKKERS, M. J. 2012. End-member modelling as an aid to diagnose remagnetization: a brief review. *In*: ELMORE, R. D., MUXWORTHY, A. R., ALDANA, M. & MENA, M. (eds) *Remagnetization and Chemical Alteration of Sedimentary Rocks*. Geological Society, London, Special Publications, **371**, first published online 22 August 2012, http://dx.doi.org/10.1144/SP371.12

DEKKERS, M. J., MATTEI, J. L., FILLION, G. & ROCHETTE, P. 1989. Grain-size dependence of the magnetic behavior of pyrrhotite during its low-temperature transition at 34 K. *Geophysical Research Letters*, **16**, 855–858.

DENNIE, D., ELMORE, R. D., DENG, J., MANNING, E. & PANNALAL, J. 2012. Palaeomagnetism of the Mississippian Barnett Shale, Fort Worth Basin, Texas. *In*:

ELMORE, R. D., MUXWORTHY, A. R., ALDANA, M. & MENA, M. (eds) *Remagnetization and Chemical Alteration of Sedimentary Rocks*. Geological Society, London, Special Publications, **371**, first published online 22 August 2012, http://dx.doi.org/10.1144/SP371.10

DÍAZ, M., ALDANA, M., COSTANZO-ÁLVAREZ, V., SILVA, P. & PEREZ, A. 2000. EPR and magnetic susceptibility studies in well samples from some Venezuelan oil fields. *Physics and Chemistry of the Earth*, **25**, 447–453.

DONOVAN, T. J., HENDRICKS, J. D. & ROBERTS, A. 1979. Aeromagnetic detection of diagenetic magnetite over oil fields. *AAPG Bulletin*, **63**, 245–248.

DONOVAN, T. J., FORGEY, R., ROBERTS, A. & ELIASON, P. T. 1984. Low-altitude aeromagnetic reconnaissance for petroleum in the Artic National Wildlife Refuge, Alaska. *Geophysics*, **49**, 1338–1353.

DUNLOP, D. J. & ARGYLE, K. S. 1991. Separating multidomain and single-domain-like remanences in pseudo-single-domain magnetites (215–540 nm) by low-temperature demagnetization. *Journal of Geophysical Research*, **96**, 2007–2017.

DUNLOP, D. J. & ÖZDEMIR, Ö. 1997. *Rock Magnetism: Fundamental and Frontiers*. Cambridge University Press, Cambridge, UK.

DUNLOP, D. J., ÖZDEMIR, Ö., CLARK, D. A. & SCHMIDT, P. W. 2000. Time-temperature relations for the remagnetization of pyrrhotite ($Fe_7S_8$) and their use in estimating paleotemperatures. *Earth Planetary Science Letters*, **176**, 107–116.

EDEL, J. B. & SCHNEIDER, J. L. 1995. The Late Carboniferous to Early Triassic geodynamic evolution of Variscan Europe in the light of magnetic overprints in Early Permian rhyolites from the northern Vosges (France) and the central Black Forest (Germany). *Geophysical Journal International*, **122**, 858–876.

ELLIOTT, W. C., OSBORN, S., O'BRIEN, V., ELMORE, R. D., ENGEL, M. H. & WAMPLER, M. 2006*a*. A comparison of K–Ar ages of diagenetic Illite and the age implications of a remagnetization in the Cretaceous Marias River Shale, Disturbed Belt, Montana. *Journal Geochemical Exploration*, **89**, 92–95.

ELLIOTT, W. E., BASU, A., WAMPLER, J. M., ELMORE, R. D. & GRATHOFF, G. 2006*b*. Comparison of K-Ar Ages of diagenetic illite-smectite to the age of a chemical remanent magnetization (CRM): an example from the Isle of Skye, Scotland. *Clays and Clay Minerals*, **54**, 314–323.

ELMORE, R. D. 2001. A review of paleomagnetic data on the timing and origin of multiple fluid-flow events in the Arbuckle Mountains, Southern Oklahoma. *Petroleum Geoscience*, **7**, 223–229.

ELMORE, R. D. & CRAWFORD, L. 1990. Remanence in authigenic magnetite: testing the hydrocarbon-magnetite hypothesis. *Journal Geophysical Research*, **95**, 4539–4549.

ELMORE, R. D. & LEACH, M. C. 1990. Paleomagnetism of the Rush Springs Sandstone, Cement, Oklahoma: implications for dating hydrocarbon migration, aeromagnetic exploration, and understanding remagnetization mechanisms. *Geology*, **18**, 124–127.

ELMORE, R. D., DUNN, W. & PECK, C. 1985. Absolute dating of a diagenetic event using paleomagnetic analysis. *Geology*, **15**, 558–561.

ELMORE, R. D., ENGEL, M. H., CRAWFORD, L., NICK, K., IMBUS, S. & SOFER, Z. 1987. Evidence for a relationship between hydrocarbons and authigenic magnetite. *Nature*, **325**, 428–430.

ELMORE, R. D., LONDON, D., BAGLEY, D. & GAO, G. 1993a. Remagnetization by basinal Fluids: testing the hypothesis in the Viola Limestone, southern Oklahoma. *Journal Geophysical Research*, **98**, 6237–6254.

ELMORE, R. D., IMBUS, S., ENGEL, M. & FRUIT, D. 1993b. Hydrocarbons and magnetizations in magnetite. *In*: AÏSSAOUI, D. M., MCNEILL, D. F. & HURLEY, N. F. (eds) *Applications of Paleomagnetism to Sedimentary Geology*. Society Economic Paleontologists and Mineralogists, Tulsa, Special Publications, **49**, 181–191.

ELMORE, R. D., CATES, K., GAO, G. & LAND, L. 1994. Geochemical constraints on the origin of secondary magnetizations in the Cambro-Ordovician Royer Dolomite, Arbuckle Mountains, southern Oklahoma. *Physics of the Earth and Planetary Interiors*, **85**, 3–13.

ELMORE, R. D., BANERJEE, S., CAMPBELL, T. & BIXLER, G. 1998. Paleomagnetic dating of ancient fluid-flow events and paleoplumbing in the Arbuckle Mountains, Southern Oklahoma. *In*: PARNELL, J. (ed.) *Dating and Duration of Fluid Flow Events and Rock-Fluid Interaction*. Geological Society, London, Special Publications, **144**, 9–25.

ELMORE, R. D., KELLEY, J., EVANS, M. & LEWCHUK, M. 2001. Remagnetization and orogenic fluids: testing the hypothesis in the central Appalachians. *Geological Journal International*, **144**, 568–576.

ELMORE, R. D., PARNELL, J., ENGEL, M. H., BARON, M., WOODS, S., ABRAHAM, M. & DAVIDSON, M. 2002. Paleomagnetic dating of fluid-flow events in dolomitized rocks along the Highland Boundary Fault, central Scotland. *Geofluids*, **2**, 299–314.

ELMORE, R. D., DULIN, S., ENGEL, M. H. & PARNELL, J. 2006a. Remagnetization and fluid flow in the Old Red Sandstone along the Great Glen Fault, Scotland. *Journal of Geochemical Exploration*, **89**, 96–99.

ELMORE, R. D., FOUCHER, J. L.-E., EVANS, M., LEWCHUK, M. & COX, E. 2006b. Remagnetization of the Tonoloway Formation and the Helderberg Group in the Central Appalachians; testing the origin of syntilting magnetizations. *Geophysical Journal International*, **166**, 1062–1076.

ELMORE, M., ENGEL, D., HOOD, R. & PARNELL, J. 2010. Paleomagnetic dating of fracturing using breccia veins in Durness Group carbonates, NW Scotland. *Journal Structural Geology*, **32**, 1933–1942, http://dx.doi.org/10.1016/j.jsg.2010.05.011

EMMERTON, S., MUXWORTHY, A. R. & SEPHTON, M. A. 2012. Magnetic characterization of oil sands at Osmington Mills and Mupe Bay, Wessex Basin, UK. *In*: ELMORE, R. D., MUXWORTHY, A. R., ALDANA, M. & MENA, M. (eds) *Remagnetization and Chemical Alteration of Sedimentary Rocks*. Geological Society, London, Special Publications, **371**, first published online 26 June 2012, http://dx.doi.org/10.1144/SP371.6

ENKIN, R. J. 2003. The direction-correction tilt test: an all-purpose tilt/fold test for paleomagnetic studies. *Earth Planetary Science Letters*, **212**, 151–166.

ENKIN, R. J., OSADETZ, K. G., BAKER, J. & KISILEVSKY, D. 2000. Orogenic remagnetizations in the front ranges and inner foothills of the southern Canadian Cordillera: chemical harbinger and thermal handmaiden of Cordilleran deformation. *Geological Society of America Bulletin*, **112**, 929–942.

EVANS, M. A. & ELMORE, R. D. 2006. Fluid control of localized mineral domains in limestone pressure solution structures. *Journal Structural Geology*, **28**, 284–301.

EVANS, S. C., ELMORE, R. D., DENNIE, D. & DULIN, S. A. 2012. Remagnetization of the Alamo Breccia, Nevada. *In*: ELMORE, R. D., MUXWORTHY, A. R., ALDANA, M. & MENA, M. (eds) *Remagnetization and Chemical Alteration of Sedimentary Rocks*. Geological Society, London, Special Publications, **371**, first published online 22 August 2012, http://dx.doi.org/10.1144/SP371.8

FÀBREGA, C., PARCERISA, D., THIRY, M., FRANKE, C. & GÒMEZ-GRAS, D. 2012. *Albitization profiles related to the Variscan basement a case study of Catalan Coastal Ranges and Eastern Pyrenees (NE Iberia)*. Abstract volume, GEOFLUIDS VII – International Conference IFP Energies nouvelles, Rueil-Malmaison (France), June 6–8, 103–107.

FONT, E., TRINDADE, R. I. F. & NEDELEC, A. 2006. Remagnetization in bituminous limestones of the Neoproterozoic Araras Group (Amazon craton): hydrocarbon maturation, burial diagenesis, or both? *Journal Geophysical Research*, **111**, 17.

FONT, E., RAPALINI, A. E., TOMEZZOLI, R. N., TRINDADE, R. I. F. & TOHVER, E. 2012. Episodic Remagnetizations related to tectonic events and their consequences for the South America Polar Wander Path. *In*: ELMORE, R. D., MUXWORTHY, A. R., ALDANA, M. & MENA, M. (eds) *Remagnetization and Chemical Alteration of Sedimentary Rocks*. Geological Society, London, Special Publications, **371**, first published online 22 August 2012, http://dx.doi.org/10.1144/SP371.7

FOOTE, R. S. 1996. Relationship of near-surface magnetic anomalies to oil-and gas-producing area. *In*: SCHUMACHER, D. & ABRAMS, M. (eds) *Hydrocarbon Migration and Its Near-Surface Expression*. AAPG, Tulsa, Memoir, **66**, 111–126.

FRANKE, C., GOMEZ-GRAS, D. *ET AL.* 2010. Paleomagnetic age constrains and magnetomineralogic implications for the Triassic paleosurface in Europe. *Geophysical Research Abstracts*, **12**, EUG2010-7858.

FRUIT, D., ELMORE, R. D. & HALGEDAHL, S. 1995. Remagnetization of the Folded Belden Formation, northwest Colorado. *Journal of Geophysical Research*, **100**, 15009–15024.

GARVEN, G. 1995. Continental-scale groundwater flow and geological processes. *Annual Review of Earth and Planetary Sciences*, **24**, 89–117.

GAY, S. P. JR. 1992. Epigenetic versus syngenetic magnetite as a cause of magnetic anomalies. *Geophysics*, **57**, 60–68.

GEISSMAN, J. W. & HARLAN, S. S. 2002. Late Paleozoic remagnetization of Precambrian crystalline rocks along the Precambrian/Carboniferous nonconformity, Rocky Mountains: a relationship among deformation, remagnetization, and fluid migration. *Earth and Planetary Science Letters*, **203**, 905–924.

GILL, J. D., ELMORE, R. D. & ENGEL, M. H. 2002. Chemical remagnetization and clay diagenesis: testing the hypothesis in the Cretaceous sedimentary rocks of northwestern Montana. *Physics and Chemistry of the Earth*, **27**, 1131–1139.

GILLETT, S. L. & KARLIN, R. E. 2004. Pervasive late Paleozoic-Triassic remagnetization of the miogeoclinal carbonate racks in the Basin and Range and vicinity, SW USA: regional results and possible tectonic implications. *Physics of the Earth and Planetary Interiors*, **141**, 95–120.

GONZÁLEZ, F., ALDANA, M., COSTANZO-ÁLVAREZ, V., DÍAZ, M. & ROMERO, I . 2002. An integrated rock magnetic and EPR study in soil samples from a hydrocarbon prospective area. *Physics and Chemistry of the Earth*, **27**, 193–199.

GOSE, W. A. & KYLE, R. J. 1993. Paleomagnetic dating of sulfide mineralization and cap-rock formation in Gulf Coast salt domes. *In*: AÏSSAOUI, D. M., MCNEILL, D. F. & HURLEY, N. F. (eds) *Applications of Paleomagnetism to Sedimentary Geology*. Society of Economic Paleontologists and Mineralogists, Tulsa, Special Publications, **49**, 157–166.

GUZMÁN, O., COSTANZO-ÁLVAREZ, V., ALDANA, M. & DÍAZ, M. 2011. Study of magnetic contrasts applied to hydrocarbon exploration in the Maturin sub-basin (Eastern Venezuela). *Studia Geophysica et Geodaetica*, **55**, 359–376.

HAMILTON, M., ELMORE, R. D., WEAVER, B. & DULIN, S. 2012. *Paleomagnetic and Petrologic investigation of Long Mountain Granite, Wichita Mountains, Oklahoma.* Abstract volume, GEOFLUIDS VII – International Conference IFP Energies nouvelles, Rueil-Malmaison (France), June 6–8, 79–82.

HARLAN, S. S., GEISSMAN, J. W., SNEE, L. W. & REYNOLDS, R. L. 1996. Late Cretaceous remagnetization of Proterozoic mafic dikes, southern Highland Range, southwestern Montana: a paleomagnetic and 40Ar/39Ar study. *Geological Society of America Bulletin*, **108**, 653–668.

HELLER, F. 1979. Palaeomagnetism of Upper Cretaceous limestones from the Münster Basin, Germany. *Journal Geophysics*, **46**, 413–427.

HERRERO-BERVERA, E. & HELSLEY, C. E. 1983. Paleomagnetism of a polarity transition in the Lower (?) Triassic Chugwater Formation, Wyoming. *Journal Geophysical Research*, **88**, 3506–3522.

HIRT, A. M., LOWRIE, W. & PFIFFNER, O. A. 1986. A paleomagnetic study of the tectonically deformed red beds of the Lower Glarus Nappe Complex, eastern Switzerland. *Tectonics*, **5**, 723–732.

HIRT, A. M., BANIN, A. & GEHRING, U. A. 1993. Thermal generation of ferromagnetic minerals from iron-enriched smectites. *Geophysical Journal International*, **115**, 1161–1168.

HIRT, A. M., EVANS, K. F. & ENGELDER, T. 1995. Correlation between magnetic anisotropy and fabric for Devonian shales on the Appalachian Plateau. *Tectonophysics*, **247**, 121–132.

HOUSEN, B. A. & MUSGRAVE, R. J. 1996. Rock-magnetic signature of gas hydrates in accretionary prism sediments. *Earth and Planetary Science Letters*, **139**, 509–519.

HUDSON, M. R., REYNOLDS, R. L. & FISHMAN, N. S. 1989. Synfolding magnetization in the Jurassic Preuss Sandstone, Wyoming-Idaho-Utah thrust belt. *Journal Geophysical Research*, **94**, 13681–13705.

JACKSON, M. 1990. Diagenetic sources of stable remanence in remagnetized Paleozoic cratonic carbonates: a rock magnetic study. *Journal Geophysical Research*, **95**, 2753–2761.

JACKSON, M. & SUN, W. W. 1992. The rock magnetic fingerprint of chemical remagnetization in mid-continental Paleozoic carbonates. *Geophysical Research Letters*, **19**, 781–784.

JACKSON, M. & SWANSON-HYSELL JACKSON, N. 2012. Rock magnetism of remagnetized carbonate rocks: another look. *In*: ELMORE, R. D., MUXWORTHY, A. R., ALDANA, M. & MENA, M. (eds) *Remagnetization and Chemical Alteration of Sedimentary Rocks*. Geological Society, London, Special Publications, **371**, first published online 26 June 2012, http://dx.doi.org/10.1144/SP371.3

JACKSON, M. J., MCCABE, C., BALLARD, M. M. & VAN DER VOO, R. 1988. Magnetite authigenesis and diagenetic paleotemperatures across the northern Appalachian Basin. *Geology*, **16**, 592–595.

JACKSON, M., ROCHETTE, P., FILLION, G., BANERJEE, S. & MARVIN, J. 1993. Rock magnetism of remagnetized Paleozoic carbonates; low-temperature behavior and susceptibility characteristics. *Journal of Geophysical Research*, **98**, 6217–6225.

JOHNSON, R. J., VAN DER VOO, R. & LOWRIE, W. 1984. Paleomagnetism and late diagenesis of Jurassic carbonates from the Jura mountains, Switzerland and France. *Geological Society of America Bulletin*, **95**, 478–488.

KATZ, B., ELMORE, R. D., ENGEL, M. H. & LEYTHAEUSER, D. 1996. Paleomagnetism of the Jurassic Asphaltkalk-deposits, Holzen, northern Germany. *Geophysical Journal International*, **127**, 305–310.

KATZ, B., ELMORE, R. D., COGOINI, M. & FERRY, S. 1998. Widespread chemical remagnetization: orogenic fluids or burial diagenesis of clays? *Geology*, **23**, 603–606.

KATZ, B., ELMORE, R. D., ENGEL, M. H., COGOINI, M. & FERRY, S. 2000. Associations between burial diagenesis of smectite, chemical remagnetization and magnetite authigenesis in the Vocontian Trough of SE-France. *Journal Geophysical Research*, **105**, 851–868.

KENNEDY, M. J., PEVEAR, D. R. & HILL, R. J. 2002. Mineral surface control of organic carbon in black shale. *Science*, **295**, 657–660.

KENT, D. V. 1985. Thermoviscous remagnetization in some Appalachian limestones. *Geophysical Research Letters*, **12**, 805–808.

KENT, D. V. & OPDYKE, N. D. 1985. Multicomponent magnetizations from the Mississippian Mauch Chunk Formation of the central Appalachians and their tectonic implications. *Journal Geophysical Research*, **90**, 5371–5383.

KILGORE, B. & ELMORE, R. D. 1989. A study of the relationship between hydrocarbon migration and precipitation of authigenic magnetic minerals in the Triassic Chugwater formation, southern Montana. *Geological Society of America Bulletin*, **101**, 1280–1288.

KLIGFIELD, R. & CHANNELL, J. E. T. 1981. Widespread remagnetization of Helvetic limestones. *Journal Geophysical Research*, **86**, 1888–1900.

KLIGFIELD, R. W., LOWRIE, W., HIRT, A. & SIDDENS, A. W. B. 1983. Effect of progressive deformation on remanent magnetization of Permian red beds from the Alps Maritimes (France). *Tectonophysics*, **97**, 59–85.

KODAMA, K. P. 1988. Remanence rotation due to rock strain during folding and the stepwise application of the fold test. *Journal Geophysical Research*, **93**, 3357–3371.

LARRASOAÑA, J. C., ROBERTS, A. P., MUSGRAVE, R. J., GRÀCIA, E., PIÑERO, E., VEGA, M. & MARTÍNEZ-RUIZ, F. 2007. Diagenetic formation of greigite and pyrrhotite in hydrate marine sedimentary systems. *Earth and Planetary Science Letters*, **261**, 350–366.

LEACH, D. L., BRADLEY, D., LEWCHUK, M. T., SYMONS, D. T. A., DE MARSILY, G. & BRANNON, J. 2001. Mississippi Valley-type lead-zinc deposits through geologic time: implications from recent age-dating research. *Mineralium Deposita*, **36**, 711–740.

LEWCHUK, M. T. & SYMONS, D. T. A. 1995. Age and duration of Mississippi Valley-type (MVT) Pb-Zn-Ba-F mineralizing events. *Geology*, **23**, 233–236.

LEWCHUK, M. T., AL-AASM, I. S., SYMONS, D. T. A. & GILLEN, K. P. 1998. Dolomitization of Mississippian carbonates in the Shell Waterton gas field, southwestern Alberta: insights from paleomagnetism, petrography, and geochemistry. *Canadian Society Petroleum Geology Bulletin*, **46**, 387–410.

LEWCHUK, M. T., EVANS, M. & ELMORE, R. D. 2003. Synfolding remagnetization and deformation: results from Paleozoic sedimentary rocks in West Virginia. *Geophysical Journal International*, **152**, 266–279.

LOUCKS, V. & ELMORE, R. D. 1986. Absolute dating of dedolomitization and the origin of magnetization in the Morgan Creek Limestone, central Texas. *Geological Society American Bulletin*, **97**, 486–496.

LOWRIE, W. 1990. Identification of ferromagnetic minerals in a rock by coercivity and unblocking temperature properties. *Geophysical Research Letters*, **17**, 159–162.

LU, G., MCCABE, C., HANER, J. S. & FERRELL, R. E. 1991. A genetic link between remagnetization and potassic metasomatism in the Devonian Onondaga formation, northern Appalachian basin. *Geophysical Research Letters*, **18**, 2047–2050.

MACHEL, H. G. 1995. Magnetic mineral assemblages and magnetic contrasts in diagenetic environments – with implications for studies of palaeomagnetism, hydrocarbon migration and exploration. *In*: TURNER, P. & TURNER, A. (eds) *Palaeomagnetic Applications in Hydrocarbon Exploration and Production*. Geological Society, London, Special Publications, **98**, 9–29.

MACHEL, H. G. & BURTON, E. A. 1991. Chemical and microbial processes causing anomalous magnetization in environments affected by hydrocarbon seepage. *Geophysics*, **56**, 598–605.

MANNING, E. B. & ELMORE, R. D. 2012. Rock magnetism and identification of remanence components in the Marcellus Shale, Pennsylvania. *In*: ELMORE, R. D., MUXWORTHY, A. R., ALDANA, M. & MENA, M. (eds) *Remagnetization and Chemical Alteration of Sedimentary Rocks*. Geological Society, London, Special Publications, **371**, first published online 3 September 2012, http://dx.doi.org/10.1144/SP371.9

MCCABE, C. & ELMORE, R. D. 1989. The occurrence and origin of Late Paleozoic remagnetization in the sedimentary rocks of North America. *Reviews of Geophysics*, **27**, 471–493.

MCCABE, C. & CHANNELL, J. E. T. 1994. Late Paleozoic remagnetization in limestone of the Craven Basin (northern England) and rock magnetic fingerprint of remagnetized sedimentary carbonate. *Journal Geophysical Research*, **99**, 4603–4612.

MCCABE, C., VAN DER VOO, R., PEACOR, D. R., SCOTESE, C. R. & FREEMAN, R. 1983. Diagenetic magnetite carries ancient yet secondary remanence in some Paleozoic sedimentary carbonates. *Geology*, **11**, 221–223.

MCCABE, C., SASSEN, R. & SAFFER, B. 1987. Occurrence of secondary magnetite within biodegraded crude oil. *Geology*, **15**, 7–10.

MCCABE, C., JACKSON, M. & SAFFER, B. 1989. Regional patterns of magnetite authigenesis in the Appalachian basin: implications for the mechanism of late Paleozoic remagnetization. *Journal of Geophysical Research*, **94**, 10429–10443.

MENA, M. & WALTHER, A. M. 2012. Rock magnetic properties of drill cutting from a hydrocarbon exploratory well and their relationship to hydrocarbon presence and petrophysical properties. *In*: ELMORE, R. D., MUXWORTHY, A. R., ALDANA, M. & MENA, M. (eds) *Remagnetization and Chemical Alteration of Sedimentary Rocks*. Geological Society, London, Special Publications, **371**, first published online 1 October 2012, http://dx.doi.org/10.1144/SP371.14

MIDDLETON, M. W. & SCHMIDT, P. W. 1980. Paleothermometry of the Sydney Basin. *Journal of Geophysical Research*, **87**, 5351–5359.

MILLER, J. D. & KENT, D. V. 1986. Synfolding and prefolding magnetizations in the Upper Devonian Catskill formation of eastern Pennsylvania: implications for the tectonic history of Acadia. *Journal of Geophysical Research*, **91**, 12791–12803.

MILLER, J. D. & KENT, D. V. 1988. Regional trends in the timing of Alleghenian remagnetization in the Appalachians. *Geology*, **16**, 588–591.

MOREAU, M. G., ADER, M. & ENKIN, R. J. 2005. The magnetization of clay-rich rocks in sedimentary basins: low temperature experimental formation of magnetic carriers in natural samples. *Earth and Planetary Science Letters*, **230**, 193–210.

NICK, K. & ELMORE, R. D. 1990. Paleomagnetism of the Cambrian Royer Dolomite and Pennsylvanian Collings Ranch Conglomerate, southern Oklahoma: an early Paleozoic magnetization and non-pervasive remagnetization by weathering. *Geological Society America Bulletin*, **102**, 1517–1525.

O'BRIEN, V. J., MORELAND, K. M., ELMORE, R. D., ENGEL, M. H. & EVANS, M. A. 2007. Origin of orogenic remagnetizations in Mississippian carbonates, Sawtooth Range, Montana. *Journal of Geophysical Research*, **112**, http://dx.doi.org/10.1029/2006JB004699

OLIVER, R. 1986. Fluids expelled tectonically from orogenic belts: their role in hydrocarbon migration and other geologic phenomena. *Geology*, **14**, 99–102.

OLIVER, J. 1992. The spots and stains of plate tectonics. *Earth Science Reviews*, **32**, 77–106.

PARNELL, J., BARON, M., DAVIDSON, M., ELMORE, D. & ENGEL, M. 2000. Dolomitic breccia veins as evidence for extension and fluid flow in the Dalradian of Argyll. *Geological Magazine*, **137**, 447–462.

PIERCE, J. W., GOUSSEV, S. A., CHARTERS, R. A., AMBER-CROMBIE, H. J. & DEPAOLI, G. R. 1998. Intrasedimentary magnetization by vertical fluid flow and exotic geochemistry. *The Leading Edge*, **17**, 89–92.

PLASTER-KIRK, L., ELMORE, R. D., ENGEL, M. H. & IMBUS, S. W. 1995. Paleomagnetic investigation of organic-rich lacustrine deposits, Middle Old Red Sandstone, Scotland. *Scottish Journal of Geology*, **31**, 97–105.

PREEDEN, U., PLADO, J., MERTANEN, S. & PUURA, V. 2008. Multiply remagnetized Silurian carbonate sequence in Estonia. *Estonian Journal of Earth Sciences*, **57**, 170–180.

PREEDEN, U., MERTANEN, S., ELMINEN, T. & PLADO, J. 2009. Secondary magnetizations in shear and fault zones in southern Finland. *Tectonophysics*, **479**, 203–213.

PULLAIAH, G., IRVING, E., BUCHAN, K. L. & DUNLOP, D. J. 1975. Magnetization changes caused by burial and uplift. *Earth and Planetary Science Letters*, **28**, 133–143.

REYNOLDS, R. L., GOLDHABER, M. B. & TUTTLE, M. L. 1993. Sulfidization and magnetization above hydrocarbon reservoirs. *In*: AÏSSAOUI, D. M., MCNEILL, D. F. & HURLEY, N. F. (eds) *Applications of Paleomagnetism to Sedimentary Geology*. Society Economic Geologists & Mineralogists, Tulsa, Special Publications, **49**, 167–179.

RICORDEL, C., PARCERISA, D., THIRY, M., MOREAU, M.-G. & GOMEZ-GRAS, D. 2007. Triassic magnetic overprints related to albitization in granites from the Morvan massif (France). *Palaeogeography, Palaeoclimatology, Palaeoecology*, **251**, 268–282.

RIJAL, M. L., APPEL, E., PETROVSKÝ, E. & BLAHA, U. 2010. Change of magnetic properties due to fluctuations of hydrocarbon contaminated groundwater in unconsolidated sediments. *Environmental Pollution*, **158**, 1756–1762.

RIPPERDAN, R. L., RICIPUTI, L., COLE, D., ELMORE, R. D., BANERJEE, S. & ENGEL, M. H. 1998. SIMS measurement of oxygen isotope ratios in authigenic magnetites from the Belden Formation, Colorado. *Journal Geophysical Research*, **103**, 21015–21024.

ROBERTS, A. P. & WEAVER, R. 2005. Multiple mechanisms of remagnetization involving sedimentary greigite (Fe3S4). *Earth and Planetary Science Letters*, **231**, 263–277, http://dx.doi.org/10.1016/j.epsl.2004.11.024

ROBERTS, A. P., CUI, Y. & VERSOUB, K. L. 1995. Wasp-waisted hysteresis loops: mineral magnetic characteristics and discrimination of component in mixed magnetic systems. *Journal Geophysical Research*, **100**, 17900–17924.

ROCHETTE, P., FILLION, G., MATTÉI, J.-L. & DEKKERS, M. J. 1990. Magnetic transition at 30–34 Kelvin in pyrrhotite: insight into a widespread occurrence of this mineral in rocks. *Earth and Planetary Science Letters*, **98**, 319–328.

ROCHETTE, P., FILLION, G. & DEKKERS, M. J. 2011. Interpretation of low-temperature data part 4: the low-temperature magnetic transition of monoclinic pyrrhotite. *The IRM Quarterly*, **21**, 7–10.

ROWAN, C. J., ROBERTS, A. P. & BROADBENT, T. 2009. Reductive diagenesis, magnetite dissolution, greigite growth and paleomagnetic smoothing in marine sediments: a new view. *Earth Planetary Science Letters*, **277**, 223–235.

SAFFER, B. & MCCABE, C. 1992. Further studies of carbonate remagnetization in the northern Appalachian basin. *Journal of Geophysical Research*, **97**, 4231–4248.

SALMON, E., EDEL, J. B., PIQUE, A. & WESTPHAL, M. 1988. Possible origins of Permian remagnetizations in Devonian and Carboniferous limestones from the Moroccan Anti-Atlas (Tafilalet) and Meseta. *Physics of the Earth and Planetary Interiors*, **52**, 339–351.

SAUNDERS, D., BURSON, K. & THOMPSON, C. 1991. Observed relation of soil magnetic susceptibility and soil gas hydrocarbon analyses to subsurface hydrocarbon accumulations. *AAPG Bulletin*, **55**, 344–353.

SHIPUNOV, S. V. 1997. Synfolding magnetization: detection, testing and geological applications. *Geophysical Journal International*, **130**, 405–410.

SHIVE, P. N., STEINER, M. B. & HUYCKE, D. T. 1984. Magnetostratigraphy, paleomagnetism, and remanence acquisition in the Triassic Chugwater Formation of Wyoming. *Journal of Geophysical Research*, **89**, 1801–1815.

STAMATAKOS, J. & KODAMA, K. P. 1991. Flexural flow folding and the paleomagnetic fold test: an example of strain reorientation of remanence in the Mauch Chunk Formation. *Tectonics*, **10**, 807–819.

STAMATAKOS, J., HIRT, A. M. & LOWRIE, W. 1996. The age and timing of folding in the central Appalachians from paleomagnetic results. *Geological Society of America Bulletin*, **108**, 815–829.

SUK, D., VAN DER VOO, R. & PEACOR, D. R. 1990. Scanning and transmission electron microscope observations of magnetite and other iron phases in Ordovician carbonates from east Tennessee. *Journal Geophysical Research*, **95**, 12327–12336.

SYMONS, D. T. A., SANGSTER, D. F. & LEACH, D. L. 1996. Paleomagnetic dating of Mississippi Valley-type Pb-Zn-Ba deposits. *In*: SANGSTER, D. F. (ed.) *Carbonate-Hosted Lead-Zinc Deposits*. Society of Economic Geologists, Colorado, Special Publications, **4**, 515–526.

SYMONS, D. T. A., ENKIN, R. & CIOPPA, M. T. 1999. Paleomagnetism in the Western Canada Sedimentary Basin: dating fluid flow and deformation events. *Bulletin of Canadian Petroleum Geology*, **47**, 534–547.

SYMONS, D. T. A., PANALAL, S. J., COVENEY, R. M. JR. & SANGSTER, D. F. 2005. Paleomagnetism of Late Paleozoic strata and mineralization in the Tri-State lead-zinc ore district. *Economic Geology*, **100**, 295–309.

SZABO, E. & CIOPPA, M. T. 2012. Multiple magnetizations in Ordovician–Devonian carbonates in the Williston Basin (Manitoba, Canada). *In*: ELMORE, R. D., MUXWORTHY, A. R., ALDANA, M. & MENA, M. (eds) *Remagnetization and Chemical Alteration of Sedimentary Rocks*. Geological Society, London, Special Publication, **371**, first published online 26 June 2012, http://dx.doi.org/10.1144/SP371.5

TARLING, D. H. 1985. Palaeomagnetic studies of the Orcadian Basin. *Scottish Journal of Geology*, **21**, 261–273.

TAUXE, L. 2005. Lectures in Paleomagnetism. World Wide Web Address: http://earthref.org/MAGIC/books/Tauxe/2005/

TOHVER, E., WEIL, A. B., SOLUM, J. G. & HALL, C. M. 2008. Direct dating of carbonate remagnetization by

40Ar/39Ar analysis of the smectite–illite transformation. *Earth Planetary Science Letters*, **274**, 524–530.

TORSVIK, T. H., LOVLIE, R. & STORRETVEDT, K. M. 1983. Multicomponent magnetization in the Helmsdale Granite, North Scotland: geotectonic implications. *Tectonophysics*, **98**, 111–129.

TORSVIK, T. H., STURT, B. A., SWENSSON, E., ANDERSEN, T. B. & DEWEY, J. F. 1992. Palaeomagnetic dating of fault rocks: evidence for Permian and Mesozoic movements and brittle deformation along the extensional Dalsfjord Fault, western Norway. *Geophysical Journal International*, **109**, 565–580.

URBAT, M., DEKKERS, M. J. & KRUMSIEK, K. 2000. Discharge of hydrothermal fluids through sediment at the Escanaba Trough, Gorda Ridge (ODP Leg 169): assessing the effects on the rock magnetic signal. *Earth and Planetary Science Letters*, **176**, 481–494.

VAN DER PLUIJM, B. 1987. Grain-scale deformation and the fold test – evaluation of synfolding remagnetization. *Geophysical Research Letters*, **14**, 155–157.

VAN DER VOO, R. 1993. *Paleomagnetism of the Atlantic, Tethys and Iapetus Oceans*. Cambridge University Press, Cambridge.

VAN DER VOO, R. & SCOTESE, C. R. 1981. Paleomagnetic evidence for a large (c. 2000 km) sinistral offset along the Great Glen Fault during the Carboniferous. *Geology*, **4**, 177–180.

VAN DER VOO, R. & TORSVIK, T. H. 2012. The history of remagnetization of sedimentary rocks: deceptions, developments and discoveries. *In*: ELMORE, R. D., MUXWORTHY, A. R., ALDANA, M. & MENA, M. (eds) *Remagnetization and Chemical Alteration of Sedimentary Rocks*. Geological Society, London, Special Publications, **371**, first published online 26 June 2012, http://dx.doi.org/10.1144/SP371.2

VERWEY, E. J. W. 1939. Electronic conduction of magnetite ($Fe_3O_4$) and its transition point at low-temperature. *Nature*, **144**, 327–328.

WALDHÖER, M. & APPEL, E. 2006. Intersections of remanence small circles: new tools to improve data processing and interpretation in palaeomagnetism. *Geophysical Journal International*, **166**, 33–45.

WEAVER, R., ROBERTS, A. P. & BARKER, A. J. 2002. A late diagenetic (syn-folding) magnetization carried by pyrrhotite; implications for paleomagnetic studies from magnetic iron sulphide-bearing sediments. *Earth and Planetary Science Letters*, **200**, 371–386.

WEIL, A. & VAN DER VOO, R. 2002a. The evolution of the paleomagnetic fold test as applied to complex geologic situations, illustrated by a case study from northern Spain. *Physics and Chemistry of the Earth*, **27**, 1223–1235.

WEIL, A. B. & VAN DER VOO, R. 2002b. Insights into the mechanism for orogen-related carbonate remagnetization from growth of authigenic Fe-oxide: a SEM and rock magnetic study of Devonian carbonates from northern Spain. *Journal Geophysical Research*, **107**(B4), 2063, http://dx.doi.org/10.1029/2001JB 000200

WOODS, S., ELMORE, R. D. & ENGEL, M. 2002. Paleomagnetic dating of the smectite-to-illite conversion: testing the hypothesis in Jurassic sedimentary rocks, Skye, Scotland. *Journal of Geophysical Research*, **107**, 2091, http://dx.doi.org/10.1029/2000JB000053

XU, S. & MERRILL, R. T. 1992. Stress, grain size, and magnetic stability of magnetite. *Journal of Geophysical Research*, **97**, 4321–4329.

XU, W., VAN DER VOO, R. & PEACOR, D. R. 1998. Electron microscopic and rock magnetic study of remagnetized Leadville carbonates, central Colorado. *Tectonophysics*, **296**, 333–362.

ZECHMEISTER, M. S., PANNALAL, S. & ELMORE, R. D. 2012. A multidisciplinary investigation of multiple remagnetizations within the Southern Canadian Cordillera, SW Alberta and SE British Columbia. *In*: ELMORE, R. D., MUXWORTHY, A. R., ALDANA, M. & MENA, M. (eds) *Remagnetization and Chemical Alteration of Sedimentary Rocks*. Geological Society, London, Special Publications, **371**, first published online 22 August 2012, http://dx.doi.org/10.1144/SP371.11

ZEGERS, T. E., DEKKERS, M. J. & BAILLY, S. 2003. Late Carboniferous to Permian remagnetization of Devonian limestones in the Ardennes: role of temperature, fluids, and deformation. *Journal Geophysical Research*, **108**, 2357, http://dx.doi.org/10.1029/2002JB002213

ZWING, A., CLAUER, N., LIEWIG, N. & BACHTADSE, V. 2009. Identification of remagnetization processes in Paleozoic sedimentary rocks of the northeast Rhenish Massif in Germany by K/Ar dating and REE tracing of authigenic illite and Fe oxides. *Journal of Geophysical Research*, **114**, B06104, http://dx.doi.org/10.1029/2008JB006137

# The history of remagnetization of sedimentary rocks: deceptions, developments and discoveries

ROB VAN DER VOO[1,2]* & TROND H. TORSVIK[2,3,4,5]

[1]*Department of Earth and Environmental Sciences, University of Michigan, Ann Arbor, MI 48109-1005, USA*

[2]*Center for Advanced Study, Norwegian Academy of Science and Letters, Drammensveien 78, N 0271 Oslo, Norway*

[3]*Department of Physics, University of Oslo, Postboks 1048 Blindern 0316, Oslo, 7 Norway*

[4]*Centre for Geodynamics, Norwegian Geological Survey, Leiv Erikssons vei 39, 7491 Trondheim, Norway*

[5]*School of Geosciences, University of the Witwatersrand, Wits 2050, South Africa*

**Corresponding author (e-mail: voo@umich.edu)*



**Abstract:** Remagnetizations have been recognized ever since magnetizations in rocks were demonstrably shown to have been acquired at a much later time than the formation or deposition of the rocks themselves. There was mention of remagnetizations as early as the 1950s, and in the 1960s the concept was frequently hypothesized as an explanation for repetitions and loops in apparent polar wander paths. In this paper, remagnetization features and processes are organized by magnetic carrier: hematite, magnetite, Fe-sulphides and goethite. Selected case histories are presented which are chosen in order to reveal important diagnostics, although many origins of remagnetizations are still obscure or incompletely known.


It is not clear when the first mention of the phenomenon known as 'remagnetization' appeared in the literature; clearly, solid-state physicists knew long ago that a heating and cooling cycle in the presence of a weak field applied to magnetic minerals, such as the iron-oxides, could produce a new magnetization which either overprinted an existing remanence or which replaced it entirely if the temperature exceeded the Curie or Néel point. Jaeger (1957) refers to a study of the Torridonian red beds (Irving & Runcorn 1957) and mentions that these authors have shown that 'coarse-grained, ill-sorted sediments may be expected to show a wide dispersion in directions, but if they have been remagnetized after heating above their Curie points they should show a narrow dispersion'. The first mention of remagnetization in the title of an article that we could find was in a discussion by Graham (1961) of lightning currents.

There is also a study by Clegg *et al.* (1957) of Triassic red beds in Spain, in which they found magnetizations with directions (not corrected for tilt of the strata) that clustered around that of the present-day geomagnetic field. Tilt corrections would produce disparate groupings, and they concluded that 'since the mean magnetic inclination is approximately 25° steeper than that of [previously studied] British rocks, while the mean latitude of the Spanish sites is some 10° farther south than that of the British, the final interpretation of the results remains inconclusive'. Although the word 'remagnetization' was not used, there is a clear link to the concept inherent in their cautious wording. It is worth noting that thermal, chemical or alternating field demagnetizations were not then carried out on these red beds, so that a primary remanence may well have been hidden by a larger present-day field overprint.

Following these somewhat tentative mentions of doubts that arose about the time of acquisition of a remanence, the evidence for remagnetizations became increasingly abundant even although the full extent of the problem would not be recognized until the 1980s. Remagnetizations were abundantly documented (e.g. Creer 1962, 1968; Chamalaun & Creer 1963; Storetvedt 1968) in the 1960s, however. Creer's 'remagnetization hypothesis' deserves mention here as it will play a role in discussions that follow of remagnetizations carried by magnetite in carbonate rocks. Noticing that many magnetization directions in lower Palaeozoic rocks resembled those in upper Palaeozoic rocks, Creer proposed that the lower Palaeozoic rocks that were located in low latitudes were remagnetized in a moist, tropical, weathering and soil-forming environment, in a process called lateritization (also written as


*From*: ELMORE, R. D., MUXWORTHY, A. R., ALDANA, M. M. & MENA, M. (eds) 2012. *Remagnetization and Chemical Alteration of Sedimentary Rocks*. Geological Society, London, Special Publications, **371**, 23–53.
First published online June 26, 2012, http://dx.doi.org/10.1144/SP371.2

laterization, see Creer 1968). A similar process was subsequently echoed for western Australia in Cenozoic times instead of the late Palaeozoic (Schmidt & Embleton 1976).

Since those early days, remagnetization has been documented with increasing frequency. Entering the word in a GeoRef search (on 6 September 2011) yields 409 references that contain 'remagnetization' in the title and some 1672 journal articles that contain the word anywhere in the text. This is our initial stumbling block because the concept is poorly represented by its name: strictly defined, a remagnetization is a remanence that is restored in a substance that had been demagnetized. Palaeomagnetists, however, generally do not know whether a remagnetized rock previously possessed another magnetization that was more or less fully erased afterwards. A previously existing remanence can only be detected when a partial remagnetization is evident. The custom has therefore taken hold to call any secondarily acquired remanence a remagnetization, but it must be noted that some workers prefer to adhere to the term 'secondary magnetization'. We shall see later that there are also fuzzy boundaries to this definition of remagnetizations, when we will briefly examine delayed acquisition of remanence in sedimentary rocks.

Because remagnetizations are common and abundant (as seen in the GeoRef numbers in the paragraph above), a review like this one has to be very selective. Practicing palaeomagnetists hoping to see their favourite discovery of a secondary remanence mentioned in the following pages may be rather disappointed, because space limitations prevent us from making this a comprehensive compilation. For excellent reviews that are more complete than this history of selected remagnetizations, we refer to McCabe & Elmore (1989) and Butler's electronic version of his out-of-print 1992 book (to be found at http://www.geo.arizona.edu/Paleomag/book/).

The organization of this review is along lines that divide the carriers of magnetization; after discussing the great red-bed controversy and the various observations of hematite as a secondary mineral, it will be the turn of magnetite, followed by Fe-sulphides and goethite. Most of the case histories presented will involve sedimentary rocks; metamorphic and igneous rocks are not specifically ignored, but will not be covered in detail in order to avoid rendering this review too long and to stay within the scope of this special publication.

## Hematite, red beds

Exposures of red beds of late Palaeozoic and Mesozoic ages are abundant in the western USA, notably in and around the Colorado Plateau and Wyoming.

Many of these formations were palaeomagnetically studied in the 1950s and then increasingly in the 1960s and early 1970s with thermal demagnetization using a limited series of rather regular steps, arguably 'cleaning' away the overprints (if any) of demonstrably much younger age. Some of the motivations for these studies were to document reversal stratigraphies for time intervals after the long Kiaman Reversed Superchron (KRS), as well as for determinations of palaeopoles that could help refine the apparent polar wander path (APWP) of Laurentia. The latter paralleled efforts underway in Europe and South America where red beds were equally abundant, or in Australia where late Palaeozoic igneous rocks appeared to be suitable. One additional reason that red beds were primary targets for palaeomagnetic studies is that they could readily be measured with the instruments available in the 1960s, such as astatic and fluxgate spinner magnetometers.

The increasing knowledge of red-bed characteristics, such as the demagnetization behaviour and remanence features of the Triassic Moenkopi and Chugwater formations, caused palaeomagnetists to become convinced that their favourite formation and its remanence characteristics revealed generic and widely applicable features. This led to rather serious disagreements and debates, in the literature and at meetings, between two opposing camps. The situation came to a head in 1979, when a Penrose Conference was organized in Durango, Colorado, by Gene Shoemaker; participants in that meeting found themselves in a position to gain first-hand insights into what the scientists believed to be true rather than into what the palaeomagnetic observations actually revealed. As palaeomagnetist Jim Channell (then from the UK and Switzerland) thought during the Penrose Conference: 'Wow, so this is how they do science over here'! (Channell, pers. comm., March 7, 2011).

One side of the controversial debate provided documentation that hematite occurred as specular hematite in Arizona's Moenkopi Formation outcrops; the workers inferred that this hematite was detrital, and hence carried a depositional remanent magnetization (DRM) dating back to the age of the strata and thus primary in origin (Elston & Purucker 1979; Steiner 1983). Magnetostratigraphic columns, including polarity transitions recorded by a metre-thick section of strata in the Chugwater Formation, could be matched over distances up to a kilometre (Herrero-Bervera & Helsley 1983; Shive et al. 1984).

The other side in the debate observed that the red beds contained multiple components of magnetization, superposed on each other as if the magnetizations were acquired over durations of millions of years and carried by chemical precipitates of hematite (Roy & Park 1972; Larson & Walker 1975, 1982; Turner 1980; Larson *et al.* 1982). Of course, such protracted processes could produce coherent magnetostratigraphic records only in gross outline, whereas shorter-term geomagnetic variations (rapid reversal sequences or polarity transitions themselves) would have little or no chance of being faithfully reproduced. This could well be called the chemical remanent magnetization (CRM) hypothesis.

In perfect hindsight, it seems wondrous to observe that principal component analysis (PCA) of multivectorial remanence, established by careful stepwise progressive thermal and/or chemical demagnetization, would have helped overcome the barriers and entrenched convictions; PCA was however not yet employed in those years by the scientists, who typically only demagnetized their rocks with blanket treatments. Ironically, it was already well known that hematite's two principal modes of occurrence (pigment and specularite) typically had contrasting laboratory unblocking temperature spectra (Fig. 1a; Collinson 1974), with the prevailing one of these two modes determining how the demagnetization of a whole-rock sample would proceed (e.g. open circles in Fig. 1a). In addition, the response of the two modes of hematite to chemical leaching is diagnostically different. A decade later, compelling examples of the thermal unblocking of pigment and specularite remanence would become routine in PCA of carefully chosen demagnetization steps and trajectories; note, for instance, the apex (turning point) of the Zijderveld diagram in Figure 1b with temperatures between 550 and 620 °C marking the isolation of detrital specularite after removal of abundant pigment (in steps between 350 and 550 °C).

Contrast this with the data in the stereonet of Figure 2 which show northward shifts upon selective destructive demagnetization, in this case by removal through sanding of red silt layers in a sample (with pigment presumably) leaving the coarser (specularite containing?) sandy layers to contribute progressively more to the remanence (Larson *et al.* 1982) as more and more silt material is removed. The remaining remanence can, in retrospect, be attributed to detrital grains (and therefore a primary remanence), but this does not appear to have been recognized as such at the time by the authors of the experiment.

Two other aspects of the red-bed controversy eventually established some credibility for the hypothesis that red beds could contain an early-acquired DRM, besides those already mentioned (Herrero-Bervera & Helsley 1983; Shive *et al.* 1984). The CRM hypothesis (Larson & Walker 1982) was not to be outdone either, and it has turned out to be a valid way of looking at some

other remagnetization problems (discussed below). The two aspects referred to are illustrated in Figures 3 and 4. Specularite grains can be rather equidimensional and, if so, they will have a tendency to roll down a short distance along a foreset slope in cross-bedded sequences (Fig. 3d). The more they roll the more their magnetization directions deviate from the ambient field, resulting in a correlation between inclination and foreset bedding angle (Fig. 3a, c). The deviation of the magnetization direction shows that the grains were already magnetized when they were deposited, in a positive confirmation of the DRM hypothesis. The irony is, of course, that the magnetization direction of the

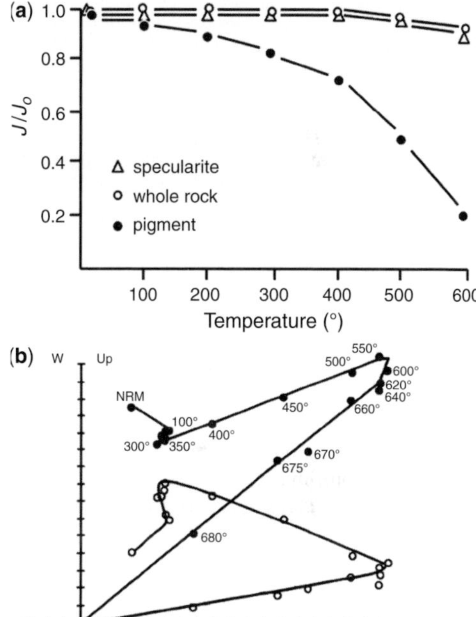

**Fig. 1.** (a) The normalized magnetization of hematite pigment and specularite contained in red beds, as a function of temperature, as well as an example of the resulting whole-rock behaviour (reprinted from Collinson 1974 with permission from Wiley & Sons). (b) An example of a thermal demagnetization (Carboniferous Mauch Chunk red beds; Kent & Opdyke 1985; McCabe & Elmore 1989) which shows very different unblocking and directional trajectories for the pigment (350–550 °C) and for (presumably coarser) hematite grains with unblocking between 640 and *c.* 680 °C, in accord with the predicted behaviour from (a). The pigment carries a late Palaeozoic remagnetization direction (SSE and shallowly up), whereas the coarser hematite carries an Early Carboniferous primary remanence.

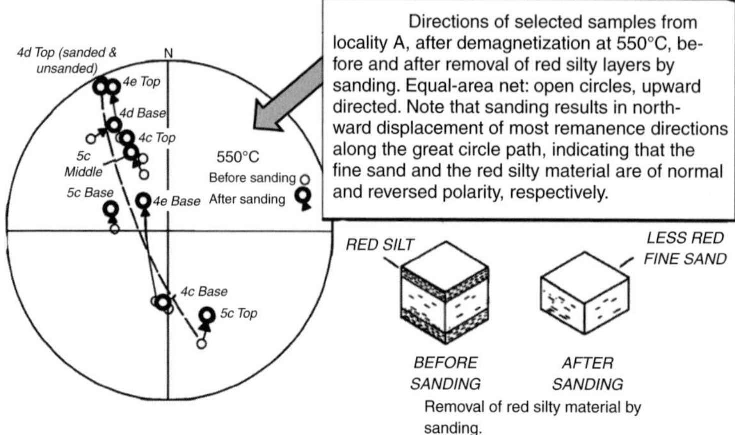



**Fig. 2.** Changes in magnetization direction upon removal by sanding of red silty layers of Triassic Moenkopi red beds, taken as evidence for superposed dual polarity remanences, one of them (or both?) being a remagnetization. Note that vector subtraction has not been carried out, so it cannot be ascertained whether the vector removed is Triassic or present-day field or of any other age (from Larson *et al.* 1982).

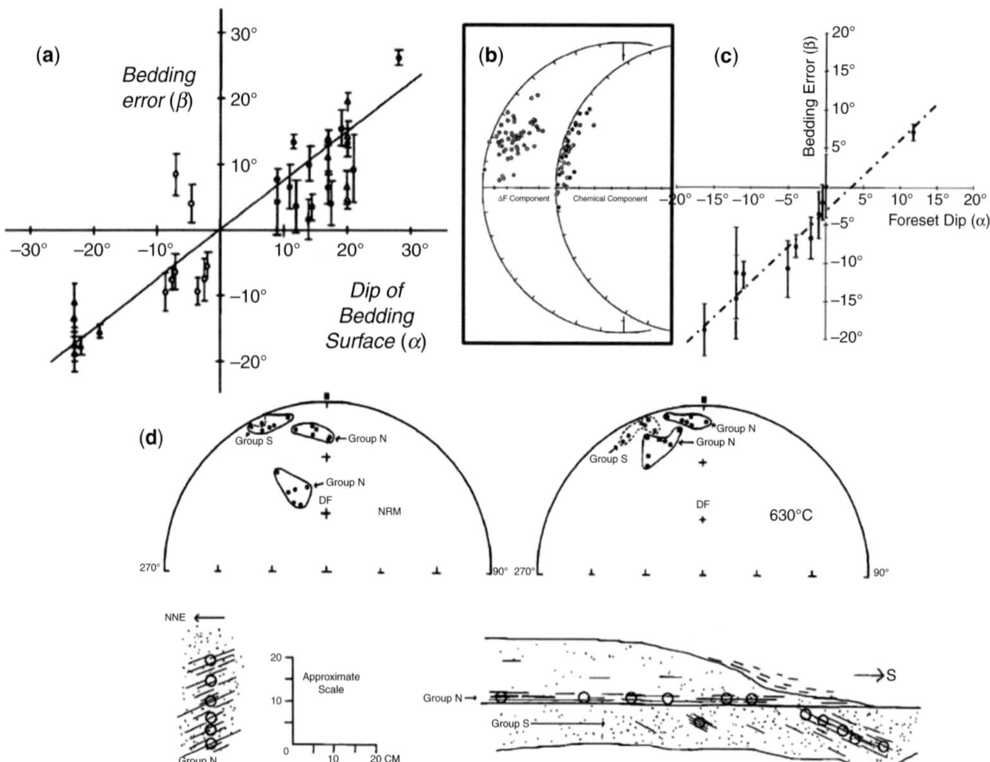



**Fig. 3.** (**a**) Grains rolling down foreset beds in cross-bedded Triassic Moenkopi red beds produce an inclination error that correlates one-on-one with foreset slope (Elston & Purucker 1979). (**b**) Magnetization directions isolated by AF or chemical demagnetization of *c.* 1.05 Ga Copper Harbor red beds (Elmore & Van der Voo 1982). (**c**) As in (a) but then for the Copper Harbor foreset beds (Elmore & Van der Voo 1982). (**d**) Directions of sample groups show a shift upon thermal demagnetization at 630 °C and locations of samples in foreset beds (Elston & Purucker 1979).

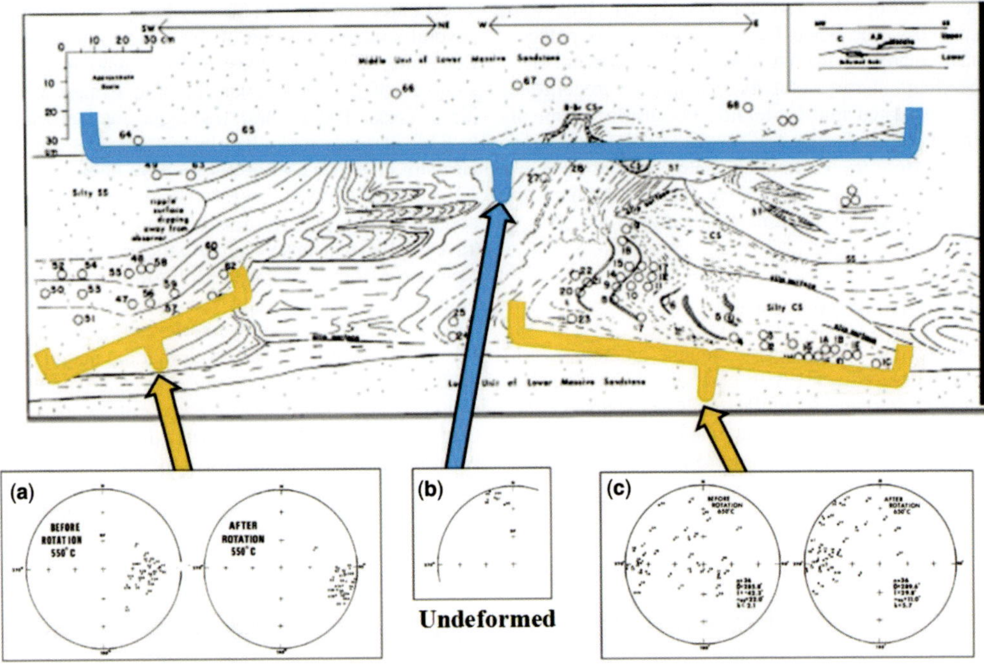

**Syn-sedimentary slump of silty sandstone**   **Syn-sedimentary slump of clay stone**

**Fig. 4.** Undeformed Moenkopi red beds overlie syn-sedimentary slumps that have been palaeomagnetically investigated by Purucker *et al.* (1980). (**a**) Directions from silty sandstone samples from southwest side of the outcrop. (**b**) Sample directions from the overlying undeformed bed. (**c**) Sample directions from the slump in hematitic claystone on the outcrop's east side.

samples is deflected from that of the ambient field, resulting in one form of inclination error.

Figure 4 illustrates data from a couple of syn-sedimentary slumps below an undisturbed level of Moenkopi red beds in Arizona (Purucker *et al.* 1980). The randomization of the magnetization directions in the slumps constrains the timing of the acquisition of the remanence to nearly immediately after deposition.

Erroneous interpretations of red bed (re-)magnetizations in the Central Appalachians and the United Kingdom were being published at about the same time as the Durango Penrose Conference, and they are illustrative enough to be mentioned here as well. The palaeomagnetism of Devonian (Catskill) red beds in the Central Appalachians had been studied by Kent & Opdyke (1978) and Van der Voo *et al.* (1979). The results were logically (but wrongly) considered to be a valid palaeomagnetic record for the Laurentian craton during the Devonian. It seemed that Carboniferous results from the Barnett Formation in Texas (Kent & Opdyke 1979), Devonian limestone results in Arizona (Elston & Bressler 1977) and the Columbus Limestone directions in Ohio (Martin 1975) all supported the

emerging palaeolatitude pattern derived from the Catskill rocks, which located the equator as running approximately from Virginia to Arizona. Outboard of the craton in New England, Devonian and Carboniferous results were obtained from what was labelled 'Acadia', but these showed a significant (*c.* 20°) palaeolatitude difference with the craton (Kent 1982). Van der Voo & Scotese (1981) pointed out that Europe and Laurentia showed a similar palaeolatitude difference, but considered that the Scottish Lewisian Basement and the Archean–Proterozoic basement in Greenland should remain strongly coupled. This led them to suggest that the boundary between Laurentia and Europe was located to the south of the Scottish Lewisian, and they proposed the Great Glen Fault as an obvious candidate. Indeed the same 20° palaeolatitude difference was revealed by a comparison of the Devonian in the Orcadian Basin in northernmost Scotland and that in Great Britain to the south of the fault. A flurry of comments and replies followed and eventually the 20° displacement of the 'Acadia Terrane' and the Great Glen hypothesis was rejected, even by its original authors (e.g. Kent & Opdyke 1985, based on results such as those of Fig. 1b; Van der Voo 1993). This was especially true when Early Carboniferous palaeomagnetic results from the Deer Lake Group in Newfoundland showed that Acadia had not undergone the postulated mid-Carboniferous displacement of 20° (Irving & Strong 1984), whereas the Devonian results from the Laurentian craton and from Scotland north of the Great Glen Fault were demonstrated to be Permian remagnetizations.

## Magnetite in carbonates

The advent of cryogenic magnetometers, with their increased sensitivity and rapid measurement systems (Goree & Fuller 1976), expanded the target rock types for palaeomagnetic study in the early 1970s to include limestones and dolomites.

The Laurentian interior quickly became a target of several studies, and one of the earlier studies was that of McElhinny & Opdyke (1973) who measured samples of the Ordovician Trenton Limestone in New York. The natural remanent magnetization (NRM) of these rocks had been studied earlier by Graham (1956) who did not employ demagnetization; the results were therefore rather meaningless and directed nearly vertically down, which would have implied that these warmer-water deposits would have come from a polar area. McElhinny and Opdyke obtained very similar and very steep NRM directions (Fig. 5a) and employed alternating field (AF) demagnetization, with the result that the magnetization directions moved during demagnetization to southerly and shallow positive inclination positions (Fig. 5a). Because they asserted that these downward directions were significantly different to the

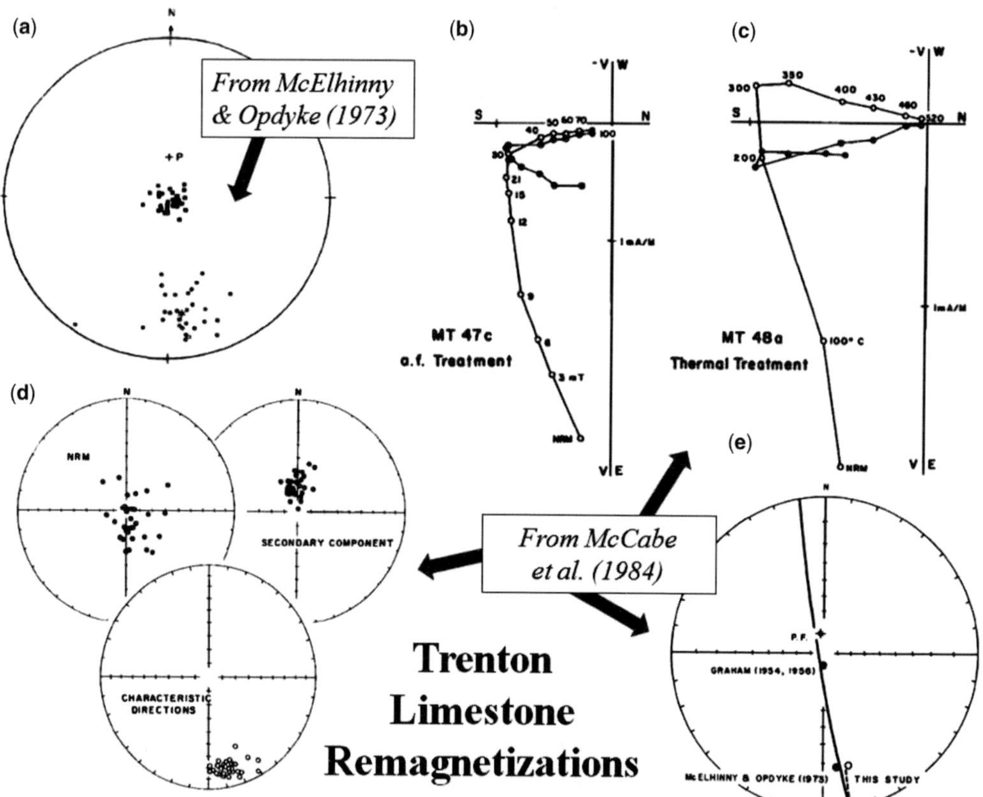


**Fig. 5.** Successive palaeomagnetic results from the Ordovician Trenton limestone in New York. (**a**) Near-vertical directions characterize the undemagnetized NRMs. SSE/shallow down sample directions are obtained in AF demagnetization (McElhinny & Opdyke 1973). (**b**) AF demagnetization (McCabe *et al.* 1984). (**c**) Thermal demagnetization (McCabe *et al.* 1984) showing greater effectiveness in isolating the ChRM directions than AF demagnetization. (**d**) Sample directions of the NRM, directions of secondary, present-day field overprints and directions of the ChRM. (**e**) Mean directions of present-day field overprints, Graham's original total NRM directions, the AF demagnetized directions of McElhinny & Opdyke (1973) and the thermally demagnetized directions of McCabe *et al.* (1984).

(shallow negative inclination) late Palaeozoic directions in New York, they argued that these limestones constituted evidence that the remagnetization hypothesis discussed earlier (Creer 1968) could be discounted. This tale, however, does not end here. Another decade later McCabe and colleagues (1984) resampled the Trenton Limestone, this time employing thermal demagnetization as the principal means of isolating the characteristic magnetization component. The differences between the latter with AF demagnetization results can be seen in Figure 5b and c and by comparing Figure 5a and 5d. The thermal demagnetization succeeds much more effectively in removing the steeply downward direction that had been a source of discouragement to John W. Graham in the 1950s. Figure 5e sums it all up and shows the mean directions of the three successive studies (Graham 1956; McElhinny & Opdyke 1973; McCabe *et al.* 1984), which in the end contained compelling evidence for remagnetization after all even if not befitting the context of Ken Creer's hypothesis. Subsequent studies have concentrated on subsurface samples of the Trenton strata and have confirmed (Suk *et al.* 1993*b*; Garner & Cioppa 2006) what was concluded earlier by McCabe *et al.* (1984) about a Permian remagnetization age.

The successful removal of the Trenton remanence during AF or thermal demagnetization, with characteristic coercivities or laboratory unblocking temperatures, collectively suggests that the remanence carrier in these dark-grey fossiliferous limestones is magnetite. The mean direction reported by McCabe *et al.* (1984) is similar to clearly postfolding, but Permian-like, magnetizations in limestones from the Valley and Ridge Province of the West Virginia Appalachians (Chen & Schmidt 1984). This similarity suggests that this is a remagnetization. This possibility had not yet occurred to most palaeomagnetists in the 1970s, however; limestone palaeopoles were reported as reflecting primary magnetizations carried by magnetite by Martin (1975) on rocks from Ohio, by Elston & Bressler (1977) for Arizona, by Kent & Opdyke (1979) for Texas and by Scott (1979) for Arkansas. Only Steiner (1973) had diagnosed the remanence unblocked below 400 °C in the Arbuckle carbonates as a partial overprint, aided in that interpretation by a fold test, but her conclusion was otherwise largely unheeded in other studies of carbonate rocks during this decade.

The notion that prevailed in the 1970s is well summarized by the excellent review of McCabe &

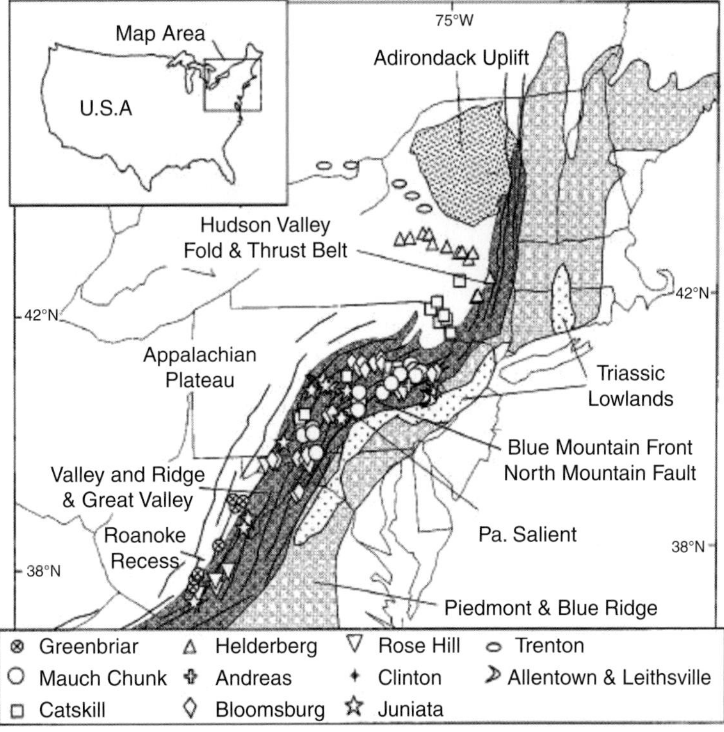


**Fig. 6.** Map of the New York, Pennsylvania and Virginia Appalachians with sampling sites of red beds and carbonate rocks. From Stamatakos *et al.* (1996).

Elmore (1989), when they state 'carbonate magnetizations residing in magnetite had usually been interpreted as an early-acquired remanence' and that 'it was thought that magnetite in sediments must have a detrital origin'. Fabric studies also seemed to support the primary nature of the remanence (McElhinny & Opdyke 1973).

However, application of the fold test came to the rescue in an unexpected fashion. While planning the chapters of his dissertation at the University of Chicago, Chris Scotese proposed to collect palaeomagnetic samples from the Siluro-Devonian rocks in the Helderberg escarpment in New York, as well as from the Valley and Ridge Province in Virginia and Pennsylvania, and measure them at the University of Michigan. He collected from both limbs of folds and found that the *in situ* directions of the characteristic remanent magnetizations (ChRMs) were generally south-directed with shallow (upwards or downwards) magnetizations. Upon tilt correction, the directions changed to shallow downwards or shallow upwards, respectively, bypassing each other in the process at about midpoint (i.e. at about 50% tilt correction). The most reasonable explanation for this was to have the magnetizations acquired during the folding (Scotese *et al.* 1982; McCabe *et al.* 1983; Scotese 1985). An alternate explanation was proffered by van der

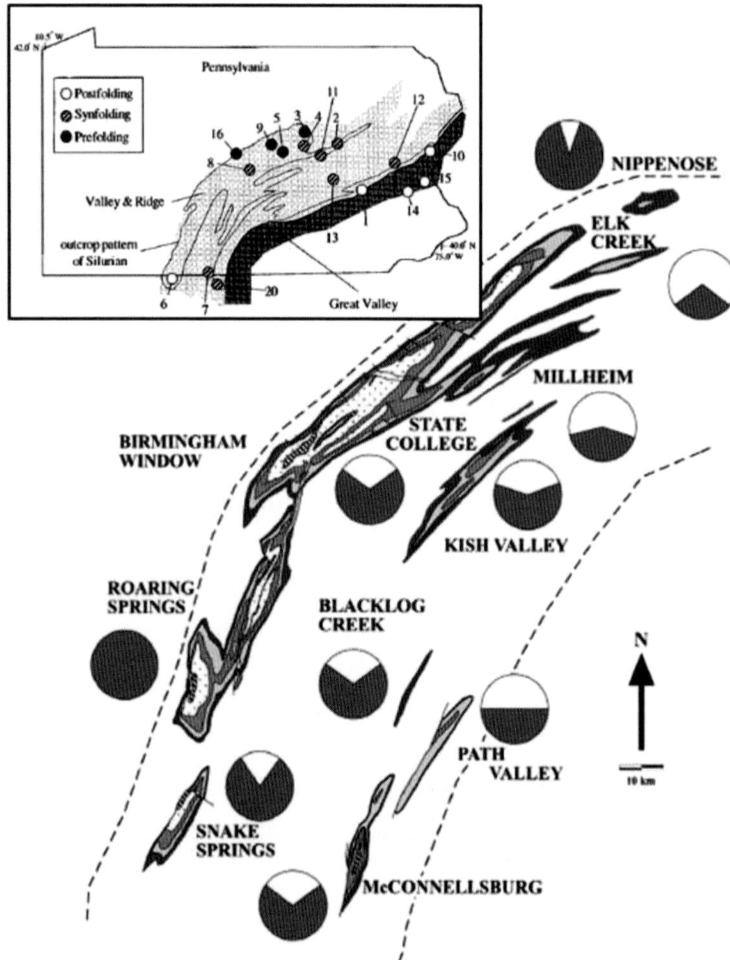


**Fig. 7.** Inset: map of the Pennsylvania salient with a gradation from southeast to northwest of post-folding, syn-folding and pre-folding red-bed remagnetizations (Stamatakos *et al.* 1996). Main map shows the locations of the fold tests carried out on the limbs of Valley and Ridge fold structures in Cambro-Ordovician remagnetized carbonates (reprinted from Cederquist *et al.* 2006 with permission of Elsevier), with the pie diagram shading indicating the percent unfolding needed to bring the results from paired limbs into agreement. The east to west pattern mimics that of the red beds.

Pluijm (1987) and Kodama (1988), suggesting that the grain-scale strain involved in folding could be rotating the remanence to appear as syn-folding. However, tests of the alternatives have generally provided results in favour of a real syn-folding magnetization acquisition.

A plethora of palaeomagnetic studies in the Central Appalachians between upstate New York and southern Virginia has succeeded in giving us a reasonably good understanding of magnetizations in carbonate rocks (as well as red beds). Figure 6 (from Stamatakos *et al.* 1996) shows the sampling site distribution by the mid 1990s, whereas Figure 7 illustrates the temporal relationship between folding and magnetization acquisition for the Pennsylvania salient. The red-bed results (inset of Fig. 7) show that in the easternmost zone of the Valley and Ridge Province the magnetizations are post-folding (open circles). In the westward zone however, the magnetizations appear to be acquired during folding, followed in turn by pre-folding

magnetizations in the frontal folds (fully filled symbols). Syn- to pre-folding results in Cambro-Ordovician carbonates in the same areas mimic this situation, with the pie-diagrams of Figure 7 showing percent unfolding needed to bring results from fold limbs into agreement (Cederquist *et al.* 2006). In Figure 8, the temporal and spatial development of the post-, syn- and pre-folding results is portrayed; it appears that folding was protracted, whereas magnetization acquisition must have been relatively fast. Some selected examples of the multiple fold tests of the carbonate-hosted magnetizations are given in Figure 9 (from Cederquist *et al.* 2006). The upward-pointing triangles mark the level at which the value of the concentration parameter $k$ (Fisher 1953) is located in relation to statistical significance. It can be seen that three of the four examples in Figure 9 satisfy the significance requirement.

Although the folded strata of the Central Appalachians have revealed the temporal–spatial

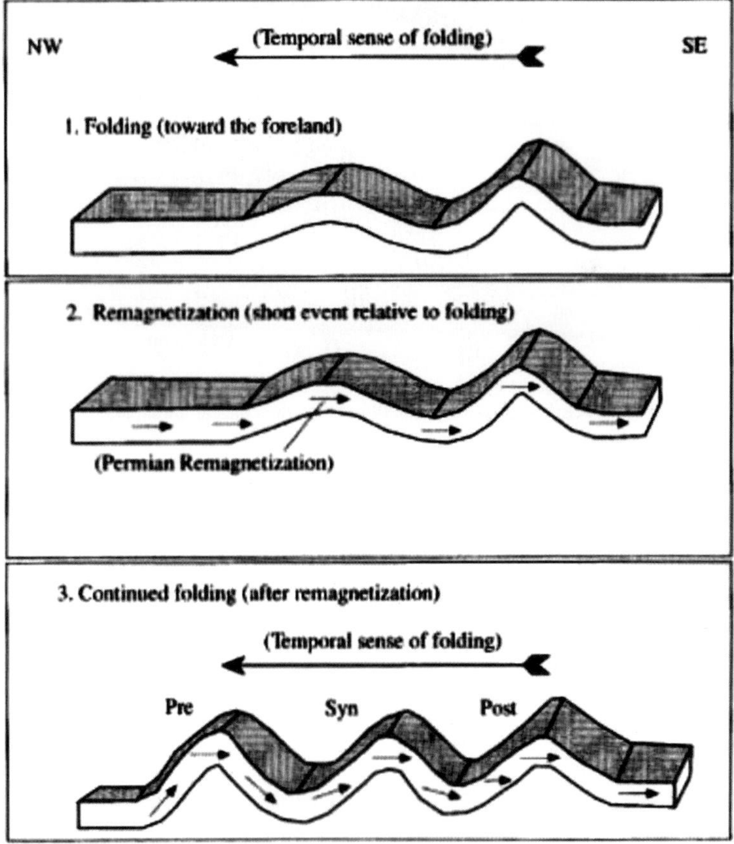



**Fig. 8.** Slow progressive folding, coupled with fast remanence acquisition, is portrayed during the remagnetization event thought to be responsible for the patterns of Figure 7 (from Stamatakos *et al.* 1996).

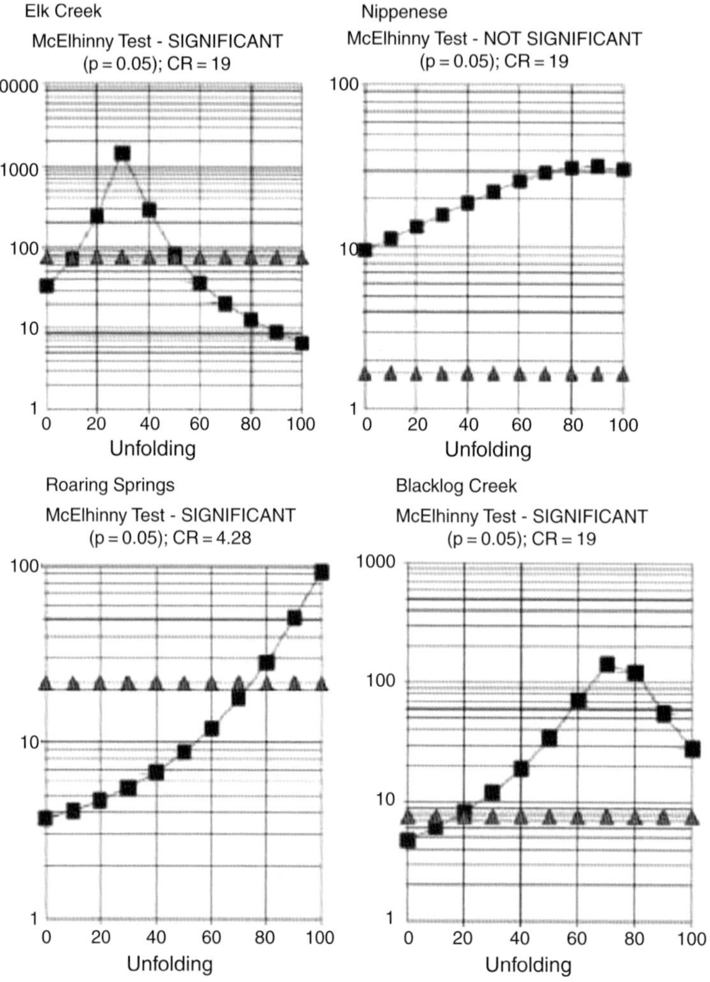


**Fig. 9.** Selected examples of fold test results in Cambro-Ordovician carbonates of the Pennsylvania salient (reprinted from Cederquist *et al.* 2006 with permission of Elsevier). Plots show the concentration parameter *k* versus percent unfolding.

relationships of remagnetization processes, it appears that the cratonic interior also has plenty to offer in this respect. Remagnetizations in the flat-lying Ordovician Trenton Limestone have already been mentioned, but carbonates in Ohio (Martin 1975), Michigan (Suk *et al.* 1993*b*), Wisconsin (Kean 1981), Missouri (Wisniowiecki *et al.* 1983), Illinois and other mid-western states (Lu *et al.* 1990) have all contributed to an overall image of continent-wide remagnetization that occurred in late Palaeozoic time (we will discuss the age estimates in a later section). The cause of this pattern remains one of the more puzzling aspects today, although it is tempting to invoke the contemporaneity of the remagnetization event with the Alleghenian–Ouachita orogeny as having

something to do with it. Other orogenies in North America with different ages – Taconic (Ordovician), Sevier (Cretaceous–Eocene), Laramide (Cretaceous–Eocene), Ellesmerian (Devonian), etc. – and other continents have their own remagnetization associations. A few examples may suffice: Devonian and Lower Carboniferous limestones in the Belgian Ardennes contain Permo-Carboniferous remagnetizations that resemble those in the Central Appalachians (e.g. Molina-Garza & Zijderveld 1996; Zegers *et al.* 2003); Permian–Triassic remagnetizations have been found in the Basin and Range Province (Gillett & Karlin 2004); Palaeozoic carbonates in Colorado carry Permian as well as late Mesozoic remagnetizations (Xu *et al.* 1998); lower Palaeozoic limestones near the

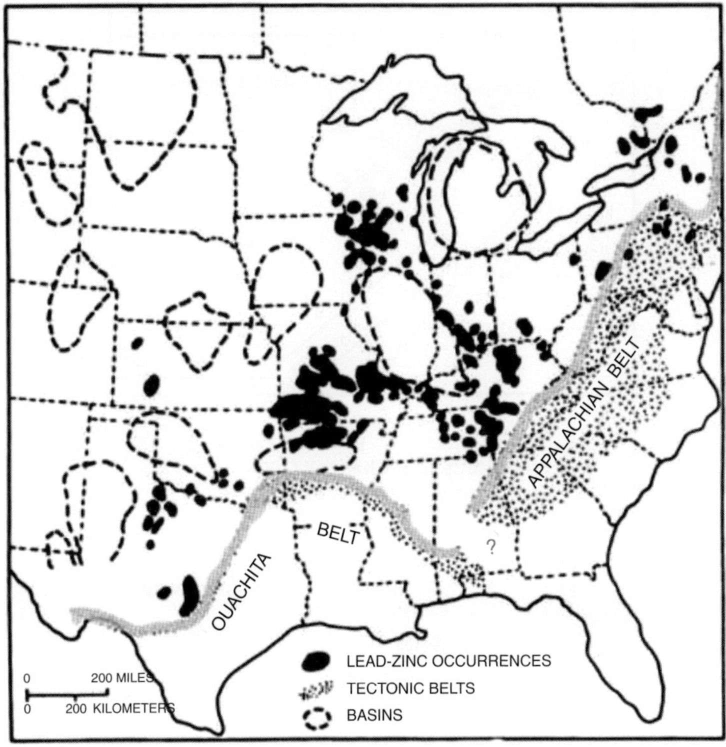


**Fig. 10.** Distribution of Mississippi-Valley-Type deposits in Laurentia (Oliver 1986) which occur in intracratonic basins in similar places as remagnetizations in platform carbonates (not shown), hinting that fluid migration is responsible for both the occurrences of MVTs and remagnetized carbonates.

Taconic orogenic belt in Vermont show remagnetizations of likely Ordovician age (Tucker & Kent 1988); Devonian limestones in Cantabria, Spain, carry remagnetizations of Permian age (Weil & Van der Voo 2002); late Early Permian syntectonic remagnetizations have been documented in the Tunas Formation, Sierras Australes of Argentina (Tomezzoli 2001). Many, many more examples deserve to be added, but cannot for lack of space.

The association between orogeny and event age has also been proposed for the emplacement of Mississippi-Valley-Type (MVT) deposits, typically in the form of lead-zinc ores (Oliver 1986) as well as hydrocarbon migration and stress-related bedding-parallel shortening as recorded by calcite twinning (Craddock & van der Pluijm 1989). Collectively, this has been recognized by Oliver (1986) as possibly caused by fluids expelled from the orogenic belts at the cratonic margin (Fig. 10). However, despite many efforts and studies of remagnetizations in a variety of settings, the squeegee hypothesis (as it is often called) has not become better substantiated than it was in the late 1980s (e.g. Evans *et al.* 2000). A somewhat related discovery showing secondary magnetite to be associated

with solid bitumen in quarries in Illinois and Mississippi (McCabe *et al.* 1987) has remained an isolated observation.

## Rock magnetism and imaging of tiny magnetite grains

This review would not be complete without a discussion of magnetite's mineralogic characteristics in carbonate rocks, given that some progress has been made in characterizing the likely grain sizes and morphologies and perhaps even the ancestry of magnetic grains in a few cases. Figure 11 shows plots of hysteresis parameters (so-called 'Day plots'; Day *et al.* 1977) with the different symbols representing remagnetized limestones from North America and Europe (McCabe & Channell 1994; Elmore *et al.* 2006). The locations of these data points are unique and deviate from the usual mixing lines of single-domain (SD) and multidomain (MD) hysteresis parameters. This is attributed to a substantial contribution of superparamagnetic grains, that is, grains of ferromagnetic materials that are too small to compete with the

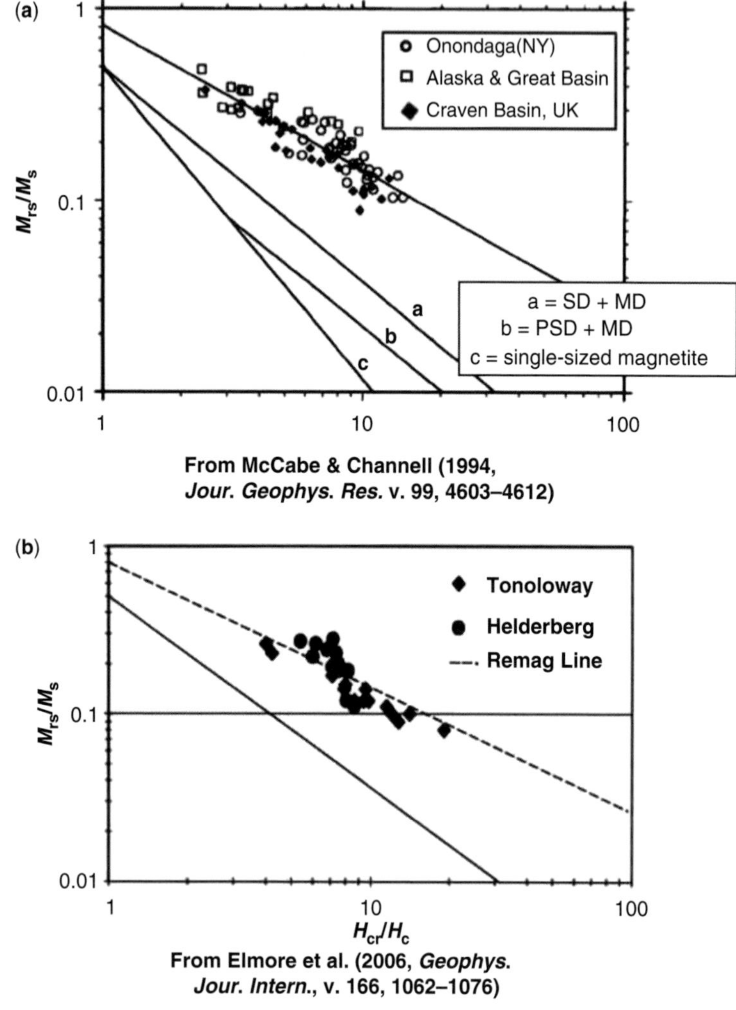



**From McCabe & Channell (1994,**
***Jour. Geophys. Res.* v. 99, 4603–4612)**

**From Elmore et al. (2006, *Geophys.***
***Jour. Intern.*, v. 166, 1062–1076)**

**Fig. 11.** Day plots (Day *et al.* 1977) that illustrate the unique hysteresis parameters are associated with remagnetized carbonates. Nanometer-scale magnetite, too small to carry a remanence at room temperature, is thought to contribute substantially to the distributions of hysteresis data. From (**a**) McCabe & Channell (1994) and (**b**) reprinted from Elmore *et al.* (2006) with permission of Wiley & Sons.

thermal energy at room temperature and therefore behave like paramagnetic substances. A manifestation of that grain size aspect in the hysteresis plot itself is shown in Figure 12a (top) and shows a constriction of the hysteresis loop at very low fields, which quite appropriately is called wasp-waisted. Figure 12b compares wasp-waisted responses with the typical SD, pseudo-single-domain (PSD) and MD fields in the Day plot (now in log-scale mode). The superparamagnetic grains, as inferred because they are often not easy to locate and image even with transmission electron microscopy (TEM), have apparently not grown enough to carry stable remanence; other grains

most likely accomplished SD status in the authigenic processes involved in late Palaeozoic remagnetization, however (Weil & Van der Voo 2002). A correlation between ferromagnetic and paramagnetic susceptibilities (Fig. 13) lends credence to the hypothesis of authigenic grain growth of nanometre-sized magnetite (McCabe *et al.* 1989).

An alternative to grain growth and thus to the acquisition of a chemical remanent magnetization, whether in response to fluid migration or not, has been proposed by Kent (1985) who argued that existing grains in these limestones were subjected to conditions that allow long-term thermoviscous growth of the remanence. Néel theory of SD

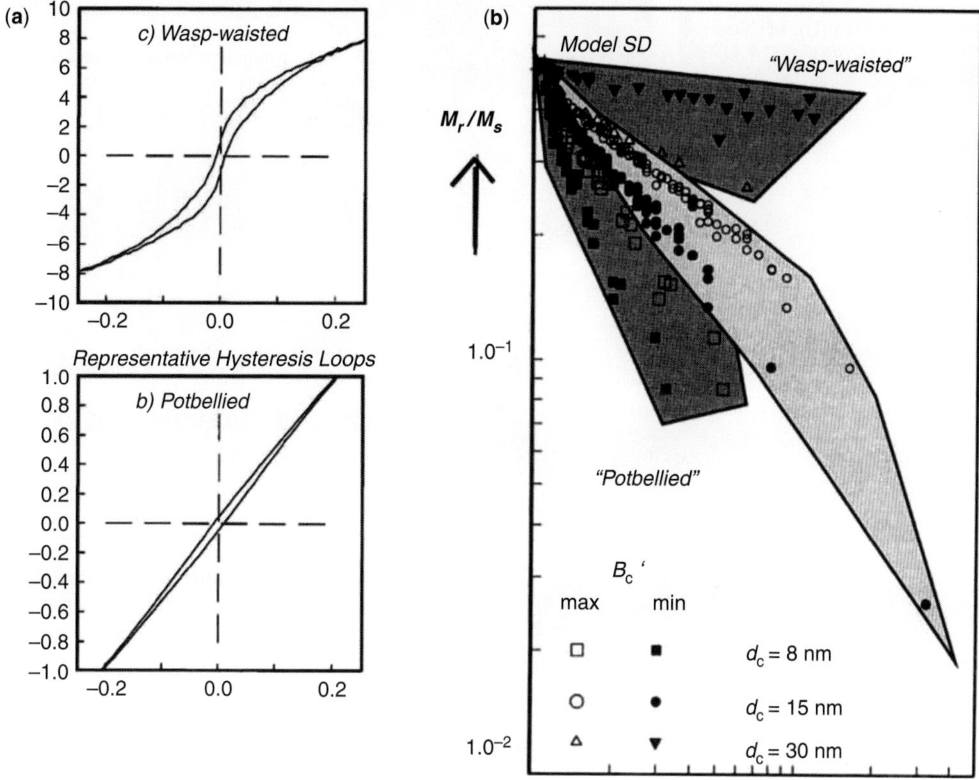



**Fig. 12.** Diagrams from Tauxe *et al.* (1996*a*) that illustrate various hysteresis parameters and characteristics. The remagnetized carbonate field has hysteresis plots that are wasp-waisted (i.e. restricted at the origin), attributed to superparamagnetic (SP) grains.

grains forms the basis for calculations that equate long durations of exposure at moderate temperatures to much shorter durations of thermal treatment (in the laboratory) at higher temperatures (Pullaiah *et al.* 1975). To test this idea, Kent astutely selected

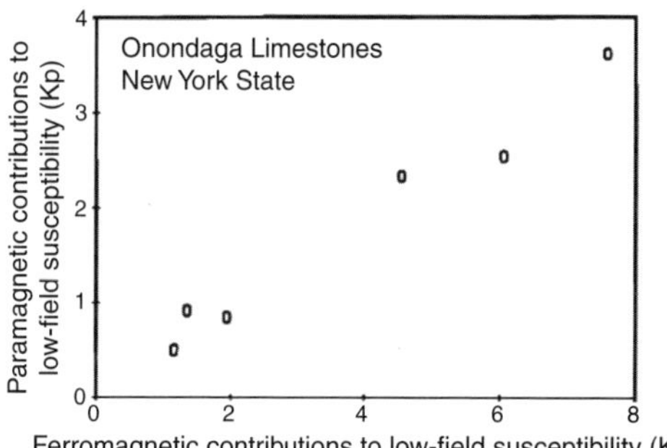



**Fig. 13.** Diagram of paramagnetic versus ferromagnetic susceptibility indicating progressive growth in both types of minerals during remanence acquisition (from McCabe *et al.* 1989).

limestone cobbles in drumlins, for which overprint magnetizations could be characterized in terms of unblocking temperatures of up to 275 °C and which are grouped around the present-day geomagnetic field direction. Because of their transport by glaciers at about 10 ka it is known how long the cobbles have been in the drumlins, and hence how much time the samples had in order to acquire a viscous remanent magnetization (VRM). This VRM, Kent explained, required 'thermal activation that is considerably more potent than the theory used by Pullaiah *et al.* (1975) allows'. He concluded that a thermoviscous origin for the Permian remagnetization component was entirely possible and even probable.

This suggestion made it all the more opportune to be able to examine the magnetite in different carbonate rocks with scanning electron microscopy, and McCabe *et al.* (1983) had already published a first attempt. They found spherules and framboids of pure iron-oxide composition, with sometimes hollow interiors as shown in Figure 14 (right-hand side). Later work by Lu *et al.* (1990) and Suk

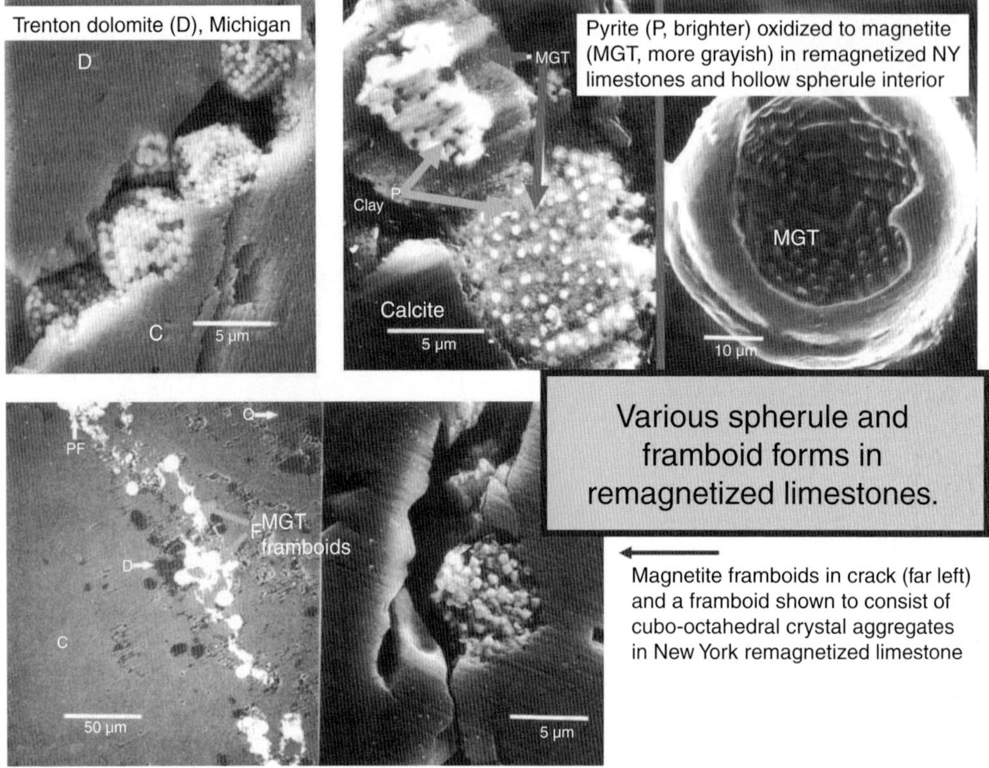



**Fig. 14.** Scanning electron microscope images of spherules, framboids and cubo-octahedral crystallites occurring in cracks and spaces in carbonate rocks from New York State, with occasional pyrite (P) relicts that survived oxidation to magnetite (MGT). D = dolomite, C = calcite, Q = quartz. Figure from Suk *et al.* (1990, 1993*a*).

*et al.* (1990, 1993*a*) found similar magnetite spherules in extracts and notably also *in situ* in cracks and spaces between calcite grains (Fig. 14), as well as in association with sulphides. In a study employing transmission electron microscopy, such associations were clarified (Fig. 15): Fe-sulphide cores (likely pyrite) were surrounded by rims of Fe-oxide, the cubic grain shape of the original pyrite being preserved. Identification of some of the sulphide as pyrrhotite and of the magnetite could be accomplished with selected area electron diffraction (SAED) (Fig. 15). The oxidation of Fe-sulphide to new magnetite grains is counter to the idea of thermoviscous remanence acquisition, so that most contributions on remagnetizations in magnetite-bearing carbonate rocks have since adopted the authigenesis model. There is, however, a little snake in the grass.

The spherules are sufficiently large to actually yield meaningful hysteresis curves (e.g. Fig. 16a) and many of them yield results that range from PSD to MD (Fig. 16b); clearly these results do not resemble the data in Figure 11. Xu *et al.* (1994) hence maintained that the spherules are fully capable of carrying a remanence, but the hysteresis characteristics are very different from those of the bulk rock. In all likelihood, these spherules are responsible instead for the typical large present-day overprint (see Fig. 5) that has been known of ever since the study of McElhinny & Opdyke (1973). What carries the late Palaeozoic remagnetizations is therefore still not well resolved, partly due to the absence of TEM observations of the tiny nanometre-sized grains that are hypothesized to have grown in so many of these remagnetized carbonate rocks.

## Ages of remagnetizations carried by magnetite

With the abundant remagnetization data from the Appalachians, analyses have been performed to examine trends (either across-strike or along-strike) of the ages of the events that gave rise to them (Miller & Kent 1988). New data have become available for North American formations in the last 20

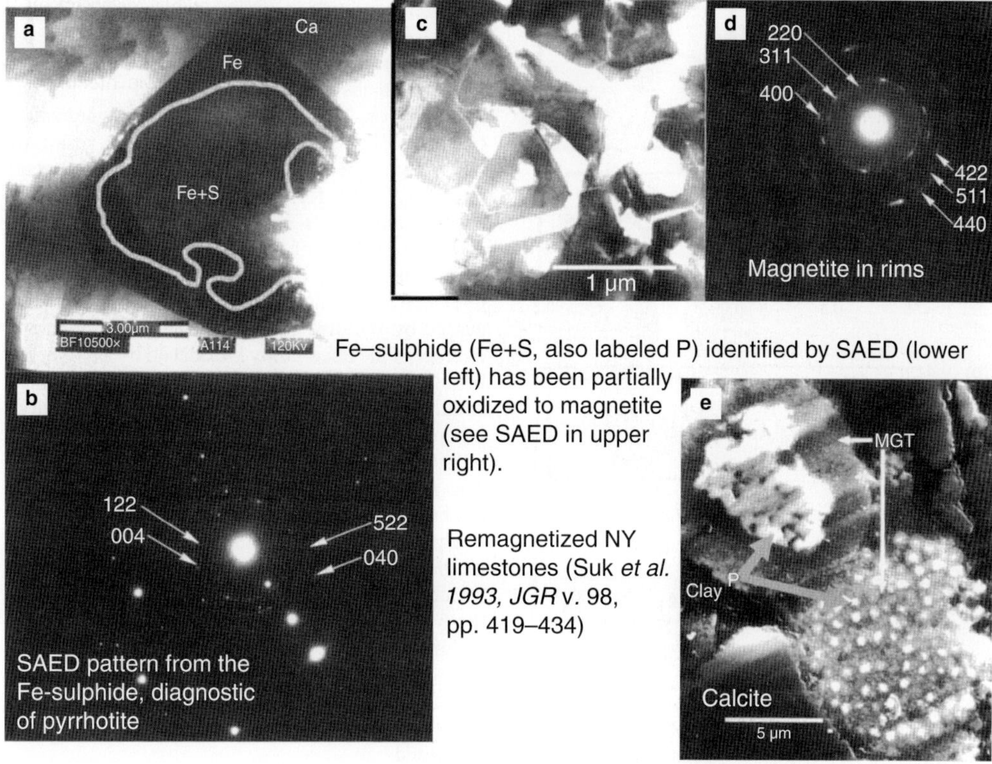



**Fig. 15.** Transmission electron microscope images of (**a**, **c**) a thinned cubo-octahedral crystal with a Fe-sulphide core and magnetite rim, as confirmed by selected-area electron diffraction (SAED) as (**b**) pyrrhotite and (**d**) magnetite. (**e**) This scanning electron microscope image shows pyrite and magnetite relationships inside a framboid, where pyrite (P) is the brighter material. Figure from Suk *et al.* (1993*a*).

years and, for this article, we have summarized results of carbonate-hosted remagnetizations as well as secondary remanences in three occurrences of ironstones (Table 1). Some 20 new results have been published in the last decade alone.

It is worth repeating in this section that remagnetizations can also take place in other time intervals than the late Palaeozoic and that documentation of contemporaneity exists for remagnetizations acquired in the Ordovician, Triassic, Cretaceous and Palaeogene due to orogenies close in time, such as Taconic, Sonoman, Sevier and Laramide in North America; other continents are less well studied but appear to support the correlative trends.

Analyses of across-strike trends have already been discussed (Figs 6–8), whereas Figure 17 shows an analysis along-strike spanning some 12° of latitude from New York to Alabama. Abbreviations are added to allow identification of selected data points. The ages of the remagnetizations are, of course, not directly known, unless one wishes to take K–Ar age determinations on albite as proxies (Hearn & Sutter 1985; see also Tohver

*et al.* 2008). Age estimates can also be obtained by comparing palaeopoles from the remagnetizations with a reference APWP for Laurentia for late Palaeozoic time (Figs 18 & 19). The trend in Figure 17 is rather clear; the more southern sites (Alabama, Georgia, Tennessee) were apparently remagnetized earlier than those of the Central Appalachians, a conclusion in agreement with the earlier publication by Miller & Kent (1988).

The reference APWP for Laurentia has evolved over time as new data became available; four versions are shown (1981, 1993 and two versions constructed from a new compilation in 2011; Fig. 19). The two versions of 2011 differ from each other in that in the last frame, a correction for hypothetical inclination shallowing has been applied to characteristic magnetization directions in detrital sedimentary rocks, assuming a flattening factor $f = 0.6$ (which is a rather conservative value). Also plotted is the large cluster of palaeopoles calculated from remagnetizations of late Palaeozoic age in Laurentian rocks (Table 1) with their cones of 95% confidence; these data are repeated for all

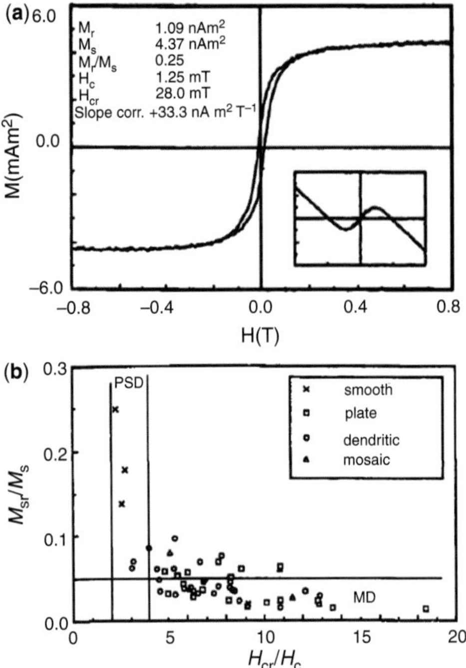



**Fig. 16.** (**a**) Hysteresis plot for a single spherule, as in Figure 14 top right, extracted from a New York carbonate rock. (**b**) Day plot of hysteresis parameters for a large collection of individual spherules with different surface characteristic (smooth, platey, dendritic, mosaic-like), showing that they are pseudo-single domain (PSD) and unlike the characteristics of bulk remagnetized carbonate samples (see Fig. 11). Figure from Xu *et al.* (1994).

four frames. Even though the reference APWPs have changed significantly with time, the average orthogonal great-circle distance between the remagnetization poles and each reference path has not changed appreciably over the last three decades. The largest shift in the reference APWPs has been for the Carboniferous segment (300–320 Ma), where corrections for inclination shallowing have had a more pronounced effect on palaeomagnetic poles assumed to represent primary magnetizations (Bilardello & Kodama 2010). We also suspect that the sampling areas of these formations in the Canadian Maritime Provinces may have undergone some vertical-axis rotations during the Alleghenian assembly of Pangea, resulting in palaeopoles that are somewhat displaced.

Using the shortest distance from a remagnetization palaeopole to the reference path is one way to estimate the age of the remagnetization event. To effectuate this, a projected inclination of the remagnetization has been calculated from each palaeopole using a common site at 40°N, 265°E, and these inclinations (inc* in Table 1) can be calibrated age-wise with the inclinations of the reference palaeopoles (Fig. 18). It must be emphasized that these estimates cannot be considered very precise; for instance, a given syn-folding magnetization (e.g. the Elk Creek or Blacklog Creek results in Fig. 9) may yield a clear peak at some percentage unfolding (30% for Elk Creek, 70% for Blacklog Creek), but this is based on the assumption that equal proportions of tilting took place in each fold limb. An analysis technique to verify this (called intersection of small circles, or ISC) has been discussed by Waldhör & Appel (2006); it may help to overcome the uncertainty introduced by asymmetric tilting of the limbs. The underlying principle is that a small circle produced by incremental tilt corrections inherently contains the appropriately corrected (but unknown) magnetization direction for each entry. The intersection of multiple small circles will be the best estimate of the desired direction of magnetization, as it must be the one in common. The method does not always work, for example when structural settings are complex. Regardless of this, the overall gross outcomes of corrections assuming symmetry between limbs are not likely to be very unrealistic and the plot of age estimates along the Appalachian Belt in the eastern USA (Fig. 17) has sufficient data entries to be representative of a real trend.

If it seems to the reader that remagnetizations are just a major nuisance standing in the way of palaeomagnetic research to obtain a palaeopole for the targeted age span selected from the available rocks, the impression is of course mostly correct but not entirely. Despite the fact that direct dating is challenging (if not impossible) and age estimates are at best only constrained by tests, secondary magnetizations have at times contributed impressively with new knowledge of geological processes that would have been nearly impossible to obtain otherwise. If a good reference APWP is available for the block or craton in question, and if a remagnetization can be associated with a particular event, it is a useful dating tool for that event (e.g. Symons & Stratakos 2002; Pannalal *et al.* 2004; Symons *et al.* 2005). Applications to tectonics are still quite feasible. For instance, through a set of palaeomagnetic studies of Devonian and Lower Carboniferous carbonate rocks in Cantabria, Spain, a strongly curved orocline has become spectacularly documented with bending constrained to earliest Permian–latest Carboniferous (Weil *et al.* 2000, 2001). Not a single primary magnetization was identified in these carbonate rocks but rotations happened during and after a second phase of folding and well after the first phase, requiring complicated analyses of the structures. This was however made straightforward by the palaeomagnetic

**Table 1.** *Late Palaeozoic remagnetizations in Laurentia. REFNO is the reference number in the compilation (Global Paleomagnetic Data Base) of McElhinny & Lock (1996)*

| Rock formation | Dec | Inc | $a_{95}$ | $G_{lat}$ | $G_{lon}$ | $P_{lat}$ | $P_{lon}$ | Dec* | Inc* | Age est. | REFNO or Reference |
|---|---|---|---|---|---|---|---|---|---|---|---|
| Helderberg | 168.4 | −4.2 | 4.2 | 39.1 | 280.8 | −51.9 | 299.5 | 159.4 | −13.2 | 256.0 | Elmore et al. (2006) |
| Helderberg–Onondaga | 165.0 | −10.0 | 2.4 | 42.6 | 285.4 | −50.1 | 309.2 | 153.1 | −15.9 | 256 | 1526 |
| Helderberg–Onondaga | 166.5 | −10.4 | 3.8 | 42.5 | 285.0 | −50.9 | 306.6 | 155.0 | −15.5 | 256 | Kent (1985) |
| Trenton, NY, Que, Ont | 166.7 | −17.9 | 3.1 | 45.0 | 285.0 | −53.0 | 306.9 | 155.9 | −19.1 | 256 | 1681 |
| Martin Formation, AZ | 158.4 | −12.7 | 2.2 | 33.5 | 249.0 | −55.7 | 289.2 | 166.5 | −15.5 | 256 | 1044 |
| St George Formation, NFL | 167.9 | −9.2 | 7.0 | 48.5 | 301.1 | −44.9 | 318.2 | 145.1 | −14.5 | 256 | 1928 |
| Muav Limestone, AZ | 157.0 | −16.3 | 3.0 | 36.0 | 248.0 | −55.0 | 289.6 | 166.1 | −14.4 | 256 | 1044 |
| West Virginia carbonates | 167.1 | −5.0 | 2.5 | 39.5 | 281.3 | −51.2 | 302.1 | 157.6 | −13.4 | 256 | Lewchuk et al. (2003) |
| Helderberg | 163.0 | −3.6 | 5.2 | 39.3 | 280.2 | −49.4 | 306.9 | 154.0 | −13.3 | 258 | Elmore et al. (2001) |
| Knox–host, TN | 152.1 | 1.3 | 2.6 | 37.0 | 273.0 | −44.9 | 315.5 | 146.6 | −12.3 | 260 | 3248 |
| Temple Butte Formation, AZ | 153.2 | −15.7 | 1.8 | 36.0 | 248.0 | −52.6 | 294.9 | 162.3 | −12.2 | 260 | 1044 |
| Patterson Creek Mntn, WVA | 166.6 | −3.3 | 5.5 | 39.5 | 281.3 | −50.2 | 302.5 | 156.9 | −11.9 | 260 | Lewchuk et al. (2002) |
| Appalachian Carbonates, TN | 161.6 | −4.5 | 4.7 | 43.0 | 282.0 | −46.3 | 311.6 | 149.7 | −11.5 | 261 | 2418 |
| Tennessee Carbonates | 157.9 | 2.0 | 5.6 | 36.0 | 274.0 | −47.8 | 307.9 | 152.6 | −11.3 | 262 | 2299 |
| Leithsville Formation, PA | 174.9 | −7.2 | 12.8 | 40.7 | 284.8 | −52.6 | 293.2 | 163.2 | −11.4 | 262 | 1762 |
| Viola Limestones | 165.0 | 0.1 | 7.1 | 34.5 | 262.8 | −52.7 | 288.6 | 165.9 | −9.8 | 263 | Elmore et al. (1993) |
| Trenton & Black River, MI | 170.5 | −10.9 | 7.6 | 42.2 | 275.1 | −52.4 | 291.0 | 164.4 | −10.2 | 263 | 3112 |
| Arbuckle Carbonates, OK | 150.2 | −1.9 | 3.3 | 34.2 | 262.8 | −46.6 | 309.2 | 151.3 | −10.2 | 263 | 205 |
| Ironstones IA, IL, WI | 152.0 | −12.0 | 16.0 | 42.6 | 271.1 | −45.4 | 312.0 | 149.0 | −10.3 | 263 | 1673 |
| Trenton subsurface, Ontario | 152.3 | −12.3 | 8.7 | 42.1 | 277.5 | −52.4 | 291.0 | 164.4 | −10.2 | 263 | Garner & Cioppa (2006) |
| Oriskany Spring Section | 163.2 | 1.1 | 5.2 | 39.3 | 280.2 | −47.1 | 305.4 | 153.7 | −8.4 | 267 | Elmore et al. (2001) |
| Cambro–Ordov. limestones, NE | 178.0 | −6.0 | 9.5 | 40.9 | 282.5 | −52.3 | 286.4 | 167.1 | −8.3 | 267 | Cederquist et al. (2006) |
| Barnett Formation Texas | 156.3 | 5.8 | 3.0 | 31.1 | 261.3 | −49.1 | 299.3 | 158.3 | −8.2 | 267 | 1441 |
| Leadville Dolomite, CO | 156.3 | −12.5 | 12.7 | 39.3 | 253.7 | −50.7 | 292.8 | 162.8 | −7.8 | 267 | Xu et al. (1998) |
| Tonoloway | 167.6 | 1.4 | 5.0 | 39.1 | 280.8 | −48.5 | 299.4 | 158.0 | −7.1 | 269 | Elmore et al. (2006) |

*(Continued)*

**Table 1.** *Continued*

| Rock formation | Dec | Inc | $a_{95}$ | $G_{lat}$ | $G_{lon}$ | $P_{lat}$ | $P_{lon}$ | Dec* | Inc* | Age est. | REFNO or Reference |
|---|---|---|---|---|---|---|---|---|---|---|---|
| Cambro-Ordov. limestones, SW | 171.0 | 0.1 | 8.1 | 40.1 | 281.9 | −48.8 | 296.0 | 160.1 | −5.8 | 271 | Cederquist *et al.* (2006) |
| Greenbrier Ls., PA | 157.9 | 8.4 | 7.7 | 38.0 | 279.6 | −43.2 | 310.5 | 148.6 | −5.6 | 271 | 1758 |
| MVT-hosting limestones, WI | 151.6 | −7.3 | 6.3 | 42.8 | 269.7 | −43.5 | 310.5 | 148.8 | −6.1 | 271 | Pannalal *et al.* (2004) |
| MVT-hosting dolomites, WI | 154.8 | −7.6 | 3.5 | 42.8 | 269.7 | −45.1 | 306.7 | 152.0 | −5.9 | 271 | Pannalal *et al.* (2004) |
| Leadville Formation, CO | 148.0 | −11.0 | 8.6 | 39.0 | 254.0 | −46.0 | 303.0 | 154.7 | −4.9 | 272 | 1678 |
| Knox−ore, TN | 168.8 | 3.7 | 3.5 | 37.0 | 273.0 | −49.8 | 290.5 | 163.8 | −5.1 | 272 | 3248 |
| Peerless Formation, CO | 138.5 | −11.3 | 4.1 | 38.9 | 255.1 | −40.0 | 315.0 | 144.0 | −4.1 | 273 | 1538 |
| Viburnum MVTs, SE MO | 157.3 | 4.0 | 3.3 | 37.5 | 268.8 | −45.2 | 302.1 | 154.8 | −2.9 | 274 | 3254 |
| Kindblade Formation, OK | 149.0 | 5.0 | 4.4 | 34.4 | 262.8 | −43.0 | 308.0 | 150.1 | −3.3 | 274 | 1847 |
| Royer Dolomite, OK | 147.1 | 3.8 | 3.7 | 34.4 | 262.8 | −42.0 | 310.0 | 148.3 | −3.2 | 274 | 2289 |
| Columbus Ls., OH | 164.2 | 4.0 | 2.0 | 40.0 | 277.0 | −45.5 | 299.8 | 156.4 | −2.0 | 278 | 188 |
| Bonneterre Dolomite, MO | 154.0 | 7.0 | 5.1 | 37.6 | 269.8 | −43.0 | 306.0 | 151.3 | −1.8 | 278 | 1267 |
| Oriskany Rocks Section | 164.6 | 8.3 | 11.4 | 39.3 | 280.2 | −44.2 | 301.9 | 154.5 | −1.0 | 282 | Elmore *et al.* (2001) |
| Allentown Dolomite, PA | 173.4 | 4.8 | 10.2 | 40.7 | 284.7 | −46.6 | 294.4 | 163.0 | −1.0 | 282 | 1762 |
| Nevada carbonates | 150.3 | −10.8 | 13.3 | 39.5 | 245.2 | −46.7 | 291.2 | 162.4 | 0.4 | 287 | Gillett & Karlin (2004) |
| St Joe limestone, AR | 146.7 | 9.1 | 11.9 | 36.0 | 267.0 | −38.8 | 311.6 | 145.5 | 0.9 | 290 | 1217 |
| Red Mountain ironstones, AL | 149.6 | 19.0 | 3.4 | 33.5 | 273.3 | −38.0 | 312.4 | 144.5 | 1.5 | 292 | 1532 |
| Taum Sauk Ls., MO | 147.0 | 10.0 | 4.6 | 37.5 | 269.5 | −37.3 | 313.1 | 143.7 | 2.0 | 295 | 1284 |
| Red Mountain ironstones, AL | 150.0 | 20.0 | 3.5 | 33.8 | 273.8 | −38.0 | 312.0 | 144.8 | 1.8 | 295 | 1477 |
| Knox Grp., TN | 157.0 | 18.0 | 7.7 | 36.0 | 276.3 | −39.9 | 306.3 | 149.6 | 3.3 | 298 | 1304 |
| Tri-State MVT district | 146.9 | 9.8 | 3.5 | 37.0 | 265.3 | −37.9 | 308.9 | 146.8 | 4.7 | 302 | Symons *et al.* (2005) |
| Nolichucky Formation, TN | 161.0 | 20.0 | 5.0 | 36.1 | 276.1 | −40.0 | 300.0 | 153.9 | 7.7 | 311 | 1464 |
| Chickamauga Group | 154.5 | 26.2 | 3.9 | 34.6 | 276.3 | −35.9 | 307.4 | 146.8 | 9.3 | 315 | Hnat *et al.* (2009) |
| Knox Mascot-Jefferson, TN | 159.2 | 26.0 | 4.1 | 36.0 | 276.5 | −34.0 | 308.7 | 144.9 | 11.3 | 316 | Symons & Stratakos (2002) |
| Central MO Barite district | 153.5 | 17.6 | 4.8 | 38.5 | 267.5 | −36.6 | 301.0 | 151.6 | 13.0 | 320 | 2693 |

$G_{lat}$, $G_{lon}$ are the mean latitude and longitude coordinates of the sampling area; $P_{lat}$, $P_{lon}$ are the coordinates of the palaeopole; dec* and inc* are the declination and inclination values calculated from the palaeopole for a common site at 40°N, 265°E; Age est. is an approximate age estimated from the inc* value, in comparison with the reference APWP (Van der Voo 1993).

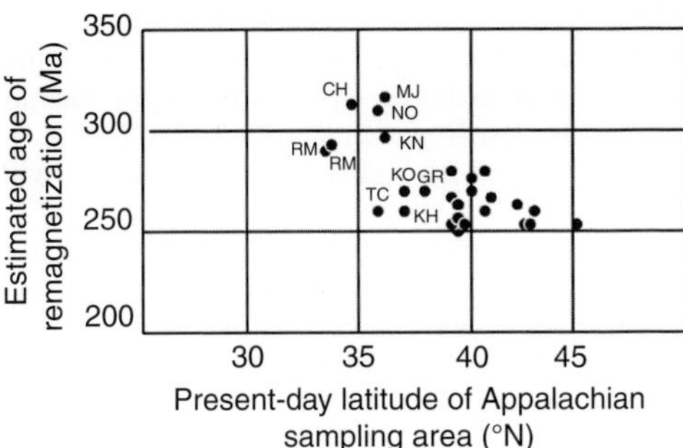



**Fig. 17.** Estimated age of the remagnetizations in the carbonate rocks from the Appalachians, plotted as a function of present-day latitude. Data from Table 1.

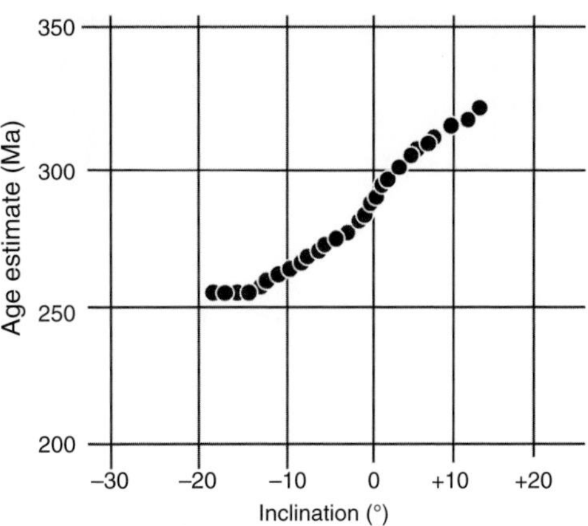



**Fig. 18.** Age estimates of remagnetization directions in carbonate rocks can be calibrated by comparing their inclinations with those (as shown) from the reference APWP (Fig. 19), all calculated for a common site of 40 °N, 265 °E.

results and the well-known interference patterns of the fold axes.

## Delayed remanence acquisition mechanisms, magnetite in loess and Fe-sulphide-magnetite relationships

Redeposition experiments have demonstrated that a sedimentary sequence acquires a geologically stable remanence with some delay (e.g. Fig. 20 from Løvlie 1974; Løvlie & Torsvik 1984) and typically with a somewhat flattened inclination (Tauxe & Kent 1984; Zhao & Roberts 2010). The latter is relevant for palaeopole calculations with the Geocentric Axial Dipole (GAD) hypothesis, as we have seen in the discussion of Figure 19 where a correction has been applied (assuming $f = 0.6$) to account for the flattening in the detrital sedimentary contributions to the latest of the APWPs. The delay is of importance for detailed magnetostratigraphy, reversal transition studies and investigations of short geomagnetic field events within longer chrons. As such,

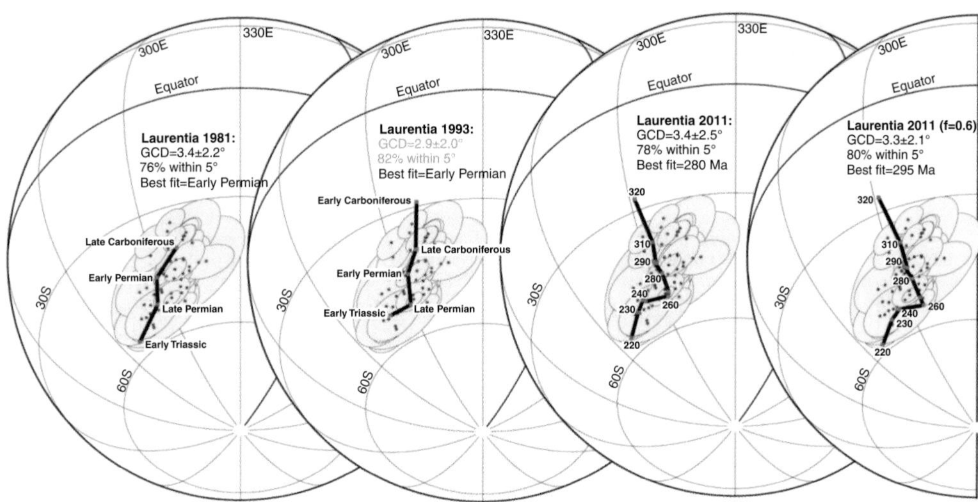



**Fig. 19.** Reference APWPs (south poles) for Laurentia in three generations (1981, 1993, 2011), with the last plot on the right based on Laurentian poles that have been corrected for inclination shallowing in detrital sedimentary rocks (assuming a flattening $f = 0.6$). The remagnetizations in carbonate rocks have given the poles identified by their blue cones of 95% confidence (Table 1) as background and the same for all four plots. GCD = Mean great-circle distance orthogonal to APWP. The older APWPs are replotted from Van der Voo (1981) and Van der Voo (1993).

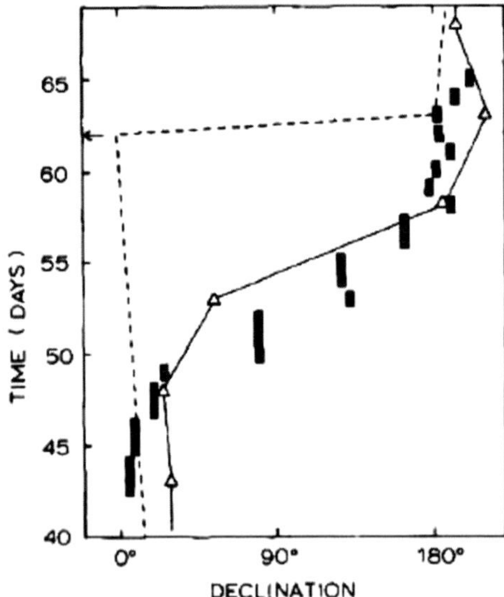

**Fig. 20.** In a redepositioning experiment (reprinted from Løvlie 1974 with permission from Elsevier) the ambient field was switched from near-zero to 180° at a time some 62 days before the end of the experiment, represented by the dashed line. The acquired and measured magnetization declination (open triangles connected by a line for the whole core, solid bars for AF demagnetized disks) is delayed by 5–15 days. During the last 45 days the declinations all remain parallel.

delayed magnetization acquisition borders on being a remagnetization, depending on whether an earlier magnetization is being replaced through a chemical reaction (such as dehydration or oxidation) or whether it is the first remanence the sediment will possess.

One of the more striking examples of such a complexity was presented by Zhou & Shackleton (1999). The well-known marine isotope data place the Matuyama–Brunhes Boundary (MBB) in stage 19 (Tauxe *et al.* 1996*b*), which is a negative peak in the $\delta^{18}O$ data (Fig. 21b) and hence a relatively warmer episode. In contrast, the MBB in Chinese loess (which by all standards is an excellent recorder of Neogene polarity reversals) is most often placed in a cold (L8) interval between the warmer soil-forming episodes of S7 and S8 (Fig. 21a). Correlation with the marine isotope stages would require that the MBB lies in palaeosol S7 instead.

The question then arises: is the MBB recorded by the loess-palaeosol sequence with a delay of some 25 000 years, or was the original windblown magnetic material (likely magnetite) oxidized in variable ways at variable times to, say, maghemite? Or is there a third (yet unrecognized) explanation to be sought in the marine record, as explored among other aspects by Liu *et al.* (2008)? Answers to some of these questions were possibly provided by the investigation by Zhou & Shackleton

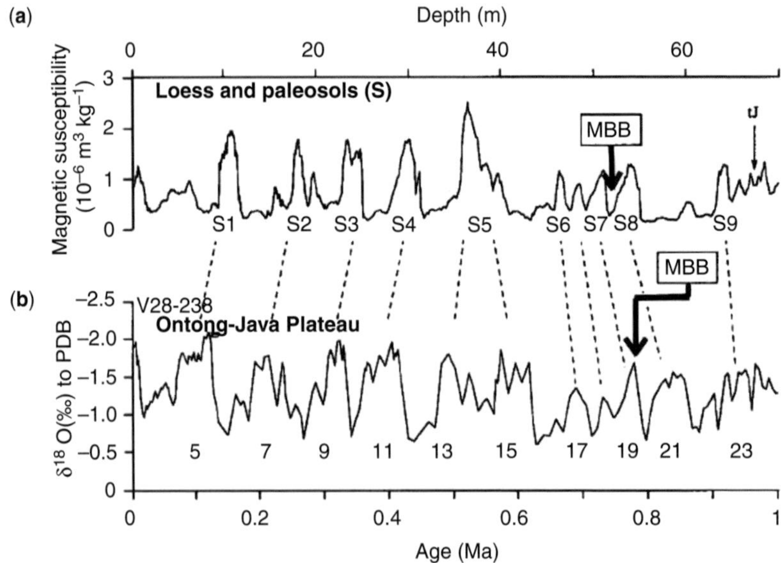

**Fig. 21.** Magnetic susceptibility, correlating with palaeosol formation (**a**) plotted for the last one million years, together with (**b**) the marine $\delta^{18}O$ record, allowing the placements of the Matuyama–Brunhes Boundary (MBB) to be compared. The relative positions of MBB are displaced from each other by some 10 000–35 000 years given its position in a cold (Loess 8) interval in China, whereas it is located in a warm (Marine Isotope Stage 19) interval in the SW Pacific. Diagram reprinted from Zhou & Shackleton (1999) with permission of Elsevier.

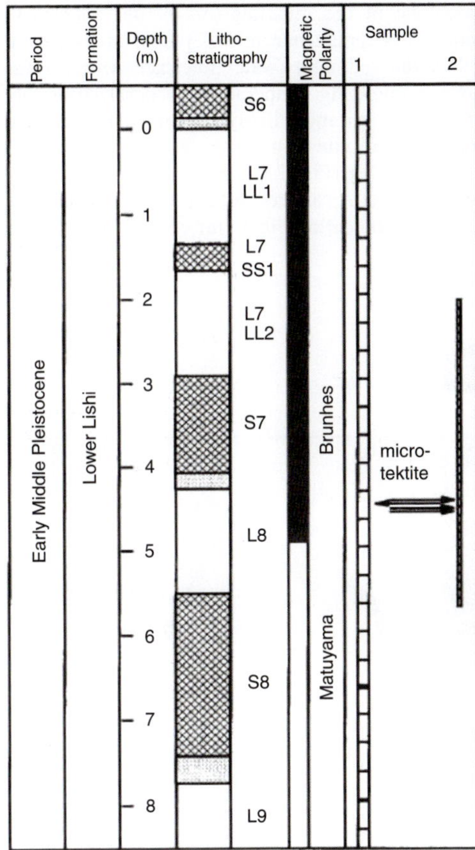



Loess at Luochan, P.R. China

**Fig. 22.** Placement of the MBB in Loess L8 falls below the occurrence of microtektites, whereas in the oceans it is placed above the microtektites. Assuming the tektites are from the same event, a delayed magnetization acquisition time of some 10 000–35 000 years is required in the loess. Diagram reprinted from Zhou & Shackleton (1999) with permission of Elsevier.

(1999), who used the occurrence of microtektites in the marine sedimentary as well as continental loess stratigraphies. They assumed that these deposits were part of the same global event, which allowed them to conclude that the MBB is placed too low (in loess L8; Fig. 22). This in turn documents that the lowest normal-polarity Brunhes remanence was acquired well after the deposition of L8, whereas it should have been placed in S7. If such delays are not constant but change on short time-scales, they may have the effect of masking events that are shorter than the delaying time of some 25 ka. This was used as an argument by Parés *et al.* (2004) to explain why the short reverse-polarity Blake Event, with a duration of less than 10 ka at about 115 ka, is seen only sporadically in even the thickest loess sections near Lanzhou, China (Fang *et al.* 1997).

There is general consensus that delayed magnetization acquisition is common in sedimentary rocks, especially in cases of Post-Depositional Remanent Magnetizations (PDRMs), and that loess may acquire its well-behaved remanence well after settling of the aeolian particles when pedogenesis affects the Fe-oxide grains that may alter to maghemite or other secondary minerals. In the latter case the remanence would be a Chemical Remanent Magnetization (CRM), fitting the definition of a remagnetization. The estimate of Zhou & Shackleton of a delay of the order of some 25 ka however remains controversial. Liu *et al.* (2008), for example, find a shift of only about 20 cm downward for the magnetic record of the MBB. Future work should examine the assumption that the tektite provinces have the same age.

Channell *et al.* (1982) concluded that a delay in remanence acquisition after carbonate precipitation also can occur in the mineral hematite. They observed that 'the depth below reversal

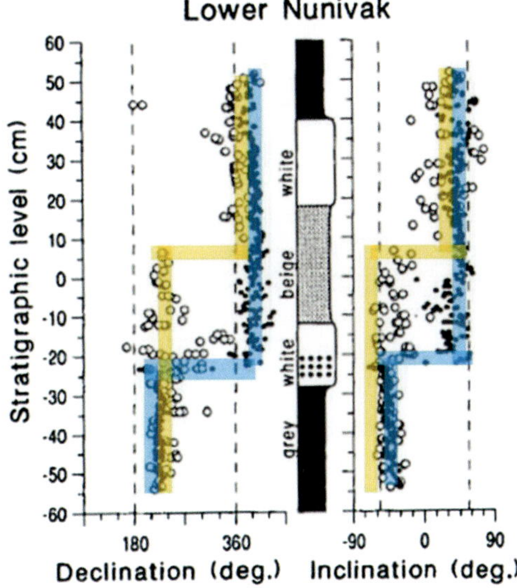



**Fig. 23.** Directions, represented by their declination and inclination, are recorded by magnetite (closed circles) and an Fe-sulphide (pyrrhotite or greigite, open circles) during a transition from reverse to normal polarity. The stratigraphic level at which these minerals show that the reversal is placed at about $-23$ cm for magnetite and at $+10$ cm for the sulphide, illustrating that one or both of these remanences are either delayed or are remagnetized. Blue and yellow highlighting bars are employed to enhance the visibility of these transitions. Plot reprinted from Van Hoof & Langereis (1991) with permission of Nature Publishing Group.

boundaries to which haematite with post-reversal magnetization can occur is estimated to be about 60 cm (after compaction), and is equivalent to a time of about $10^5$ years for these particular sediments'.

A comparable process (but one that demonstrably involves two different mineral carriers) affected marine marls in the southern Mediterranean coast of Sicily, studied by palaeomagnetists from Utrecht University (Van Hoof & Langereis 1991). In this case, one of the minerals (magnetite) appears to acquire its remanence at a depth of about 30 cm below the sediment–water interface, whereas the lower-unblocking-temperature mineral (possibly maghemite, but more likely a magnetic Fe-sulphide such as pyrrhotite or greigite), acquired its magnetization earlier (Fig. 23). The magnetization directions in both carriers reflect a 35° clockwise rotation, indicating that the entire remanence

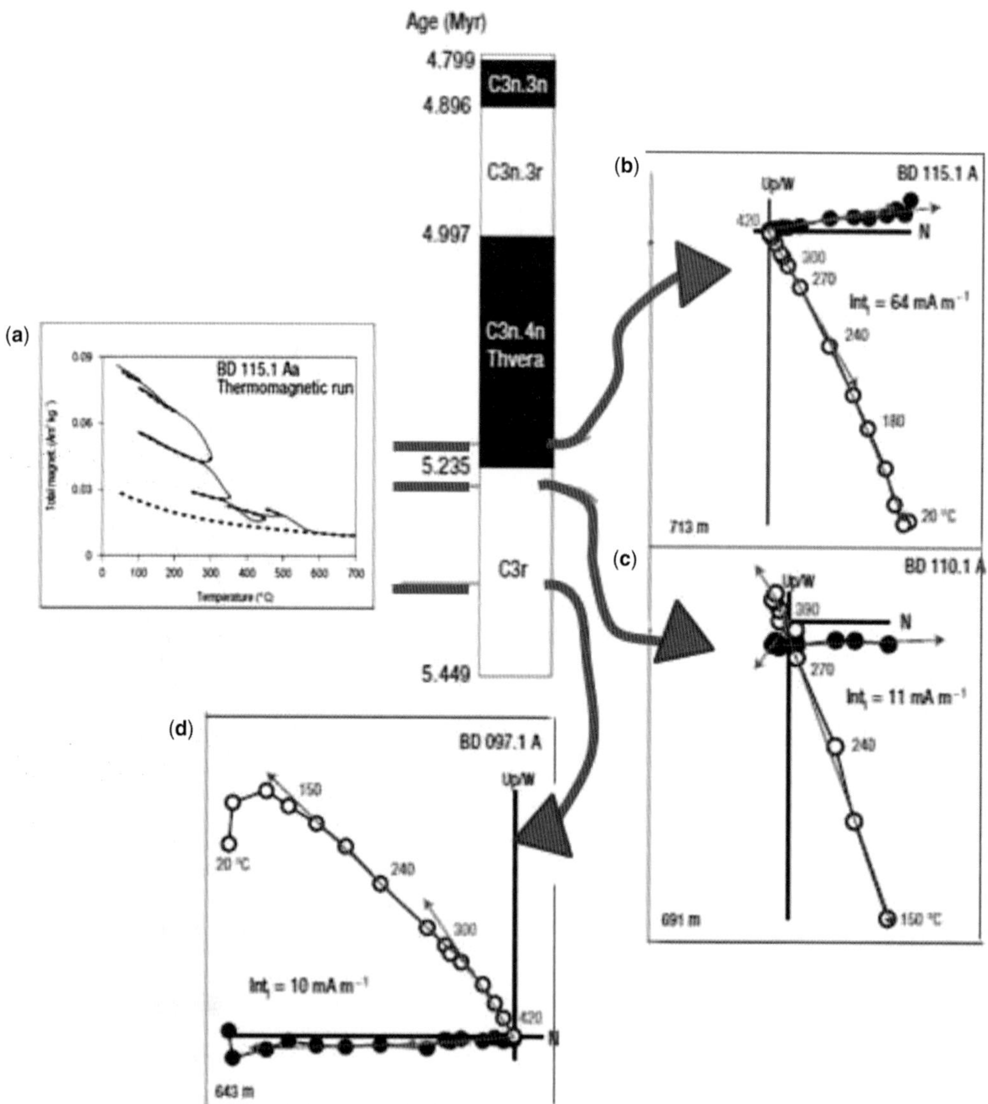


**Fig. 24.** Demagnetization diagrams (reprinted from Vasiliev *et al.* 2008 with permission of the Nature Publishing Group) of samples located near the reversal boundary between Chron C3n.4n and C3r in Pliocene claystones from Romania, in which the remanence carriers are two types of greigite. One has all the characteristics of being of intracellular bacterial origin and is therefore called a magnetosome; it has the higher-temperature unblocking. The other magnetic carrier is low-temperature greigite, likely of diagenetic origin.

predates the rotation and is not of recent vintage. An unexpected aspect of this situation is related to the usual assumption that a sulphide typically forms as a secondary phase, whereas the magnetite would normally be detrital. The Sicilian marls however had magnetite acquire its remanence after the sulphides (Fig. 23), making the magnetite a secondary mineral carrying a remagnetization; we can now see why this particular result is allotted some paragraphs in this review. Once again, however, there is more.

The more recent discovery of similar mineralogic occurrences and behaviour in sedimentary rocks from Romania (Vasiliev *et al.* 2008) may turn the above argument upside down. Pliocene claystones were observed to contain nanometre-scale SD greigite, identified by its demagnetization characteristics (Fig. 24) and by transmission electron microscopy (Fig. 25). The authors make a convincing case for these crystals being magnetosomal (i.e. from bacteria) in origin which, they state, 'would place

them among the oldest greigite magnetofossils identified so far'. The crystals appear to carry a primary magnetization signal, acquired between *c.* 5.3 and 2.6 Ma.

The importance of this observation for the Sicilian marls is yet to be determined; if the Fe-sulphides in these marls are also derived from bacterial magnetosomes, then they are primary carriers of the remanence with magnetite being secondary and therefore carrying a remagnetization.

## Pyrrhotite and greigite

Despite these very recent developments suggesting a primary nature of the remanence carried by Fe-sulphide, remagnetizations in greigite ($Fe_3S_4$) are often demonstrably present and relatively common having been detected in Neogene marine sediments in New Zealand, Taiwan, offshore Oman and California (Rowan & Roberts 2006, 2008; Rowan

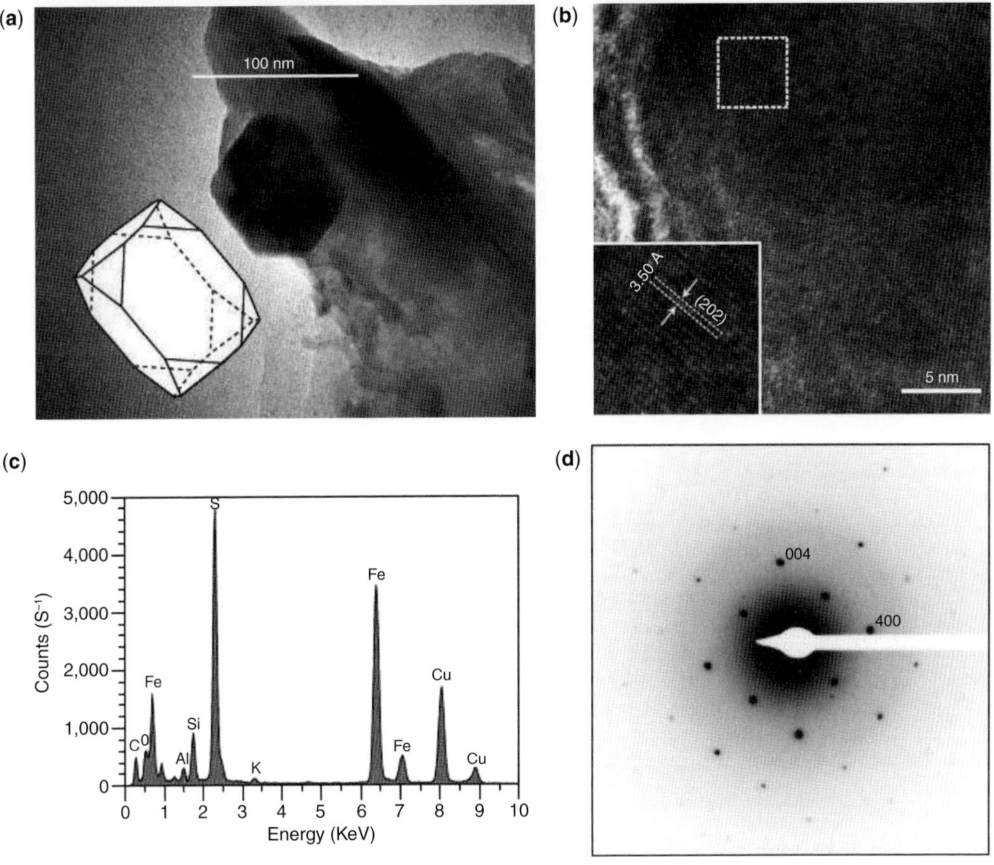



**Fig. 25.** Transmission electron microscopic evidence for a magnetosome origin of the greigite in the claystones studied by Vasiliev *et al.* (2008). Reprinted with permission of the Nature Publishing Group.

*et al.* 2009). Moreover, they have been known to carry dual magnetic polarities of secondary origin (Jiang *et al.* 2001; Sagnotti *et al.* 2005) or to have acquired a syn-folding remanence (Weaver *et al.* 2002). Likewise, monoclinic pyrrhotite ($Fe_7S_8$) seems to be more common as a late diagenetic or younger mineral and to be generally associated with unambiguously remagnetized rocks.

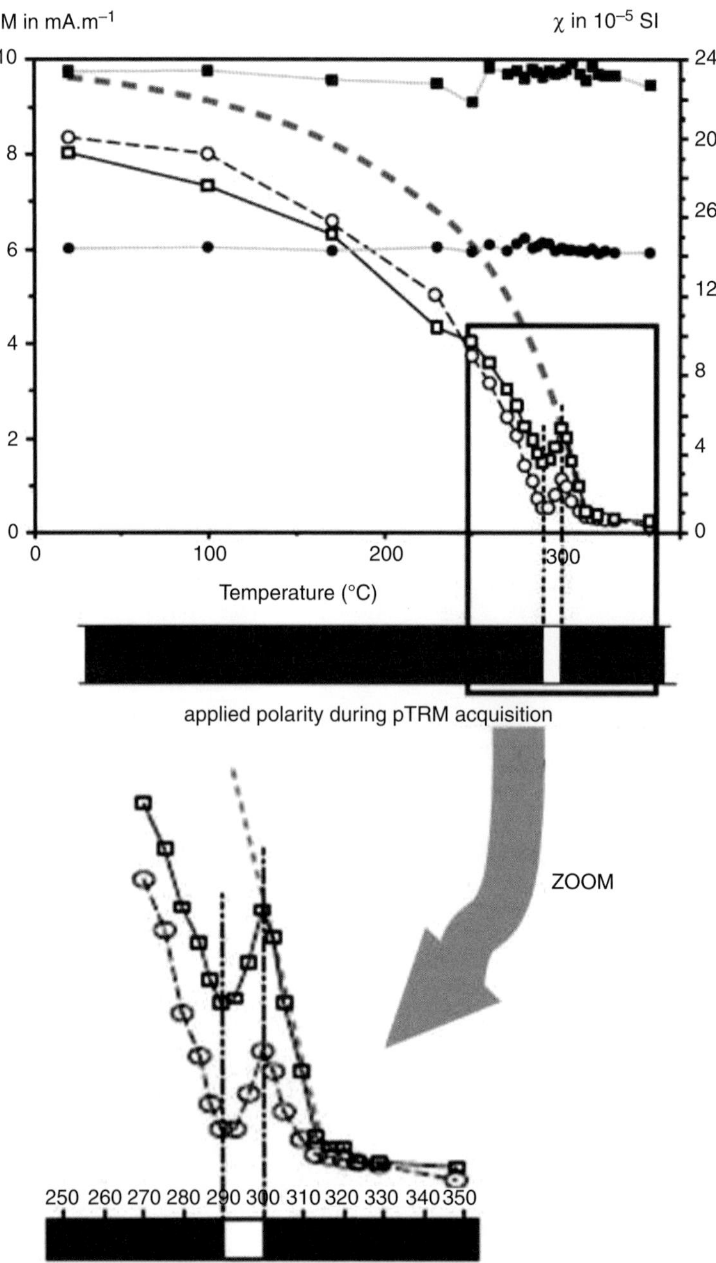


**Fig. 26.** Experiments designed to test the acquisition of a partial thermoremanent magnetization (pTRM) over a narrow temperature interval (290–300 °C) which, upon thermal demagnetization, is manifested by a zig-zag pattern in which the remanence decays, grows (290–300 °C) and then decays to zero. Figure reprinted from Crouzet *et al.* (2001*b*) with permission of Wiley & Sons.

Roberts & Weaver (2005) presented five mechanisms by which rocks can become remagnetized by growth of greigite: (1) on the surfaces of early diagenetic framboidal and nodular pyrite; (2) within cleavages of detrital sheet silicate grains; (3) on the surfaces of authigenic clays (smectite, illite); (4) on the surfaces of siderite; or (5) on the surface of gypsum that resulted from earlier

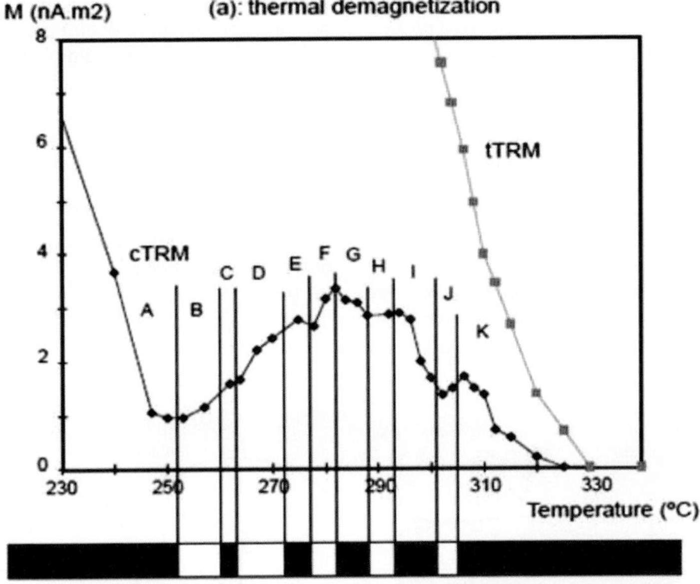


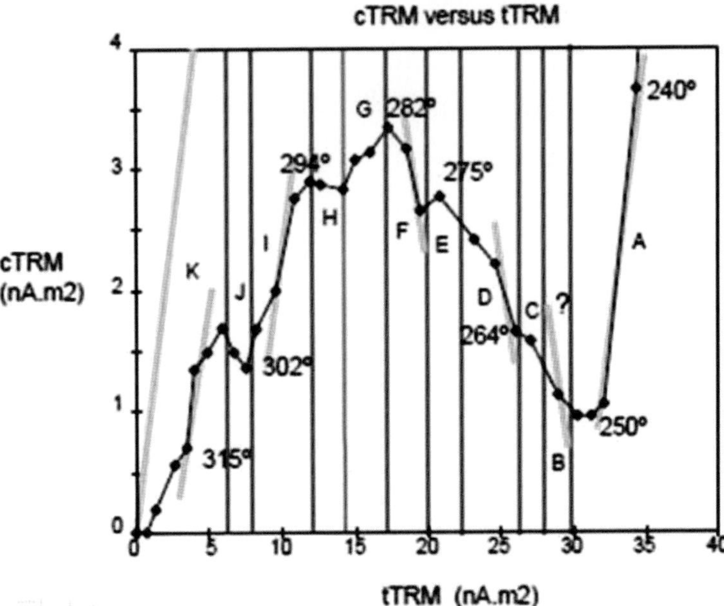


**Fig. 27.** Five brief reverse-polarity intervals are employed to test multiple pTRM acquisitions and how these would be recognized if they were to be naturally acquired. Such a behaviour is recognized in metamorphosed limestones in the Western Alps of France. Figure reprinted from Crouzet *et al.* (2001*b*) with permission of Wiley & Sons.

oxidation of nodular pyrite. They accept that it is possible that other mechanisms exist as well. They conclude that 'sediments containing greigite should be routinely suspected of remagnetization'.

The relatively low laboratory unblocking temperatures of pyrrhotite (less than $c.$ 350 °C) allow it to be thermally remagnetized under low-grade metamorphic conditions, or it may also be a product of hydrothermal fluid/rock interactions such as postulated for the Oslo Graben by Dominguez *et al.* (2011). In this case, pyrrhotite carries a remagnetization by virtue of being superposed on a (secondary) remanence in magnetite in early Palaeozoic limestones inside the rift structure. Pyrrhotite has also been identified as the carrier of remagnetizations in the Himalayas (Crouzet *et al.* 2003; Schill *et al.* 2003, 2004).

Muttoni (1995) has observed that wasp-waisted hysteresis loops can be produced by pyrrhotite and magnetite co-existing in grey to very dark limestone samples of the remagnetized Triassic Prezzo

Formation. In his study, an increasing degree of constriction in the hysteresis loop could be correlated with increasing presence of pyrrhotite as the higher-coercivity mineral.

Dual polarity remagnetizations carried by pyrrhotite in carbonate rocks in the Western Alps of France were shown, by careful thermal demagnetization, to be acquired over a sequence of magnetic polarity intervals attending slow, post-tectonic cooling/uplift of these rocks in the Miocene (Crouzet *et al.* 1999, 2001*a*). The pyrrhotite grains in these rocks are SD and are apparently able to record successive independent and antiparallel partial thermoremanent magnetizations (pTRMs). Crouzet & coworkers (2001*b*) attempted to estimate the temperature at which each reversal occurs during the post-metamorphic cooling. They referred to this method as thermopalaeomagnetism, and designed experiments to examine the sequence of polarity chrons recorded as a function of cooling temperature (Figs 26 & 27). The application of a

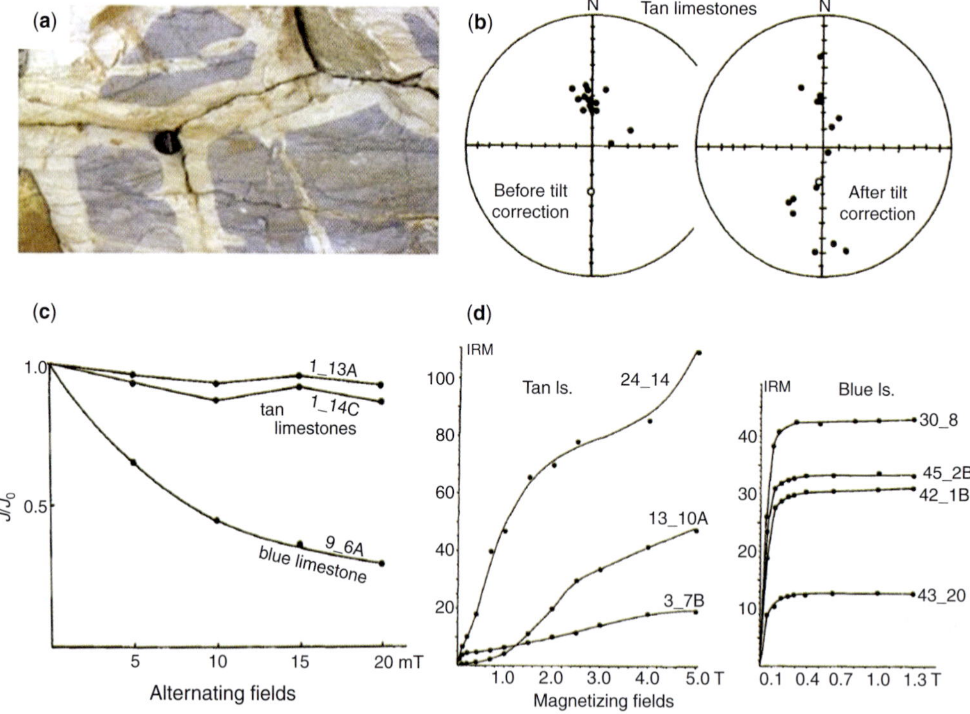


**Fig. 28.** (**a**) Bluish limestone in the Jura Mountains of Switzerland is shown here with fracture-controlled alteration patterns that bleached the limestone adjacent to the fractures and spared the bluish interior. In the process magnetite was replaced as the magnetic carrier by goethite, resulting in the prevailing tan colour of such limestones throughout the area. (**b**) A negative fold test, a late Miocene–Pliocene age of folding and the presence of a reversed site-mean combine to document that the goethite carries a post-folding, pre-Brunhes remanence of likely Pliocene age. (**c**) Characteristic coercivities of goethite (tan limestones) and magnetite (blue limestones). (**d**) Acquisition of isothermal remanent magnetizations of goethite and magnetite-bearing samples, showing extremely high coercivities (in excess of 3 Tesla) for goethite. Figure from Johnson *et al.* (1984).

short reversed laboratory field lets the sample acquire a reverse-polarity TRM in the single short interval 290–300 °C (Fig. 26). Subsequent thermal demagnetization show that the pTRM is blocked over a brief interval in which the remanence appears to grow before it is fully removed in the last treatments above 300 °C. Next they selected five short intervals (Fig. 27a) of reverse polarity at the following temperatures: 305–301 °C, 293–288 °C, 282–277 °C, 272–263 °C and 260–252 °C with subsequent thermal demagnetization of the resulting composite TRM (labelled 'cTRM'). Good correspondence is observed in these experiments between magnetization blocking (cTRM) and laboratory unblocking (tTRM) temperatures (Fig. 27) in a Thellier–Thellier experiment (Crouzet *et al.* 2001*b*). The authors conclude that the results of their experiments show that the rocks examined are able to record a complex sequence of successive pTRMs of alternating polarities acquired over narrow temperature intervals and therefore short intervals of time.

## Goethite

Many carbonate rocks have been exposed to hydration and oxidation processes during their long histories, and this implies that oxyhydroxides can commonly form at the expense of magnetite or Fe-sulphide that were deposited or formed earlier in the rock's history. Heller (1978, 1979) reported goethite (FeOOH) in his studies of Mesozoic limestones in Germany. He established much of what we know of its rock magnetic characteristics, which are characterized by low unblocking temperatures (<150 °C) and very high maximum coercivities (≫3 T). Because of its low-temperature unblocking, the magnetization directions carried by goethite are not always well defined in thermal demagnetization if the temperature increments are large, but in most cases there is little of intrinsic value to be learned anyway. However, in some situations, an unusual story is revealed as we present below.

In summer 1978, William Lowrie of ETH-Zürich hosted the first author for an extended visit. During this visit they had many pleasurable excursions to the Jura Mountains between Basel and Lausanne, which allowed them to collect classic Jurassic limestone sequences in the hopes of uncovering a polarity reversal history. Rock types included shales and limestones, with the latter often having a micritic and tan-coloured appearance; laboratory treatments revealed the presence of goethite (Johnson *et al.* 1984), subsequently confirmed by Gehring *et al.* (1991) in intercalated oolitic ironstones. A few samples of bluish, more

marly, limestones however contained magnetite. In a second field season in 1979, Lowrie and Van der Voo stumbled on a rare occurrence that explained the tan colour as a secondary development (Fig. 28a), where the blue interiors survived alteration penetrating from cracks and joints inward. The contrasting behaviour of the magnetization carried by magnetite and goethite in the carbonate rocks is also illustrated in Figure 28. Interestingly, although post-folding (Fig. 28b), the goethite remanence contains a reverse-polarity remanence but obviously not of Jurassic age. Carbonate rocks of Jurassic age in southeast France have also yielded reverse-polarity remanence with a direction of magnetization that indicates this to be a remagnetization of likely Cenozoic age (Dekkers & Rochette 1992). In contrast to its post-folding nature in the Jura limestones, the goethite in the Swiss oolitic ironstones contains a pre- or syn-folding remanence (Gehring *et al.* 1991).

## Summary

Remagnetization, defined for the purposes of this review as a secondarily acquired remanence without necessarily requiring that a pre-existing magnetization was replaced or that such an earlier condition even existed, is abundant, common and to be treated with enormous circumspection (lest one is tempted to conclude what may simply be a temporally erroneous postulate). Notably, remagnetizations can also reveal geological events, structures or conditions that may not be uncovered with any other method.

We provide a history of the remagnetization concept, but also review several case histories. Paraphrasing the Spanish philosopher George Santayana: If one does not learn from a mistake that allowed a remagnetization to be taken for a primary remanence, one is bound to repeat it!

We thank reviewers J. W. Geissman and M. J. Dekkers for their thorough and thoughtful commentaries on the manuscript, and editor R. D. Elmore for the invitation to contribute a history of remagnetization's various aspects to this Special Publication.

## References

BILARDELLO, D. & KODAMA, K. P. 2010. Palaeomagnetism and magnetic anisotropy of Carboniferous red beds from the Maritime Provinces of Canada: evidence for shallow palaeomagnetic inclinations and implications for North American apparent polar wander. *Geophysical Journal International*, **180**, 1013–1029.

BUTLER, R. F. 1992. *Paleomagnetism: Magnetic Domains to Geologic Terranes.* Blackwell Scientific Publishing, Oxford (now available at http://www.geo.arizona. edu/Paleo mag/book/).

CEDERQUIST, D. P., VAN DER VOO, R. & VAN DER PLUIJM, B. A. 2006. Syn-folding remagnetization of Cambro-Ordovician carbonates from the Pennsylvania Salient post-dates oroclinal rotation. *Tectonophysics*, **422**, 41–54.

CHAMALAUN, F. H. & CREER, K. M. 1963. A revised Devonian pole for Britain. *Nature*, **198**, 375.

CHANNELL, J. E. T., FREEMAN, R., HELLER, F. & LOWRIE, W. 1982. Timing of diagenetic haematite growth in red pelagic limestones from Gubbio (Italy). *Earth and Planetary Science Letters*, **58**, 189–201.

CHEN, D. L. & SCHMIDT, V. 1984. Paleomagnetism of the middle Mississippian Greenbrier Group in West Virginia, USA. *Geodynamic Series*, **12**, 48–62.

CLEGG, J. A., DEUTSCH, E. R., EVERITT, C. W. F. & STUBBS, P. H. S. 1957. Some recent palaeomagnetic measurements made at Imperial College, London. *Advance Physics*, **6**, 219–231.

COLLINSON, D. W. 1974. The Role of pigment and specularite in the remanent magnetism of red sandstones, Geophys. *Journal of the Royal Astronomical Society*, **38**, 253–264.

CRADDOCK, J. P. & VAN DER PLUIJM, B. A. 1989. Late Paleozoic deformation of the cratonic carbonate cover of eastern North America. *Geology*, **17**, 416–419.

CREER, K. M. 1962. A statistical enquiry into the partial remagnetization of folded Old Red Sandstone rocks. *Journal of Geophysical Research*, **67**, 1899–1906.

CREER, K. M. 1968. Paleozoic paleomagnetism. *Nature*, **219**, 246–250.

CROUZET, C., MÉNARD, G. & ROCHETTE, P. 1999. High-precision three-dimensional paleothermometry derived from paleomagnetic data in an Alpine metamorphic unit. *Geology*, **27**, 503–506.

CROUZET, C., MÉNARD, G. & ROCHETTE, P. 2001a. Cooling history of the Dauphinoise Zone (Western Alps, France) deduced from the thermopaleomagnetic record: geodynamic implications. *Tectonophysics*, **340**, 79–93.

CROUZET, C., ROCHETTE, P. & MÉNARD, G. 2001b. Experimental evaluation of successive polarities during uplift of metasediments. *Geophysical Journal International*, **145**, 771–785.

CROUZET, C., GAUTAM, P., SCHILL, E. & APPEL, E. 2003. Multicomponent magnetization in western Dolpo (Tethyan Himalaya, Nepal): tectonic implications. *Tectonophysics*, **377**, 179–196.

DAY, R., FULLER, M. D. & SCHMIDT, V. A. 1977. Hysteresis properties of titanomagnetites: grain-size and compositional dependence. *Physics Earth and Planetary Interiors*, **13**, 260–267.

DEKKERS, M. & ROCHETTE, P. 1992. Magnetic properties of Chemical Remanent Magnetization in synthetic and natural goethite: prospects for a Natural Remanent Magnetization/Thermoremanent Magnetization ratio paleomagnetic stability test? *Journal of Geophysical Research*, **97**, 17291–17307.

DOMINGUEZ, A. R., VAN DER VOO, R. ET AL. 2011. The ~270 Ma paleolatitude of Baltica and its significance for Pangea models. *Geophysical Journal International*, **186**, 529–550.

ELMORE, R. D. & VAN DER VOO, R. 1982. Origin of hematite and its associated remanence in the Copper Harbor Conglomerate (Keweenawan), Upper Michigan. *Journal of Geophysical Research*, **87**, 10 918–10 929.

ELMORE, R. D., LONDON, D., BAGLEY, D., FRUIT, D. & GAO, G. Q. 1993. Remagnetization by basinal fluids: testing the hypothesis in the Viola Limestone, southern Oklahoma. *Journal of Geophysical Research*, **98**, 6237–6254.

ELMORE, R. D., KELLEY, J., EVANS, M. & LEWCHUK, M. T. 2001. Remagnetization and orogenic fluids: testing the hypothesis in the Central Appalachians. *Geophysical Journal International*, **144**, 568–576.

ELMORE, R. D., FOUCHER, J. L.-E., EVANS, M., LEWCHUK, M. & COX, E. 2006. Remagnetization of the Tonoloway Formation and the Helderberg Group in the Central Appalachians: testing the origin of syntilting magnetizations. *Geophysical Journal International*, **166**, 1062–1076.

ELSTON, D. P. & BRESSLER, S. L. 1977. Paleomagnetic poles and polarity zonation from Cambrian and Devonian strata in Arizona. *Earth and Planetary Science Letters*, **36**, 423–433.

ELSTON, D. P. & PURUCKER, M. E. 1979. Detrital magnetization in red beds of the Moenkopi Formation (Triassic), Gray Mountain, Arizona. *Journal of Geophysical Research*, **84**, 1653–1665.

EVANS, M. A., ELMORE, R. D. & LEWCHUK, M. T. 2000. Examining the relationship between remagnetization and orogenic fluids: central Appalachians. *Journal of Geochemical. Explortion*, **69–70**, 139–142.

FANG, X.-M., LI, J.-J. ET AL. 1997. A record of the Blake Event during the last interglacial paleosol in the western Loess Plateau of China. *Earth and Planetary Science Letters*, **146**, 73–82.

FISHER, R. A. 1953. Dispersion on a sphere. *Proceedings of the Royal Society: A*, **217**, 295–305.

GARNER, N. & CIOPPA, M. T. 2006. Late Paleozoic remagnetization of the Trenton Formation in Ordovician petroleum reservoirs of southwestern Ontario. *Journal of Geochemical Exploration*, **89**, 119–123.

GEHRING, A. U., KELLER, P. & HELLER, F. 1991. Paleomagnetism and tectonics of the Jura arcuate mountain belt in France and Switzerland. *Tectonophysics*, **186**, 269–278.

GILLETT, S. L. & KARLIN, R. E. 2004. Pervasive late Paleozoic–Triassic remagnetization of miogeoclinal carbonate rocks in the Basin and Range and vicinity, SW USA: regional results and possible tectonic implications. *Physics of the Earth and Planetary Interiors*, **141**, 95–120.

GOREE, W. S. & FULLER, M. D. 1976. Magnetometers using RF-driven squids and their applications in rock magnetism and paleomagnetism. *Review of the Geophysical Space Physics*, **14**, 591–608.

GRAHAM, J. W. 1956. Paleomagnetism and magnetostriction. *Journal of Geophysical Research*, **61**, 735–739.

GRAHAM, K. W. T. 1961. The remagnetization of a surface outcrop by lightning currents. *Geophysics Journal*, **6**, 85–102.

HEARN, P. & SUTTER, J. F. 1985. Authigenic potassium feldspar in Cambrian carbonates: evidence of Alleghanian brine migration. *Science*, **228**, 1529–1531.

HELLER, F. 1978. Rock magnetic studies of Upper Jurassic limestones from southern Germany. *Journal of Geophysics*, **44**.

HELLER, F. 1979. Palaeomagnetism of Upper Cretaceous limestones from the Münster Basin, Germany. *Journal of Geophysics*, **46**, 413–427.

HERRERO-BERVERA, E. & HELSLEY, C. E. 1983. Paleomagnetism of a polarity transition in the lower (?) Triassic Chugwater Formation, Wyoming. *Journal of Geophysical Research*, **88**, 3506–3522.

HNAT, J. S., VAN DER PLUIJM, B. A. & VAN DER VOO, R. 2009. Remagnetization in the Tennessee Salient, Southern Appalachians, USA: constraints on the timing of deformation. *Tectonophysics*, **474**, 709–722.

IRVING, E. & RUNCORN, S. K. 1957. Palaeomagnetic investigations in Great Britain, analysis of the palaeomagnetism of the Torridonian sandstone series of NW Scotland. *Philosophical Transactions of Royal Society London, Series A, No. 974*, **250**, 83–99.

IRVING, E. & STRONG, D. F. 1984. Paleomagnetism of the Early Carboniferous Deer Lake Group, western Newfoundland: no evidence for mid-Carboniferous displacement of 'Acadia'. *Earth and Planetary Science Letters*, **69**, 379–390.

JAEGER, J. C. 1957. The temperature in the neighborhood of a cooling intrusive sheet. *American Journal of Science*, **255**, 306–318.

JIANG, W. T., HORNG, C. S., ROBERTS, A. P. & PEACOR, D. R. 2001. Contradictory magnetic polarities in sediments and variable timing of neoformation of authigenic greigite. *Earth and Planetary Science Letters*, **193**, 1–12.

JOHNSON, R. J., VAN DER VOO, R. & LOWRIE, W. 1984. Paleomagnetism and late diagenesis of Jurassic carbonates from the Jura mountains, Switzerland and France. *Geological Society of America Bulletin*, **95**, 478–488.

KEAN, W. F. 1981. Paleomagnetism of the Late Ordovician Neda iron ore from Wisconsin, Iowa and Illinois. *Geophysics Research Letters*, **8**, 880–882.

KENT, D. V. 1982. Paleomagnetic evidence for post-Devonian displacement of the Avalon Platform (Newfoundland). *Journal of Geophysical Research*, **87**, 8709–8716.

KENT, D. V. 1985. Thermoviscous remagnetization in some Appalachian limestones. *Geophysics Research Letters*, **12**, 805–808.

KENT, D. V. & OPDYKE, N. D. 1978. Paleomagnetism of the Devonian Catskill red beds: evidence for motion of coastal New England—Canadian Maritime region relative to cratonic North America. *Journal of Geophysical Research*, **83**, 4441–4450.

KENT, D. V. & OPDYKE, N. D. 1979. The Early Carboniferous paleomagnetic field of North America and its bearing on the tectonics of the northern Appalachians. *Earth and Planetary Science Letters*, **44**, 365–372.

KENT, D. V. & OPDYKE, N. D. 1985. Multicomponent magnetizations from the Mississippian Mauch Chunk Formation of the central Appalachians and their tectonic implications. *Journal of Geophysical Research*, **90**, 5371–5383.

KODAMA, K. P. 1988. Remanence rotation due to rock strain during folding and the stepwise application of the fold test. *Journal of Geophysical Research*, **93**, 3357–3371.

LARSON, E. E. & WALKER, T. R. 1975. Development of chemical remanent magnetization during early stages of red bed formation in late Cenozoic sediments, Baja California. *Geological Society of America Bulletin*, **86**, 639–650.

LARSON, E. E. & WALKER, T. R. 1982. A rock magnetic study of the Lower Massive Sandstone, Moenkopi Formation (Triassic), Gray Mountain area, Arizona. *Journal of Geophysical Research*, **87**, 4819–4836.

LARSON, E. E., WALKER, T. R., PATTERSON, P. E., HOBLITT, R. P. & ROSENBAUM, J. G. 1982. Paleomagnetism of the Moenkopi Formation, Colorado Plateau: basis for long-term model of acquisition of chemical remanent magnetization in red beds. *Journal of Geophysical Research*, **87**, 1081–1106.

LEWCHUK, M. T., ELMORE, R. D. & EVANS, M. 2002. Remagnetization signature of Paleozoic sedimentary rocks from the Patterson Creek Mountain Anticline in West Virginia. *Physics and Chemistry of the Earth*, **27**, 1141–1150.

LEWCHUK, M. T., EVANS, M. & ELMORE, R. D. 2003. Synfolding remagnetization and deformation: results from Palaeozoic sedimentary rocks in West Virginia. *Geophysics Journal International*, **152**, 266–279.

LIU, Q., ROBERTS, A. P., ROHLING, E. J., ZHU, R. X. & SUN, Y. B. 2008. Post-depositional remanent magnetization lock-in and the location of the Matuyama-Brunhes geomagnetic reversal boundary in marine and Chinese loess sequences. *Earth and Planetary Science Letters*, **275**, 102–110.

LØVLIE, R. 1974. Post-depositional remanent magnetization in a re-deposited deep-sea sediment. *Earth and Planetary Science Letters*, **21**, 315–320.

LØVLIE, R. & TORSVIK, T. H. 1984. Magnetic remanence and fabric properties of laboratory-deposited hematite-bearing red sandstone. *Geophysics Research Letters*, **11**, 221–224.

LU, G., MARSHAK, S. & KENT, D. V. 1990. Characteristics of magnetic carriers responsible for late Paleozoic remagnetization in carbonate strata of the mid-continent, U.S.A. *Earth and Planetary Science Letters*, **99**, 351–361.

MARTIN, D. L. 1975. A paleomagnetic polarity transition in the Devonian Columbus limestone of Ohio: a possible stratigraphic tool. *Tectonophysics*, **28**, 125–134.

McCABE, C. & ELMORE, R. D. 1989. The occurrence and origin of late Paleozoic remagnetization in the sedimentary rocks of North America. *Reviews of Geophysics*, **27**, 471–494.

McCABE, C. & CHANNELL, J. E. T. 1994. Late Paleozoic remagnetization in limestones of the Craven Basin (northern England) and the rock magnetic fingerprint of remagnetized sedimentary carbonates. *Journal of Geophysical Research*, **99**, 4603–4612.

McCABE, C., VAN DER VOO, R., PEACOR, D. R., SCOTESE, C. R. & FREEMAN, R. 1983. Diagenetic magnetite carries ancient yet secondary remanence in some Paleozoic carbonates. *Geology*, **11**, 221–223.

McCABE, C., VAN DER VOO, R. & BALLARD, M. 1984. Late Paleozoic remagnetization of the Trenton Limestone. *Geophysics Research Letters*, **11**, 979–982.

McCabe, C., Sassen, R. & Saffer, B. 1987. Occurrence of secondary magnetite within biodegraded oil. *Geology*, **15**, 7–10.

McCabe, C., Jackson, M. & Saffer, B. 1989. Regional patterns of magnetite authigenesis in the Appalachian Basin: implications for the mechanism of late Paleozoic remagnetization. *Journal of Geophysical Research*, **94**, 10429–10443.

McElhinny, M. W. & Lock, J. 1996. IAGA paleomagnetic databases with access. *Surveys in Geophysics*, **17**, 575–591.

McElhinny, M. W. & Opdyke, N. D. 1973. Remagnetization hypothesis discounted: a paleomagnetic study of the Trenton Limestone, N. Y. State. *Geological Society of America Bulletin*, **84**, 3697–3708.

Miller, J. D. & Kent, D. V. 1988. Regional trends in the timing of Alleghenian remagnetization in the Appalachians. *Geology*, **16**, 588–591.

Molina-Garza, R. S. & Zijderveld, J. D. A. 1996. Paleomagnetism of Paleozoic strata, Brabant and Ardennes massifs, Belgium: implications of prefolding and postfolding Late Carboniferous secondary magnetizations for European apparent polar wander. *Journal of Geophysical Research*, **101**, 15799–15818.

Muttoni, G. 1995. "Wasp-waisted" hysteresis loops from a pyrrhotite and magnetite-bearing remagnetized Triassic limestone. *Geophysical Research Letters*, **22**, 3167–3170.

Oliver, J. 1986. Fluids expelled tectonically from orogenic belts: their role in hydrocarbon migration and other geologic phenomena. *Geology*, **14**, 99–102.

Pannalal, S. J., Symons, D. T. A. & Sangster, D. F. 2004. Paleomagnetic dating of Upper Mississippi Valley zinc-lead mineralisation, WI, USA. *Joural of Applied Geophysics*, **56**, 135–153.

Parés, J. M., Van der Voo, R., Yan, M. & Fang, X. M. 2004. After the dust settles: why is the Blake event imperfectly recorded in Chinese loess? *In*: Channell, J. E. T. & Kent, D. V., Lowrie, W. & Meert, J. G. (eds) *AGU Monograph on Timescales of the Paleomagnetic Field*. AGU, Boulder, 191–204.

Pullaiah, G., Irving, E., Buchan, K. L. & Dunlop, D. J. 1975. Magnetization changes caused by burial and uplift. *Earth and Planetary Science Letters*, **28**, 133–143.

Purucker, M. E., Elston, D. P. & Shoemaker, E. M. 1980. Early acquisition of characteristic magnetization in red beds of the Moenkopi Formation (Triassic), Gray Mountain, Arizona. *Journal of Geophysical Research*, **85**, 997–1012.

Roberts, A. P. & Weaver, R. 2005. Multiple mechanisms of remagnetization involving sedimentary greigite (Fe(3)S(4)). *Earth and Planetary Science Letters*, **231**, 263–277.

Roy, J. L. & Park, J. K. 1972. Red beds: DRM or CRM? *Earth and Planetary Science Letters*, **17**, 211–216.

Rowan, C. J. & Roberts, A. P. 2006. Magnetite dissolution, diachronous greigite formation, and secondary magnetizations from pyrite oxidation: Unravelling complex magnetizations in Neogene marine sediments from New Zealand. *Earth and Planetary Science Letters*, **241**, 119–137.

Rowan, C. J. & Roberts, A. P. 2008. Widespread remagnetizations and a new view of Neogene tectonic rotations within the Australia-Pacific plate boundary zone, New Zealand. *Journal of Geophysical Research*, 113, Article Number: B03103, doi: 10.1029/ 2006JB004594.

Rowan, C. J., Roberts, A. P. & Broadbent, T. 2009. Reductive diagenesis, magnetite dissolution, greigite growth and paleomagnetic smoothing in marine sediments: a new view. *Earth and Planetary Science Letters*, **277**, 223–235.

Sagnotti, L., Roberts, A. P. et al. 2005. Apparent magnetic polarity reversals due to remagnetization resulting from late diagenetic growth of greigite from siderite. *Geophysics Journal International*, **160**, 89–100.

Schill, E., Appel, E., Godin, L., Crouzet, C., Gautam, P. & Regmi, K. R. 2003. Record of deformation by secondary magnetic remanences and magnetic anisotropy in the Nar-Phu Valley (central Himalaya). *Tectonophysics*, **377**, 197–209.

Schill, E., Appel, E., Crouzet, C., Gautam, P., Wehland, F. & Staiger, M. 2004. Oroclinal bending v. regional significant clockwise rotations in the Himalayan arc: constraints from secondary pyrrhotite remanences. *Special Paper Geological Society of America*, **383**, 73–85.

Schmidt, P. W. & Embleton, B. J. J. 1976. Palaeomagnetic results from sediments of the Perth Basin, Western Australia, and their bearing on the timing of regional lateritisation. *Palaeogeography, Palaeoclimatolology, Palaeoecology*, **19**, 257–273.

Scotese, C. R. 1985. *The assembly of Pangea: Middle and late Paleozoic paleomagnetic results from N. America*, PhD thesis, University of Chicago, Chicago, Illinois.

Scotese, C. R., Van der Voo, R. & McCabe, C . 1982. Paleomagnetism of the Upper Silurian and Lower Devonian carbonates of New York State: evidence for secondary magnetizations residing in magnetite. *Physics Earth and Planetary Interiors*, **30**, 385–395.

Scott, G. R. 1979. Paleomagnetic studies of the Early Carboniferous St. Joe limestone, Arkansas. *Journal of Geophysical Research*, **84**, 6277–6285.

Shive, P. N., Steiner, M. B. & Huycke, D. T. 1984. Magnetostratigraphy, paleomagnetism, and remanence acquisition in the Triassic Chugwater Formation of Wyoming. *Journal of Geophysical Research*, **89**, 1801–1815.

Stamatakos, J. A., Hirt, A. M. & Lowrie, W. 1996. The age and timing of folding in the central Appalachians from paleomagnetic results. *Geological Society of America Bulletin*, **108**, 815–829.

Steiner, M. B. 1973. Late Paleozoic partial remagnetization of Ordovician rocks from southern Oklahoma. *Geological Society of America Bulletin*, **84**, 341–346.

Steiner, M. B. 1983. Detrital remanent magnetization in hematite. *Journal of Geophysical Research*, **88**, 6523–6539.

Storetvedt, K. M. 1968. On remagnetization problems in palaeomagnetism. *Earth and Planetary Science Letters*, **4**, 107–112.

Suk, D., Peacor, D. R. & Van der Voo, R. 1990. Replacement of pyrite framboids by magnetite in limestone and implications for paleomagnetism. *Nature*, **345**, 611–613.

Suk, D., Van der Voo, R. & Peacor, D. R. 1993a. Origin of magnetite responsible for remagnetization of early Paleozoic limestones of New York State. *Journal of Geophysical Research*, **98**, 419–434.

Suk, D., Van der Voo, R., Peacor, D. R. & Lohmann, K. C. 1993b. Late Paleozoic remagnetization and its carrier in the Trenton and Black River carbonates from the Michigan Basin. *Journal of Geology*, **101**, 795–808. .

Symons, D. T. A. & Stratakos, K. K. 2002. Paleomagnetic dating of Alleghanian orogenesis and mineralisation in the Mascot-Jefferson City zinc district of East Tennessee, USA. *Tectonophysics*, **348**, 51–72.

Symons, D. T. A., Pannalal, S. J., Coveney, R. M. Jr. & Sangster, D. F. 2005. Paleomagnetism of late Paleozoic strata and mineralization in the Tri-State lead–zinc ore district. *Economical Geology*, **100**, 295–309.

Tauxe, L. & Kent, D. V. 1984. Properties of a detrital remanence carried by haematite from study of modern river deposits and laboratory redeposition experiments. *Journal of the Royal Astronomical Society*, **76**, 543–561.

Tauxe, L., Mullender, T. A. T. & Pick, T. 1996a. Potbellies, wasp-waists, and superparamagnetism in magnetic hysteresis. *Journal of Geophysical Research*, **101**, 571–583.

Tauxe, L., Herbert, T., Shackleton, N. J. & Kok, Y. S. 1996b. Astronomical calibration of the Matuyama–Brunhes boundary: consequences for magnetic remanence in marine carbonates and the Asian loess sequences. *Earth and Planetary Science Letters*, **140**, 133–146.

Tohver, E., Weil, A. B., Solum, J. G. & Hall, C. M. 2008. Direct dating of carbonate remagnetization by $^{40}$Ar/$^{39}$Ar analysis of the smectite–illite transformation. *Earth and Planetary Science Letters*, **274**, 524–530.

Tomezzoli, R. N. 2001. Further palaeomagnetic results from the Sierras Australes fold and thrust belt, Argentina. *Geophysics Journal International*, **147**, 356–366.

Tucker, S. & Kent, D. V. 1988. Multiple remagnetizations of the lower Paleozoic limestones from the Taconics of Vermont. *Geophysics Research Letters*, **15**, 1251–1254.

Turner, P. 1980. *Continental Red Beds*. Elsevier, Amsterdam, Developments in Sedimentology, **29**.

Van der Pluijm, B. A. 1987. Grain-scale deformation and the fold test – evaluation of synfolding remagnetization. *Geophysics Research Letters*, **14**, 155–157.

Van der Voo, R. 1981. Paleomagnetism of North America: a brief review. *In*: McElhinny, M. W. & Valencio, D. A. (eds) *Paleoreconstruction of the Continents*. American Geophysical Union, Washington, DC, Geodynamics Series, **2**, 159–176.

Van der Voo, R. 1993. *Paleomagnetism of the Atlantic, Tethys and Iapetus Oceans*. Cambridge University Press, Cambridge.

Van der Voo, R., French, A. N. & French, R. B. 1979. A paleomagnetic pole position from the folded Upper Devonian Catskill red beds, and its tectonic implications. *Geology*, **7**, 345–348.

Van der Voo, R. & Scotese, C. R. 1981. Paleomagnetic evidence for a large (c. 2000 km) sinistral offset along the Great Glen Fault during the Carboniferous. *Geology*, **9**, 583–589.

Van Hoof, A. A. M. & Langereis, C. G. 1991. Reversal records in marine marls and delayed acquisition of remanent magnetization. *Nature*, **351**, 223–225.

Vasiliev, I., Franke, C., Meeldijk, J. D., Dekkers, M. J., Langereis, C. G. & Krijgsman, W. 2008. Putative greigite magnetofossils from the Pliocene Epoch. *Nature Geoscience*, **1**, 782–786.

Waldhör, M. & Appel, E. 2006. Intersections of remanence small circles: new tools to improve data processing and interpretation in palaeomagnetism. *Geophysics Journal International*, **166**, 33–45.

Weaver, R., Roberts, A. P. & Barker, A. J. 2002. A late diagenetic (syn-folding) magnetization carried by pyrrhotite: implications for paleomagnetic studies from magnetic iron sulphide-bearing sediments. *Earth and Planetary Science Letters*, **200**, 371–386.

Weil, A. B. & Van der Voo, R. 2002. Insights into the mechanism for orogen-related carbonate remagnetization from growth of authigenic Fe-oxide: a SEM and rock magnetic study of Devonian carbonates from northern Spain. *Journal of Geophysical Research*, **107**, 2063, doi 10.1029/2001JB000200.

Weil, A. B., Van der Voo, R., van der Pluijm, B. A. & Parés, J. M. 2000. The formation of an orocline by multiphase deformation: a paleomagnetic investigation of the Cantabria-Asturias Arc (northern Spain). *Journal of Structural Geology*, **22**, 735–756.

Weil, A. B., Van der Voo, R. & van der Pluijm, B. A. 2001. Oroclinal bending and evidence against the Pangea megashear: the Cantabria-Asturias Arc (northern Spain). *Geology*, **29**, 991–994.

Wisniowiecki, M., Van der Voo, R., McCabe, C. & Kelly, W. C. 1983. A Pennsylvanian paleomagnetic pole from the mineralized Late Cambrian Bonneterre Formation, SE Missouri. *Journal of Geophysical Research*, **88**, 6540–6548.

Xu, W., Van der Voo, R. & Peacor, D. R. 1994. Are magnetite spherules capable of carrying stable magnetizations? *Geophysics Research Letters*, **21**, 517–520.

Xu, W., Van der Voo, R. & Peacor, D. R. 1998. Electron microscopic and rock magnetic study of remagnetized Leadville carbonates, central Colorado. *Tectonophysics*, **296**, 333–362.

Zegers, T. E., Dekkers, M. J. & Bailly, S. 2003. Late Carboniferous to Permian remagnetization of Devonian limestones in the Ardennes: role of temperature, fluids, and deformation. *Journal of Geophysical Research*, **108**, 2357, doi: 10.1029/2002JB 002213

Zhao, X. & Roberts, A. P. 2010. How does Chinese loess become magnetized? *Earth and Planetary Science Letters*, **292**, 112–122.

Zhou, L. P. & Shackleton, N. J. 1999. Misleading positions of geomagnetic reversal boundaries in Eurasian loess and implications for correlation between continental and marine sedimentary sequences. *Earth and Planetary Science Letters*, **168**, 117–130.

# Episodic Remagnetizations related to tectonic events and their consequences for the South America Polar Wander Path

E. FONT[1]*, A. E. RAPALINI[2], R. N. TOMEZZOLI[2], R. I. F. TRINDADE[3] & E. TOHVER[4]

[1]*IDL-Faculdade de Ciências da Universidade de Lisboa, Edifício C8,
Campo grande 1749-016, Lisbon, Portugal*

[2]*Departamento de Ciencias Geológicas, Laboratorio de Paleomagnetismo
Daniel A.Valencio, Instituto de Geociencias Básicas, Aplicadas y Ambientales
de Buenos Aires (IGEBA), FCEN, Universidad de Buenos Aires*

[3]*Instituto de Astronomia, Geofísica e Ciências Atmosféricas,
Universidade de São Paulo, São Paulo, Brazil*

[4]*School of Earth & Environment, University of Western Australia,
Crawley, 6009, Western Australia, Australia*

*Corresponding author (e-mail: font_eric@hotmail.com)*


**Abstract:** The South American record of remagnetizations is linked to specific events of its tectonic history stretching back to Precambrian times. At the Ediacaran–Cambrian time interval (570–500 Ma), the final stages of the western Gondwana assemblage led to remagnetization of Neoproterozoic carbonates within the São Francisco–Congo Craton and at the border of the Amazon Craton, along the Araguaia–Paraguay–Pampean Belt. From the late Permian to early Triassic, the San Rafaelic orogeny and the emplacement of the Choiyoi magmatic province was responsible for widespread remagnetizations in Argentina and Uruguay. Cretaceous remagnetization has also been documented in Brazil and interpreted to result from magmatism and fault reactivations linked to the opening of the South Atlantic Ocean. We present a review of these widespread remagnetization events principally based on palaeomagnetic data and, when available, on rock magnetic and radiogenic isotope age data. This study gives an overview of the geographical distribution of the remagnetization events in South America, and provides important clues to better understand the geodynamic evolution of the South American plate at these times. In addition, magnetic mineralogy data for the different case studies presented here constrain the physical–chemical mechanisms that led to partial or total resetting of magnetic remanences in sedimentary rocks.


Global plate tectonics reconstructions using palaeomagnetism are based on the record of the natural remanent magnetization (NRM) in rocks at the time of their formation. Consequently, secondary remanent magnetizations acquired late in the geological history of sedimentary and igneous rocks have long been considered as an obstacle for palaeogeographic reconstructions based on palaeomagnetism. Over the past two decades, a better understating of physical–chemical mechanisms responsible for magnetic resetting and the genesis of authigenic magnetic carriers, in addition to improvements in radiogenic isotope dating methods, permit the use of remagnetized rocks as a tool for the study of large-scale geological processes. Widespread remagnetization events linked to large-scale tectonic processes, such as formation of orogenic belts, deformation and metamorphism, are well documented in the literature for North America

(e.g. McCabe & Elmore 1989; Elmore *et al.* 1993; Banerjee *et al.* 1997; Xu *et al.* 1998) and Europe (e.g. McCabe & Channell 1994; Katz *et al.* 2000; Jordanova *et al.* 2001; Weil & van der Voo 2002; Gong *et al.* 2009) but are still poorly constrained for the South American continent (D'Agrella-Filho *et al.* 2000; Trindade *et al.* 2004; Rapalini & Sánchez-Bettucci 2008; Tohver *et al.* 2010; Font *et al.* 2011) despite the important role they plays in global plate-tectonic reconstructions.

From Precambrian times until the end of the Cretaceous, the South American plate has experienced a significant number of tectonic events: the accretion of crustal blocks, opening and closing of ocean basins and large-scale magmatic and metamorphic processes. In early Cambrian times, during the final stages of assemblage of western Gondwana, the closure of the Clymene Ocean that separated the Amazon Craton from the São Francisco and Rio de

*From*: ELMORE, R. D., MUXWORTHY, A. R., ALDANA, M. M. & MENA, M. (eds) 2012. *Remagnetization and Chemical Alteration of Sedimentary Rocks*. Geological Society, London, Special Publications, **371**, 55–87.
First published online August 22, 2012, http://dx.doi.org/10.1144/SP371.7

la Plata blocks was apparently responsible for widespread remagnetizations along the Paraguay, Pampean and Dom Feliciano belts (e.g. D'Agrella-Filho *et al.* 2000; Trindade *et al.* 2003, 2004, 2006; Font *et al.* 2006; Rapalini & Sánchez-Bettucci 2008; Tohver *et al.* 2010, 2011). During the Permian, southern South America was affected by the San Rafaelic deformation and other approximately coeval tectonic events for which several genetic models have been proposed. Among them is accretionary tectonics, collision of the North Patagonian block and the contemporary to slightly younger felsic magmatic activity of the large Choiyoi province (Azcuy & Caminos 1987; Llambías & Sato 1995; Ramos 2008; Vaughan & Pankhurst 2008; Tomezzoli 2009; Rapalini *et al.* 2010). The timing of this orogenic event coincides with wide-scale remagnetizations in Argentina (Valencio *et al.* 1980; Rapalini & Tarling 1993; Tomezzoli & Vilas 1999; Rapalini *et al.* 2000; Tomezzoli 2001) and Uruguay (Rapalini & Sánchez-Bettucci 2008). Finally, a large-scale deformational event occurred during the Cretaceous in eastern Brazil due to intense volcanism and fault reactivation linked to the opening of the South Atlantic Ocean (Font *et al.* 2011), which also left an important imprint in the magnetic record of older rocks.

The purpose of the present contribution is two-fold: (1) to provide a review of the time and location of widespread remagnetization events that have affected South America since the Precambrian, and highlight evidence of their direct relationship with large-scale plate-tectonics events; and (2) to give a synthesis of the nature and origin of the physical–chemical mechanisms that lead to the partial or total magnetic resetting in these rocks. Beforehand, a synthesis of the apparent polar wander path (APWP) of South America, based on updated palaeomagnetic pole compilations, is presented and discussed to provide a useful palaeomagnetic background for comparison with the remagnetized poles described further.

## APWPs of South America at Ediacaran–Cambrian, late Palaeozoic and Cretaceous times

Isotopic, palaeomagnetic and geochronological data suggest that the final assemblage of Gondwana was a long process spanning the Neoproterozoic and early Cambrian times, resulting from successive diachronic collisions involving major and minor blocks (Stern 2002; Cordani *et al.* 2003; Meert 2003; Trindade *et al.* 2006; Meert & Lieberman 2008; Vaughan & Pankhurst 2008; Cordani *et al.* 2009; Gray *et al.* 2009; Tohver *et al.* 2011). The APWP

of Gondwana is anchored by a few palaeomagnetic poles for the 570–500 Ma interval (Fig. 1b, Table 1) (Tohver *et al.* 2006; Trindade *et al.* 2006; Moloto-A-Kenguemba *et al.* 2008), and the older portion of this period is probably marked by ongoing convergence of different cratonic blocks. No individual South American craton can be constrained by palaeomagnetic poles between the age of the Marinoan glaciation in the Amazon Craton (*c.* 635 Ma; Trindade *et al.* 2003; Font *et al.* 2005) and the age of the Nola dyke (571 ± 6 Ma; Moloto-A-Kenguemba *et al.* 2008) and the Sierra de Animas two poles (579 ± 2 Ma; Sanchez-Bettuci & Rapalini 2002; Oyhantçabal *et al.* 2007). For ages older than 580 Ma there are only two palaeomagnetic poles available, which are not well constrained. The first, the Playa Hermosa pole (Sánchez-Bettucci & Rapalini 2002), has a coherent position in the APWP of Gondwana (Fig. 1b) but is of low quality due to: (1) absence of radiometric isotope ages that may constrain the depositional age of the sediments; (2) low number of data (2 sites, 6 samples); (3) absence of field tests; and (4) absence of flattening inclination correction. However, new geochronological, magnetic fabric and palaeomagnetic results (Lossada *et al.* 2011) seem to ratify the pole position and its proposed age (around 595 Ma). The second pole from this interval is the Campo Alegre pole (D'Agrella-Filho & Pacca 1988) which has long been used for palaeogeographic reconstructions. This has recently been called into question however by its geographic proximity with the widespread remagnetization that affected the Itajaí basin during Permian or Cretaceous times (see below). The remanent magnetization of the Campo Alegre rocks is very stable and coherent, being carried by both magnetite and hematite, but only an incomplete baked contact test is reported in the original study of D'Agrella-Filho & Pacca (1988) which is insufficient to ascertain its primary nature. Re-evaluating the tilt correction applied to these rocks, it is observed that the *in situ* site mean characteristic component of the Campo Alegre rocks lies very close to the remagnetized Itajaí virtual geomagnetic poles (VGPs), suggesting that both results reflect a contemporaneous remagnetization (Font *et al.* 2011). At the other end of the APWP, starting from 525 Ma, a well-defined APWP for Gondwana can be traced along northern Africa until the early–mid-Ordovician (*c.* 500 Ma, see compilations by Meert 2003 and Trindade *et al.* 2006; Fig. 1c; Table 1).

The South American APWP for the time period from the Carboniferous (*c.* 350 Ma) until the Permo-Triassic (*c.* 250 Ma) is the subject of numerous controversies and has been at the core of the different Pangea models (i.e. Pangea A v. Pangea B, e.g. Wegener 1912; Bullard *et al.* 1965; Irving 1977;

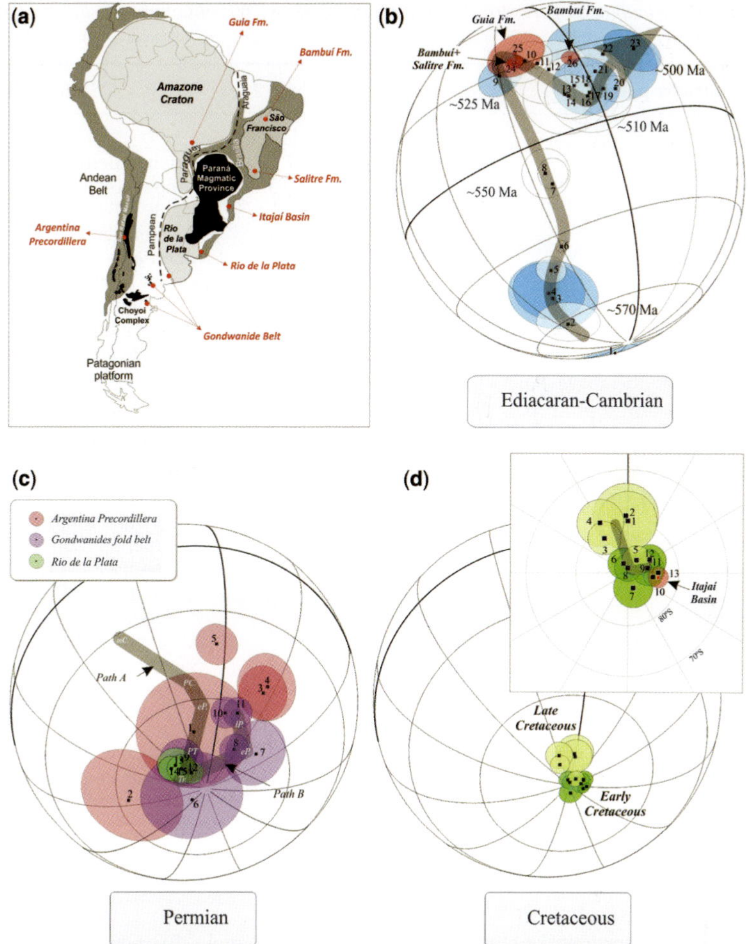



**Fig. 1.** (**a**) Geological map of South America (modified from Tohver *et al.* 2011) showing the location of the studied areas; Apparent Polar Wander Paths for the South American plate at (**b**) Ediacaran–Cambrian (poles from South America are highlighted in blue; see text for details); (**c**) Permian, showing a short Path A (Brandt *et al.* 2009) and a more conventional Path B (Tomezzoli 2009); (**d**) Cretaceous times (Font *et al.* 2009). Selected palaeomagnetic poles are referenced in Tables 1–3. Positions of the remagnetized formations synthetized in this work are also indicated: (b) Bambuí, Salitre and Guia Formations (in red); (c) Argentina Precordillera (in pink), Gondwanides Fold Belt (in purple) and Rio de la Plata (in green); and (d) Itajaí Basin (in green).

Morel & Irving 1981; Torsvik & Van der Voo 2002; Muttoni *et al.* 2003, 2009; Van der Voo & Torsvik 2004; Brandt *et al.* 2009; Domeier *et al.* 2011; Yuan *et al.* 2011). Ever since the publication of the first Late Palaeozoic APWP for South America (e.g. Vilas 1981), it has been observed that Late Carboniferous–Late Permian poles plot along a large loop. This trend has been subsequently confirmed by numerous palaeomagnetic studies (see Rapalini *et al.* 2006; Tomezzoli *et al.* 2009 for recent reviews; Fig. 1c). Most but not all of these results derive from the study of clastic sedimentary rocks.

This loop indicates a relatively low-latitude position for the western areas of Gondwana in the Early Permian, which gives rise to an apparently large overlap of continental crust with southern Laurasia in a Wegener type Pangea A reconstruction. This palaeogeographic overlap was eliminated by Irving's (1977) proposal of a different Permian configuration for Pangaea (Pangea B) that transposed Gondwana to the east relative to Laurrussia. In order to restore Pangaea to its known configuration at the time of break-up (reconstructed from ocean floor magnetic anomalies), the Pangea model requires a

**Table 1.** *Selected palaeomagnetic poles for the calibration of the APWP of Gondwana during Ediacaran–Cambrian times*

| Selected palaeomagnetic poles | | $P_{lat}$ | $P_{long}$ | $\alpha_{95}$ | References |
|---|---|---|---|---|---|
| *Gondwana APWP* | | | | | |
| 1 | Praya Hermoza | −75 | 182 | 12 | Sánchez-Bettucci & Rapalini (2002) |
| 2 | Nola Dolerite | −62 | 305 | 10.5 | Moloto-Kenguemba *et al.* (2008) |
| 3 | Sierra de las Animas II | −49 | 312 | 15 | Sánchez-Bettucci & Rapalini (2002) |
| 4 | Los Barrientos | −47 | 313 | 12 | Rapalini (2006) |
| 5 | Sinyai dolerite | −40 | 321 | 5 | Meert & Van der Voo (1996) |
| 6 | Mirbat SS | −34 | 330 | 2 | Kempf *et al.* (2000) |
| 7 | Arumbera | −14 | 336 | 5 | Kirschvink (1978) |
| 8 | Todd River | −10 | 335 | 8 | Kirschvink (1978) |
| 9 | Itabaina dykes | 29 | 330 | 8 | Trindade *et al.* (2006) |
| 10 | Adma diorites | 31 | 341 | 9 | Morel (1981) |
| 11 | Ntonya Ring | 28 | 345 | 2 | Briden (1968); Briden *et al.* (1993) |
| 12 | Hawker Group | 24 | 348 | 13 | Klootwijk (1980) |
| 13 | Pertaoorta Group | 12 | 351 | 8 | Klootwijk (1980) |
| 14 | Madagascar vigration | 13 | 350 | 14 | Meert *et al.* (2003) |
| 15 | Kangaroo Island | 15 | 354 | 12 | Klootwijk (1980) |
| 16 | Billy Creek | 11 | 358 | 14 | Klootwijk (1980) |
| 17 | Carion Granite | 14 | 358 | 11 | Meert *et al.* (2001) |
| 18 | Juiz de Fora Complex | 10 | 357 | 10 | D'Agrella-Filho *et al.* (2004) |
| 19 | Giles Creek | 11 | 3 | 10 | Klootwijk (1980) |
| 20 | Sor Rondane | 10 | 7 | 5 | Zijderveld (1968) |
| 21 | Lower Lake Frome | 18 | 3 | 10 | Klootwijk (1980) |
| 22 | Sierras de la Animas I | 24 | 9 | 19 | Sánchez-Bettucci & Rapalini (2002) |
| 23 | Piquete Formation | 24 | 22 | 10 | D'Agrella-Filho *et al.* (1986) |
| *Remagnetized poles* | | | | | |
| 24 | Bambui + Salitre C | 32 | 337 | 3 | Trindade *et al.* (2004) |
| 25 | Araras Group B | 36 | 338 | 10 | Trindade *et al.* (2003) |
| 26 | Bambui B | 26 | 357 | 3 | D'Agrella-Filho *et al.* (2004) |

Remagnetized palaeomagnetic poles of the Guia, Bambuí and Salitre formations are also shown. Palaeomagnetic poles were rotated using the Euler pole from Reeves *et al.* (2004) for South America: Lat = 43.017°N, Long = 329.935°E, Angle = 58.842°.

very long (3000 km or more) strike-slip displacement between Gondwana and Laurasia at some time during the Permian and/or the Triassic (Irving 2004). This palaeoreconstruction has attracted occasional support from palaeomagnetic studies from different areas of Gondwana such as Africa (Bachtadse *et al.* 2002), Adria (Muttoni *et al.* 2003) and Iran (Torcq *et al.* 1997). Amongst geologists, the enormous amount of displacement between Gondwana and Laurasia required by the Pangaea B configuration has been a source of scepticism. Several tests have demonstrated that large-scale and long-standing non-dipolar fields are not a plausible explanation for the apparent overlap between northern and southern continents (Muttoni *et al.* 2003, 2009; Tauxe 2005). Thus, palaeomagnetic objections to the Pangaea B configuration are related to artefacts or biases in the palaeomagnetic database, for example incomplete demagnetization, overprinted results, incorrectly dated rocks or the inclination error in sedimentary rocks (e.g. Rochette & Vandamme 2001; Weil *et al.* 2001; Torsvik & Van der Voo 2002). Recent palaeomagnetic results

in Permian rocks for South America (e.g. Brandt *et al.* 2009; Domeier *et al.* 2011; Yuan *et al.* 2011; Fig. 1c; Table 2) have been in favour of a much shorter APWP path for South America between the Late Carboniferous and the Early Triassic that would be more compatible with a Pangea A reconstruction throughout the whole time span. To consider this proposal, most (if not all) the previously available poles must however be discarded, no matter whether they correspond to pre-tectonic, syn-tectonic or post-tectonic magnetizations. The short APWP is also incompatible with some poles based on magmatic rocks, which are intrinsically devoid of inclination-shallowing effects (Conti & Rapalini 1990; Tomezzoli *et al.* 2005, 2009). Which APWP is the most reliable still remains an open question.

The Cretaceous APWP for South America is comparatively well established (Ernesto *et al.* 2002; Ernesto 2007; Somoza & Zaffarana 2008; Font *et al.* 2009) and is consistent with its North America counterpart after adequate rotation, at least for the mid- to late Cretaceous interval

**Table 2.** *Mean palaeomagnetic poles of the Path A (Brandt* et al. *2009) and Path B (Tomezzoli 2009) illustrated in Figure 1c*

| Selected palaeomagnetic poles | | $P_{lat}$ | $P_{long}$ | $\alpha_{95}$ | References |
|---|---|---|---|---|---|
| *Mean palaeomagnetic poles* | | | | | |
| **Path A** (Brandt *et al.* 2009) | | | | | |
| Middle Permian–early Triassic | | −80 | 311 | 6.9 | |
| Early Permian | | −62.4 | 347.6 | 8.1 | |
| Permian–Carboniferous | | −54.3 | 341 | 12.4 | |
| Middle Carboniferous | | −31.6 | 317.5 | 8.3 | |
| **Path B** (Tomezzoli 2009) | | | | | |
| Triassic | | 4 | −81 | 279.5 | |
| Late Permian | | – | −72 | 39 | |
| Early Permian | | 7 | −64 | 14 | |
| *Remagnetized poles* | | | | | |
| 1 | San Juan Formation | −70.2 | 337.8 | 19.2 | Rapalini & Tarling (1993) |
| 2 | La Flecha Formation | −63.8 | 244.6 | 18 | Rapalini & Astini (2005) |
| 3 | Ponón Trehue | −53.4 | 25 | 9 | Truco & Rapalini (1996) |
| 4 | Alcaparrossa | −50.8 | 25.8 | 11 | Rapalini & Tarling (1993) |
| 5 | Hoyada Verde | −41.9 | 356.2 | 7 | Rapalini *et al.* (1989) |
| 6 | Gonzalves Chaves | −84 | 216 | 17 | Tomezzoli & Vilas (1997) |
| 7 | San Roberto | −70 | 49 | 11 | Tomezzoli *et al.* (2006) |
| 8 | Tunas II | −74 | 26 | 5 | Tomezzoli & Vilas (1999) |
| 9 | Sierra Grande | −77 | 311 | 7 | Rapalini (1998) |
| 10 | Rio Curaco | −64 | 5 | 5 | Tomezzoli *et al.* (2006) |
| 11 | Tunas I | −63 | 14 | 5 | Tomezzoli & Vilas (1999) |
| 12 | Cerro Victoria | −82.6 | 309.3 | 3.9 | Rapalini & Sánchez-Bettucci (2008) |
| 13 | Yerbal | −77 | 298.4 | 5.9 | Rapalini & Sánchez-Bettucci (2008) |
| 14 | Rocha | −76.6 | 291 | 4.2 | Rapalini & Sánchez-Bettucci (2008) |
| 15 | Sierras Bayas Group | −79.3 | 300.9 | 4.8 | Valencio *et al.* (1980) |

Remagnetized palaeomagnetic poles of the Gondwanides Belt, the Argentina Precordillera and the Rio de la Plata Craton are also indicated.

(Tarduno & Smirnov 2001; Somoza & Zaffarana 2008). Future refinements to this section of the APWP should address age determinations and more complete geographic distribution for the mid- to late Cretaceous interval (Table 3). The compilation of Font *et al.* (2009) is based solely on poles from magmatic rocks, in order to minimize uncertainties in the palaeomagnetic record of sedimentary rocks due to problems of shallowing and time of magnetization (Fig. 1d; Table 3). The majority of high-quality palaeomagnetic poles are from the Paraná Magmatic Province of Brazil, with a few poles from other regions of South America. Of the thirteen palaeomagnetic poles selected by Font *et al.* (2009), eight are from Brazil (Table 3), four are from Argentina (Geuna & Vizán 1998; Geuna *et al.* 2000) including two from Patagonia (Butler *et al.* 1991; Somoza & Zaffarana 2008) and one is from Paraguay (Ernesto *et al.* 1996). The APWP of South America is better defined for the early Cretaceous interval from *c.* 135 to

*c.* 125 Ma, where high-quality palaeomagnetic data based on a large number of independent sites are available from well-dated rocks by $^{40}Ar/^{39}Ar$ methods (see review in Ernesto 2007). Of particular interest are the Serra Geral Formation and the Ponta Grossa dykes from the Paraná Magmatic Province, dated at 133–132 and 131–129 Ma, respectively (Ernesto *et al.* 1990, 1999; Renne *et al.* 1992, 1996); and the Central Alkaline Magmatic Province, Brazil dated at 127–130 Ma (K–Ar method; Velázquez *et al.* 1992; Ernesto *et al.* 1996). From early- to mid-Cretaceous, South American poles cluster near the geographic pole (Fig. 1d). From mid- to late Cretaceous, the South American plate experienced southward continental drift associated with clockwise rotations (Ernesto *et al.* 2002). The transition is well constrained in the APWP by the high-quality palaeomagnetic pole from the Cabo Magmatic Province, NE Brazil ($Q = 5$), well dated at $102 \pm 1$ Ma (Font *et al.* 2009).

**Table 3.** *Selected palaeomagnetic poles for the calibration of the APWP of South America at the Cretaceous (modified from Font* et al. *2009)*

| Formation | Pole number | Age (Ma) | Polarity | N | Long. (°E) | Lat. (°N) | References |
|---|---|---|---|---|---|---|---|
| *South America (North Pole)* | | | | | | | |
| Patagonian Basalts | 1 | 64–79 | N/R | 18 | 178.4 | 78.7 | Butler *et al.* (1991) |
| Itatiaia and Passa Quatro comb. | 2 | 70–71 | N/R | 18 | 180.0 | 79.5 | Montes-Lauar *et al.* (1995) |
| Poços de Caldas Complex | 3 | 84 | N/R | 36 | 145.7 | 82.2 | Montes-Lauar *et al.* (1995) recalc. |
| São Sebastião Dykes | 4 | 80–90 | N | 26 | 151.3 | 79.0 | Montes-Lauar *et al.* (1995) |
| San Bernardo | 5 | 85–98 | N | 9 | 215.0 | 87.0 | Somoza & Zaffarana (2008) |
| Cabo Magmatic Province, Brazil | 6 | 100 | N | 24 | 155.9 | 87.9 | This paper |
| Cerro Barcino, Argentina | 7 | 112–130 | N | 16 | 339.0 | 87.0 | Geuna *et al.* (2000) |
| Florianópolis Dykes, SE Brazil | 8 | 119–128 | N/R | 65 | 183.3 | 89.1 | Raposo *et al.* (1998) |
| Central Alkaline Province | 9 | 127–130 | N/R | 75 | 242.3 | 85.4 | Ernesto *et al.* (1996) |
| Ponta Grossa Dykes, SE Brazil | 10 | 129–131 | N/R | 115 | 238.5 | 84.5 | Ernesto *et al.* (1999) |
| Serra Geral Formation | 11 | 133–132 | N/R | 392 | 269.2 | 84.1 | Ernesto *et al.* (1990, 1999) |
| Cordoba Province, Argentina | 12 | 133–115 | N/R | 55 | 255.9 | 86.0 | Geuna & Vizán (1998) |
| Northeastern Brazil Magmatism | 13 | 125–145 | N/R | 44 | 277.6 | 85.2 | Ernesto *et al.* (2002) |
| *Mean South America Poles* | | | | | | | |
| Upper Cretaceous | | 65–80 | N/R | 4 | 165.2 | 80.2 | $\alpha_{95} = 3.9$; $K = 552$ |
| Middle Cretaceous | | 80–110 | N | 3 | 230.9 | 89.2 | $\alpha_{95} = 4.9$; $K = 646$ |
| Lower Cretaceous | | 110–140 | N/R | 6 | 254.8 | 86 | $\alpha_{95} = 1.9$; $K = 1152$ |
| *Remagnetized Poles* | | | | | | | |
| Itajaí Basin | | | R | 95 | 277.5 | 84 | Font *et al.* (2011) |

Position of the remagnetized Itajaí pole is also shown.

## Cambrian remagnetizations at the final assemblage of Western Gondwana

### The Guia Formation

The Araras Group in the SE Amazon Craton (Fig. 1a) presents a peculiar case where both primary and secondary magnetic remanences are preserved within a single carbonate sequence. The stable and undeformed region near the city of Mirassol d'Oeste (MO), Mato Grosso state, Brazil presents well-preserved outcrops of the lower part of the Araras Group in the Terconi Quarry. Here, basal pink microbial dolostones (i.e. the so-called cap carbonates; Hoffman *et al.* 1998; Hoffman *et al.* 2007) of the Mirassol d'Oeste Formation and bituminous limestones of the Guia Formation cover the Marinoan (*c.* 635 Ma) glacial deposits of the Puga Formation (Nogueira *et al.* 2003, 2007; Trindade *et al.* 2003; Font *et al.* 2010).

The basal pink dolostone records a stable magnetization carried by hematite that yielded a palaeomagnetic pole, $P_{long} = 283.7°$, $P_{lat} = 83.5°S$ ($N = 51$; $dp = 3.1°$; $dm = 5.3°$; palaeolatitude $= 21.2°$), for the Amazon Craton at the end of the Neoproterozoic (Trindade *et al.* 2003; Font *et al.* 2005, 2010). The presence of geomagnetic reversals (Fig. 2a) and a positive fold test (Trindade, R., unpublished data) suggests a primary, detrital origin for the magnetization. The presence of at least four reversals within the 20 m of dolostones observed by Trindade *et al.* (2003) and Font *et al.* (2010) also suggests sedimentation over a much longer time period than that predicted by the original 'snowball Earth' hypothesis of Hoffman *et al.* (1998). In contrast, the bituminous limestones of the Guia Formation record a secondary, post-folding remagnetization (Fig. 2b; Trindade *et al.* 2003; Tohver *et al.* 2010). Rock magnetic properties are typical of those of remagnetized carbonates (Font

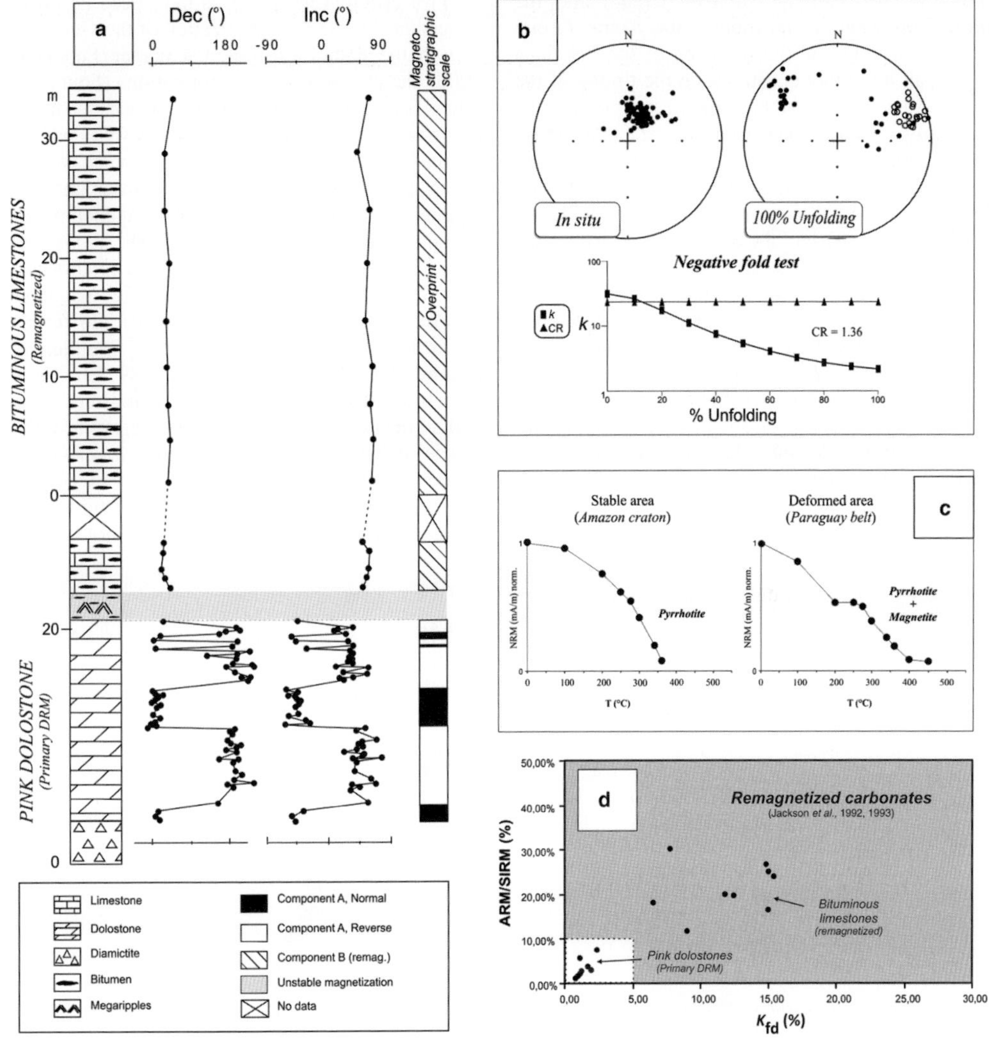



**Fig. 2.** (**a**) Magnetostratigraphic column of the Mirassol d'Oeste and Guia formations (Araras Group) showing primary (with geomagnetic reversals; A component) and secondary (B component) characteristic remanent magnetizations, respectively. (**b**) Negative fold test (McElhinny 1964) of the Guia limestones in the Paraguay Fold and Thrust Belt. Above: stereographic projections in geographic and tilt-corrected coordinates showing a divergence of magnetic directions after 100% unfolding; below: dispersion parameter $k$ v. percentage of unfolding (CR represent the critical ratio at which $k$ values become significant at the 95% confidence level for the number of data entries). (**c**) Thermal demagnetization curves of Guia limestones in stable (Amazon Craton) and deformed (Paraguay Belt) areas showing pyrrhotite and a mixture of pyrrhotite and magnetite as principal magnetic carriers, respectively. (**d**) Dominance of superparamagnetic particles in the remagnetized Guia limestones is indicated by high values of ARM/SIRM and $K_{fd}$ (%), whereas the latter are absent in the preserved MO dolostones (typical values for remagnetized carbonates are from Jackson *et al.* 1992, 1993).

*et al.* 2006) and indicate magnetite and iron sulphides (mostly pyrrhotite) as principal magnetic carriers in these rocks (Fig. 2c, d).

This secondary palaeomagnetic direction from the central portion of the Paraguay Belt reported by Trindade *et al.* (2003) yields a palaeomagnetic pole

($P_{long} = 326.6°$; $P_{lat} = 33.1°N$; $N = 74$; dp $= 3.3°$; dm $= 3.6°$; palaeolatitude $= 36.4°$) that plots on the APWP for West Gondwana (Fig. 1b) close to the high-quality 525 Ma Itabaiana pole of the Borborema Province, NE Brazil (Trindade *et al.* 2006). This observation indicates the palaeogeographic proximity between the Amazon Craton and the proto-Gondwana at the time of the Araras Group remagnetization and the intrusion of the Itabaiana dykes, an important constraint on the timing of the final collision between these two cratons during the assembly of Gondwana (Trindade *et al.* 2003, 2006). Further study along the entire >90° arc of the curved Paraguay Belt demonstrates a clear correlation between the declination of this secondary magnetization and the strike of the belt (Fig. 3; Tohver *et al.* 2010). This positive strike test, equivalent to a vertical axis fold test, indicates the secondary, oroclinal nature of the Paraguay Belt curvature. Thus, the >90° bend of the Paraguay Belt was created by large-scale tectonic rotations (principally of the *c.* 200 km long, east–west limb of the belt) after 525 Ma.

The Araras Group offers an excellent opportunity to study the interplay between carbonate rocks, hydrocarbons and the physio-chemical conditions that led to the complete resetting of their primary remanence.

Two different mechanisms have been proposed to account for the remagnetization of the Guia Formation: the maturation of organic matter (Font *et al.* 2006) and clay mineral transformations (Font *et al.* 2006; Tohver *et al.* 2010). In the first case, the maturation of hydrocarbons by microbial activity in sedimentary rocks and particularly in carbonates results in an anaerobic iron-reducing environment that favours the growth of authigenic magnetite and ferrous sulphide (e.g. Seewald 2001, 2003; Kao *et al.* 2004). The first evidence for a relationship between hydrocarbons and authigenic magnetite was reported by Elmore *et al.* (1987) and further confirmed by others workers (e.g. Benthien & Elmore 1987; Kilgore & Elmore 1989; Elmore & Crawford 1990; Suk *et al.* 1990; Blumstein *et al.* 2004; Font *et al.* 2006). Models for magnetite authigenesis in hydrocarbon-rich sediments invoke reducing conditions, induced by the release of $H_2S$ from hydrocarbon maturation, leading to the transformation of primary iron oxides (e.g. Reynolds *et al.*

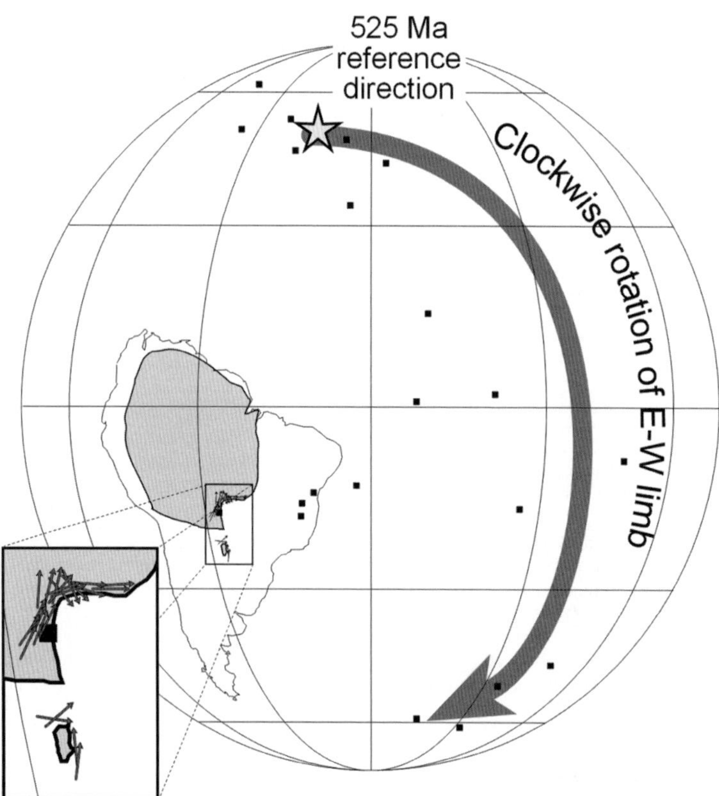



**Fig. 3.** Position of site mean VGPs from the Araras Group carbonates demonstrating divergence from the 525 Ma Itabaiana reference pole. Inset shows how the declination of site level directions varies with regional strike of the Paraguay Belt, with easterly declinations common in the east–west-trending portion of the belt, northeasterly declinations in the NE-trending hinge zone and northerly directions in the north–south-trending domain.

1990, 1991; Seewald, 2003; Kao *et al.* 2004). Once dissolved $Fe^{2+}$ is available, authigenic magnetite can precipitate inorganically. However, the presence of spherical magnetite and framboids of pyrite/magnetite in several case studies suggested that microbial activity also plays an important role in the formation of new magnetic phases (e.g. Elmore *et al.* 1987; Elmore & Leach 1990; Brothers *et al.* 1996). Magnetite is not the solely magnetic carrier that can be produced by hydrocarbon–reservoir rock interactions. It may be partially or totally substituted by ferromagnetic iron sulphides, such as pyrrhotite and greigite, which contribute significantly to the chemical remanent overprint (e.g. Dekkers 1990; Rochette *et al.* 1990; Krs *et al.* 1992; Xu *et al.* 1998; Dunlop 2000; Zegers *et al.* 2003; Font *et al.* 2006). Greigite and pyrrhotite are generally metastable in sedimentary rocks, tending to transform into pyrite (i.e. pyritization), but may be preserved in (Total Organic Carbon) TOC-poor and reactive Fe-rich environments, where effective removal of reduced sulphur by precipitation of intermediate iron sulphides prevented pyritization (Kao *et al.* 2004). The presence of pyrrhotite in the remagnetized Araras Group rocks is evidenced magnetically by its characteristic Curie temperature at *c.* 280–320 °C (Rochette *et al.* 1990) (Fig. 2c). Scanning Electron Microscopy (SEM) coupled to Energy Dispersive Spectra (EDS) identified numerous iron sulphides in close association with bitumen, namely pyrrhotite and pyrite that precipitated in voids and microcracks of the rocks (Fig. 4a–d). Relics of pyritized bacteria are represented by framboids (Machel & Burton 1991; Fig. 4e).

The second mechanism for carbonate remagnetization relies on the link between clay mineral maturity and the presence of magnetite as a secondary by-product (McCabe *et al.* 1989; Katz *et al.* 1998, 2000; Gill *et al.* 2002; Woods *et al.* 2002; Blumstein *et al.* 2004; Moreau *et al.* 2005; Tohver *et al.* 2008). In the case of the Paraguay Belt, there appears to be a link between the host minerals of the stable

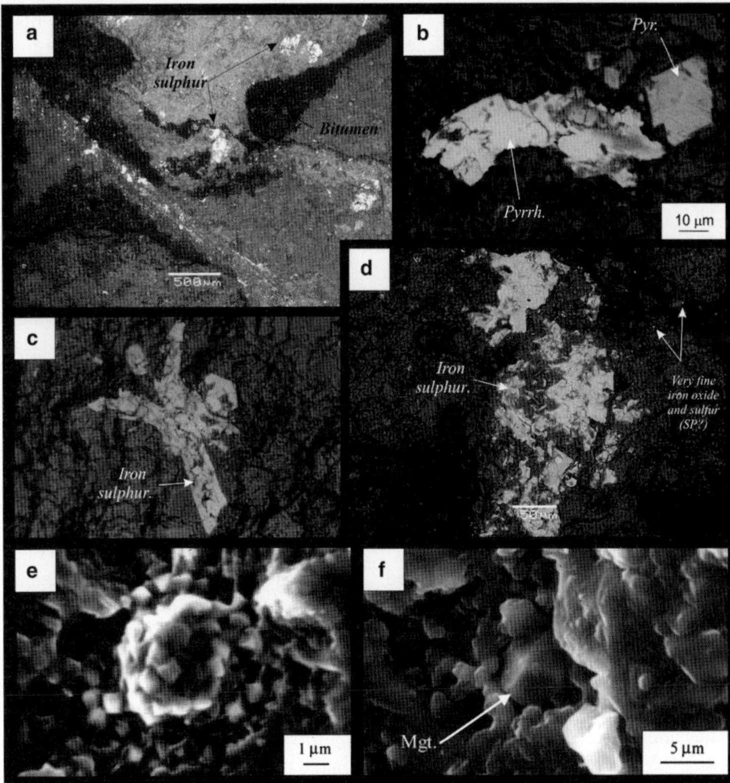



**Fig. 4.** Scanning Electronic Microscopy (SEM) photographs showing: (**a**) association of authigenic iron sulphurs with bitumen and fractures; (**b**) euhedral pyrite (diamagnetic) and probably pyrrhotite (ferromagnetic); (**c**) iron sulphur filling voids; (**d**) presence of large (>100 μm) iron sulphur and very fine (SP) iron oxides and sulphurs; (**e**) euhedral secondary magnetite; and (**f**) pyrite framboid.

magnetization and the site-level tectonic environment. For example, carbonate rocks located in the stable, that is cratonic, area (Terconi Quarry) contain essentially pyrrhotite whereas those located in the deformed area (Paraguay Belt) show a mixture of pyrrhotite and authigenic magnetite (Fig. 2c). Font *et al.* (2006) interpreted the lack of remanence signal in the upper levels of the dolostone (Fig. 2a) and the complete remagnetization of overlying limestones as a result of the higher organic matter content in these layers. In the more deformed portion of the Paraguay Belt, the higher metamorphic grade (higher illite crystallinity values of Alvarenga 1990) suggests the operation of the second clay-related mechanism of remagnetization. Here, authigenic magnetite occurs either as large euhedral crystals (i.e. palaeomagnetically irrelevant) or very fine, probably superparamagnetic, minerals (Fig. 4d, f). In their study of these remagnetized carbonates, Tohver *et al.* (2010) separated different size fractions of insoluble clay for $^{40}Ar/^{39}Ar$ encapsulation dating. The maximum age of remagnetization is indicated by the ages of the finest grain size aliquots, where the concentration of authigenic illite would be expected to be highest. The *c.* 530 Ma maximum age of remagnetization from three different sites of remagnetized limestone agrees with the *c.* 525 Ma age inferred from the aforementioned overlap in the palaeomagnetic pole position determined from the Araras Formation and the primary Itabaiana pole.

## The Bambuí case

The São Francisco Basin covers *c.* 300 000 km$^2$ of the São Francisco Craton and comprises clastic and carbonate sediments of Neoproterozoic age, with a basal glacial unit (Macaúbas Group in the south, Bebedouro Formation in the north) covered by a thick carbonate succession (Bambuí Group in the south, Salitre Formation in the north) (Karfunkel & Hoppe 1988). The basal glacial unit comprises diamictites, clast-supported conglomerates, sandstones and pelites deposited in glacio-marine environments. The upper unit comprises mostly carbonate rocks with pelitic intercalations deposited along an extensive platform that comprised not only the craton margins but most of its inner part (Fig. 1a). The succession has experienced moderate to weak deformation inside the cratonic area as far-field response of the intense deformation that occurred along the encircling Brasiliano fold belts between 600 and 520 Ma (Inda & Barbosa 1978).

The maximum age of sedimentation for both the Bebedouro and Macaúbas glacial units were obtained from the U–Pb sensitive high-resolution ion microprobe (SHRIMP) dating of detrital zircons from both units with ages around 900 Ma

(Pedrosa-Soares *et al.* 2000), which are corroborated by similar ages obtained for dykes that cut across the basement of the Neoproterozoic succession but do not reach it. Several attempts to date the upper carbonate succession using Pb–Pb and U–Pb methods were unsuccessful (Babinski *et al.* 1999; D'Agrella-Filho *et al.* 2000). Most ages cluster within 550–515 Ma with a peak at 520 Ma. Only two $^{207}Pb/^{206}Pb$ dates provide ages older than 650 Ma, thus approaching the sedimentation age of the carbonates at $686 \pm 69$ Ma (Babinski *et al.* 1999) and $740 \pm 22$ Ma (Babinski *et al.* 2007).

Palaeomagnetic data for the Bambuí Group and the Salitre Formation were reported by D'Agrella-Filho *et al.* (2000) and Trindade *et al.* (2004), respectively. In both regions, carbonate successions presented a very stable characteristic magnetic component with single polarity and tightly grouped directions across sedimentary sections hundreds of metres thick. In both regions, which are more than 600 km apart, directions were remarkably similar, comprising three magnetic components with similar palaeomagnetic poles obtained from magnetic components carried by similar magnetic carriers. The more stable components, B and C, are carried by monoclinic pyrrotite and magnetite, respectively. The corresponding palaeomagnetic poles coincide with reference poles for 520 Ma (Fig. 1b), which by its turn is the peak age obtained using Pb/Pb methods for the carbonates and Pb–Zn mineralization in the area, which is thus interpreted as the age of fluid migration across the whole São Francisco Craton (Trindade *et al.* 2004). In Salitre Formation rocks, in the northern sector of the basin, an additional component (named D) was obtained at very high unblocking temperatures, characteristic of hematite.

Samples from both regions show all the rock magnetic characteristics typical of remagnetized carbonates. The hysteresis cycles are always wasp-waisted in shape, resulting from the mixture of soft and hard coercivity fractions. In addition, all samples show anomalously high $H_c$ (coercivity) and $H_{cr}$ (coercivity remanence) ratios when plotted in the Day *et al.* (1977) diagram (Fig. 5a), following the trend of remagnetized carbonates proposed by Jackson (1990). This trend has been interpreted by Dunlop (2002) as a result of the mixture of stable single-domain magnetite with superparamagnetic grains. Lowrie–Fuller Cisowski tests were also performed in both collections (Fig. 5b). In this test, samples are alternating field (AF) demagnetized, are then given an anhysteretic remanent magnetization (ARM) at a peak field of 200 mT, demagnetized again and finally given a stepwise isothermal remanent magnetization (IRM) up to 200 mT, which is subsequently AF demagnetized. The resulting curves show the same behaviour presented in other

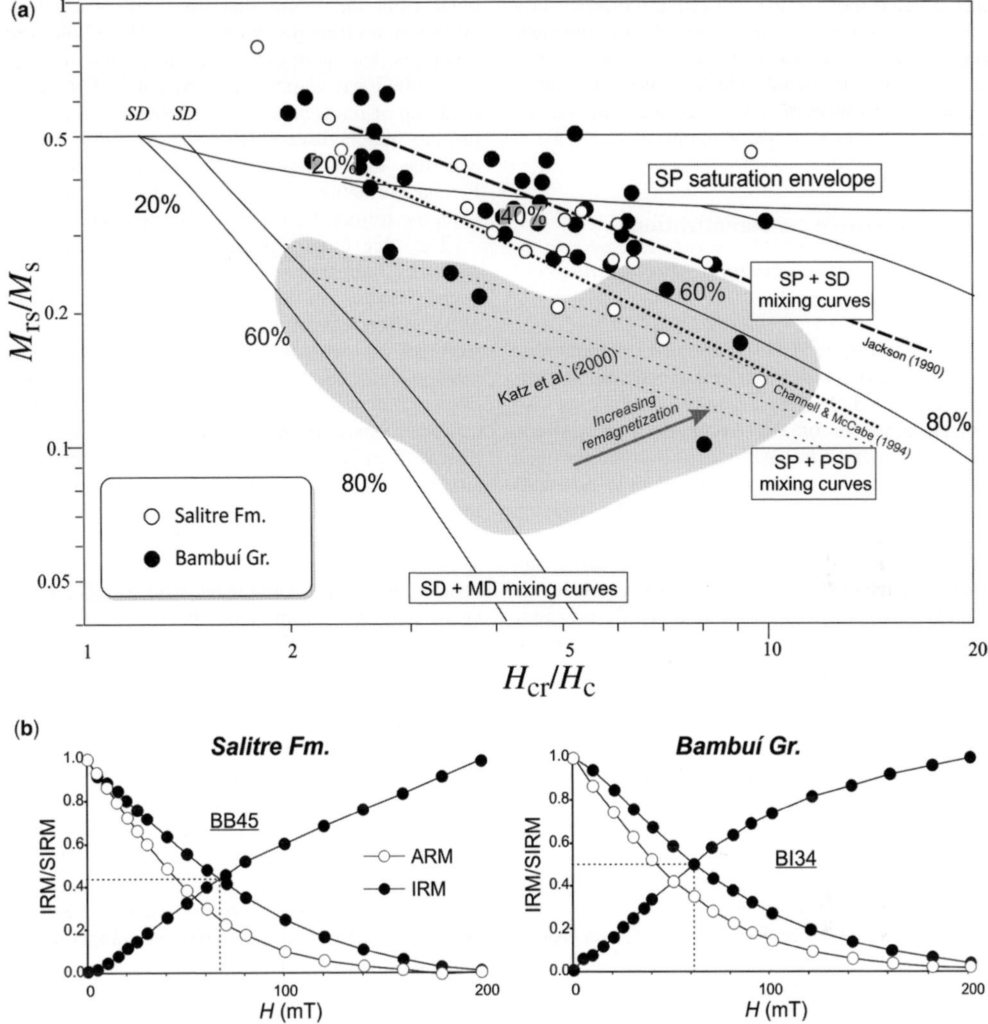



**Fig. 5.** (**a**) Hysteresis data from the Bambuí Group and Saltire Formation (Trindade *et al.* 2004) plotted on the theoretical unmixing diagram of Dunlop (2002) and showing a mixture of SD + SP typical of remagnetized carbonates (Jackson 1990). (**b**) Lowrie-Fuller (Johnson *et al.* 1975) and Cisowski (1981) tests of representative samples from the Bambuí Group and the Salitre Formation. Results are similar to the contradictory behaviour observed in other remagnetized carbonates (Jackson 1990; McCabe & Channell 1994; Huang & Opdyke 1996).

remagnetized successions (McCabe & Channel 1994; Huang & Opdyke 1996). In these curves, the symmetrical behaviour of IRM acquisition and demagnetization curves indicates that single-domain particles are non-interacting (Cisowski 1981). This contrasts with the fact that ARM is weaker than IRM to AF demagnetization, a behaviour more akin to that of multi-domain magnetite grains (Johnson *et al.* 1975) but also typical of remagnetized carbonates (McCabe & Channell 1994). Given the coincidence in magnetic directions, unblocking temperatures and magnetic carriers across a broad cratonic region, the single-polarity and tightly grouped nature of magnetic components and the rock magnetic properties typical of remagnetized carbonates, we interpreted these rocks to have been affected by a widespread remagnetization event. The age of the event can be constrained by both the coincidence of the Bambuí and Salitre poles with reference Cambrian poles for the Gondwana supercontinent, and the 520 Ma Pb/Pb ages obtained systematically in the two regions. Since this age corresponds to the final assembly of the Gondwana supercontinent, we have suggested that a large-scale brine migration has affected most of the São Francisco Basin promoting pervasive remagnetization of the carbonates, resetting of their Pb isotopic system and also a concentrating of base metals which formed widespread Pb–Zn mineralization found all along the region (e.g. Misi & Veizer 1998; Monteiro *et al.* 1999).

## Late Palaeozoic remagnetizations

### The San Rafaelic remagnetization of the Argentine Precordillera

Rapalini & Tarling (1993) reported for the first time the presence of a widespread remagnetization affecting areas of the exposed Cambro-Ordovician carbonate platform of the Argentine Precordillera. This was inferred to have occurred in the Permian on the basis of the corresponding pole positions and their exclusively reversed polarity (i.e. Kiaman Reverse Polarity Superchron). The main carrier of this remagnetization is the Early Ordovician San Juan Limestone (e.g. Benedetto 1998), which is largely exposed along the Eastern and Central Precordillera in the provinces of San Juan, La Rioja and Mendoza. Rapalini & Tarling (1993) included within the same remagnetizing event the syntectonic magnetization observed in the Carboniferous Hoyada Verde clastic sediments (Bobbio *et al.* 1990) as well as their reinterpretation of the original Alcaparrosa Formation pole (Vilas & Valencio 1978) which, according to the authors, does not reflect the original remanence but a syn-tectonic (44% unfolding) magnetization during Permian times. The original work of Rapalini & Tarling (1993) was continued by Truco & Rapalini (1996) who found a syn-tectonic magnetization affecting the Middle Ordovician limestones of the Ponón Trehué Formation, exposed in the San Rafael Block, some 400 km south from the previous study zone. More recently, Rapalini *et al.* (2000) and Rapalini & Astini (2005) carried out new palaeomagnetic studies at different localities along the Eastern and Northern Precordillera, where several localities showed a similar magnetic component to that found in the previous study. This allowed the study area affected by the remagnetization to be expanded both geographically (San Rafael Block, Eastern and Northern Precordillera) as well as stratigraphically along the carbonatic sequence. Some areas were however found to be unaffected by the remagnetization, which was apparently due to a lithological control. Rocks that showed no evidence of remagnetization were characterized by either a primary remanence or unstable or incoherent magnetizations.

Figure 6a, b shows the distribution of localities where the remagnetizing event has been found as well as those where it is apparently absent. It is clear from the figure that no single geographic pattern arises from this distribution. The lithological control is clear in the extreme north of Precordillera, where the Early Cambrian Cerro Totora Formation made up of red sandstones and siltstones records a primary magnetization (Rapalini & Astini 1998), and the limestones of the Late Cambrian La Flecha Formation, exposed less than 3 km away from the former, have been remagnetized.

The Argentine Precordillera is part of the Andean Chain that was uplifted, mainly by east-verging thrusts, in the Late Tertiary (Jordan *et al.* 1983; Allmendinger *et al.* 1990). As such, 'syntectonic' or 'post-tectonic' magnetizations may have undergone Andean tilting, which in most cases is not well constrained. In the case of the calcareous rocks exposed in the Eastern Precordillera, a large tilting of around 40° of the whole sequence since remanence acquisition was inferred from the remanent magnetization directions (Rapalini *et al.* 2000). In other cases (Rapalini & Tarling 1993; Truco & Rapalini 1996; Rapalini & Astini 2005), very minor if any significant Andean tilting has been observed from pole positions (Fig. 1c; Table 2), the distribution of which is consistent with magnetization being acquired during the Early–Late Permian. Three units (Hoyada Verde, Alcaparrosa and Ponón Trehue Fms) present a syntectonic magnetization that is close to Early Permian reference directions. All these units were affected at that time by a major deformational phase that affected the Precordillera and other neighbouring geologic provinces in Argentina called the San Rafaelic tectonic phase (Azcuy & Caminos 1987), hence the term 'San Rafaelic Remagnetization' (Table 2). Rapalini & Astini (2005) have speculated on some progressive migration of the remagnetizing 'front'. According to their consideration, the units located towards the west (i.e. The Hoyada Verde and the Alcaparrosa Formation) underwent syntectonic magnetizations during folding at c. 300–285 Ma. Meanwhile, units located more to the east were remagnetized at a later time. The La Flecha Formation exposed in the Northern Precordillera was apparently remagnetized not before c. 263 Ma as inferred from its pole position and the presence of mixed polarities. In any case, there seems to be little doubt on the relation of this event and deformation assigned to the San Rafaelic tectonic phase.

Detailed and systematic rock-magnetic studies on the remagnetized rocks mentioned above are lacking. Rapalini *et al.* (2000) performed hysteresis studies on samples from the La Flecha, La Silla and San Juan formations from Eastern Precordillera (Fig. 6c). When plotted together with unmixing theoretical curves for carbonates (Dunlop 2002), unremagnetized limestones fall within the single- and multi-domain (SD + MD) which is characteristic of primary magnetic mineralogy. Remagnetized carbonates however show a typical trend from SD + MD to pseudo-single- and single-domain (PS + SD) mixture. Geuna & Escosteguy (2006) have recently performed a magnetic mineralogy study on the Late Ordovician Alcaparrosa Formation, originally studied by Vilas & Valencio (1978) and re-interpreted by Rapalini & Tarling (1993) as an Early Permian syn-tectonic remagnetization. The new study reveals that the magnetization in the basaltic pillow lavas and dolerites of this unit is mainly carried by pyrrhotite with minor and sporadic amounts of magnetite. Geuna & Escosteguy (2006) were able to prove that the remanence is secondary and related to neoformation of magnetic minerals in the alteration halo associated with the intrusion of a small porphyritic body (Pórfiro Alcaparrosa) that has been dated as 270 Ma (Sillitoe 1977). The natural remanence and magnetic susceptibility is significantly higher inside the contact aureole surrounding the intrusion, and decreases progressively away from the contact aureole. A few kilometres from the aureole, the Alcaparrosa volcanic and sedimentary rocks are paramagnetic with unstable remanence. Geuna & Escosteguy (2006) interpreted that the original magnetic minerals in these basic volcanic rocks were altered by ocean floor metamorphism. Secondary pyrrhotite and sporadic magnetite in the rocks are almost exclusively found inside or very close to the alteration halo of the intrusive body.

Rapalini & Tarling (1993) proposed that the San Rafaelic remagnetization found in the Argentine Precordillera was a chemical remagnetization linked to tectonically expelled fluids from the orogen towards the foreland ('the migrating fluids model' of Oliver 1986). Conversely, Geuna & Escosteguy (2006) suggested that other sources of remagnetization were also active at those times and cast doubts on the validity of a single and simple model to explain the whole remagnetizing event. However, detailed rock magnetic studies are poorly documented for these remagnetized rocks, making the identification of the corresponding remagnetizing processes difficult to establish. Future magnetic mineralogy investigations in these formations will probably provide better insights to study the mechanisms of remagnetization in the Argentine Precordillera.

## Remagnetizations in the Gondwanides Fold Belt

In the late Palaeozoic, the southern margin of Gondwana was characterized by a large orogenic belt that

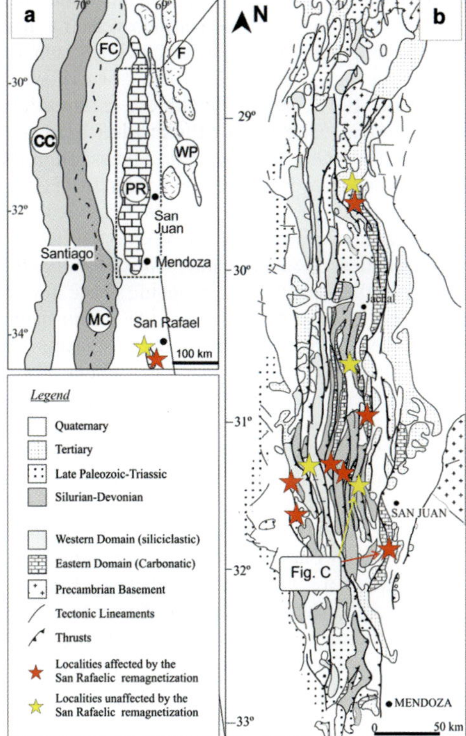


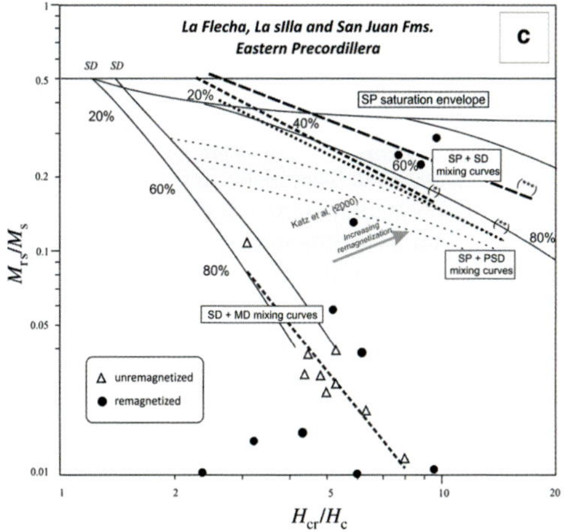


**Fig. 6.** (**a, b**) Distribution of remagnetized (red stars) and unremagnetized (yellow stars) rocks from the Argentina Precordillera. CC, Cordillera de la Costa; MC, Cordillera Principal; FC, Cordillera Frontal; PR, Precordillera; F, Famatina; WP, Sierras Pampeanas (modified from Rapalini & Astini 2005). (**c**) Hysteresis data from La Flecha, La Silla and San Juan formations (Rapalini *et al.* 2000), plotted on the theoretical unmixing diagram of Dunlop (2002) and showing a typical SD + MD mixture for unremagnetized carbonates and a SD–SP trend for remagnetized limestones. Envelope curves typical of remagnetized carbonates are also shown (*, Trindade *et al.* 2004; **, Channel & McMabe 1994; and ***, Jackson 1990; see graph curves for asterisks).

extended nearly 13 000 km across South America, southern Africa, Antarctica and eastern Australia. This belt was originally defined by Du Toit (1937) as the 'Samfrau Geosyncline' and is also known as the 'Gondwanides Belt' (Keidel 1916). The Sierras Australes Fold and Thrust Belt of the Buenos Aires province (Argentina), also widely known as Sierra de la Ventana Fold Belt (e.g. Harrington 1947; Fig. 7), has been interpreted as the result of crustal collision between Patagonia and southern South America (Ramos 2008) or as the result of accretionary tectonics (e.g. Cawood & Buchan 2007). In the last decade, large palaeomagnetic studies were carried out on the youngest (i.e. Permian) clastic sedimentary units of the stratigraphic sequence exposed along this belt and, in particular, in the Sierra Australes (Tomezzoli 2001), the Carapacha Basin (Tomezzoli *et al.* 2006) and the North Patagonian Massif (Rapalini 1998).

*The Sierras Australes.* The Tunas Formation has a broad geographical extension in the region and is the youngest and less-deformed unit of the Carboniferous–Permian Pillahuincó Group. It is composed of 1200 m of silicified fine pale-green sandstones, lithic feldsarenites interbedded with red clay-siltstones and shale which were part of a deltaic complex prograding towards the NE (Harrington 1947). A variable degree of tectonic deformation is observed along the succession of the Tunas Formation, with higher deformation at the base than at the top (e.g. Harrington 1947; Buggisch 1987).The palaeofloral associations of *Glossopteris* attest to a Permian age for these rocks (Archangelsky & Cúneo 1984). Some vitric tuffaceous interbeds of the upper half of Tunas basin were dated at 274 $\pm$ 10 Ma (U–Pb, Tohver *et al.* 2007).

For the palaeomagnetic study, more than 600 specimens were collected from 55 sites in 8 localities in the eastern parts of the belt (Sierra de las Tunas to the north and Sierra de Pillahuincó to the south; Tomezzoli 1997; Fig. 7a). Magnetic patterns were similar in all localities and showed well-defined unblocking temperatures between 550 and 680 °C and a gradual quasi-linear decay towards the origin. The isolated characteristic remanent magnetization (ChRM) shows positive (downwards) inclinations and good within-site directional consistency ($\alpha_{95} < 15°$ and $k > 20$) in at least 44 sites. Exclusive reverse polarity suggests magnetization acquired during the Kiaman Reverse Polarity Superchron (Opdyke & Channell 1996). High unblocking temperatures suggest hematite as the main magnetic carrier. A syn-tectonic origin for the remagnetization is suggested by a negative fold test (McFadden 1990) where minor scattering in magnetic directions is reached after partial unfolding (Fig. 7). At the base of the sequence

from the Sierra de las Tunas (San Carlos, Toro Negro and Golpe de Agua localities; Tomezzoli & Vilas 1999), the results of the fold test provides best grouping at an average of 42% unfolding (Fig. 7b). Younger sequences from the Sierra de Pillahuincó (Dos de Mayo, Las Lomas-La Susana, Arroyo Paretas and Las Mostazas localities; Tomezzoli 2001) however demonstrated a magnetization acquired with an average of 90% unfolding (Fig. 7c). Magnetization was likely acquired during the Kiaman Reverse Polarity Superchron, based on the fact that only a reversed magnetic polarity is observed in these rocks. A significant duration is envisioned for the remagnetization event (i.e. syn-tectonic magnetic overprint proceeding from the base to the top), estimated to *c.* 20 Ma (Fig. 7d; Tomezzoli 1999). These results are grouped into two palaeomagnetic poles called Tunas I (Tomezzoli & Vilas 1999) and Tunas II (Tomezzoli 2001) that are consistent with previous poles from South America assigned to the Permian (Fig. 1c). The Early Permian deformation responsible for the magnetization described here may correspond to the activity of the San Rafaelic orogenic phase already mentioned (Azcuy & Caminos 1987). The beginning of the San Rafaelic orogenic phase according to Llambías & Sato (1995) in the Cordillera Frontal of Argentina was during the Asselian (earliest Permian). The positions of Tunas I and II poles follow a consistent trend within their stratigraphic position in the Tunas Formation. According to a traditional Late Palaeozoic APWP for South America (Brandt *et al.* 2009; Tomezzoli 2009; Domeier *et al.* 2011), an Early Permian age for Tunas I and a Permian age for Tunas II is likely (Fig. 1c).

No detailed rock magnetic studies were carried out for these rocks. Demagnetization patterns strongly suggested hematite as the principal remanence carrier. In the Tunas Formation, some beds have red spots both in the pigment and detrital grains which are interpreted as diagenetic hematite. These rocks have not been significantly metamorphosed, having suffered only moderate diagenetic changes (Cobbold *et al.* 1986; Buggisch 1987). The magnetization is believed to be mainly syndepositional to syndiagenetic. This suggests that deformation occurred very soon after deposition of the Tunas Formation started and continued during sedimentation. Deformation appears to be diachronous, being older in the western areas of the Tunas depositional basin while deposition was starting to take place in eastern areas (Fig. 7c). For that reason, the degree of deformation diminishes substantially both stratigraphically (towards younger levels) and geographically (eastwards). This is probably why magnetization is truly syn-tectonic in the western Sierra de las Tunas (lower levels) and almost pre-tectonic in the Sierra de Pillahuincó

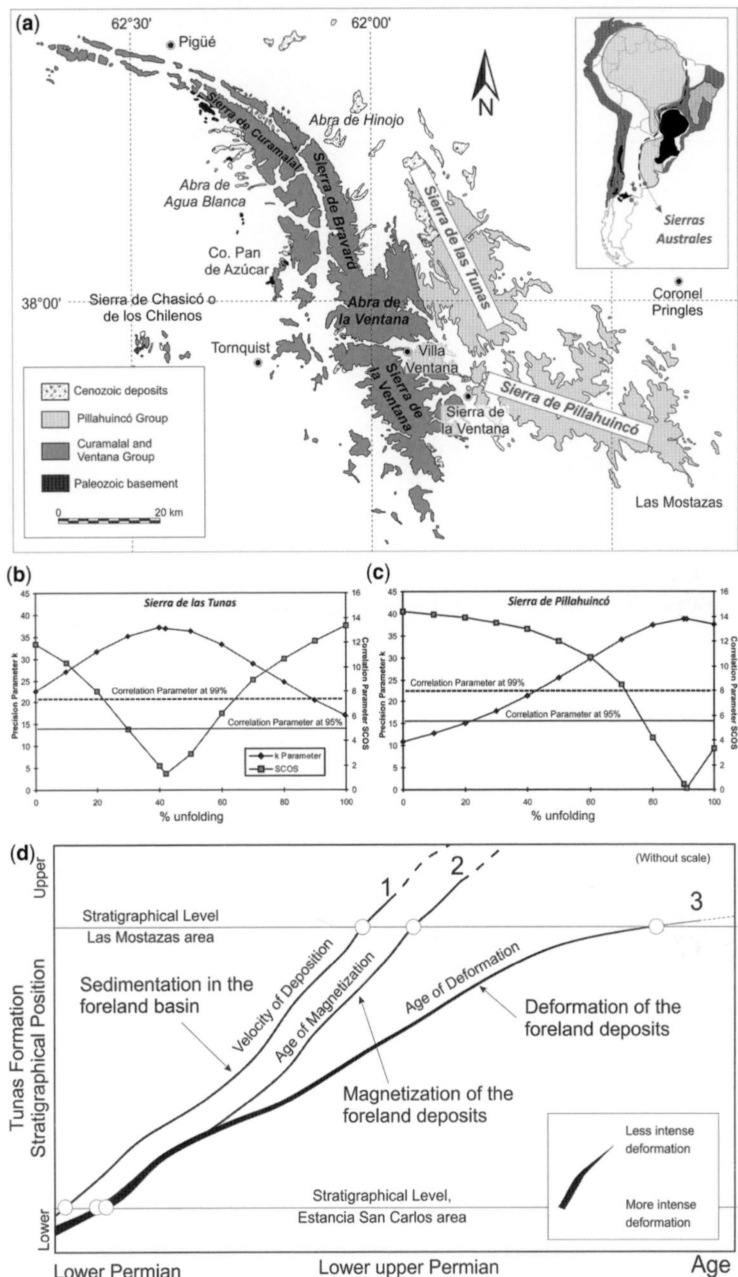



**Fig. 7.** (**a**) Geological map of the Sierras Australes, Argentina. Plots of the statistical parameters $k$ (Fisher 1953) and the SCOS (correlation parameter of McFadden 1990) for the Tunas Formation in the (**b**) Sierra de las Tunas and (**c**) Sierra de Pillahuincó v. percentage unfolding (Tomezzoli & Vilas 1999; Tomezzoli 2001). Both units showed a syn-folding origin for the remagnetization. (**d**) Schematic relationship of sedimentation, magnetization and deformation for different stratigraphic levels of the Tunas Formation (Tomezzoli 1999).

(upper levels). It must be noted that the Sierra de Pillahuincó retains the same orientation of the structures as the Sierra de las Tunas, with wider folds and smoother topography, indicating that they were affected in different degrees by the same tectonic process.

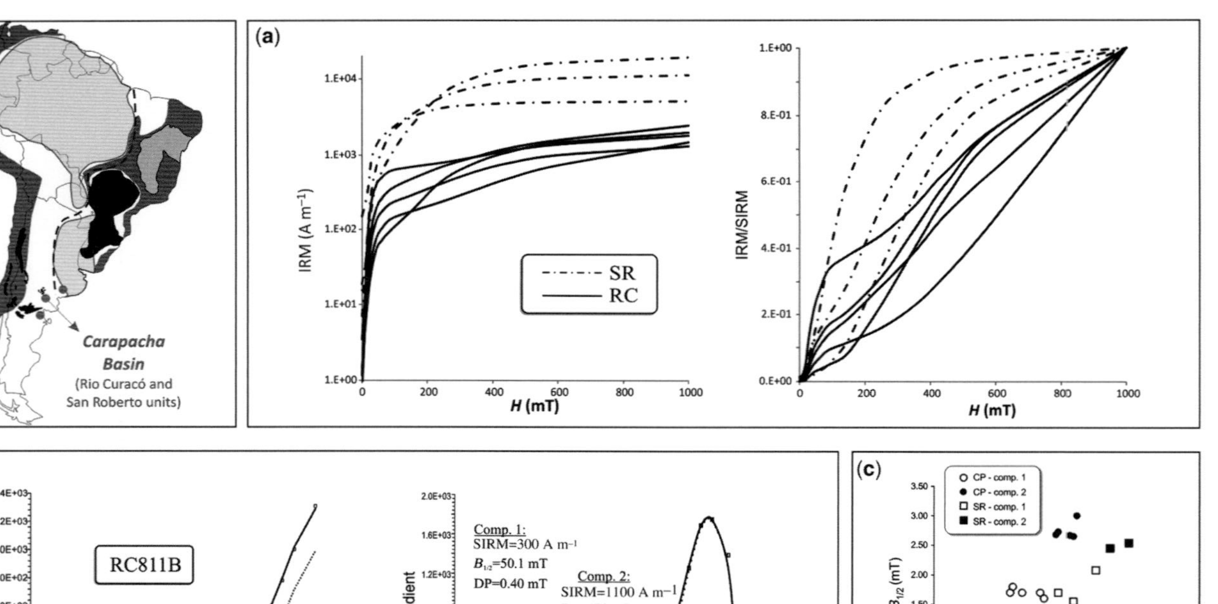

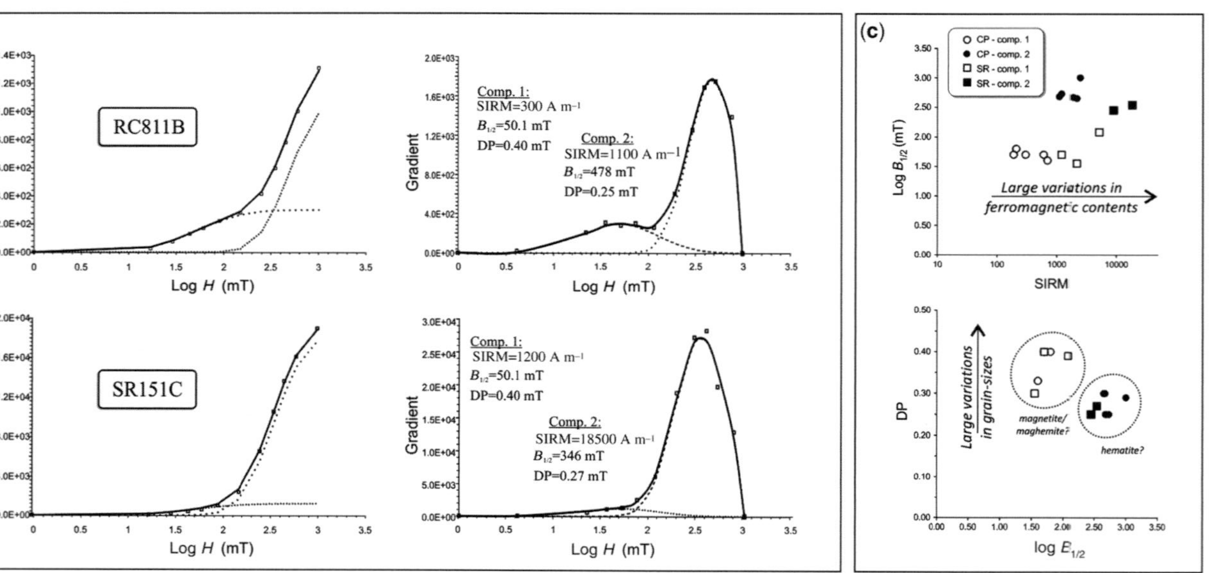

*The Carapacha Basin.* The Carapacha Basin is a continental half-graben of Permian age located in southern La Pampa province, central Argentina (Melchor 1999; Fig. 8), along the trend of the Gondwanides Belt, some 500 km west from the Sierras Australes. The basin filling is up to 630 m thick and entirely composed of fluvial and subordinate lacustrine deposits assigned to the Carapacha Formation for which a mid-Permian age have been attributed from the presence of *Glossopteris* flora (Melchor & Césari 1997). The base of the Carapacha Formation is not exposed and the uppermost part is intruded by andesites and small rhyolitic dykes, believed to be associated with Permian volcanic rocks of the Choiyoi Group. Due to the large regional distribution of these eruptive units and to the paucity of precise radiogenic isotope age data, their regional distribution is hard to establish. Basement rocks include Upper Cambrian–Lower Devonian metamorphic rocks and Late Palaeozoic granite and orthogneisses (Cerro de los Viejos Complex) that crop out in south-eastern La Pampa province. Magnetic fabrics of the Cerro de los Viejos Complex are characterized by two distinct foliations dated between $280.4 \pm 2.3$ Ma and $261 \pm 13$ Ma (Tomezzoli *et al.* 2003).

The Carapacha Formation comprises red, brown or grey arkosic or lithic mudstones, siltstones and fine-grained sandstones and massive or laminated and scarce conglomerates. It has been divided into two members separated by an unconformity: the Urre-Lauquen Member, which crops out along the Río Curacó as gently folded strata, and the homoclinal sequence of the Calencó Member (Melchor 1999). However, poor exposure of the rocks precludes a more definite determination of the tectonic structure here.

For the palaeomagnetic study, more than 250 specimens were collected from 24 sites at the two localities, with different structural attitudes (Tomezzoli *et al.* 2006). Most samples carry a single component of magnetization with positive inclination and good statistical values ($\alpha_{95} < 15°$ and $k > 20$ for site-based mean directions). Stepwise unfolding of remanence directions showed a better clustering before un-tilting, that is, negative fold test, suggesting a post-folding age for the magnetization in the Carapacha basin (Tomezzoli *et al.* 2006). Unblocking temperatures of *c.* 680 °C

suggested hematite as the principal magnetic carrier in these rocks (Fig. 9). In some samples, however, unblocking temperatures ranging from 450 to 580 °C were suggestive of magnetite or titanomagnetite. A mixture of hematite and magnetite corroborates IRM curves that remain unsaturated at 1 T (Fig. 8; Tomezzoli *et al.* 2006). IRM data was treated by the cumulative log-Gaussian function (Robertson & France 1994) using the software of Kruiver *et al.* (2001). Results show a clear contribution of hematite (65–95%) over magnetite (5–25%) in both the Rio Curacó and San Roberto units. Large values of saturation IRM (SIRM) and dispersion parameter (DP) suggest a heterogeneous (probably secondary) population of magnetic carriers in terms of concentration and/or grain size (Fig. 8).

The two different palaeomagnetic poles calculated from the Río Curacó (lower member) and San Roberto (upper member) are located at different positions in the APWP of South America (Fig. 1c). The San Roberto deposits that are less deformed occupy a younger position in the APWP while those from the Río Curacó deposits, that are older and more deformed, fall within an older section of the APWP (Tomezzoli *et al.* 2006; path B of Fig. 1c). The apparent age difference between the two poles, as judged by their relative positions along the APWP, is large enough to suggest a long period of several million years for the remagnetization phases. These results agree with those obtained in the same area from the Sierra Chica volcanic units (Tomezzoli *et al.* 2009) and for the Sierras Australes mentioned above.

Geological evidence indicates that sedimentation in Sierra de la Ventana was partially coeval with deformation and that this deformation advanced from west to east (Tomezzoli & Vilas 1999). The San Roberto pole presented here is thought to represent a younger remagnetization age, equivalent to the Tunas II palaeomagnetic pole (Tomezzoli 2001). Since the direction of the maximum compression is from SW to NE, it is logical that the age of the deformation is older to the west than to the east as seen in the Sierras Australes from the Buenos Aires Province where the time-transgression of the deformation is clearly evidenced by the palaeomagnetic study. Overall, the whole remagnetization process is broadly coeval with the San Rafael orogenic phase (Azcuy &

---

**Fig. 8.** IRM data for the Rio Curacó and San Roberto units, Carapacha Basin (Tomezzoli *et al.* 2006). (**a**) Raw (left) and normalized (right) IRM curves of SR (San Roberto units) and RC (Rio Curacó) samples. Data show that most samples did not reach saturation at 1 T, suggesting the resence of hematite. Differences in the shape of the IRM curves indicate wide ranges of ferromagnetic content and coercivity spectra. (**b**) Cumulative log-Gaussian treatment of representative samples using the software of Kruiver *et al.* (2001). Results show a characteristic bimodal distribution of coercivity spectra corresponding to a mixture of a low (magnetite?) and high (hematite?) coercive ferromagnetic phase. (**c**) Log $B_{1/2}$ v. SIRM and DP v. log $B_{1/2}$ plots showing large variations in ferromagnetic content and grain size.

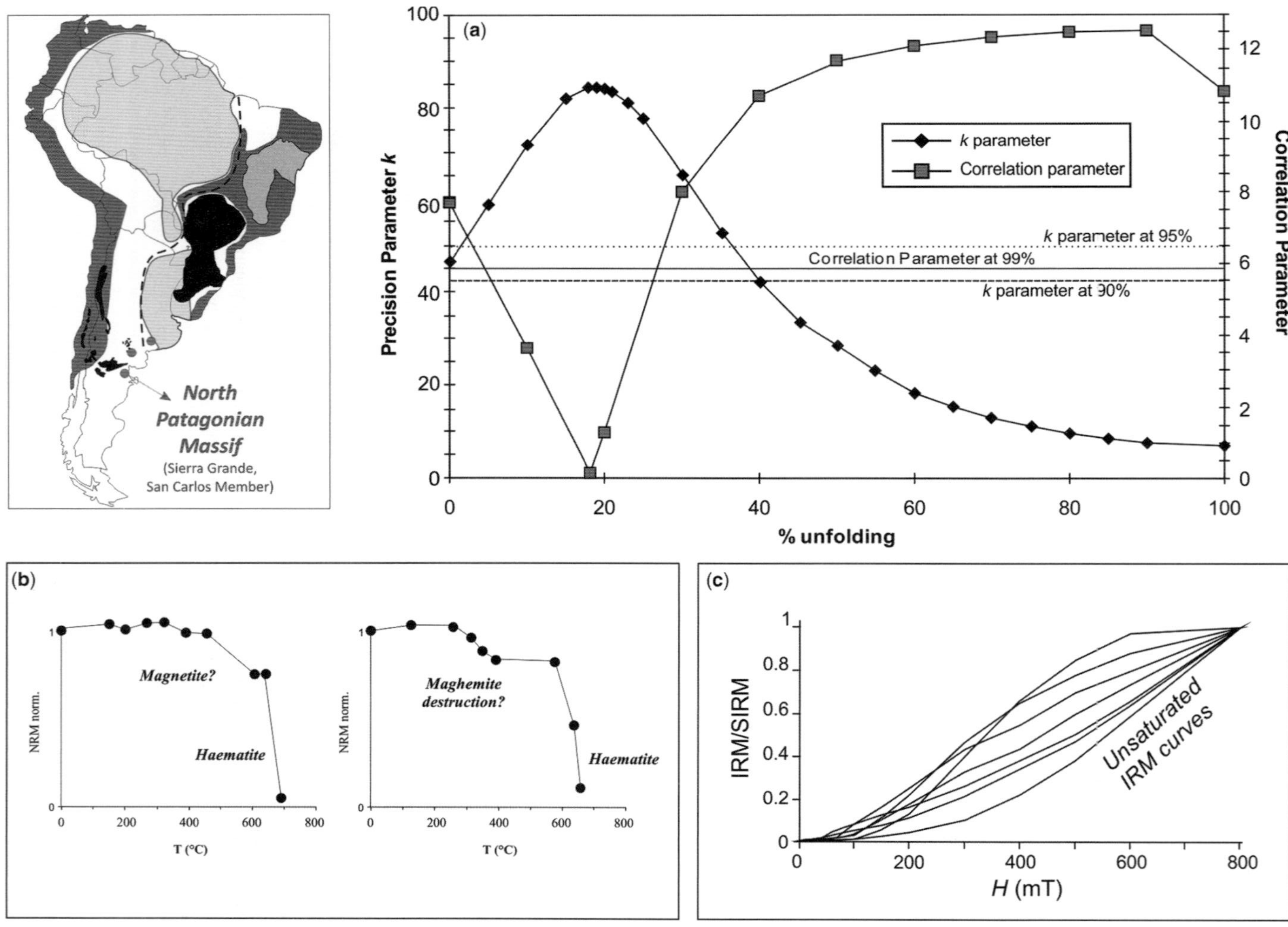
North
Patagonian
Massif
(Sierra Grande,
San Carlos Member)
(a)
Precision Parameter k
Correlation Parameter
k parameter
Correlation parameter
k parameter at 95%
Correlation Parameter at 99%
k parameter at 90%
% unfolding
(b)
NRM norm.
Magnetite?
Haematite
T (°C)
NRM norm.
Maghemite
destruction?
Haematite
T (°C)
(c)
IRM/SIRM
Unsaturated
IRM curves
H (mT)

Caminos 1987), which may link the magnetization characteristics in La Pampa and Sierra de la Ventana to those found in the Precordillera and San Rafael Block. However, this is still speculation and a thorough investigation of the processes responsible for this regional feature is needed.

*The North Patagonian Massif.* Rapalini (1998) investigated the palaeomagnetism of the Siluro-Devonian clastic sedimentary rocks of the Sierra Grande Formation, exposed in the northeastern corner of the North Patagonian Massif (41.6°S, 65.3°W, Fig. 9). This unit is well known for the presence of two ferriferous horizons called the Rosales (San Carlos Member) and the Alfaro (Herrada Member) horizons (Zanettini 1981). Near the town of Sierra Grande, these horizons have been affected by thermal metamorphism caused by the intrusion of a Late Palaeozoic granitic body which led to the formation of secondary magnetite that was commercially exploited for many years in the Hiparsa mine. Rapalini & Vilas (1991) presented the first palaeomagnetic data for these rocks, originally interpreting palaeomagnetic remanences from the Rosales horizon as primary, an assessment that was later discarded (Rapalini 1998). The upper Herrada Member however preserves a likely primary palaeomagnetic direction. Preliminary results on sandstones of the San Carlos Member, not including the Rosales horizon, showed a probable Permian remagnetization. A more detailed study by Rapalini (1998) on the sandstones of the San Carlos Member concluded that these rocks have been syn-tectonically remagnetized. Eighty-eight samples from 13 sites located along a syncline–anticline structure on these rocks presented a characteristic remanence with the best directional grouping obtained after 19% of partial unfolding (Fig. 9a). The partially corrected remanence yielded a pole position for the syn-tectonic magnetization of the Sierra Grande Formation at 77.3°S, 310.7°E (dp = 7.7°, dm = 6.6°, $N = 13$). The position of this palaeopole is coincident with late Early–early Late Permian poles of South America (Fig. 1c), giving a minimum age for deformation in this area. The exclusive reverse polarity found in the San Carlos Member was interpreted by the author as suggesting acquisition of remanence before the end of the Kiaman Superchron, which is older than 265 Ma (Gradstein *et al.* 2004; Ogg *et al.* 2008).

Detailed rock magnetic or petrographic studies have not been carried out on these rocks. However, high unblocking temperatures over 650 °C and the ineffectiveness of AF demagnetization strongly suggest hematite as the principal magnetic carrier of the remanence in the San Carlos Member, with probable minor additions of magnetite/maghemite (Fig. 9b). The presence of oolitic hematite has been confirmed in samples located few tens of kilometres from the metamorphic zone. Normalized IRM acquisition curves (IRM/SIRM) show a lack of low-coercivity fraction and unsaturated state at 1 T (Fig. 9c), which supports the theory that remanence is essentially carried by hematite. However, the presence of magnetite has been reported by Zanettini (1981) in the lower Rosales horizon. Detailed rock magnetic or petrographic studies have not been carried out on these rocks.

The syn-tectonic magnetization found in the Sierra Grande sediments may be caused by the same regional tectonic process evidenced from thrusting and folding in the North Patagonian Massif (von Gosen 2002, 2003; López de Luchi *et al.* 2010) and the Sierras Australes Fold Belt (Harrington 1947; von Gosen *et al.* 1990; Tomezzoli & Cristallini 1998) in the Permian. As such, the remagnetization of the Sierra Grande sediments may be associated directly with authigenic precipitation of iron oxides (mainly hematite) during deformation. Tomezzoli *et al.* (2010) have recently found that Early Ordovician granitoids exposed in the Sierra Grande area with little or no evidence of internal deformation have also been remagnetized in the Permian, thus suggesting that deformation is not the sole cause of remagnetization. Widespread Permian magmatism is observed in the North Patagonian Massif (see Pankhurst *et al.* 2006) and, in particular, in the Sierra Grande area where Varela (2009) have recently dated different plutons at 263 ± 9, 262 ± 6 and 260 ± 3 Ma by Rb–Sr isochrones. Considering that the Sierra Grande Formation has been severely overprinted by contact metamorphism with Permian plutons in some areas, chemical remagnetization is evidently linked to increased thermal gradient and fluids associated with the intrusions. Since deformation and magmatism were relatively contemporaneous, the syn-tectonic nature of the magnetization should not be mistaken for 'syndeformational'; in other words, deformation may not be the sole cause of magnetic memory resetting in the studied rocks.

## The Río de la Plata Craton

The presence of a widespread remagnetization during the Permian and/or Triassic in the Rio de la Plata Craton has been recently reported in the

---

**Fig. 9.** (**a**) Negative fold test (McFadden 1990) of the San Carlos Member, Sierra Grande, North Patagonian Massif (Rapalini 1998); (**b**) thermal demagnetization diagram; and (**c**) IRM curves of the San Carlos sediments showing dominance of an antiferromagnetic fraction as principal magnetic carrier of the secondary magnetization (Rapalini 1998).

limestones from the Cerro Victoria Formation (Late Ediacaran–Early Cambrian), the late Ediacaran clastic Yerbal Formation and the Ediacaran clastic Rocha Formation (Rapalini & Sanchez-Bettucci 2008). These units are exposed in different tectonostratigraphic terranes of Uruguay. While the Cerro Victoria and Yerbal Formations are exposed in the Nico Perez terrane, the Rocha Formation belongs to the Punta del Este (or Cuchilla Dionisio terrane, e.g. Basei *et al.* 2005). In all cases, a well-defined dual polarity magnetization that does not pass the fold test indicated a post-tectonic remagnetization (Fig. 10). Rapalini & Sanchez-Bettucci (2008) also claimed that this secondary direction is indistinguishable from that found in the La Tinta Formation from the Tandilia region of Argentina (Valencio *et al.* 1980), thus invalidating the use of the La Tinta Formation palaeomagnetic pole for late Precambrian palaeogeographic reconstructions. As shown in Figure 1c, the pole positions belonging to the Cerro Victoria, Yerbal, Rocha and La Tinta Formation coincides within their uncertainties. The new palaeomagnetic pole from the Sierras Bayas Group (ex-La Tinta Formation) is also shown, with no significant discrepancy from the other. The post-tectonic nature of the magnetization of these rocks indicates a dual-polarity secondary magnetization that is probably latest Permian–Early Triassic in age. Because the APWP for South America since late Permian times shows very slow and short displacements the pole positions

cannot be unambiguously assigned to the Permo-Triassic, and the latest Cretaceous–Early Tertiary portion of the APWP also presents a viable, alternative age for the remagnetization. A Late Permian–Early Triassic age is however favoured from the number of syn-tectonic magnetizations of such age found along the Gondwanides Belt (already mentioned) and from the recent report of an age of $254 \pm 7$ Ma for diagenetic processes in the Sierras Bayas Group (Zalba *et al.* 2007). Conversely, Rapalini & Sánchez-Bettucci (2008) reported a less conspicuous Cambrian remagnetization (see above) in the latest Ediacaran–Cambrian Polanco Formation and a couple of sites from the Cerro Victoria Formation.

No detailed rock magnetic studies have yet been carried out on these rocks, limiting the identification of the remagnetization processes. Additionally, the rocks on the craton do not show any macroscopic evidence of deformation during the Late Palaeozoic. Preliminary magnetic mineralogy data included IRM acquisition curves that showed a dominant signal from hematite consistent with their NRM demagnetization behaviour, suggesting a CRM origin for the remagnetization. A CRM overprint agrees well with uplift and inversion of the sedimentary column which led to the formation of a very large fore-deep (Claromecó fore-deep) towards the NE (e.g. Ramos 2008) which could expell fluids from the orogen towards the foreland, similar to that described by Oliver (1986, 1992) for eastern

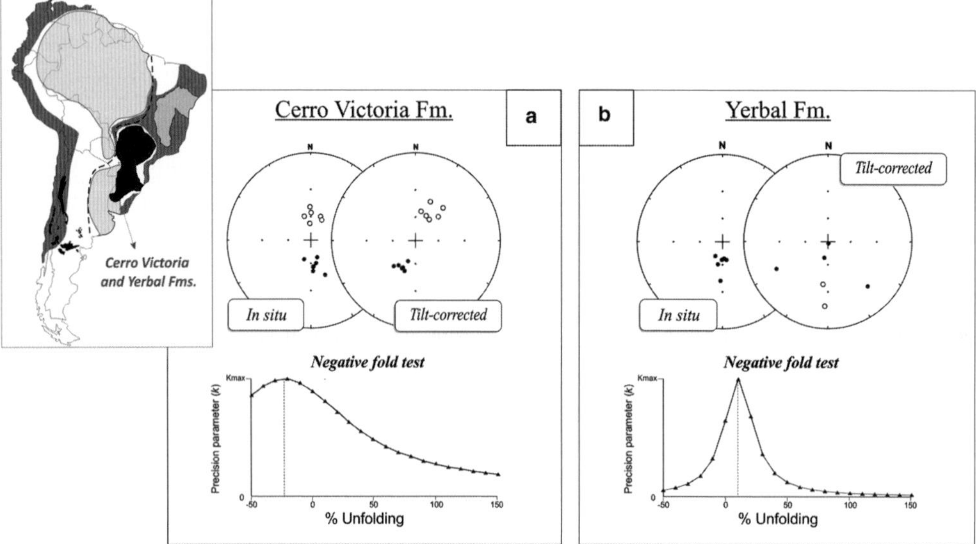



**Fig. 10.** Fold test (McFadden 1990) of the (**a**) Cerro Victoria and (**b**) Yerbal formations, Rio de la Plata, Uruguay (Rapalini & Sánchez-Bettucci 2008). Above: stereographic projections in geographic and tilt-corrected coordinates; below: dispersion parameter $k$ v. percentage of unfolding. In both cases, data show a higher dispersion $k$ after tilt correction and gave a negative (i.e. post-folding) response in the test.

North America. However, as in the case of the San Rafaelic remagnetization, a possible mineralogical control on the remagnetization is evident. Rapalini (2006) have reported Ediacaran–Cambrian red clastic sediments in the Tandilia system that have escaped remagnetization. Results of similar studies on the calcareous and clastic sediments of the Sierras Bayas Group (Augusto Rapalini, pers. comm., 2011) also indicate both hematite and magnetite in the remagnetized as well as in the unremagnetized samples. These observations highlight the importance of future detailed rock magnetic studies to unravel the nature and origin of the remagnetizing events in this area.

## Cretaceous overprint link to the opening of the Atlantic Ocean: the Itajaí Basin case

The Itajaí Basin is located in the state of Santa Catarina, Brazil, and borders the Dom Feliciano orogenic belt which is coeval with the collision that sutured the Congo, Kalahari and Rio de la Plata cratons at the end of the Neoproterozoic and Early Cambrian. It was interpreted as a collision-related foreland basin (Gresse *et al.* 1996; Rostirolla *et al.* 1999). This syn-orogenic origin has been questioned since sedimentary features indicate a quiescent environment (Almeida *et al.* 2010). The Itajaí Group is composed of detrital sediments intruded by granites and capped by rhyolites. Its age was set between $563 \pm 3$ Ma, corresponding to the age of the sandstone deposition, and $549 \pm 4$ Ma for the intrusion of rhyolites (U–Pb dating, Guadagnin *et al.* 2010).

Rhyolites and sandstones share the same mean characteristic remanent magnetization and the coordinates of the corresponding palaeomagnetic pole ($P_{long} = 277.5°E$, $P_{lat} = 84.0°S$, $\alpha 95 = 2.0$) plot close to the current geographic pole (Font *et al.* 2011). Once rotated to African coordinates using the 'tight fit' Euler poles (Trindade *et al.* 2006), the Itajaí pole (IT) is located far away from the APWP of Gondwana for the 570–500 Ma interval (Fig. 1d); this suggests that the remagnetization took place well after the final assemblage of the Gondwana supercontinent. The Itajaí pole is statistically more similar to the Lower Cretaceous poles than to the Permian portion of the APWP, suggesting that remagnetization took place at this later time (Fig. 1d).

The secondary origin for the magnetization of the Itajaí Group was suggested by a negative fold test (McElhinny 1964) performed on the sandstones (Fig. 11a; Font *et al.* 2011). It matches results of the Fuller *et al.* (2002) diagram that shows a concave-upwards NRM:IRMs diagram typical of remagnetized rocks (Fig. 11b). Principal magnetic carriers are represented by a mixture of authigenic magnetite ($T_{Curie} = 580$ °C), maghemite ($T_{destruction} = 350$ °C), goethite ($T_C = 100$ °C) and hematite ($T_{Néel} = 680$ °C) (Fig. 11c). The presence of superparamagnetic minerals was evidenced by the frequency-dependent susceptibility ($K_{fd}$) which takes values within the range 4–11%. SEM analysis indicates that most of the sandstone samples contain strongly altered iron oxides frequently associated with hydrothermal elements such as Ba and Mn, suggesting severe chemical alteration by fluid circulation (Fig. 12). Maximum demagnetization temperatures for the ChRMs of both rhyolites and sandstones are too high for these components to have been acquired by thermoviscous processes (Pullaiah *et al.* 1975). Alternatively, evidence of hydrothermal circulation via fault reactivation suggests a chemical origin for the remagnetization.

The Rio de la Plata Craton is bounded everywhere by strike-slip faults of late Neoproterozoic–early Palaeozoic ages originated by sinistral transpression during the closure of the southern Adamastor Ocean (e.g. Oyhantçabal *et al.* 2011; Rapela *et al.* 2011). During the Mesozoic opening of the South Atlantic, reactivation of ancient NE–SW faults in the Dom Feliciano Belt, such as the Perimbó Shear Zone (PSZ) that separates the Itajaí Basin from the Neoproterozoic Brusque Complex and the Major Gercino Shear Zone (MGSZ), resulted in extensional deformation (Moulin *et al.* 2010; Passarelli *et al.* 2010). These events may have facilitated hydrothermal circulation, leading to the formation of Au-rich quartz veins and Pb, Zn and Cu mineralization in the Itajaí Basin and playing an important role in the remagnetization processes. Early and Late Cretaceous dykes related to reactivation of WNW–ESE faults are compatible with this NE–SW extension and are related to the opening of the South Atlantic Ocean (R. Almeida, pers. comm.).

## Summary

Until the last decades, the paucity of palaeomagnetic data in South America severely limited our capacity to reconstruct the palaeogeography of the South American plate in crucial phases of its evolution, whether in the context of the Gondwana assemblage (e.g. Meert & Powel 2001) or in that of the Pangaea (e.g. Ernesto 2007; Brandt *et al.* 2009; Font *et al.* 2009; Tomezzoli 2009; Domeier *et al.* 2011; Yuan *et al.* 2011). Significant improvements have since been made and a globally well-defined APWP is currently available and presented here for the Precambrian–Ediacaran (570–500 Ma), the Permo-Triassic (300–250 Ma) and the Cretaceous (140–65 Ma) interval. However, few reliable

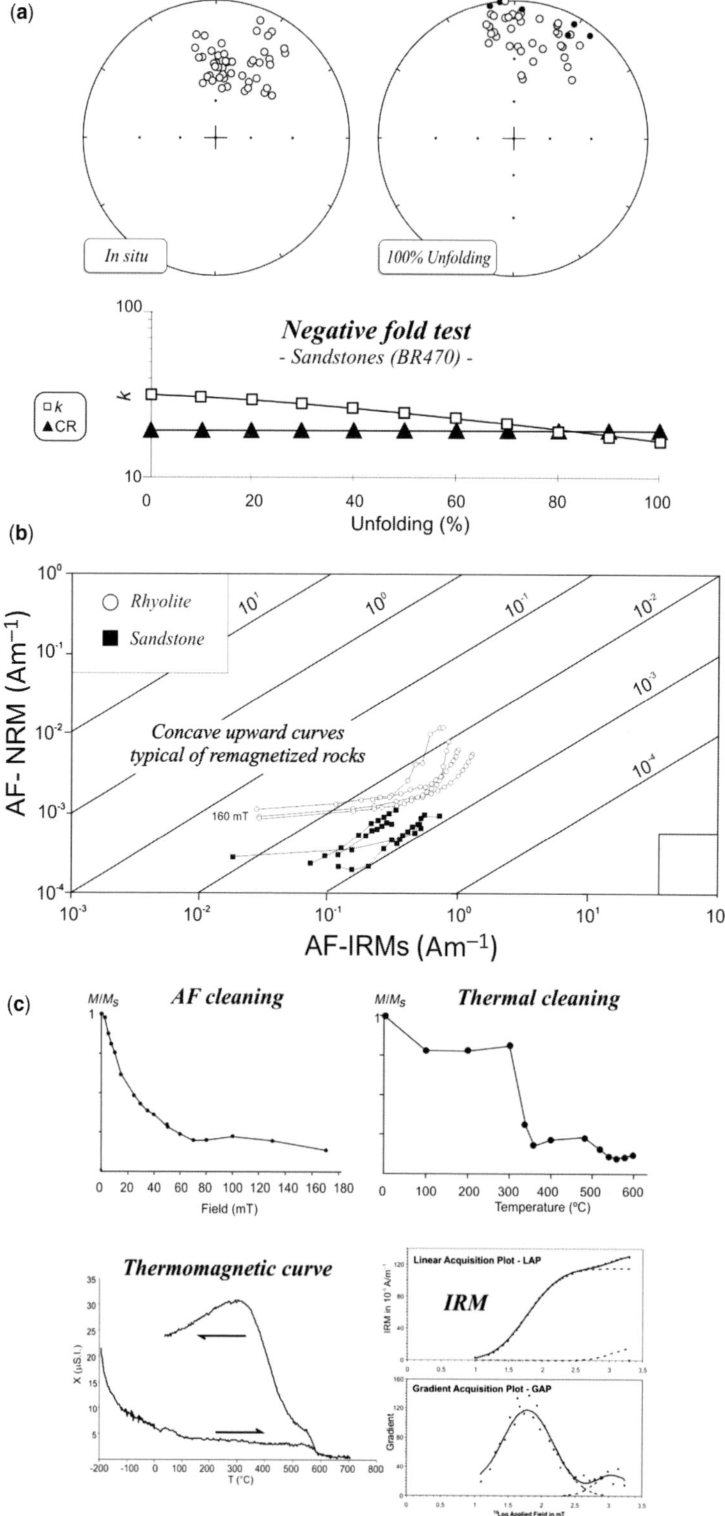

(a)
In situ
100% Unfolding
Negative fold test
- Sandstones (BR470) -
k
CR
Unfolding (%)
(b)
AF- NRM (Am⁻¹)
Rhyolite
Sandstone
Concave upward curves
typical of remagnetized rocks
160 mT
AF-IRMs (Am⁻¹)
(c)
M/Ms
AF cleaning
Field (mT)
M/Ms
Thermal cleaning
Temperature (ºC)
Thermomagnetic curve
X (µSI)
T (°C)
Linear Acquisition Plot - LAP
IRM in 10³ A/m
IRM
Gradient Acquisition Plot - GAP
Gradient
Log Applied Field in mT

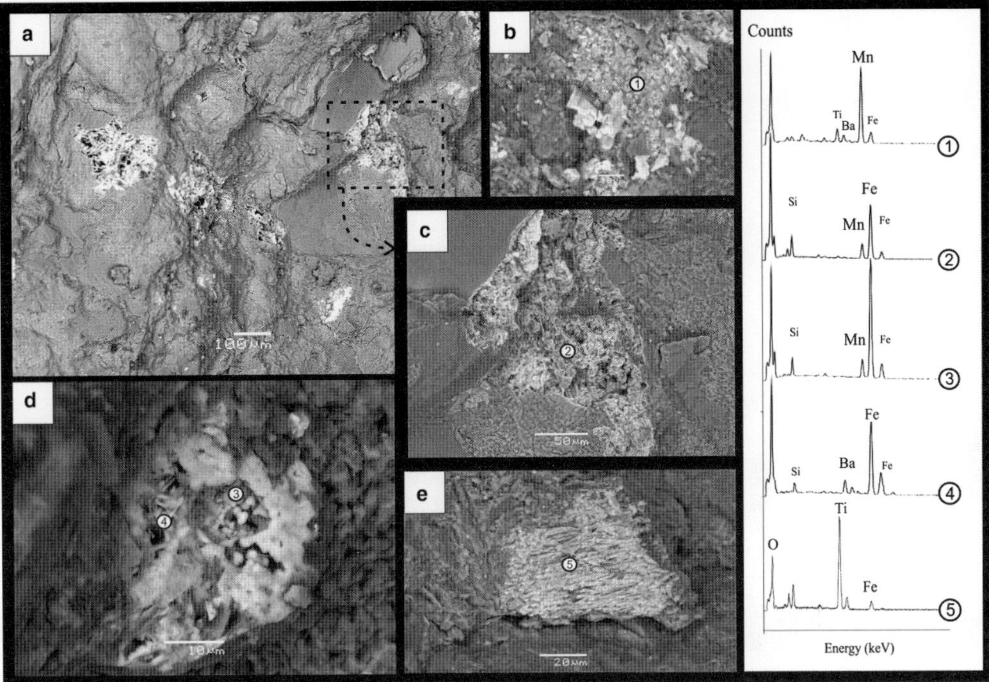

**Fig. 12.** SEM photographs and corresponding Energy Dispersive Spectra (EDS) of the remagnetized sandstones of the Itajaí Basin showing: (**a–d**) large (>100μm) and severely altered iron oxides associated with hydrothermal elements such as Mn and Ba; (**e**) Ti-rich altered iron oxide (hematite?).

palaeomagnetic poles exist between these time intervals and further palaeomagnetic investigations are still needed to reconstruct the missing pieces of the puzzle. One of the principal limitations resides in the episodic occurrence of widespread remagnetizations linked to major plate-tectonics events. A review of where, when and how these remagnetizations led to partial or total magnetic resetting of South American rocks is therefore summarized here (Fig. 13).

At least three widespread remagnetization events linked to global plate tectonics are identified in South America from Precambrian to Cretaceous times. Early Cambrian remagnetization of the neoproterozoic carbonates from the Amazon (Guia Formation) and Congo-São Francisco (i.e. Bambuí and Salitre formations) cratons, now well dated at

*c.* 520 Ma, marked the closure of the Clymene Ocean and the formation of the Paraguay and Araguaia thrust and fold belts at the final stages of the west Gondwana assemblage (Fig. 1; D'Agrella-Filho *et al.* 2000; Trindade *et al.* 2003, 2004; Tohver *et al.* 2010, 2011). Two remagnetized poles of possible Cambrian age from the Cerro Victoria and Polanco formations are also documented in Uruguay (Rapalini & Sánchez-Bettucci 2008), suggesting that the Cambrian remagnetization also spread to the Dom Feliciano orogen at the eastern margin of the Rio de la Plata Craton. The age of the Paraguay Belt overlaps with that of the Pampean Orogeny, suggesting a coeval closure for the Clymene Ocean separating Amazonia from the São Francisco and Rio de Plata cratons (Tohver *et al.* 2010). The Guia, Bambuí and Salitre

---

**Fig. 11.** (**a**) Stereographic projections of magnetic data from the neoproterozoic sandstones of the Itajaí Basin in geographic and tilt-corrected coordinates (above) and dispersion parameter *k* v. percentage of unfolding. Results give a negative (i.e. post-folding) response in the fold test (McElhinny 1964). (**b**) Diagram of AF demagnetization of NRMs v. AF demagnetization of IRMs. The concave-upwards shape of the curves is typical of remagnetized rocks (Fuller *et al.* 2002). (**c**) Above: AF and thermal demagnetization; below: thermomagnetic curve and IRM data of representative samples of the Itajaí sandstones. Data show a mixture of magnetite and hematite (and/or goethite).

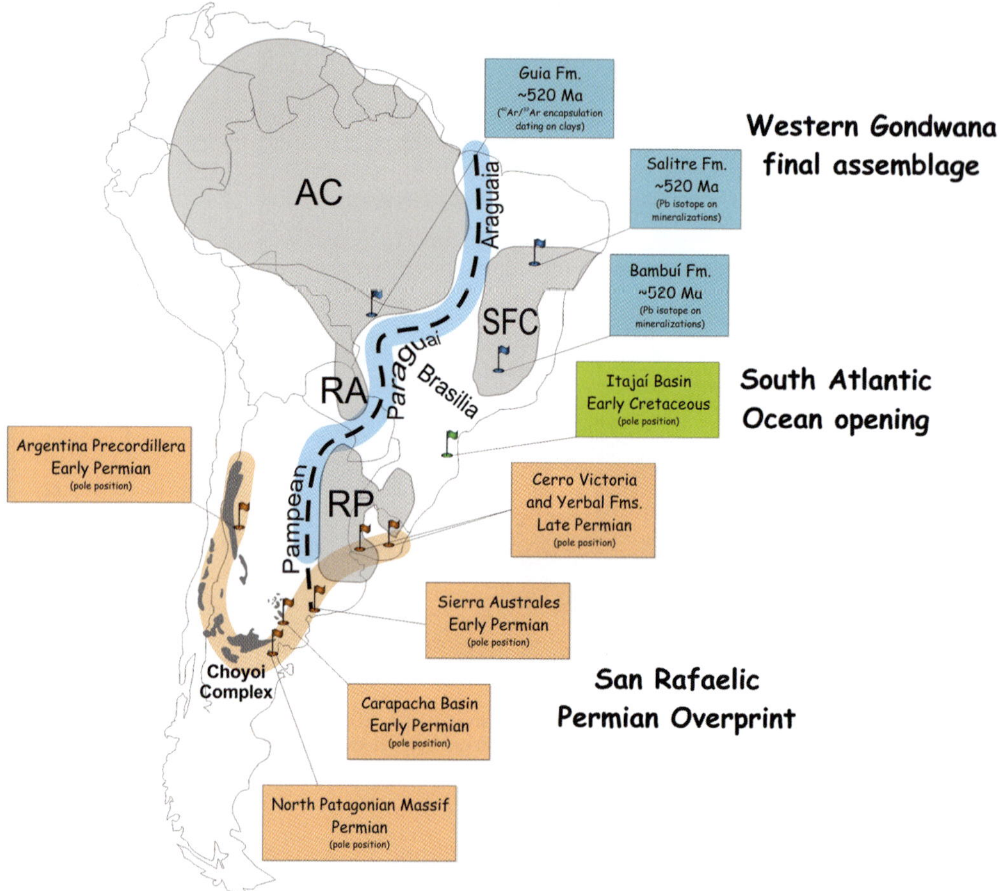


**Fig. 13.** Summary of remagnetized areas in South America showing age and dating methods of remagnetized poles.

formations are all composed of carbonates that are known to be easily prone to remagnetization through acquisition of a chemical remanent magnetization. In all cases, the secondary origin of the remanence is proved by negative fold tests. A CRM origin for the magnetization is well evidenced here by rock magnetic properties and mineralogical features where authigenic magnetite, and pyrrhotite in the case of hydrocarbon-rich sediments (i.e. Guia Formation), carried the secondary magnetization. The most probable remagnetizing mechanism is suggested here to be linked to smectite–illite transformation even if organic matter maturation by sulphato-reducing bacteria played an important role in the non-deformed Guia Limestones. Indeed, the smectite–illite transformation liberates free iron for latter magnetite authigenesis (e.g. Hirt *et al.* 1993; Katz *et al.* 2000; Gill *et al.* 2002; Zegers *et al.* 2003; Blumstein *et al.* 2004; Moreau *et al.*

2005; Tohver *et al.* 2008, 2010). There is some experimental evidence for a magnetite-producing reaction related to the breakdown of Fe-rich smectite to form Fe-poor illite and Fe-oxides (Hirt *et al.* 1993; Cama *et al.* 2000). Presumably, the liberation of Fe from smectite breakdown in a sulphur-rich environment would result in the formation of sulphides such as pyrrhotite in lieu of magnetite. Indeed, tests for the correlation between the age of remagnetization and the age of illitization have shown some recent success (Tohver *et al.* 2008; Zwing *et al.* 2009; Tohver *et al.* 2010). The genetic link between fine-grained illitic clay and its authigenic by-product magnetite provides a clear example of an amenable target for geochronology (Tohver *et al.* 2008; Zwing *et al.* 2009; Tohver *et al.* 2010). The age of the finest-grained material, dominated by authigenic illite (polytype 1M$_d$), is determined by $^{40}$Ar/$^{39}$Ar encapsulation dating.

Since this neoformed population is created by the same reactions that create authigenic magnetite, this technique can directly establish the age of remagnetization or, conversely, test for the likelihood of a primary magnetization.

The Permian was a period of intense volcanism along the continental margin of SW Gondwana where the Choiyoi magmatic belt emplaced a huge volume of rhyolites and granites in Chile and Argentina (Llambías *et al.* 2003). This period of intense volcanism is mostly coeval with the San Rafael Orogeny that affected the SW margin of Gondwana (Kleiman & Japas 2009) and was responsible for widespread remagnetization that affected the Rio de la Plata Craton (Rapalini & Sánchez-Bettucci 2008) from southern Brazil to central Argentina (Sierra Australes: Tomezzoli & Vilas 1999; Tomezzoli 2001; Sierra Chica: Tomezzoli *et al.* 2009). For instance, the age of the remagnetization in Argentina and Uruguay is estimated by the similarity of the corresponding poles with younger references, whereas no detailed magnetic mineralogy and/or radiometric dating for the magnetic overprint still exist. The case studies presented here however show good examples of remagnetization where lithology plays an important role. Indeed, in the Precordillera, the remagnetization is principally associated with carbonates of the La Flecha and San Juan formations whereas sandstones and siltstones of the Cerro Totora Formation preserved a primary magnetization in the same geographic area (Rapalini & Astini 1998). The distinction between preserved and remagnetized rocks is evidenced here by hysteresis data plotted together with theoretical unmixing curves for remagnetized carbonates (Fig. 6c; Dunlop 2002). The typical trend from SD + MD to SD + SP (single domain to superparamagnetic) is interpreted to have resulted from the CRM overprint. In the San Roberto and Rio Curacó formations in the Carapacha Basin, systematic dominance of hematite contents over magnetite as well as large SIRM and DP values point to a CRM acquisition via precipitation of authigenic magnetic carriers (Fig. 8). Similar interpretations have been made for the San Carlos Member sandstones of the Sierra Grande Formation, in the North Patagonian Massif, based on dominance of hematite over magnetite (Fig. 9).

The Cretaceous was a period of unusually active tectonics where ocean crust formation rate and off-ridge volcanism were greater than at any time since and where shallow and deep connections between the South Atlantic and North Atlantic ocean basins opened (e.g. Larson 1991; Poulsen *et al.* 2001; Phipps Morgan *et al.* 2004; Eagles 2007; Moulin *et al.* 2010). In southern Brazil, the accommodation of the extensional deformation resulting from the opening of the South Atlantic Ocean involved the

reactivation of ancient faults and the genesis of a considerable quantity of ore deposits (Biondi *et al.* 1992, 2001; Basei *et al.* 2000; Passarelli *et al.* 2010). The best example is the Major Gercino Shear Zone (MGSZ), a Proterozoic lithospheric-scale discontinuity within the Dom Feliciano Belt which extended from southern Brazil to Uruguay (e.g. Passarelli *et al.* 2010). The main transpressive phase of the MGSZ was constrained by U–Pb analysis from multi-crystal zircon fractions at $614 \pm 2$ and $609 \pm 16$ Ma; recent K–Ar ages from biotites in mylonites however indicate ages of 206 and 230 Ma (Passarelli *et al.* 2010). The Perimbó Shear Zone (PSZ) is another large-scale fault zone which limits the Itajaí Basin from the Neoproterozoic Brusque Complex (Rostirolla *et al.* 2003). A fluid circulation via fault reactivation scenario is an excellent candidate to account for the remagnetization of the Neoproterozoic Itajaí Basin (Font *et al.* 2011). Contrary to the previous case, the Itajaí Basin is composed of rhyolites and sandstones, for which remagnetization proxies are still badly known and where no direct evidence between the age of the remagnetization and the nature and origin of the processes that led to total magnetic resetting are documented. Nevertheless, the position of the palaeomagnetic pole plots close to the Lower Cretaceous referenced poles (Fig. 1; Table 3). A negative fold test indicated a post-folding age for the remagnetization (Fig. 11). Principal magnetic carriers are authigenic magnetite, hematite (pigmentary) and goethite that are ubiquitously associated with hydrothermal elements such as barium and manganese (Fig. 12). In addition, taking into account that maximum demagnetization temperatures for the ChRMs of both rhyolites and sandstones *c.* 680 °C (hematite) are too high for these components to have been acquired uniquely by thermoviscous overprint (Pullaiah *et al.* 1975), a CRM origin is here preferred. However, more insights are needed to clearly state the links between faults reactivation, volcanism and fluid circulation of this area in the context of the South Atlantic opening.

In conclusion, the review of remagnetized South America formations presented here indicated at least three widespread remagnetization events linked to major plate tectonics, namely:

- the *Ediacaran–Cambrian remagnetization* that affected the Paraguay and Dom Feliciano belts and for which radiometric ages of 520–500 Ma marked the collision of the Amazon, Congo-São Francisco, Kalahari and Rio de la Plata cratons at the final stages of the western Gondwana assembly;
- the *San Rafaelic Permian Overprint* that affected Argentina and Uruguay during the emplacement of the Choiyoi Magmatic Province; and

- the *Lower Cretaceous remagnetization* associated with the South Atlantic opening that affected the Itajaí Basin bordering the Dom Feliciano Belt by fault reactions and fluid circulation.

In most cases, the origin of the remagnetization both in sedimentary (carbonates, siliciclastics) or magmatic rocks is interpreted as a chemical remanent magnetization. In the remagnetized carbonates of the Guia, Bambuí and Salitre formations, models invoked organic matter (hydrocarbon) maturation, smectite–illite transformations and migration of mineralizing fluids through basement faults. The genetic link between fine-grained illitic clay and its authigenic by-product magnetite provides a clear example of an amenable target for geochronology. However, in the case of the San Rafaelic Permian overprint and the Cretaceous remagnetizations, the ages of the remagnetization are only constrained by pole position and further detailed magnetic mineralogy analysis, coupled with radiometric dating of authigenic magnetic carriers, is needed to better constrain the timing and duration of these widespread remagnetizations.

We first thank D. Elmore for his invitation and support with bibliographic references. We also thank M. Ernesto for numerous discussions about palaeomagnetism of cretaceous magmatism in South America and J. Miguel Miranda and C. Lee from the IDL institute (Portugal) for technical and administrative supply. Additional funding was provided by PIP-CONICET 2828, UBACYT X220, X183, PICT-1074, Pest-OE/CTE/LA0019/2011-IDL, PTDC/CTE-AST/117298/2010 and by PTDC/CTE-GIX/110205/2009.

# References

ALLMENDINGER, R. W., FIGUEROA, D., SNYDER, D., BEER, J., MPODOZIS, C. & ISACKS, B. L. 1990. Foreland shortening and crustal balancing in the Andes at 30°S latitude. *Tectonics*, **9**, 789–809.

ALMEIDA, R. P., JANIKIAN, L., FRAGOSO-CESAR, A. R. S. & FAMBRINI, G. L. 2010. The Ediacaran to Cambrian Rift System of Southeastern South America: tectonic implications. *The Journal of Geology*, **118**, 145–161.

ALVARENGA, C. J. S. 1990. *Phénomènes sédimentaires, structuraux et circulation de fluides développés à la transition chaîne-craton*. Exemple de la chaîne Paraguay d'âge protérozoïque supérieur, Mato Grosso, Brésil. PhD thesis, Université Aix-Marseille III, France.

ARCHANGELSKY, S. & CUNEO, R. 1984. Zonacion del Permico continental de Argentina sobre labase de sus plantas fosiles. *In*: DEL CARMEN PERRILLAT, M. (ed.) *Memoria Congreso Latinamericano de Paleontologia*, **3**, 143–153.

AZCUY, C. L. & CAMINOS, R. 1987. Diastrofismo. *In*: ARCHANGELSKY, S. (eds) *El Sistema Carbonífero en la República Argentina*. Academia Nacional de Ciencias Córdoba, Córdoba, 239–251.

BABINSKI, M., VAN SCHMUS, W. R. & CHEMALE, F. JR. 1999. Pb–Pb dating and Pb isotopic geochemistry of Neoproterozoic carbonate rocks from the São Francisco basin, Brazil: implications for the mobility of Pb isotopes during tectonism and metamorphism. *Chemical Geology*, **160**, 175–199.

BABINSKI, M., VIEIRA, L. C. & TRINDADE, R. I. F. 2007. Direct dating of the Sete Lagoas cap carbonate (Bambuí Group, Brazil) and implications for the Neoproteozoic glacial events. *Terra Nova*, **19**, 401–406.

BACHTADSE, V., ZANGLEIN, R., TAIT, J. & SOFFEL, H. 2002. Palaeomagnetism of the Permo/Carboniferous (280 Ma) Jebel Nehoud ring complex, Kordofan, Central Sudan. *Journal of African Earth Sciences*, **35**, 89–97.

BANERJEE, S., ELMORE, R. D. & ENGEL, M. H. 1997. Chemical remagnetization and burial diagenesis: testing the hypothesis in the Pennsylvanian Belden formation, Colorado. *Journal of Geophysical Research*, **102**, 24 825–24 841.

BASEI, M. A. S., SIGA, O. JR., MASQUELIN, H., HARARA, O. M., REIS NETO, J. M. & PRECIOZZI, F. 2000. The Dom Feliciano Belt of Brazil and Uruguay and its foreland domain the Rio de la Plata Craton: framework, tectonic evolution and correlation with similar provinces of Southwestern Africa. *In*: CORDANI, U. G., MILANI, E. J., THOMAZ FILHO, A. & CAMPOS, D. A. (eds) *Tectonic Evolution of South America*. Brazilian Academy of Science, Brazil, 311–334.

BASEI, M. A. S., FRIMMEL, H. E., NUTMAN, A. P., PRECIOZZI, F. & JACOB, J. 2005. A connection between the Neoproterozoic Dom Feliciano (Brazil/Uruguay) and Gariep (Namibia/South Africa) orogenic belts – evidence from a reconnaissance provenance study. *Precambrian Research*, **139**, 195–221.

BENEDETTO, J. L. 1998. Early Palaeozoic brachiopods and associated shelly faunas from western Gondwana: their bearing on the geodynamic history of the pre-Andean margin. *In*: PANKHURST, R. J. & RAPELA, C. W. (eds) *The Proto-Andean Margin of Gondwana*. Geological Society, London, Special Publications, **142**, 57–83.

BENTHIEN, R. H. & ELMORE, R. D. 1987. Origin of magnetization in the Phosphoria Formation at Sheep Mountain, Wyoming: a possible relationship with hydrocarbon. *Geophysical Research Letters*, **14**, 323–326.

BIONDI, J. C., SCFFICKET, G. & BUGALHO, A. 1992. Processos mineralizadores em bacias tardi-orogênicas 1. influência das estruturas rígidas na geração dos depósitos da minepar e do Ribeirão da Prata, grupo Itajaí (sc). *Revista Brasileira de Geociências*, **22**, 275–288.

BIONDI, J. C., BARTOSZECK, M. K. & VANZELA, G. A. 2001. Controles geológicos e geomorfológicos dos depósitos de caulim da bacia de Campo Alegre (SC). *Revista Brasileira de Geociências*, **3**, 13–20.

BLUMSTEIN, A. M., ELMORE, R. D. & ENGEL, M. H. 2004. Paleomagnetic dating of burial diagenesis in Mississippian carbonates, Utah. *Journal of Geophysical Research*, **109**, 1, doi: 10.1029/2003JB002698.

BOBBIO, M. L., RAPALINI, A. E. & VILAS, J. F. 1990. Estudio paleomagnético preliminar de la Formación Hoyada Verde, Precordillera de San Juan: un ejemplo de remagnetización sintectónica. *Revista Geológica de Chile*, **17**, 191–200.

BRANDT, D., ERNESTO, M., ROCHA-CAMPOS, A. C. & DOS SANTOS, P. R. 2009. Paleomagnetism of the Santa Fé Group, central Brazil: implications for the late Paleozoic apparent polar wander path for South America. *Journal of Geophysical Research*, **114**, 02101, doi: 10.1029/2008JB005735.

BRIDEN, J. C. 1968. Paleomagnetism of the Ntonya ring structure, Malawi. *Journal of Geophysical Research*, **73**, 725–733.

BRIDEN, J. C., MCCLELLAND, E. & REX, D. C. 1993. Proving the age of a paleomagnetic pole: the case of the Ntonya Ring structure, Malawi. *Journal of Geophysical Research*, **98**, 1743–1749.

BROTHERS, L. A., ENGEL, M. H. & ELMORE, R. D. 1996. The late diagenetic conversion of pyrite to magnetite by organically complexed ferric iron. *Chemical Geology*, **130**, 1–14.

BUGGISCH, W. 1987. Stratigraphy and very low grade metamorphism of the Sierras Australes de la Provincia de Buenos Aires (Argentina) and implications in Gondwana correlation. *Zentralbatt für Geologie Paläontologie*, **1**, 819–837.

BULLARD, E. C., EVERETT, J. E. & SMITH, A. G. 1965. A symposium on continental drift IV. The fit of the continents around the Atlantic. *Philosophical Transaction of Royal Society of London Serie A*, **258**, 41–51.

BUTLER, R. F., HERVÉ, F., MUNIZAGA, F., BECK, M. E. JR, BURMESTER, R. F. & OVIEDO, E. S. 1991. Paleomagnetism of the Patagonian plateau basalts, southern Chile and Argentina. *Journal of Geophysical Research*, **96**, 6023–6034.

CAMA, J., GANOR, J., AYORA, C. & LASAGA, C. A. 2000. Smectite dissolution kinetics at 80 °C and pH 8.8. *Geochimica et Cosmochimica Acta*, **64**, 2701–2717.

CAWOOD, P. A. & BUCHAN, C. 2007. Linking accretionary orogenesis with supercontinent assembly. *Earth-Science Reviews*, **82**, 217–256.

CHANNELL, J. E. T. & MCCABE, C. 1994. Comparison of magnetic hysteresis parameters of unremagnetized and remagnetized limestones. *Journal of Geophysical Research*, **99**, 4613–4623.

CISOWSKI, S. 1981. Interacting vs. non-interacting single domain behavior in natural and synthetic samples. *Physics of the Earth and Planetary Interiors*, **26**, 56–62.

COBBOLD, P. R., MASSABIE, A. C. & ROSSELLO, E. A. 1986. Hercynian wrenching and thrusting in the Sierras Australes fold belt, Argentina. *Hercynica*, **2**, 135–148.

CONTI, C. & RAPALINI, A. E. 1990. Paleomagnetismo de la Formacion Choique Mahuida, aflorante en la sierra homonima, provincia de La Pampa, Republica Argentina. *11th Congreso Geológico Argentino, San Juan, Argentina, Actas*, **2**, 235–238.

CORDANI, U. G., D'AGRELLA-FILHO, M. S., TEIXEIRA, W. & TRINDADE, R. I. F. 2003. Tearing up Rodinia: the Neoproterozoic palaeogeography of South American cratonic fragments. *Terra Nova*, **15**, 350–359.

CORDANI, U. G., TEIXEIRA, W., D'AGRELLA-FILHO, M. S. & TRINDADE, R. I. F. 2009. The position of the Amazonian Craton in supercontinents. *Gondwana Research*, **15**, 396–407.

D'AGRELLA-FILHO, M. S. & PACCA, I. 1988. Paleomagnetism of the Itajaí and Bom Jardim Group from Southern Brazil. *Geophysical Journal International*, **93**, 365–376.

D'AGRELLA-FILHO, M. S., PACCA, I. & SATO, K. 1986. Paleomagnetism of metamorphic rocks from the Piquete Region—Ribeira Valley, Southeastern Brazil. *Revista Brasileira de Geofísica*, **4**, 79–84.

D'AGRELLA-FILHO, M. S., BABINSKI, M., TRINDADE, R. I. F., VAN SCHMUS, W. R. & ERNESTO, M. 2000. Simultaneous remagnetization and U-Pb isotope resetting in Neoproterozoic carbonates of the Sao Francisco craton, Brazil. *Precambrian Research*, **99**, 179–196.

D'AGRELLA-FILHO, M. S., RAPOSO, M. I. B. & EGYDIO-SILVA, M. 2004. Paleomagnetic Study of the Juiz de Fora Complex, SE Brazil: implications for Gondwana. *Gondwana Research*, **7**, 103–113.

DAY, R., FULLER, M. D. & SCHMIDT, V. A. 1977. Magnetic hysteresis properties of synthetic titanomagnetites. *Journal of Geophysical Research*, **81**, 873–880.

DEKKERS, M. J. 1990. Magnetic monitoring of pyrrhotite alteration during thermal demagnetization. *Geophysical Research Letters*, **17**, 779–782.

DOMEIER, M., VAN DER VOO, R., TOHVER, E., TOMEZZOLI, R., VIZÁN, H., TORSVIK, T. H. & KIRSHNER, J. (2011) New Late Permian paleomagnetic data from Argentina: refinement of the apparent polar wander path of Gondwana. *Geochemistry Geophysics Geosystems*, **12**, Q07002, doi: 10.1029/2011GC003616.

DU TOIT, A. 1937. *Our Wandering Continents*. Oliver and Boyd, Edinburgh.

DUNLOP, D. 2002. Theory and application of the Day plot (Mrs/Ms v. Hcr/Hc): 2. Application to data for rocks, sediments, and soils. *Journal of Geophysical Research*, **107**, B3, doi: 10.1029/2001JB000487.

DUNLOP, D. J., ÖZDEMIR, Ö., CLARK, D. A. & SCHMIDT, P. W. 2000. Time-temperature relations for the remagnetization of pyrrhotite (Fe7S8) and their use in estimating paleotemperatures. *Earth and Planetary Science Letters*, **176**, 107–116.

EAGLES, G. 2007. New angles on South Atlantic opening. *Geophysical Journal International*, **168**, 353–361.

ELLWOOD, B. B., BALSAM, W. L. & ROBERTS, H. H. 2006. Gulf of Mexico sediment sources and sediment transport trends from magnetic susceptibility measurements of surface samples. *Marine Geology*, **230**, 237–248.

ELMORE, R. D. & CRAWFORD, L. 1990. Remanence in authigenic magnetite: testing the hydrocarbon–magnetite hypothesis. *Journal of Geophysical Research*, **95**, 4539–4549.

ELMORE, R. D. & LEACH, M. C. 1990. Remagnetization of the Rush Springs Formation, Cement, Oklahoma: implications for dating hydrocarbons migration and aeromagnetic exploration. *Geology*, **18**, 124–127.

ELMORE, R. D., ENGEL, M. H., CRAWFORD, L., NICK, K., IMBUS, S. & SOFER, Z. 1987. Evidence for a relationship between hydrocarbons and authigenic magnetite. *Nature*, **325**, 428–430.

ELMORE, R. D., LONDON, D., BAGLEY, D., FRUIT, D. & GUOQIU, G. 1993. Remagnetization by basinal fluids: testing the hypothesis in the Viola Limestone, southern Oklahoma. *Journal of Geophysical Research*, **95**, 6237–6254.

ERNESTO, M. 2007. Drift of South American platform since early Cretaceous: reviewing the apparent polar wander path. *Geociências*, **25–1**, 83–90.

ERNESTO, M., PACCA, I. G., HIODO, F. Y. & NARDY, A. J. R. 1990. Paleomagnetism of the Mesozoic Serra Geral Formation, Southern Brazil. *Physics of the Earth and Planetary Sciences*, **64**, 153–175.

ERNESTO, M., COMIN-CHIARAMONTI, P., GOMES, C. B., PICCIRILLO, E. M., CASTILLO, A. M. C. & VELÁZQUEZ, J. C. 1996. Paleomagnetic data from the Central Alkaline Province, Eastern Paraguay. *In*: COMIN-CHIARAMONTI, P. & GOMES, C. B . (eds) *Alkaline Magmatism in Central-Eastern Paraguay. Relationships with coeval magmatism in Brazil.* Editora da Universidade de São Paulo, São Paulo, Brazil, 85–102.

ERNESTO, M., RAPOSO, M. I. B., MARQUES, L. S., RENNE, P. R., DIOGO, L. A. & DE MIN, A. 1999. Paleomagnetism, geochemistry and 40Ar/39Ar dating of the Northeastern Paraná Magmatic province: tectonic implications. *Journal of Geodynamics*, **28**, 321–340.

ERNESTO, M., BELLIENI, G. *ET AL.* 2002. Paleomagnetic, geochronological and geochemical constraints on time and duration of the Mesozoic igneous activity in Northeastern Brazil. *In*: HAMES, W. *ET AL.* (eds) *The Central Atlantic Magmatic Province: Insights From Fragments of Pangea.* Geophysical Monograph Series, **136**, 129–149. AGU, Washington, DC, doi: 10.1029/136GM07.

FISHER, R. A. 1953. Dispersion on a sphere. *Proceedings Royal Society of London, Series A*, **217**, 295–305.

FONT, E., TRINDADE, R. I. F. & NÉDÉLEC, A. 2005. Detrital remanent magnetization in haematite-bearing Neoproterozoic Puga cap dolostone, Amazon craton: a rock magnetic and SEM study. *Geophysical Journal International*, **163**, 1–10, doi: 10.1111/j.1365-246X.2005.02776.

FONT, E., TRINDADE, R. I. F. & NÉDÉLEC, A. 2006. Remagnetization in bituminous limestones of the Neoproterozoic Araras Group (Amazon Craton): hydrocarbon maturation, burial diagenesis or both? *Journal of Geophysical Research*, **111**, B06204, doi: 10.1029/2005JB004106.

FONT, E., ERNESTO, M., SILVA, P. M. F., CORREIA, P. B. & NASCIMENTO, M. A. L. 2009. Paleomagnetism, Rock magnetism and AMS study of the Cabo Magmatic Province, NE Brazil. *Geophysical Journal International*, **179**, 905–922, doi: 10.1111/j.1365-246X.2009.04333.x.

FONT, E., NÉDÉLEC, A., TRINDADE, R. I. F. & MOREAU, C. 2010. Slow or fast ice melting at the snowball Earth Aftermath? the cap dolostone record. *Palaeogeography, Palaeoclimatology, Palaeoecology*, **295**, 215–225, doi: 10.1016/j.palaeo.2010.05.039.

FONT, E., YOUBI, N., FERNANDES, S., EL HACHIMI, H., KRATINOVA, Z. & HAMIM, Y. R. 2011. Revisiting the magnetostratigraphy of the Central Atlantic Magmatic Province from Morocco. *Earth and Planetary Science Letters*, **309**, 302–331, doi: 10.1016/j.epsl.2011.07.007.

FULLER, M., KIDANE, T. & ALI, J. 2002. AF demagnetization characteristics of NRM, compared with anhysteretic and saturation isothermal remanence: an aid in the interpretation of NRM. *Physics and Chemistry of the Earth*, **27**, 1169–1177.

GEUNA, S. & ESCOSTEGUY, L. 2006. Mineralogía magnética de la Formación Alcaparrosa (Ordovícico) y del Pórfiro pérmico que la intruye, Calingasta, provincia de San Juan. *Avances en Mineralogía, Metalogenia y Petrología 2006*, Asociación Mineralógica de Argentina, Universidad de Buenos Aires, 97–104.

GEUNA, S. E. & VIZÁN, H. 1998. New Early Cretaceous palaeomagnetic pole from Córdoba Province (Argentina): revision of previous studies and implications for the South American database. *Geophysical Journal International*, **135**, 1085–1100.

GEUNA, S. E., SOMOZA, R., VIZAN, H., FIGARI, E. G. & RINALDI, C. A. 2000. Paleomagnetism of Jurassic and Cretaceous rocks in central Patagonia: a key to constrain the timing of rotations during the breakup of southwestern Gondwana? *Earth and Planetary Sciences Letters*, **181**, 145–160.

GILL, J. D., ELMORE, R. D. & ENGEL, M. H. 2002. Chemical remagnetization and clay diagenesis: testing the hypothesis in the Cretaceous sedimentary rocks of northwestern Montana. *Physics and Chemistry of the Earth*, **27**, 1131–1139.

GONG, Z., VAN HINSBERGEN, D. J. J. & DEKKERS, M. J. 2009. Diachronous pervasive remagnetization in northern Iberian basins during Cretaceous rotation and extension. *Earth and Planetary Science Letters*, **284**, 292–301.

GRADSTEIN, F. M., OGG, J. G. & SMITH, A. G. 2004. *A Geologic Time Scale 2004.* Cambridge University Press, Cambridge, UK.

GRAY, D. R., FOSTER, D. A., MEERT, J. G., GOSCOMBE, B. D., ARMSTRONG, R. A., TROUW, R. A. J. & PASSCHIER, C. W. 2009. A Damara orogen perspective on the assembly of southwestern Gondwana. *In*: PANKHURST, R. J., TROUW, R. A. J., BRITO NEVES, B. B. & DE WIT, M. J. (eds) *West Gondwana: Pre-Cenozoic Correlations Across the South Atlantic Region.* Geological Society, London, Special Publications, **294**, 257–278, doi: 10.1144/SP294.14.

GRESSE, P. G., CHEMALE, F., SILVA, L. C., WALRAVEN, F. & HARTMANN, L. A. 1996. Late to post-orogenic basins of the Pan-African-Brasiliano collision orogen in southern Africa and southern Brazil. *Basin Research*, **8**, 157–171.

GUADAGNIN, F., CHEMALE, F. JR. *ET AL.* 2010. Depositional age and provenance of the Itajaí Basin, Santa Catarina State, Brazil: Implications for SW Gondwana correlation. *Precambrian Research*, **180**, 156–182, doi: 10.1016/j.precamres.2010.04.002.

HARRINGTON, H. J. 1947. *Explicación de las Hojas Geológicas 33 m y 34 m, Sierras de Curamalal y de la Ventana, Provincia de Buenos Aires.* Servicio Nacional de Minería y Geología. Boletín 61.

HIRT, A. M., BANIN, A. & GEHRING, U. A. 1993. Thermal generation of ferromagnetic minerals from iron-enriched smectites. *Geophysical Journal International*, **115**, 1161–1168.

HOFFMAN, P. F., KAUFMAN, A. J., HALVERSON, G. P. & SCHRAG, D. P. 1998. A Neoproterozoic snowball Earth. *Science*, **281**, 1342–1346.

HOFFMAN, P. F., HALVERSON, G. P., DOMACK, E. W., HUSSON, J. M., HIGGINS, J. A. & SHRAG, D. P. 2007. Are basal Ediacaran 635 Ma post-glacial 'cap dolostones' diachronous? *Earth and Planetary Science Letters*, **258**, 114–131.

HUANG, K. & OPDYKE, N. 1996. Severe remagnetization revealed from Triasic platform carbonates near Guiyang, Southwest China. *Earth and Planetary Science Letters*, **143**, 49–61.

INDA, H. A. & BARBOSA, J. F. 1978. *Texto explicativo para o mapa geológico do Estado da Bahia, escala 1:1.000.000*. SME/COM, Salvador, 137.

IRVING, E. 1977. Drift of the major continental blocks since the Devonian. *Nature*, **270**, 304–309.

IRVING, E. 2004. The case for Pangea B, and the intra-Pangean Megashear. *In*: CHANNELL, J. E. T., KENT, D. V., LOWRIE, W. & MEERT, J. G. (eds), *Timescales of the Paleomagnetic Field*. AGU, Washington, Geophysical Monograph, **145**, 13–27.

JACKSON, M. 1990. Diagenetic sources of stable remanence in remagnetized Paleozoic cratonic carbonates: a rock magnetic study. *Journal of Geophysical Research*, **95**, 2753–2762.

JACKSON, M., SUN, W. W. & CRADDOCK, J. P. 1992. The rock magnetic fingerprint of chemical remagnetization in midcontinental paleozoic carbonates. *Geophysical Research Letters*, **19**, 781–784.

JACKSON, M., ROCHETTE, P., FILON, G., BANERJEE, S. & MARVIN, J. 1993. Rock magnetism of remagnetized paleozoic carbonates: low-temperature behaviour and susceptibility characteristics. *Journal of Geophysical Research*, **98**, 6217–6225.

JOHNSON, H. P., LOWRIE, W. & KENT, D. V. 1975. Stability of anhysteretic remanent magnetization in fine and coarse magnetite and maghemite. *Geophysical Journal of Royal Astronomical Society*, **41**, 1–10.

JORDAN, T. E., ISACKS, B. L., ALLMENDINGER, R. W., BREWER, J. A., RAMOS, V. A. & ANDO, C. J. 1983. Andean tectonics related to the geometry of the subducted Nazca Plate. *Geological Society of America, Bulletin*, **94**, 341–361.

JORDANOVA, N., HENRY, B., JORDANOVA, D., IVANOV, Z., DIMOV, D. & BERGERAT, F. 2001. Paleomagnetism in northwestern Bulgaria: geological implications of widespread remagnetization. *Tectonophysics*, **343**, 79–92.

KAO, S. J., HORNG, C. S., ROBERTS, A. P. & LIU, K. K. 2004. Carbon-sulfur-iron relationships in sedimentary rocks from southwestern Taiwan: influence of geochemical environment on greigite and pyrrhotite formation. *Chemical Geology*, **203**, 153–168.

KARFUNKEL, J. & HOPPE, A. 1988. Late Proterozoic glaciation in central-eastern Brazil: synthesis and model. *Palaeogeography, Palaeoclimatology, Palaeoecology*, **65**, 1–21.

KATZ, B., ELMORE, R. D., COGOINI, M. & FERRY, S. 1998. Widespread chemical remagnetization: orogenic fluids or burial diagenesis of clays? *Geology*, **26**, 603–606.

KATZ, B., ELMORE, R. D., COGOINI, M., ENGEL, M. H. & FERRY, S. 2000. Associations between burial diagenesis of smectite, chemical remagnetization, and magnetite authigenesis in the Vocontian trough, SE France. *Journal of Geophysical Research*, **105**, 851–868.

KEIDEL, J. 1916. La geología de las sierras de la Provincia de Buenos Aires y sus relaciones con las montañas del Cabo y los Andes. Análes del Ministerio de Agricultura de la Nación. Dirección de Geología, Mineralogía y Minería, **9**(3), 1–78, Buenos Aires.

KEMPF, O., KELLERHALS, P., LOWRIE, W. & MATTER, A. 2000. Paleomagnetic directions in Late Precambrian glaciomarine sediments of the Mirbat Sandstone Formation, Oman. *Earth and Planetary Science Letters*, **175**, 181–190.

KILGORE, B. & ELMORE, R. D. 1989. A study of the relationship between hydrocarbon migration and the precipitation of authigenic minerals in the Triassic Chugwater Formation, southern Montana. *Geological Society of America Bulletin*, **101**, 1280–1288.

KIRSCHVINK, J. L. 1978. The Precambrian–Cambrian boundary problem: paleomagnetic directions from the Amadeus basin, Central Australia. *Earth and Planetary Science Letters*, **40**, 91–100.

KLEIMAN, L. E. & JAPAS, M. S. 2009. The Choiyoi volcanic province at 34°S-36°S (San Rafael, Mendoza, Argentina): implications for the Late Palaeozoic evolution of the southwestern margin of Gondwana. *Tectonophysics*, **473**, 283–299.

KLOOTWIJK, C. T. 1980. Early Paleozoic paleomagnetism in Australia. *Tectonophysics*, **64**, 249–332.

KRS, M., NOVÁK, F., KRSOVÁ, M., PRUNER, P., KOUKLÍKOVÁ, L. & JANSA, J. 1992. Magnetic properties and metastability of greigite-smythite mineralization in brown-coal basins of the Krušné hory Piedmont, Bohemia. *Physics of the Earth and Planetary Interiors*, **70**, 273–287.

KRUIVER, P. P., DEKKERS, M. J. & HESLOP, D. 2001. Quantification of magnetic coercivity components by the analysis of acquisition curves of isothermal remanent magnetization. *Earth and Planetary Sciences Letters*, **189**, 269–276.

LARSON, R. L. 1991. Latest pulse of Earth: evidence for a Mid-Cretaceous super plume. *Geology*, **19**, 547–550.

LLAMBÍAS, J. E. & SATO, A. M. 1995. El batolito de Colanguil: transición entre orogénesis y anorogénesis. *Revista de la Asociación Geológica Argentina*, **50**, 111–131.

LLAMBÍAS, E. J., QUENARDELLE, S. & MONTENEGRO, T. 2003. The Choiyoi Group from Central Argentina: a subalkaline transitional to alkaline association in the craton adjacent to the active margin of Gondwana continent. *Journal of South American Earth Sciences*, **16**, 243–257.

LÓPEZ DE LUCHI, M., RAPALINI, A. E. & TOMEZZOLI, R. N. 2010. Magnetic Fabric and microstructures of Late Paleozoic granotoids from the North Patagonian Massif: evidence of a collision between Patagonia and Gondwana? *Tectonophysics*, **494**, 118–137.

LOSSADA, A. C., TOHVER, E., SANCHEZ-BETTUCCI, L. & RAPALINI, A. E. 2011. *Fábrica Magnética, Paleomangetismo y geocronología de la Formación Playa Hermosa, Uruguay*. 18th Cong. Geol. Argentino, Actas in CD. Neuquén, Argentina.

MACHEL, H. G. & BURTON, E. A. 1991. Chemical and microbial processes causing anomalous magnetization in environments affected by hydrocarbon seepage. *Geophysics*, **56**, 598–605.

MCCABE, C. & CHANNELL, J. E. T. 1994. Late Paleozoic remagnetization if limestones of the Craven Basin (northern England) and the rock magnetic fingerprint of remagnetized sedimentary carbonates. *Journal of Geophysical Research*, **99**, 4603–4612.

MCCABE, C. & ELMORE, D. R. 1989. The occurrence and origin of late paleozoic remagnetization in the sedimentary rocks of north America. *Reviews of Geophysics*, **27**, 471–494.

McCABE, C., JACKSON, M. & SAFFER, B. 1989. Regional patterns of magnetite authigenesis in the Appalachian basin: implications for the mechanism of late Paleozoic remagnetization. *Journal of Geophysical Research*, **94**, 10 429–10 443.

McELHINNY, N. W. 1964. Statistical significance of the fold test in Palaeomagnetism. *Geophysical Journal of Royal Astronomical Society*, **8**, 338–340.

McFADDEN, P. L. 1990. A new fold test for palaeomagnetic studies. *Geophysical Journal International*, **103**, 163–169.

MEERT, J. G. 2003. A synopsis of events related to the assembly of eastern Gondwana. *Tectonphysics*, **362**, 1–40.

MEERT, J. G. & LIEBERMAN, B. S. 2008. The Neoproterozoic assembly of Gondwana and its relationship to the Ediacaran–Cambrian radiation. *Gondwana Research*, **14**, 5–21.

MEERT, J. G. & POWELL, C. McA. 2001. Introduction to the special volume on the assembly and break-up of Rodinia. *Precambrian Research*, **110**, 1–8.

MEERT, J. G. & VAN DER VOO, R. 1996. Paleomagnetic and 40Ar/39Ar study of the Sinyai dolerite, Kenya: implications for Gondwana assembly. *Journal of Geology*, **104**, 131–142.

MEERT, J. G., NÉDÉLEC, A., HALL, C., WINGATE, M. T. D. & RAKOTONDRAZAFY, M. 2001. Paleomagnetism, geochronology and tectonic implications of the Cambrian-age Carion granite, Central Madagascar. *Tectonophysics*, **340**, 1–21.

MEERT, J. G., NÉDÉLEC, A. & HALL, C. 2003. The stratoid granites of central Madagascar: paleomagnetism and further age constraints on Neoproterozoic deformation. *Precambrian Research*, **120**, 101–129.

MELCHOR, R. N. 1999. Redefinición estratigráfica de la Formación Carapacha (Pérmico), Provincia de La Pampa, Argentina. *Revista de la Asociación Geológica Argentina*, **54**, 99–108.

MELCHOR, R. N. & CÉSARI, S. N. 1997. Permian floras from Carapacha Basin, La Pampa Province, central Argentina. Description and importance. *Geobios*, **30**, 607–633.

MISI, A. & VEIZER, J. 1998. Neoproterozoic carbonate sequences of the Una Group, Irecê Basin Brazil: chemostratigraphy, age and correlations. *Precambrian Research*, **89**, 87–100.

MOLOTO-A-KENGUEMBA, G. R., TRINDADE, R. I. F., MONIÉ, P., NÉDÉLEC, A. & SIQUEIRA, R. 2008. A late Neoproterozoic paleomagnetic pole for the Congo craton: tectonic setting, paleomagnetism and geochronology of the Nola dike swarm (Central African Republic). *Precambrian Research*, **164**, 214–226.

MONTEIRO, L. V. S., BETTENCOURT, J. S., SPIRO, B., GRAÇA, R. & OLIVEIRA, T. F. 1999. The Vazante Zinc Mine, MG, Brazil: constraints on fluid evolution and willecmitic mineralization. *Exploration and Mining Geology*, **8**, 210–242.

MONTES-LAUAR, C. R., PACCA, I. G., MELFI, A. J. & KAWASHITA, K. 1995. Late Cretaceous alkaline complexes, southeastern Brazil: paleomagnetism and geochronology. *Earth and Planetary Science Letters*, **134**, 425–440.

MOREAU, M. G., ADER, M. & ENKIN, R. J. 2005. The magnetization of clay-rich rocks in sedimentary basins: low-temperature experimental formation of magnetic carriers in natural samples. *Earth and Planetary Science Letters*, **230**, 193–210.

MOREL, P. 1981. Paleomagnetism of a Pan-African diorite: a Late Precambrian pole for western Africa. *The Geophysical Journal of the Royal Astronomical Society*, **65**, 493–503.

MOREL, P. & IRVING, E. 1981. Palaeomagnetism and the evolution of Pangaea. *Journal of Geophysical Research*, **86**, 1858–1872.

MOULIN, M., ASLANIAN, D. & UNTERNEHR, P. 2010. A new starting point for the South and Equatorial Atlantic Ocean. *Earth-Science Review*, **98**, 1–37.

MUTTONI, G., KENT, D. V., GARZANTI, E., BRACK, P., ABRAHAMSEN, N. & GAETANI, M. 2003. Early permian pangea 'B' to Late Permian Pangea 'A'. *Earth and Planetary Science Letters*, **215**, 379–394.

MUTTONI, G., KENT, D. V. *ET AL.* 2009. Opening of the neotethys ocean and the Pangea B to Pangea A transformation during the Permian. *GeoArabia*, **14**, 17–48.

NOGUEIRA, A. C. R., RICCOMINI, C., SIAL, A. N., MOURA, C. A. V. & FAIRCHILD, T. R. 2003. Soft-sediment deformation at the base of the Neoproterozoic Puga cap carbonate (southwestern Amazon Craton, Brazil): confirmation of rapid icehouse-greenhouse transition in snowball Earth. *Geology*, **31**, 613–616.

NOGUEIRA, A. C. R., RICCOMINI, C., SIAL, A. N., MOURA, C. A. V., TRINDADE, R. I. F. & FAIRCHILD, T. R. 2007. S and Sr isotope fluctuations and paleoceanographic changes in the late Neoproterozoic Araras carbonate platform, southern Amazon craton, Brazil. *Chemical Geology*, **237**, 168–190.

OGG, J. G., OGG, G. & GRADSTEIN, F. M. 2008. *The Concise Geologic Time Scale*. Cambridge University Press, Cambridge.

OLIVER, J. 1986. Fluids expelled tectonically from orogenic belts: their role in hydrocarbon migration and other geologic phenomena. *Geology*, **14**, 99–102.

OLIVER, J. 1992. The spots and stains of plate tectonics. *Earth Science Review*, **32**, 77–106.

OPDYKE, N. D. & CHANNELL, J. E. T. 1996. *Magnetic Stratigraphy*. Academic Press, San Diego, California.

OYHANTÇABAL, P., SIEGEMUND, S., WEMMER, K., FREI, R. & LAYER, P. 2007. Post-collisional transition from calc-alkaline to alkaline magmatism during transcurrent deformation in the southernmost Dom Feliciano Belt (Braziliano–Pan-African, Uruguay). *Lithos*, **98**, 141–159, doi: 10.1016/j.lithos.2007.03.001.

OYHANTÇABAL, P., SIEGESMUND, S., WEMMER, K. & PASSCHIER, C. W. 2011. The transpressional connection between Dom Feliciano and Kaoko Belts at 580–550 Ma. *International Journal of Earth Science*, **100**, 379–390, doi: 10.1007/s00531-010-0577-3.

PANKHURST, R. J., RAPELA, C. W., FANNING, C. M. & MÁRQUEZ, M. 2006. Gondwanide continental collision and the origin of Patagonia. *Earth-Sciences Reviews*, **76**, 235–257.

PASSARELLI, C. R., BASEI, M. A. S., SIGA, O. JR, Mc REATH, I. & CAMPOS NETO, M. C. 2010. Deformation and geochronology of syntectonic granitoids emplaced in the Major Gercino Shear Zone, southeastern South America. *Gondwana Research*, **17**, 688–703.

PEDROSA-SOARES, A. C., CORDANI, U. G. & NUTMAN, A. 2000. Constraining the age of Neoproterozoic glaciation in eastern Brazil: first U–Pb (SHRIMP) data of detrital zircons. *Revista Brasileira de Geociências*, **30**, 58–61.

PHIPPS MORGAN, J., RESTON, T. J. & RANERO, C. R. 2004. Contemporaneous mass extinctions, continental flood basalts, and 'impact signals': are mantle plume-induced lithospheric gas explosions the causal link? *Earth and Planetary Sciences Letters*, **217**, 263–284.

POULSEN, C. J., BARRON, E. J., ARTHUR, M. A. & PETERSON, W. H. 2001. Response of the mid-Cretaceous global oceanic circulation to tectonic and $CO_2$ forcings. *Paleoceanography*, **16**, 576–592.

PULLAIAH, G., IRVING, E., BUCHAN, K. L. & DUNLOP, D. J. 1975. Magnetization changes caused by burial and uplift. *Earth and Planetary Science Letters*, **28**, 133–143.

RAMOS, V. A. 2008. Patagonia: a Paleozoic continent drift? *Journal of South American Earth Sciences*, **26**, 235–251.

RAPALINI, A. E. 1998. Syntectonic magnetization of mid-Palaeozoic Sierra Grande Formation: further constraints on the tectonic evolution of Patagonia. *Journal of the Geological Society, London*, **155**, 105–114.

RAPALINI, A. E. 2006. New Late Proterozoic paleomagnetic pole for the Rio de la Plata craton: implications for Gondwana. *Precambrian Research*, **147**, 223–233.

RAPALINI, A. E. & ASTINI, R. A. 1998. Paleomagnetic confirmation of the Laurentian origin of the Argentine Precordillera. *Earth and Planetary Science Letters*, **155**, 1–14.

RAPALINI, A. E. & ASTINI, R. A. 2005. La remagnetización Sanrafaélica de la Precordillera: Nuevas evidencias. *Revista de la Asociación Geológica Argentina*, **60**, 290–300.

RAPALINI, A. E. & TARLING, D. H. 1993. Multiple magnetizations in the Cambrian- Ordovician carbonate platform of the Argentine Precordillera and their tectonic implications. *Tectonophysics*, **227**, 49–62.

RAPALINI, A. E. & SANCHEZ-BETTUCCI, L. 2008. Widespread remagnetization of Late Proterozoic sedimentary units of Uruguay and the apparent polar wander path for the Rio de la Plata craton. *Geophysical Journal International*, **174**, 55–74.

RAPALINI, A. E. & VILAS, J. F. 1991. Preliminary paleomagnetic data from the Sierra Grande Formation: tectonic consequences of the first Middle Paleozoic paleopoles from Patagonia. *Journal of South American Earth Sciences*, **4**, 25–41.

RAPALINI, A. E., VILAS, J. F, BOBBIO, M. L. & VALENCIO, D. A. 1989. Geodynamic interpretations from paleomagnetic data of Late Paleozoic rocks in the Southern Andes. *In*: HILLHOUSE, J. W. (ed.) *Deep Structure and Past Kinematics of Accreted Terranes*. American Geophysical Union, Geophysical Monograph Series, **50**, 41–57.

RAPALINI, A. E., BORDONARO, O. & BERQUO, T. S. 2000. Paleomagnetic study of Cambrian–Ordovician rocks in the Eastern Precordillera of Argentina: some constraints on the Andean uplift of this block. *Tectonophysics*, **326**, 173–184.

RAPALINI, A. E., FAZZITO, S. & ORUÉ, D. 2006. A new Late Permian paleomagnetic pole for stable South America: the Independencia Group, eastern Paraguay. *Earth and Planets Space Special Volume*, **58**, 1247–1253.

RAPALINI, A. E., LOPEZ DE LUCHI, M., MARTINEZ DOPICO, C., LINCE KLINGER, F., GIMENEZ, M. E. & MARTINEZ, M. P. 2010. Did Patagonia collide against Gondwana in the Late Paleozoic? Some insights from a multidisciplinary study of magmatic units of the North Patagonian Massif. *Geologica Acta*, **8**, 349–371.

RAPELA, C. W., FANNING, C. M., CASQUET, C., PANKHURST, R. J., SPALLETI, L., POIRÉ, D. & BALDO, E. G . 2011. The Rio de la Plata craton and the adjoining Pan-African/Brasiliano terranes: Their origins and incorporation into South-West Gondwana. *Gondwana Research* (online first), **20**(4), November 2011, 673–690, doi: 10.1016/j.gr.2011.05.001.

RAPOSO, M. I. B., ERNESTO, M. & RENNE, P. R. 1998. Paleomagnetic and 40Ar/39Ar data on the Early Cretaceous dyke swarm from the Santa Catarina Island, Southern Brazil. *Earth and Planetary Science Letters*, **144**, 199–211.

REEVES, C. V., DE WIT, M. J. & SAHU, B. K. 2004. Tight reassembly of Gondwana exposes Phanerozoic shears in Africa as global tectonic players. *Gondwana Research*, **7**, 7–19.

RENNE, P. R., ERNESTO, M., PACCA, I. G., COE, R. S., GLEN, J. M., PRÉVOT, M. & PERRIN, M. 1992. The age of Paraná flood volcanism, rifting of Gondwanaland, and the jurassic-cretaceous boundary. *Science*, **6**, 975–9.

RENNE, P. R., DECKART, K., ERNESTO, M., FÉRAUD, G. & PICCIRILLO, E. M. 1996. Age of the Ponta Grossa dyke swarm (Brazil), and implications to Paraná flood volcanism. *Earth and Planetary Science Letters*, **144**, 199–211.

REYNOLDS, R. L., FISHMAN, N. S., WANTY, R. B. & GOLDHABER, M. B. 1990. Iron sulfides minerals at Cement oil field, Oklahoma: implications for magnetic detection of oil fields. *Geological Society of America Bulletin*, **102**, 368–380.

REYNOLDS, R. L., FISHMAN, N. S. & HUDSON, M. R. 1991. Sources of aeromagnetic anomalies over Cement oild field (Oklahoma), Simpson oild fiel (Alaska), and the Wyoming-Idaho-Utah thrust belt. *Geophysics*, **56**, 606–617.

ROBERTSON, D. J. & FRANCE, D. E. 1994. Discrimination of remanence-carrying minerals in mixtures using isothermal remanent magnetisation acquisition curves. *Physics of the Earth and Planetary Sciences Letters*, **82**, 223–234.

ROCHETTE, P. & VANDAMME, D. 2001. Pangea B: an artfact of incorrect paleomagnetic assumptions? *Annali di Geofisica*, **44**, 649–658.

ROCHETTE, P., FILLION, G., MATTÉI, J. L. & DEKKERS, M. J. 1990. Magnetic transition at 30–34 Kelvin in pyrrhotite: insight into a widespread occurrence of this mineral in rocks. *Earth Planetary Science Letters*, **98**, 319–328.

ROSTIROLLA, S. P., AHRENDT, A., SOARES, P. C. & CARMIGNANI, L. 1999. Basin analysis and mineral endowment of the Proterozoic Itajaí Basin, south-east Brazil. *Basin Research*, **11**, 127–142.

ROSTIROLLA, S. P., MANCINI, F., RIGOTI, A. & KRAFT, R. P. 2003. Structural styles of the intracratonic reactivation of the Perimbó fault zone, Paraná basin, Brazil. *Journal of South American Earth Sciences*, **16**, 287–300.

SÁNCHEZ-BETTUCCI, L. & RAPALINI, A. E. 2002. Paleomagnetism of the Sierra de Las Animas Complex, southern Uruguay: its implications in the assembly of western Gondwana. *Precambrian Research*, **118**, 243–265.

SEEWALD, J. S. 2001. Aqueous geochemistry of low molecular weight hydrocarbons at elevated temperatures and pressures: Constraints from mineral buffered laboratory experiments. *Geochimica et Cosmochimica Acta*, **65**, 1641–1664.

SEEWALD, J. S. 2003. Organic–inorganic interactions in petroleum-producing sedimentary basins. *Nature*, **426**, 327–333.

SILLITOE, R. H. 1977. Permo-Carboniferous, Upper Cretaceous and Miocene Porphyry Copper type mineralization in the Argentinian Andes. *Economic Geology*, **72**, 99–103.

SOMOZA, R. & ZAFFARANA, C. B. 2008. Mid-Cretaceous polar standstill of South America, motion of the Atlantic hotspots and the birth of the Andean cordillera. *Earth and Planetary Sciences Letters*, **271**, 267–277.

STERN, R. J. 2002. Crustal evolution in the East African orogeny: a neodymium isotopic perspective. *Journal of Africa Earth Sciences*, **34**, 109–117.

SUK, D., PEACOR, D. R. & VAN DER VOO, R. 1990. Replacement of pyrite framboids by magnetite in calcáreo and implications for Palaeomagnetism. *Nature*, **345**, 611–613.

TARDUNO, J. A. & SMIRNOV, A. V. 2001. Stability of the Earth with respect to the spin axis for the last 130 million years. *Earth and Planetary Science Letters*, **184**, 549–553.

TAUXE, L. 2005. Inclination flattening and the geocentric axial dipole hypothesis. *Earth and Planetary Science Letters*, **233**, 247–261.

TOHVER, E., D'AGRELLA FILHO, M. & TRINDADE, R. I. F. 2006. Paleomagnetic record of Africa and South America for the 1200–500 Ma interval, and evaluation of Rodinia and Gondwana assemblies. *Precambrian Research*, **147**, 193–222.

TOHVER, E., CAWOOD, P. A., ROSSELLO, E. A. & RICCOMINI, C. 2007. *Demise of the Gondwana ice sheet in SW Gondwana: time constraints from U–Pb zircon ages.* GEOSUR 2007, An International Congress on the Geology and Geophysics of the Southern Hemisphere, Universidad de Chile, Santiago, 163 (Abstract volume), 139–142.

TOHVER, E., WEIL, A. B., SOLUM, J. G. & HALL, C. M. 2008. Direct dating of chemical remagnetizations in sedimentary rocks, insights from clay mineralogy and 40Ar/39Ar age analysis. *Earth and Planetary Science Letters*, **274**, 524–530, doi: 10.1016/j.epsl.2008.08.002.

TOHVER, E., TRINDADE, R. I. F., SOLUM, J. G., HALL, C. M., RICCOMINI, C. & NOGUEIRA, A. C. 2010. Closing the Clymene Ocean and bending a Brasiliano belt, evidence for the Cambrian formation of Gondwana from SE Amazon craton. *Geology*, **38**, 267–270, doi:10.1130/G30510.1.

TOHVER, E., CAWOOD, P. A., ROSSELLO, E. A. & JOURDAN, F . 2011. Closure of the Clymene Ocean and formation of West Gondwana in the Cambrian: Evidence from the Sierras Australes of the southernmost Rio de la Plata craton, Argentina. *Gondwana Research* (online first), **21**(2–3), March 2012, 394–405, doi:10.1016/j.gr. 2011.04.001.

TOMEZZOLI, R. N. 1997. *Geología y Paleomagnetismo en el ámbito de las Sierras Australes de la provincia de Buenos Aires.* PhD thesis (unpublished), Universidad de Buenos Aires.

TOMEZZOLI, R. N. 1999. Edad de la sedimentación y deformación de la Formación Tunas en las Sierras Australes de la provincia de Buenos Aires (37°-39°S − 61° − 63°W). *Revista de la Asociación Geológica Argentina*, **54**, 220–228.

TOMEZZOLI, R. N. 2001. Further palaeomagnetic results from the Sierras Australes fold and thrust belt, Argentina. *Geophysical Journal International*, **147**, 356–366.

TOMEZZOLI, R. N. 2009. The apparent polar wander path for South America during the Permian–Triassic. *Gondwana Research*, **15**, 209–215.

TOMEZZOLI, R. N. & CRISTALLINI, E. O. 1998. Nuevas evidencias sobre la importancia del fallamiento en la estructura de las Sierras Australes de la Provincia de Buenos Aires. *Revista de la Asociación Geológica Argentina*, **53**, 117–129.

TOMEZZOLI, R. N. & VILAS, J. F. 1997. Paleomagnetismo y fábrica magnética en afloramientos cercanos a las Sierras Australes de la Provincia de Buenos Aires (López Lecube y González Chaves). *Revista de la Asociación Geológica Argentina*, **52**, 419–432.

TOMEZZOLI, R. N. & VILAS, J. F. 1999. Paleomagnetic constraints on age of deformation of the Sierras Australes thrust and fold belt, Argentina. *Geophysical Journal International*, **138**, 857–870.

TOMEZZOLI, R. N., MACDONALD, W. D. & TICKYJ, H. 2003. Composite magnetic fabrics from S-C granitic gneiss of Cerro de los Viejos, La Pampa province, Argentina. *Journal of Structural Geology*, **25/2**, 159–169.

TOMEZZOLI, R. N., KLEIMAN, L., SALVARREDI, J., TERRIZZANO, C. & CRISTALLINI, E. O. 2005. Relaciones estratigráicas de volcanitas del Choiyoi Inferior sobre la base de estudios paleomagnéticos, Bloque de San Rafael, Mendoza, Argentina. *XVI Congreso Geológico Argentino, Buenos Aires*, **I**, 227–232.

TOMEZZOLI, R. N., MELCHOR, R. & MACDONALD, W. D. 2006. Tectonic implications of post-folding Permian magnetizations in the Carapacha basin, La Pampa province, Argentina. *Earth and Planets Space*, **58**, 1235–1246.

TOMEZZOLI, R. N., SAINT PIERRE, T. & VALENZUELA, C. 2009. New Palaeomagnetic results from Late Paleozoic volcanic units along the western Gondwana in La Pampa, Argentina. *Earth and Planets Space*, **60**, 1–7.

TORCQ, F., BESSE, J., VASLET, D., MARCOUX, J., RICOU, L. E., HALAWANI, M. & BASAHEL, M. 1997. Paleomagnetic results from Saudi Arabia and the Permo-Triassic Pangea configuration. *Earth and Planetary Science Letters*, **148**, 553–567.

TORSVIK, T. H. & VAN DER VOO, R. 2002. Refining Gondwana and Pangea Palaeogeography: estimates of Phanerozoic (octupole) non-dipole fields. *Geophysical Journal International*, **151**, 771–794.

TRINDADE, R. I. F., FONT, E., D'AGRELLA FILHO, M. S., NOGUEIRA, A. C. R. & RICCOMINI, C. 2003. Low-latitude and multiple geomagnetic reversals in the

Neoproterozoic Puga cap carbonate, Amazon craton. *Terra Nova*, **15**, 441–446.

TRINDADE, R. I. F., D'AGRELLA-FILHO, M. S., BABINSKI, M., FONT, E. & BRITO NEVES, B. B. 2004. Paleomagnetism and geochronology of the Bebedouro cap carbonate: evidence for continental-scale Cambrian remagnetization in the São Francisco craton, Brazil. *Precambrian Research*, **128**, 83–103.

TRINDADE, R. I. F., D'AGRELLA-FILHO, M. S., EPOF, I. & BRITO NEVES, B. B. 2006. Paleomagnetism of early Cambrian Itabaiana mafic dikes (NE Brazil) and the final assembly of Gondwana. *Earth and Planetary Science Letters*, **244**, 361–377.

TRUCO, S. & RAPALINI, A. E. 1996. New evidence of a widespread Permian remagnetizing event in the central Andean zone of Argentina. *Third International Symposium on Andean Geodynamics*, St. Malo, France, Extended Abstracts, 799–802.

VALENCIO, D. A., SINITO, A. M. & VILAS, J. F. 1980. Paleomagnetism of upper Precambrian rocks of the La Tinta Formation, Argentina. *Geophysical Journal of Royal Astronomical Society*, **62**, 563–575.

VAN DER VOO, R. & TORSVIK, T. H. 2004. The quality of European permo-triassic paleopoles and its impact on Pangea Reconstructions. *In*: CHANNELL, J. E. T., KENT, D. V., LOWRIE, W. & MEERT, J. G. (eds) *Timescales of the Paleomagnetic Field*. AGU, Washington, Geophysical Monograph, **145**, 29–42.

VARELA, R. 2009. Geology and Rb–Sr geochronology of granitoids of Sierra Grande, province of Río Negro. *Revista de la Asociación Geológica Argentina (online)*, **64–2**, 275–284.

VAUGHAN, A. P. & PANKHURST, R. J. 2008. Tectonic overview of the West Gondwana margin. *Gondwana Research*, **13**, 150–162.

VELÁZQUEZ, V. F., GOMES, C. B. ET AL. 1992. Magmatismo alcalino Mesozóico na porção centro-oriental do Paraguay: aspectos geocronólogicos. *Geochimica Brasiliensis*, **6**, 23–35.

VILAS, J. F. 1981. Paleomagnetism of South American rocks and the dynamic processes related with the fragmentation of Western Gondwana. *In:* MCELHINNY, M. W. & VALENCIO, D. A. (eds) *Paleoreconstruction of the Continents*. American Geophysical Union. Geological Society of America, Geodynamics Series, **2**, 106–114.

VILAS, J. F. & VALENCIO, D. A. 1978. Paleomagnetism and K–Ar age of the Upper Ordovician Alcaparrosa Formation, Argentina. *Geophysical Journal of Royal Astronomical Society*, **55**, 143–154.

VON GOSEN, W. 2002. Polyphase structural evolution in the northeastern segment of the North Patagonian Massif (southern Argentina). *Journal of South American Earth Sciences*, **15**, 591–623.

VON GOSEN, W. 2003. Thrust tectonics in the North Patagonian massif (Argentina): implications for a Patagonian plate. *Tectonics*, **22–1**, 1005.

VON GOSEN, W., BUGGISCH, W. & DIMIERI, L. V. 1990. Structural and metamorphic evolution of the Sierras Australes (Buenos Aires Province/Argentina). *Geologische Rundschau Stuttgart*, **79/3**, 797–821.

WEGENER, A. 1912. Die Entstehung der Kontinente, Peterm. *Geologische Rundschau*, **3**, 276–292.

WEIL, A. B. & VAN DER VOO, R. 2002. Insights into the mechanism for orogen-related carbonate remagnetization from growth of authigenic Fe-oxide: a scanning electron microscopy and rock magnetic study of Devonian carbonates from northern Spain. *Journal of Geophysical Research*, **107**, doi: 10.1029/2001JB000200.

WEIL, A. B., VAN DER VOO, R. & VAN DER PLUIJM, B. A. 2001. Oroclinal bending and evidence against the Pangea megashear: the Cantabria-Asturias arc (northern Spain). *Geology*, **29**, 991–994.

WOODS, S. D., ELMORE, R. D. & ENGEL, M. H. 2002. Paleomagnetic dating of the smectite–illite conversion: testing the hypothesis in Jurassic sedimentary rocks, Skye, Scotland. *Journal of Geophysical Research*, **107**, 2091, doi: 10.1029/2000JB000053.

XU, W., VAN DER VOO, R. & PEACOR, D. R. 1998. Electron microscopic and rock magnetic study of remagnetized Leadville carbonates, central Colorado. *Tectonophysics*, **296**, 333–362.

YUAN, K., VAN DER VOO, R., BAZHENOV, M. L., BAKHMUTOV, V., ALEKHIN, V. & HENDRIKS, B. W. H. 2011. Permian and Triassic palaeolatitudes of the Ukrainian shield with implications for Pangea reconstructions Kenneth Yuan. *Geophysical Journal International*, **184**, 595–610.

ZALBA, P. E., MANASSERO, M. J., LAVERRET, E., BEAUFORT, D., MEUNIER, A., MOROSI, M. & SEGOVIA, L. 2007. Middle Permian telodiagenetic processes in neoproterozoic sequences tandilia system Argentina. *Journal of Sedimentary Research*, **77**, 525–538.

ZANETTINI, J. C. 1981. La Formación Sierra Grande (Provincia de Rio Negro). *Revista de la Asociación Geológica Argentina*, **36**, 160–179.

ZEGERS, T. E., DEKKERS, M. J. & BAILLY, S. 2003. Late Carboniferous to Permian remagnetization of Devonian limestones in the Ardennes: role of temperature, fluids, and deformation. *Journal of Geophysical Research*, **108**, doi: 10.1029/2002JB002213.

ZIJDERVELD, D. A. 1968. Natural remanent magnetizations of some instrusive rocks from the Sor Rondane mountains, Queen Maud Land, Antarctica. *Journal of Geophysical Research*, **73**, 3773–3785.

ZWING, A., CLAUER, N., LIEWIG, N. & BACHTADSE, V. 2009. Identification of remagnetization processes in Paleozoic sedimentary rocks of the northeast Rhenish Massif in Germany by K–Ar dating and REE tracing of authigenic illite and Fe oxides. *Journal of Geophysical Research*, **114**, B06104, doi: 10.1029/2008JB006137.

# Palaeomagnetism of the Mississippian Barnett Shale, Fort Worth Basin, Texas

DEVIN DENNIE*, R. D. ELMORE, JOHN DENG, EARL MANNING &
JOHARI PANNALAL

*School of Geology and Geophysics, University of Oklahoma, Norman, OK 3019, USA*

**Corresponding author (e-mail: devin.dennie@gmail.com)*

**Abstract:** A palaeomagnetic study of four oriented cores was conducted to better understand the timing of diagenetic events in the Mississippian Barnett Shale, a primary source rock and the unconventional gas reservoir in the Fort Worth Basin, Texas. Thermal demagnetization removes a present-field modern viscous remanent magnetization (VRM) as well as a chemical remanent magnetization (CRM) that has shallow inclinations and streaked south–SE-directed declinations. The VRM was used to orient the CRM data for one well and it produced a similar streak of directions. The streaking of directions could represent a mixing trend between two or more CRMs. Specimens from bedding-parallel and NE subvertical mineralized fractures contain a CRM that is interpreted to be of Pennsylvanian age and to have formed in response to burial diagenetic processes. NE- and NW-oriented vertical fractures are common to rocks that contain late Permian to Triassic CRMs. Sr and sulphur isotope results from vein minerals around NE fractures suggest the CRM could be related to fluids sourced from the Ouachita front. The SE directions in the streak could be explained if the northern part of the basin experienced a component of anticlockwise rotation of up to 20° in the Pennsylvanian.

Organic-rich shales are currently the subject of focused research largely due to the fact that they have become exploitable unconventional reservoirs of hydrocarbons. Studies (e.g. Neuzil 1994; Best & Katsube 1995; Harrington & Horseman 1999; Loucks *et al.* 2009) suggest that shale and mudstone source rocks tend to be extremely tight to natural fluid movement, with permeabilities measured in the nanodarcy to microdarcy ranges. As a result, shales require artificial fractures to be created by hydraulic stimulation in order to enhance the rock's permeability and to make economic reservoirs (Pollastro *et al.* 2007). This new interest in shales has led to additional research to better understand both the depositional and diagenetic aspects of these fine-grained rocks. One feature of these shales that has been neglected is their palaeomagnetism, which can provide information on the timing of fluid-related (Elmore *et al.* 1993, 1998) and burial diagenetic events (Katz *et al.* 2000; Blumstein *et al.* 2004; Moreau *et al.* 2005; Tohver *et al.* 2008).

The Mississippian (Chesterian) Barnett Shale in the Fort Worth Basin, Texas (Fig. 1) is the first major shale-gas play of significant importance. Further understanding of the diagenetic history of the shale is important for shale diagenesis but also in terms of hydrocarbon prospectivity and producibility. For example, Bowker (2007) and Pollastro *et al.* (2007) suggest that orogenic fluids

(e.g. Oliver 1992) exerted a major control on the thermal maturity of the Barnett Shale. Based on elevated vitrinite reflectance values in some stratigraphic intervals and along some faults, for example the Mineral Wells fault (Fig. 1), these studies hypothesized that warm fluids migrated away from the Ouachita thrust front, resulting in maturity variations in the shales that are independent of depth of burial. Recent studies by Gale *et al.* (2007) and Papazis (2005) have also presented evidence for multiple veins that record a complex history of fluid alteration, suggesting that multiple fluid events occurred in the Barnett Shale.

This study presents palaeomagnetic data, supported by geochemical and petrographic results, from the Barnett Shale in the Fort Worth Basin that bear on the origin and timing of chemical remagnetization events. The study specifically tests the hypothesis that the Barnett Shale has been altered by externally derived fluids and that these processes caused remagnetization.

Data for this study were obtained from four subsurface oriented cores in Wise and Johnson Counties, Texas (Fig. 1). Palaeomagnetic and rock magnetic studies were undertaken to identify and characterize any chemical remanent magnetizations (CRMs) and determine the absolute timing of the events. Petrographic studies were conducted to characterize and determine the origin of the diagenetic features. In addition, geochemical results

*From*: ELMORE, R. D., MUXWORTHY, A. R., ALDANA, M. M. & MENA, M. (eds) 2012. *Remagnetization and Chemical Alteration of Sedimentary Rocks*. Geological Society, London, Special Publications, **371**, 89–106.
First published online August 22, 2012, http://dx.doi.org/10.1144/SP371.10

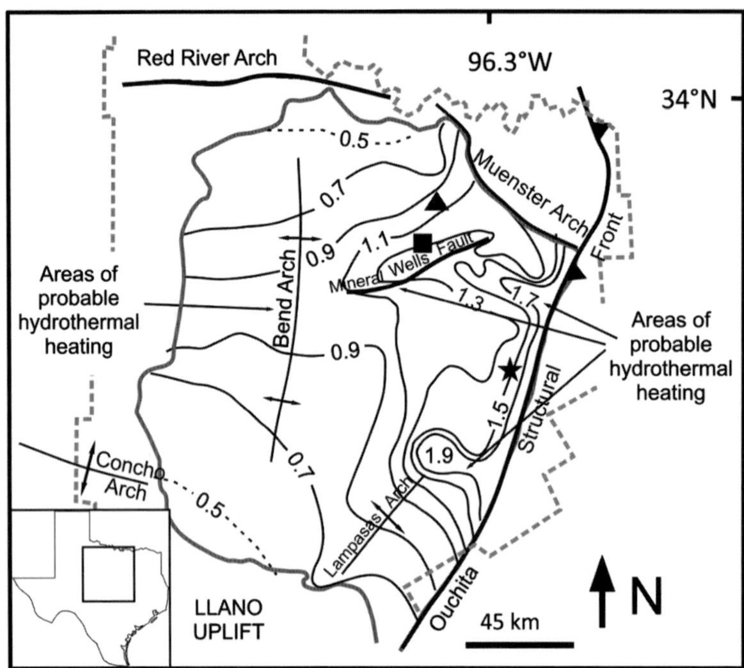



**Fig. 1.** Map of the Fort Worth Basin with lines of equal thermal maturity, as determined from a mean vitrinite isoreflectance (Ro) map for the Barnett Shale (based on data from Humble Geochemical Services, Humble, Texas). Inset map is an outline of Texas with a box which shows the area in the Fort Worth Basin map. Symbols represent the cores used in this study: the triangle is the Ernest Smith A 1; square is the Adams SW 7; and star shows the location of cores in Johnson County (Freddy Lynn 1 and Spencer Trussel 1). Modified from Pollastro *et al.* (2007).

(Dennie 2010; Deng 2011) provide evidence that the unit was diagenetically altered.

## Geological setting

The Barnett Shale was the first of now over 30 identified and potentially economically viable shale resource plays in the continental United States (Jarvie *et al.* 2007). The shale acts as both source and reservoir rock for production of natural gas. The Barnett Shale occurs in the subsurface of northern, central and western Texas, and is exposed in the Llano Uplift region of central Texas. The Barnett Shale is thickest near the Muenster Arch (over 350 m) and thins to less than 35 m over the Bend Arch and at the Llano Uplift (Pollastro *et al.* 2007; Jarvie *et al.* 2007). The basin is bordered to the north by the pre-Ouachita Muenster Arch and Red River uplifts, which are tectonically adjacent to the Southern Oklahoma Aulacogen (Thomas 1993). Palaeozoic rocks of the southern Fort Worth Basin onlap the Bend Arch to the west where they transition to carbonate rocks. The shale is in the thermal maturity window for hydrocarbon

generation in much of the Fort Worth Basin, with the most prolific areas located along the basin's 320 + km frontage with the Ouachita fold-thrust system and the Muenster Arch uplift (Bowker 2007; Jarvie *et al.* 2007).

The Fort Worth Basin began forming in the late Mississippian Period prior to the initiation of deformation associated with the Ouachita fold-thrust belt (Meckel *et al.* 1992). The Muenster Arch and Southern Oklahoma Aulacogen may have exerted an important early control on the deposition of the Barnett Shale. Reactivation of major basement reverse faults (Henry 1982) associated with the aulacogen during the early Mississippian Period (Johnson *et al.* 1988; Hale-Erlich & Coleman 1993; Erlich & Coleman 2005) may have formed a NW-oriented sub-basin parallel to the Muenster Arch, which became the Barnett Shale depositional basin prior to development of the Ouachita fold-thrust belt.

Many regional faults in the basin have a NE orientation and show north and west down displacement, although some basement faults have a NW trend (downthrown to the north and east). With initiation of down dropping of the foreland basin along the Ouachita fold-thrust belt in the Early to Mid-Pennsylvanian Period, deposition of the Pennsylvanian clastic molasse commenced. Changes in stress direction and rate of orogeny produced complex changes in fault and opening-mode fracture character as the Ouachita front evolved, including strike-slip displacements and related rotations of blocks as suggested by Erlich & Coleman (2005). These interpretations are supported by recent seismic analysis of basin faults (e.g. Sullivan *et al.* 2006).

The region contains Precambrian–Upper Cretaceous rocks (Fig. 2). Cambrian sandstones, shales and carbonate rocks unconformably overlie Precambrian granite and diorite basement (Pollastro *et al.* 2007). These are overlain by the thick Cambro-Ordovician Ellenburger Group dolomites, limestones and sandstones. A period of exposure and erosion in the Mid-Ordovician Period resulted in extensive karstification of the Ellenburger carbonates (Flippin 1982). Following this period of erosion, the Mid-Ordovician Simpson Group and Upper Ordovician Viola Group shales and limestones were deposited and were then eroded during a pre-Mississippian period of subaerial exposure.

The Barnett Shale is informally subdivided into the upper and lower Barnett Shale (Fig. 2) within the Fort Worth Basin (Pollastro *et al.* 2007). The Forestburg Lime is a muddy limestone that stratigraphically divides the Upper and Lower Barnett in NE parts of the basin. To the south of Fort Worth, the Barnett Shale is a carbonate-poor unit whereas Barnett-equivalent Chappel limestones occur in the shallower western parts of the basin (Bowker 2003; Pollastro *et al.* 2007).

The Barnett Shale is composed primarily of dark, organic- and silica–carbonate-rich mudstones. Detailed descriptions of the lithofacies and mineralogy of the unit (Papazis 2005; Jarvie *et al.* 2007; Milliken *et al.* 2007; Singh 2008; Loucks *et al.* 2009) show that the shale is heterogeneous in nature despite its relatively uniform appearance. Silica-rich organic (black) mudstone is commonly interbedded with carbonate-rich, pyritic and phosphatic mudstone. Overall, the Barnett contains less than one-third clay minerals by volume, with quartz (silica) being the largest mineral component. Carbonate content is highly variable by bed and is both depositional and diagenetic in origin. Silica is present in some zones as both very fine quartz silt and as siliceous cement. The silica cement is interpreted to be sourced from a combination of terrigenous and biogenic sources (Papazis 2005; Loucks & Ruppel 2007; Milliken *et al.* 2007). Silica diagenesis may have been a major factor in decreasing the porosity and permeability of the mudstones and siltstones during burial, as proposed in studies of other shales (e.g. Schieber *et al.* 2000).

Veins and vertical–subvertical fractures are present in the Barnett Shale and most are healed with multiple generations of calcite cement although silica, barite, sphalerite and pyrite are also present (Papazis 2005; Gale *et al.* 2007). Several workers (Bowker 2003, 2007; Pollastro *et al.* 2007) invoke externally derived fluids as a potential source of Barnett Shale variability in terms of maturity and producibility. It is not clear how the hypothesized

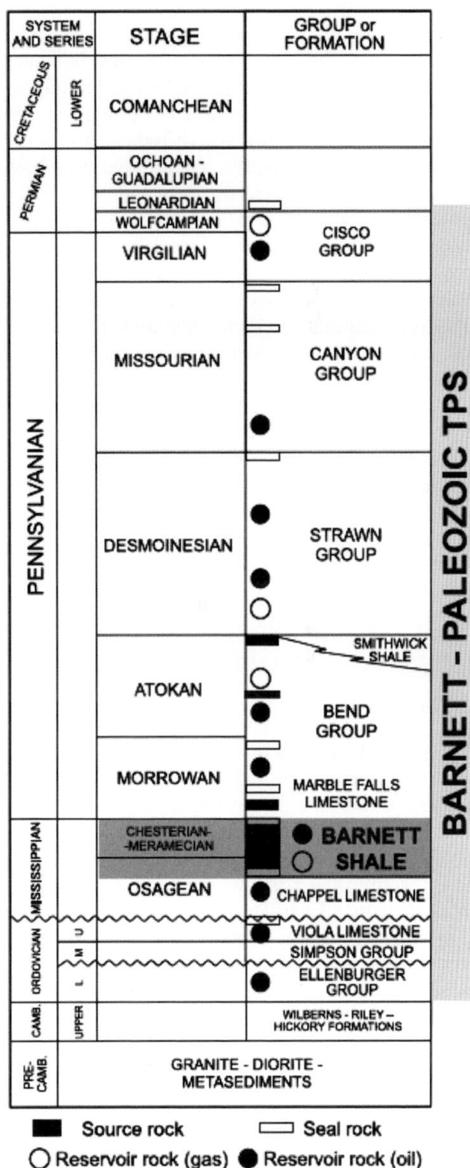


**Fig. 2.** Generalized stratigraphic column for the Fort Worth Basin. Modified from Pollastro *et al.* (2007).

regional-scale thermal alteration around faults (Fig. 1) is related to localized alteration in cores.

There have been several palaeomagnetic studies of the Barnett Shale. Howell & Martinez (1957) did not employ demagnetization techniques and the results are difficult to interpret. Kent & Opdyke (1979) sampled outcrops in central Texas and reported that that carbonate concretions in the Barnett Shale contain a characteristic magnetization with SE declinations and shallow inclinations. They interpreted the magnetization as forming early in the history of the rock and Mississippian in age. A better understanding of the apparent polar wander path (APWP) since the paper was published indicates that the pole position (latitude 49.1°N and longitude 119.2°E) for the magnetization is Permian in age (e.g. Van der Voo 1993), which indicates that the magnetization is secondary. There have also been several studies of associated units and other rocks in the region. A regional palaeomagnetic study was conducted on the outcrops of the Ellenburger carbonate rocks in central Texas and in the Franklin Mountains in west Texas (Haubold 1999). The study reported that the rocks contain a magnetization carried by authigenic magnetite, and also found a correlation between magnetization intensity, directions and lithofacies, suggesting a permeability control on remagnetization by migrating fluids (Haubold 1999). The pervasiveness of this secondary remanence decreased with distance from the Ouachita thrust in a manner consistent with an orogenic fluid event as the agent of remagnetization. Other studies have demonstrated that rocks in Oklahoma and Texas were remagnetized during the Ouachita orogeny (Lu *et al.* 1991; Elmore 2001).

## Methods

This study utilizes four oriented well cores retrieved by drill stem coring tools oriented by standard industry down-hole magnetic survey and scribe methods (Nelson *et al.* 1987). Two cores, the Devon Adams SW 7 (ASW) and Ernest Smith A 1 (ES), are from the north-western part of the basin (Wise County, TX). The ASW core contains a 92 m Barnett section (upper and lower Barnett) overlying a relict Viola Limestone interval and brecciated (karst) Ellenburger dolomite (Papazis 2005; Loucks & Ruppel 2007). The ES core (122.6 m of Barnett Shale) is located about 32 km closer to the Muenster Arch than the ASW, and lacks significant evidence of nearby regional faulting or karsting prior to or during Barnett deposition. The other two cores, the Devon Freddy Lynn 1H (FL) and Devon Spencer Trussel, are from the eastern part of the basin (Johnson County TX) near the Ouachita thrust (Fig. 1). These two cores were combined so a

complete section of the Barnett Shale (101.2 m) could be sampled and are referred to together as the South Core (SC).

Samples of representative rock types (Fig. 3) were collected for palaeomagnetic and rock magnetic analysis. In addition, to test for a connection between fractures and magnetization, samples were collected in intervals adjacent to veins in the ASW and ES cores (Fig. 3). Fracture orientations were identified from formation microimager logs (FMI) and oriented core. Samples collected from within a 3 m (10 feet) window around the fractures were grouped as one site. A total of 139 samples were analysed from the Barnett Shale cores.

The cores were sampled with a fixed variable-speed and water-cooled drill press with a non-magnetic core bit located at the Oklahoma Petroleum Information Center Core Facility, Norman, Oklahoma. Most cores were sampled normal to the corrected tool-face azimuth (corrected true north) as calculated from the reference and scribe data. Because measured wellbore plunges are steep (between 1 and 3° from vertical) and beds are near horizontal, this mode of sampling approximates a true *in situ* sample orientation.

The orientation of the ES core was more problematic because the core was slabbed to maximum bedding dip as opposed to the orientation scribes. Samples were drilled normal to the slab face as opposed to the orientation scribe lines. The samples were reoriented after measurement based on the present-day field viscous remanent magnetization (VRM) component as isolated by low-temperature thermal demagnetization (Van der Voo & Watts 1978; Suk *et al.* 1993; Van Alstine & Butterworth 2002).

The samples were cut to specimens of standardized length (2.2 cm) by removing the outer part of the sample (core) to reduce the possibility of contamination. The natural remanent magnetizations (NRMs) were measured using a 2G-Enterprises cryogenic magnetometer with DC SQUIDs (superconducting quantum interference device) in the shielded palaeomagnetic laboratory at the University of Oklahoma. The specimens were subjected to low-temperature demagnetization (LTD) by immersion in liquid nitrogen, then returned to room temperature in a zero field in order to test for and remove the unstable remanence residing in multidomain (MD) magnetite grains (Dunlop & Argyle 1991). The specimens were subjected to two LTD treatments.

The specimens were then subjected to stepwise thermal demagnetization in 22 steps from 100 °C to 580 °C using an ASC Scientific Thermal Specimen Demagnetizer. Some specimens were initially heated to 680 °C in 27 steps to test for remanence in hematite; however, the NRM was unblocked

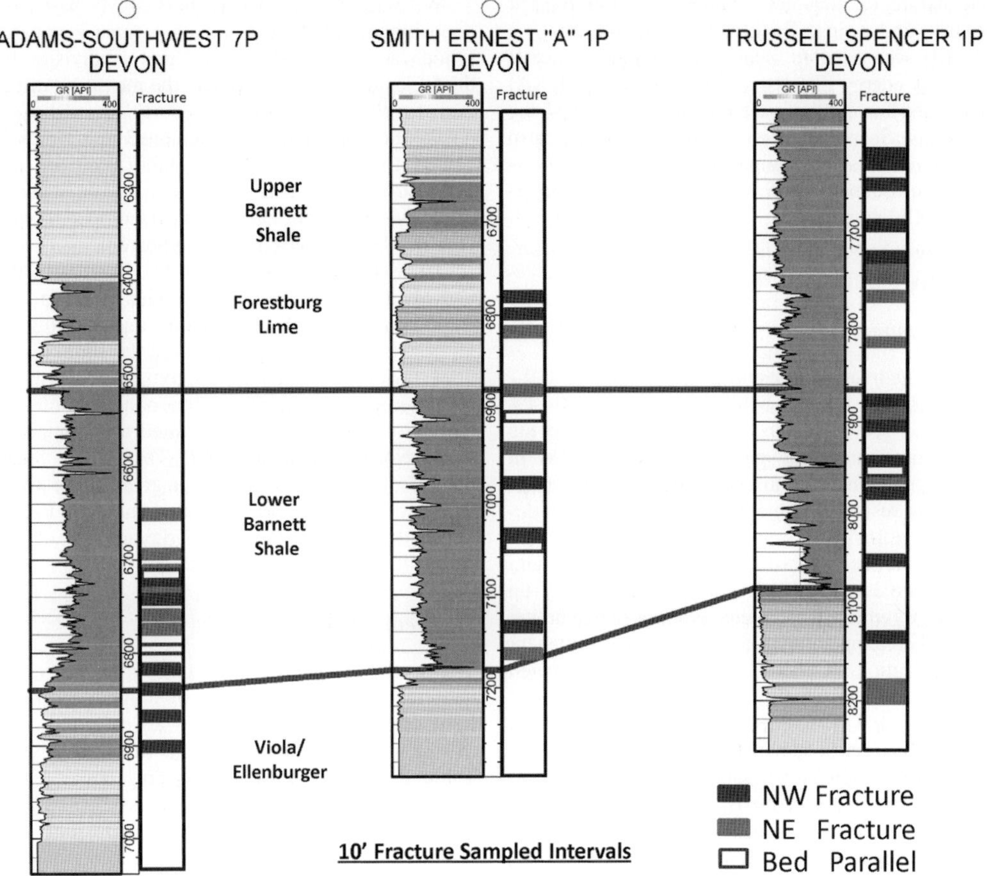



**Fig. 3.** Gamma ray curves for the three cores (darker colours are predominantly organic-rich mudstones and the lighter colours are progressively more carbonate-rich mudstones, limestones and dolostones) with a rectangle to the left which shows the locations of fractures and their orientations. Lower solid line is the contact between the Ellenburger Group and the Barnett Shale, whereas the upper dark line shows the contact between the lower and upper Barnett Shale.

below 580 °C in all specimens. Additional steps were added at low temperatures (100–200 °C) on some runs to help delineate the VRM for sample orientation purposes in the ES core.

Zijderveld (1967) diagrams were examined to determine principal components (Kirschvink 1980) of the magnetization. Specimens with maximum angular deviation (MAD) of values less than 15° were included in the calculation of the mean statistics (Fisher 1953). In order to investigate the magnetic mineralogy of the samples, coercivity spectrum analysis (Kruiver *et al.* 2001) was conducted on the isothermal remanent magnetization (IRM) acquisition data. The specimens were AF demagnetized up to 120 mT using a 2 G Automated Degaussing System and then subjected to a 26-step IRM acquisition up to 2500 mT using an ASC

Scientific Impulse Magnetizer. The IRM curves were analysed using the IRMUNMIX 2.2 software program (Heslop *et al.* 2002) and cumulative log-Gaussian (CLG) analysis was performed with IRM-CLG 1.0 (Kruiver *et al.* 2001) to model individual coercivity contributions (Heslop *et al.* 2004).

Core-samples and specimens were described with particular emphasis on recognizing diagenetic features. Matrix and vein cements were selected for petrographic analysis. Petrographic analysis using both transmitted and reflected light microscopy was conducted on 134 thin sections produced from both palaeomagnetic specimens and from samples specially cut from the core.

An automated Cameca SX50 electron probe microanalyser (University of Oklahoma Electron Microprobe Laboratory) was used to investigate the nature of diagenetic minerals in the Barnett matrix and veins. Backscatter electron imaging (BSEI) was used to image the opaque minerals whereas energy dispersive x-ray analysis (EDXA) was used for qualitative elemental analysis of the minerals. Scanning electron microscopy was also conducted on select samples using a scanning electron microscope (ZEISS DSM-960A) at the University of Oklahoma.

## Results and interpretations

### Palaeomagnetism

The LTD treatment resulted in the loss of 0.9–19.7% of the NRM on the first cycle and 0–4% after the second cycle. Multi-domain magnetite does contribute some part of the total NRM in some specimens, but most of the remanence is interpreted to reside in single-domain or pseudo-single-domain grains.

Thermal demagnetization isolated a component interpreted as a VRM at low temperatures (20 °C up to 280 °C) which is interpreted to reside in magnetite. At higher temperatures, a characteristic remanent magnetization (ChRM) with south–SE-directed declinations and shallow inclinations is removed (Fig. 4a–c). The magnetization is fully unblocked by 450 °C in most specimens (Fig. 4a–c), although maximum unblocking temperatures extend up to 480 °C in a few specimens. These results suggest that the ChRM resides in magnetite. Thirty-nine of the specimens analysed were not included in the analysis because of low magnetization intensities, unstable decay or spurious directions.

The ChRMs from the specimens in the ASW and SC cores yield south–SE declinations and generally shallow-up inclinations (Fig. 5a, b). For the ES core, the magnetization removed below 280 °C was used to orient the core. This magnetization in the ES core has a mean inclination of 68° which is close to the present-day field inclination of 62°, suggesting that it is not a vertically oriented drilling-induced magnetization (e.g. Audunsson & Levi 1989; de Wall & Worm 2001). It is interpreted as a present field VRM. The VRM declination was 'corrected' back to the present-day field direction (N4°E) and ChRM was corrected based on the VRM correction (Fig. 6a, b). Based on this analysis, the specimen directions from all cores have shallow inclinations with a streak of south–southeasterly declinations (Figs 5 & 6).

We initially subdivided the ChRMs based on depositional lithofacies and/or gamma ray parasequences (e.g. Singh 2008). Upon subdivision by the facies in the Barnett Shale, the measured directions for the site means were poorly grouped. As a result, the approach was abandoned and we concluded that the thin Barnett lithofacies are too subtle to resolve with a palaeomagnetic dataset. The ChRM directions were also compared with core depths but no trends were apparent.

To explore the streaking of directions, the ChRM directions were subdivided into groups starting with the SE directions and moving in 10° declination increments toward the south directions (Table 1). Each group consisted of 12–23 specimens (Table 1). The corresponding pole for each group of data was calculated and compared to the North American APWP. The Barnett Shale poles from all wells plot on a great circle track that starts about 20° to the east of the APWP, and moves toward the Permian–Triassic part of the path (Fig. 7).

### Magnetic mineralogy

The results of the CLG analysis suggest that magnetite is the primary carrier of magnetization in most of the carbonate rocks, with potential contributions by pyrrhotite in some samples (Fig. 8, Table 2). If pyrrhotite does carry some remanence, it has the same direction as the magnetite component because the Zijderveld diagrams for the specimens show only one component that is unblocked between 280 °C and 480 °C. The hematite that is present in some specimens (Table 2) is not interpreted to carry a stable remanence.

### Petrography

Evidence of post-depositional alteration in the Barnett Shale consists of authigenic minerals such as calcite and dolomite in the fine-grained matrix and in cemented fractures (Fig. 9a). These minerals replace allochems in the matrix. As noted by many previous workers, authigenic silica is common in many forms in the Barnett Shale (Papazis 2005; Loucks & Ruppel 2007; Milliken *et al.* 2007). Hydrocarbons are also found in inclusions in some silica and calcite cements, and as solid residue (pyrobitumen) coating some diagenetic minerals.

**Fig. 4.** Representative Zijderveld orthogonal demagnetization diagrams and normalized thermal decay curves for specimens from the (**a**, **b**) ASW and (**c**) SC cores. Closed symbols represent the horizontal projection; open symbols represent the vertical projection and do not directly show the inclination.

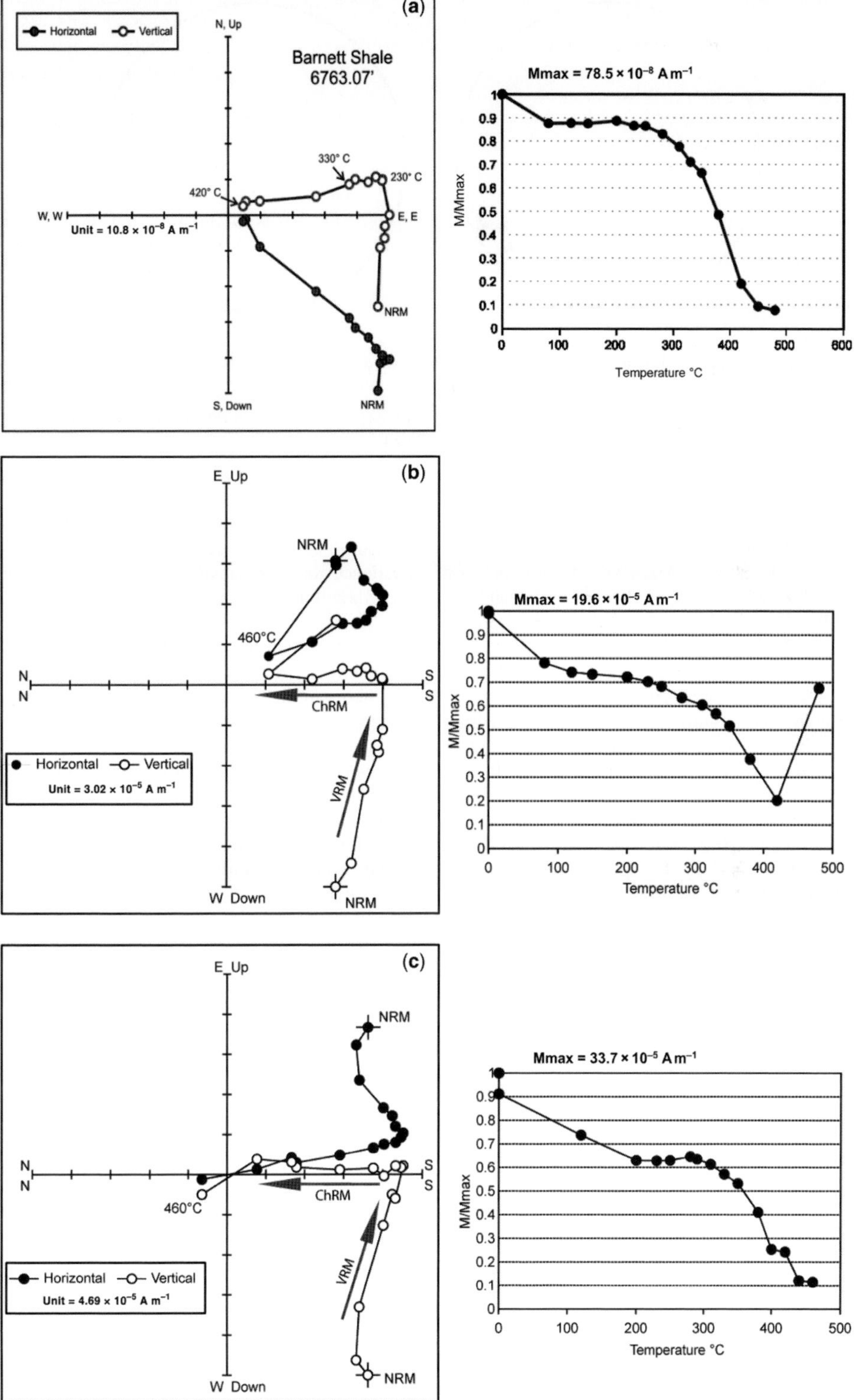
Horizontal
Vertical
N, Up
Barnett Shale
6763.07'
330° C
230° C
420° C
W, W
E, E
Unit = 10.8 × 10⁻⁸ A m⁻¹
NRM
S, Down
NRM
(a)
Mmax = 78.5 × 10⁻⁸ A m⁻¹
M/Mmax
Temperature °C
E_Up
NRM
460°C
N
N
S
S
ChRM
Horizontal
Vertical
Unit = 3.02 × 10⁻⁵ A m⁻¹
VRM
W Down
NRM
(b)
Mmax = 19.6 × 10⁻⁵ A m⁻¹
M/Mmax
Temperature °C
E_Up
NRM
N
N
S
S
460°C
ChRM
Horizontal
Vertical
Unit = 4.69 × 10⁻⁵ A m⁻¹
VRM
W Down
NRM
(c)
Mmax = 33.7 × 10⁻⁵ A m⁻¹
M/Mmax
Temperature °C

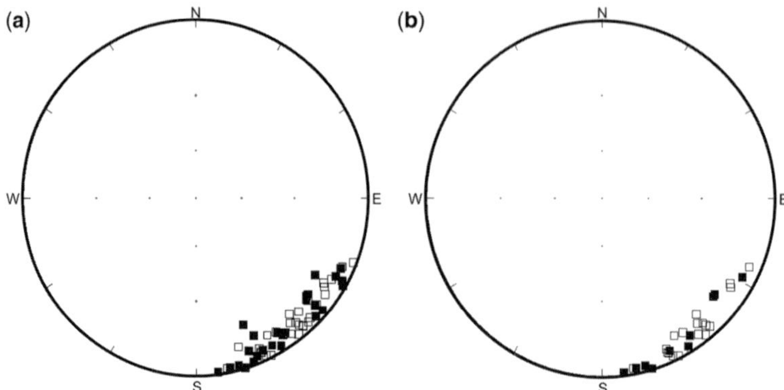

**Fig. 5.** Equal area stereonet showing distribution of specimen direction data from the Barnett Shale in the (**a**) ASW and (**b**) SC cores. Closed symbols indicate positive (lower projection) inclinations and open symbols represent negative (up) inclinations.

Sulphides are common in the Barnett Shale. Pyrite is the most common sulphide, and occurs as framboids in beds (Fig. 9b) and in dodecahedral, cubic and rarely octahedral forms replacing allochems and in fractures. Marcasite is also common in some intervals. Cubic pyrite and marcasite are probably formed late in the diagenetic sequence (Beier & Feldman 1991). Sphalerite (Fig. 9b) occurs with barite and pyrite in phosphatic intervals and in zones with a high gamma ray response.

Sulphates are common in some veins and fractures in the Barnett Shale. Barite occurs as cement in the matrix and as a fracture fill (Fig. 9c). It does not appear to be a 'scale' product of drilling. Barite-celestine is most common in subvertical or stratiform veins (Fig. 9c) where it is associated with albite, quartz, pyrite, sphalerite and calcite (Fig. 9d). Anhydrite is found in fractures with a late phase of calcite, euhedral vein-filling pyrite and celestine (Fig. 9c) as well as in the matrix. Some prior studies have suggested the sulphate can be a product of drilling via oxidation of sulphides in carbonate rocks (e.g. Dean & Ross 1976). Evidence that the anhydrite is a natural cement in the Barnett includes its location in the interior of the cores and the fact that it is cut by younger pyrite and other minerals unlikely to form as a scale during core retrieval.

Other authigenic minerals are present in carbonate- and phosphate-rich intervals and suggest burial or fluid diagenesis of the shales. For example, rare earth element (REE)-enriched (cerium, thorium- and yttrium-rich) monazite as well as authigenic illite, authigenic potassium feldspar, albite, titanite and chromite are found in some localized zones. Some zones contain abundant phosphate that appears to have been remobilized and precipitated into fractures.

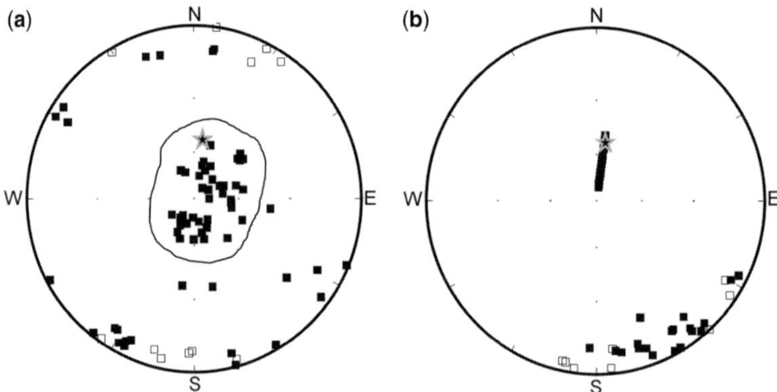

**Fig. 6.** (**a**) Unoriented, uncorrected specimen directions from the ES core, with the present-day field direction indicated by a star. (**b**) Declination-corrected directions of the VRM (circles) and ChRM (squares). The line in VRM data shows the remaining variation in uncorrected inclination over the dataset.

**Table 1.** *Ten-degree means from streaked ChRM directional data. Each pole corresponds to a separate mean starting with the SE directions and moving in 10° declination increments toward the south directions. N, number of specimens; Dec, declination; Inc, inclination; k, a measure of grouping; a95, 95% cone of confidence; $P_{lat}$, pole latitude; $P_{long}$, pole longitude; dp, dm, semiaxes of 95% cone of confidence of pole*

| Statistical site | Site mean ChRM directions | | | | | Palaeopoles | | | |
|---|---|---|---|---|---|---|---|---|---|
| | Dec° | Inc° | $a_{95}$ | k | N | $P_{lat}$ | $P_{long}$ | dp | dm |
| −130 | 122.1 | −1 | 2.9 | 163.3 | 16 | 26 | 148.3 | 1.5 | 2.9 |
| 130–140 | 135.9 | −2.1 | 2.6 | 172.6 | 19 | 37.5 | 139.2 | 1.3 | 2.6 |
| 140–150 | 143.6 | −0.8 | 3 | 108.2 | 23 | 42.5 | 131.4 | 1.5 | 2.9 |
| 150–160 | 155.7 | 2.2 | 3.8 | 89.6 | 17 | 48.5 | 116.3 | 1.9 | 3.8 |
| 160–170 | 164.9 | 7.6 | 3.8 | 129.2 | 12 | 50.4 | 102 | 1.9 | 3.8 |
| 170– | 179.9 | 7.1 | 5.1 | 66.9 | 13 | 55.4 | 78.1 | 2.6 | 5.1 |

## Fractures and ChRMs

Fractures of varying width, orientation and planarity are present in the Barnett Shale. Most fractures are vertical, tectonic, have very narrow aperatures and are filled with calcite or with calcite, silica and minor barite (Fig. 9a, c, d). Some fractures and veins, especially near known faults, have kinematic apertures >1 cm in diameter that are open and partially mineralized. Many fractures terminate

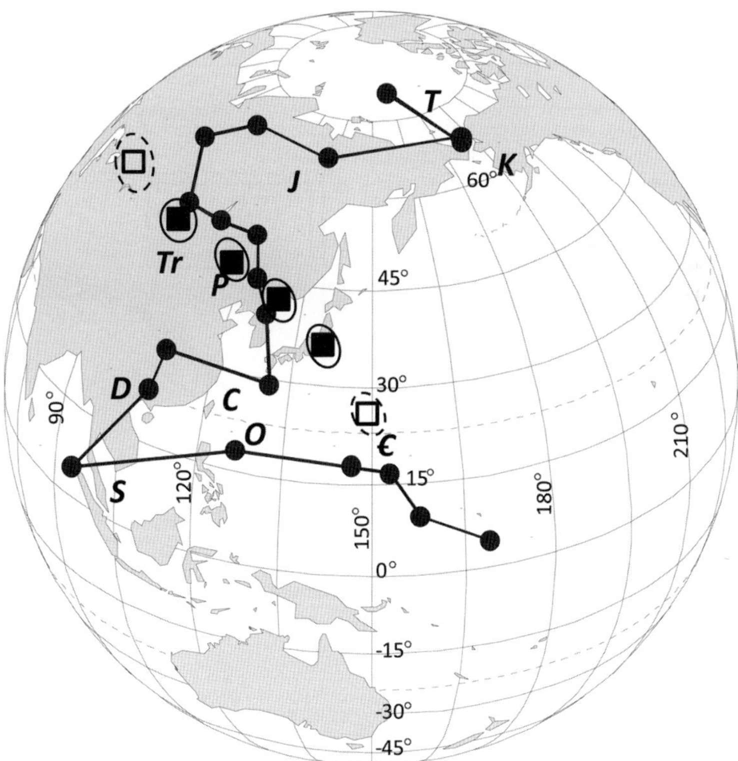


**Fig. 7.** Barnett Shale poles (with 95% error ellipses) derived from the streak of directions are plotted on the apparent polar wander path of Van der Voo (1993). Each pole corresponds to a separate mean starting with the SE directions and moving in 10° declination increments towards the south directions (Table 1). The two dashed end-member groupings include greater than 10° means but are included for completeness.

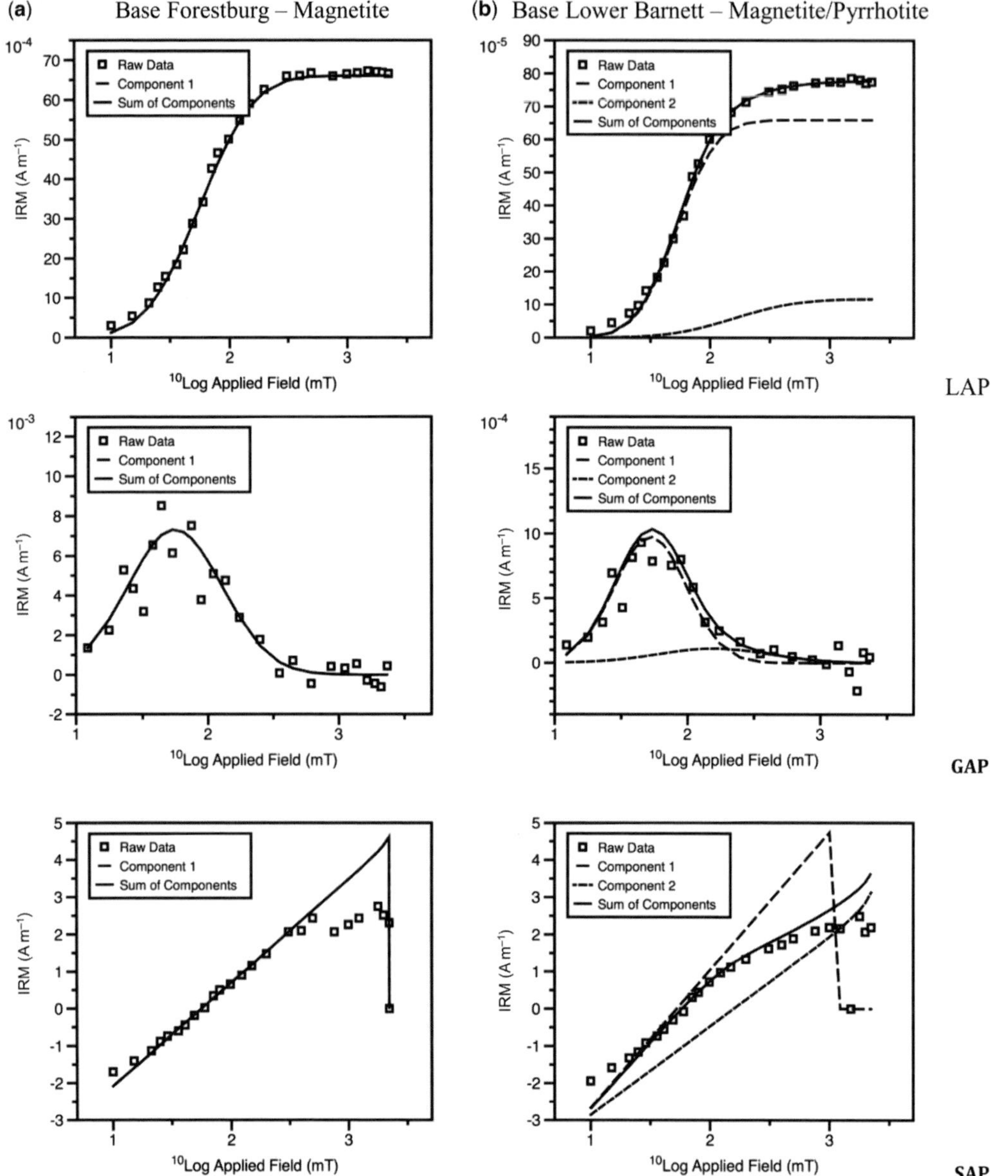



**Fig. 8.** Coercivity spectrum analysis for representative samples from the ASW core. (**a**) Base of Forestburg dominated by magnetite and (**b**) base of lower Barnett suggesting the presence of both magnetite and pyrrhotite. See Table 2 for the coercivity spectrum analysis results for all samples.

at lithologic contacts. Most are tensile- (natural opening; Gale *et al.* 2007) mode fractures although some NE-oriented fractures show evidence of shear.

The cores were subdivided into two fracture groups, ASW and ES representing Wise County and the SC representing Johnson County. The two areas are 110 km apart and the fracture orientations between the two areas are different, although internal angles between sets are similar. Four general fracture trends can be defined in the Barnett Shale as follows:

(1)   Bedding-parallel or stratiform fractures are minor in abundance but have distinctive mineral assemblages consisting of sulphate-rich

**Table 2.** *Magnetic mineral from coercivity spectrum analysis of ASW samples; SIRM: saturation isothermal remanent magnetization; $B_{1/2}$ is the field where half of the SIRM is reached*

| Sample | Depth | Components | Contribution (Percent) | SIRM (A m$^{-1}$) | $B_{1/2}$ (mT) | Magnetic mineralogy |
|---|---|---|---|---|---|---|
| 21802 | 6418.09 | 1 | 100 | 66.1 | 56 | Magnetite |
| 60612 | 6654.12 | 1 | 94 | 4.7 | 50 | Magnetite |
| | | 2 | 6 | 0.3 | 398 | Hematite |
| 71011 | 6726.1 | 1 | 85 | 6.6 | 53 | Magnetite |
| | | 2 | 15 | 1.2 | 159 | Pyrrhotite/Hematite |
| 80922 | 6782.0 | 1 | 81 | 12.8 | 53 | Magnetite |
| | | 2 | 19 | 3 | 163 | Pyrrhotite/Hematite |

cements (barite and celestine) as well as euhedral albite, quartz and pyrite. These are often found with phosphatic, pyritic intervals. They are associated with hot gamma ray signals on well logs that are identified as parasequence boundaries which are interpreted as maximum flooding surfaces (Singh 2008).

(2) A second minor fracture set consists of north–NE-trending subvertical veins that contain a similar mineralogy to the bedding-parallel fractures. These fractures show evidence of shear motion and reactivation. These fractures contain carbonate, sulphate and sulphide cements and have the most complex healing history, largest apertures and the most heterogeneous mineralogy of the fracture sets in the cores.

(3) NW fractures are narrow, vertical features and are filled with one or more generations of calcite and silica cements and only minor sulphate minerals. Hydrocarbons are found as inclusions within silica and also coating silica cements, suggesting this fracture set formed concurrently with hydrocarbon generation and migration. En echelon fracturing and fracture tip interference is seen in this set and suggests a natural hydraulic fracturing mechanism during advanced burial diagenesis.

(4) A second vertical fracture set has narrow apertures and a NE direction. Unlike NW fractures, evidence of shear motion is present in this set.

One hypothesis is that the fractures may exert a control on the CRMs by providing conduits for fluids which would be an agent of chemical remagnetization. To test this hypothesis, fracture orientation/fracture-fill data were compared with the ChRM data for specimens around veins in the ASW and ES cores (Fig. 3).

The site mean directions (Fig. 10; Table 3) from around the different fracture sets in the ASW and ES cores show a general correlation when grouped by fracture type (Table 4). The ChRM directions from the bedding-parallel fractures have SE declinations and shallow inclinations ($D = 130.2°$, $I = 4.3°$; Fig. 10a) with a pole that falls off the mid-Pennsylvanian part of the APWP (Fig. 11). Subvertical and NE fractures (Fig. 10a) contain a similar distribution of site directions ($D = 139.8°$, $I = -0.6°$) to the bedding-parallel fractures and the pole is near the late Pennsylvanian part of the path (Fig. 11). Although there are only a few specimens and the grouping is relatively poor, matrix sites (Fig. 10b; Tables 3 & 4) yield south–SE directions ($D = 151.7°, I = 11.3°$) with a pole on the late Pennsylvanian–Permian part of the path (Fig. 11). The sites around NW vertical fractures (Fig. 10c; Tables 3 & 4) show more south-directed ChRMs ($D = 159.3°, I = 3.9°$) with a Permian–early Triassic pole (Fig. 11). The vertical NE fractures (Fig. 10d) also have south–SE directions ($D = 161.2°$, $I = 9.5°$) with a late Permian–early Triassic pole (Fig. 11; Tables 3 & 4).

The palaeomagnetic data from some of the fracture sets are under-represented due to the lack of dense sampling; the interpretations should therefore be considered as preliminary. The inferred CRM ages need to be tested by additional sampling and palaeomagnetic analysis. One problem with the vein tests conducted here is that the sampling was limited by the core width and by sample spacing above and/or below a vein. More sampling above and below a fracture would partially address this issue and would yield a statistically more useful dataset.

## Discussion

The ChRMs in the Barnett Shale are interpreted to reside in magnetite, based on laboratory unblocking temperatures between 320 °C and 580 °C and the results of coercivity spectrum analysis. Rock pyrolysis and vitrinite reflectance data (e.g. Jarvie *et al.* 2007; Pollastro *et al.* 2007) indicate that palaeotemperatures in the basin are between 100 °C and 180 °C. Based on comparisons with the unblocking-temperature–relaxation-time curve of Pullaiah *et al.*

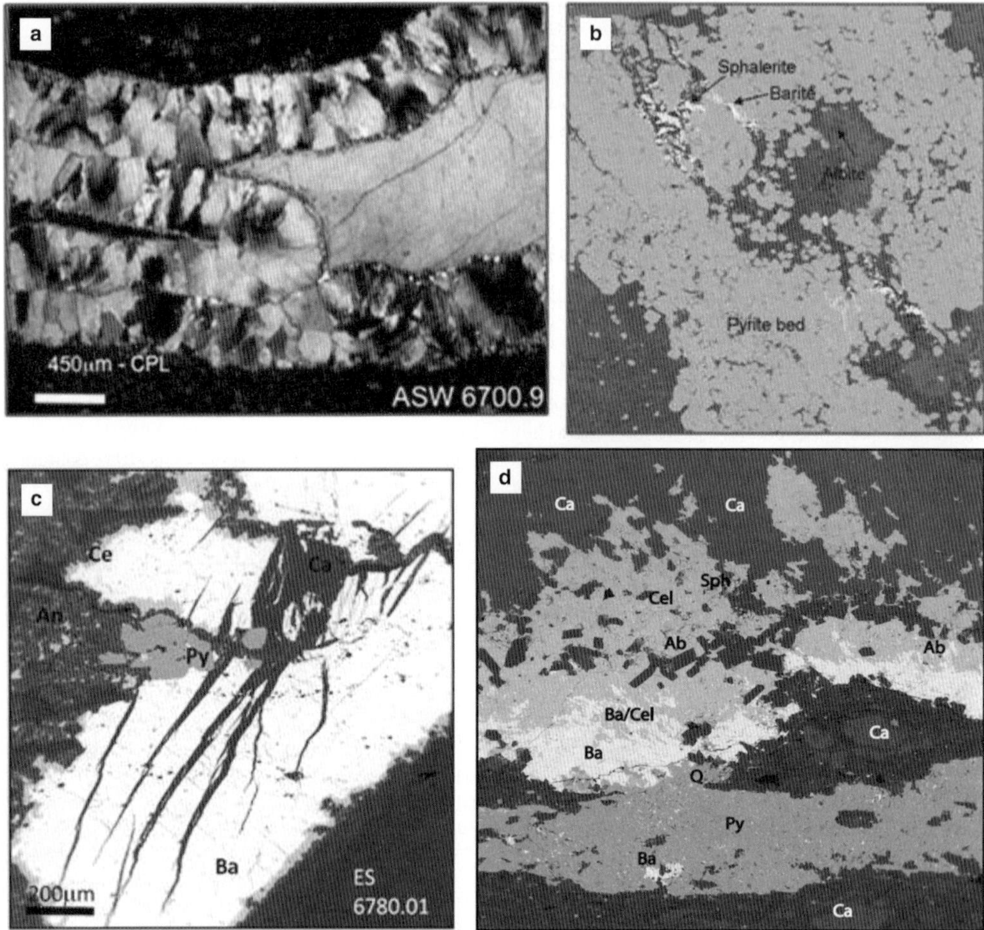



**Fig. 9.** Representative photomicrographs of diagenetic features in the Barnett Shale. (**a**) Vein with outer and inner calcite separated by a thin silica cement. (**b**) Bedded pyrite with internal fracture, filled with sphalerite and lesser Pb–Cu sulphides, and barite. Dark grey mineral is albite (up to 20% of the rock volume). Image is 395 µm × 395 µm (ASW 6603). (**c**) Multi-generational fracture in the Forestburg Lime of the ES core. Celestine (CE; outer rim) and barite (Ba) make up the bulk of the vein. Late anhydrite (An) and pyrite (Py) crosscut the vein. A late calcite cement (Ca) 'refractures' the barite, celestine and anhydrite along cleavage planes and between the barite and matrix. (**d**) SEM image of a stratiform fracture fill/deposit (ASW 6603). The sample coincides with a very high gamma ray spike, a pyritic and a phosphatic shale interval. Barite–Celestine appears to replace calcite in the vein and matrix, which contain an early generation of quartz and authigenic albite (790 × 790 µm). Minerals Ca, early calcite vein fill; Q, authigenic quartz; Ab, eudedral plagioclase; Cel, Celestine; Ba, Barite; Py, late bedded pyrite; Sph, sphalerite. (a) cross polarized light; (b–d) BSEI images.

(1975), the maximum unblocking temperature of 480 °C is too high for the ChRMs to be thermo-viscous remanent magnetizations. The ChRMs are therefore interpreted as CRMs residing in magnetite.

The directions from all specimens define a streak from *c.* 120° to 190°. Both orientation techniques produce the same distribution of directions, which suggests that the distribution is real. The poles from the more SE part of the streak lie to the east of the APWP whereas the poles for the more

southerly directions fall on the Permian–Triassic part of the path. One explanation for the more east-directed ChRMs is that an early Pennsylvanian CRM could have been acquired in a relatively short period of time that did not average out secular variation, resulting in a pole that falls off the path. Another possibility is that the area sampled by the drill cores was either locally or regionally tectonically rotated. A 15–20° anticlock-wise rotation would account for the more easterly

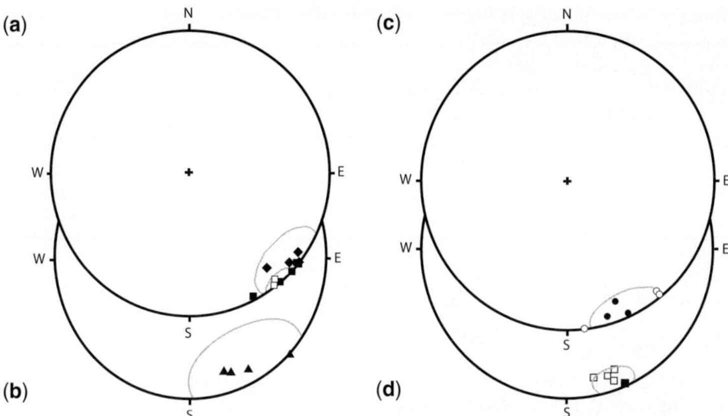

**Fig. 10.** Palaeomagnetic site means for the different fracture groupings in the ASW and ES cores. (**a**) Diamonds are bedding-parallel (stratiform) fractures and squares are NE and subvertical fractures; (**b**) triangles are matrix only; (**c**) circles are vertical NW tectonic fractures; and (**d**) squares are subvertical–vertical and NE shear fractures.

poles (Fig. 7). If the SE-oriented ChRMs are corrected for the anticlockwise offset, the corresponding pole falls on the late Pennsylvanian–early Permian part of the APWP. The Fort Worth Basin is interpreted to have formed through oblique continent–continent collision that began in the NE corner of the basin and progressed to the SW (Walper 1982). This would result in an anticlockwise 'torque' on the basement blocks as the orogenic front swung past the end of the Muenster

**Table 3.** *Results of grouping specimen data into site means around fracture sites for the ASW and ES cores Group is the facture orientation (MT, matrix; NE1, NE subvertical fractures; NE2, NE vertical fractures; BP, bedding parallel; NW, NW vertical fractures). Note that the total n is larger than the number of specimens analysed for each core because some specimens are used more than once in calculation of site means if they occurred within the 10 foot window around the fractures*

| Group | Core | Depth | Dec | Inc | $a_{95}$ | n | k |
|---|---|---|---|---|---|---|---|
| MT | ASW | 6656 | 133 | 0.4 | 5.9 | 4 | 246.3 |
| MT | ASW | 6727 | 162.8 | 17.3 | 7.5 | 2 | 1104.6 |
| MT | ASW | 6734 | 159.7 | 15 | 15.7 | 3 | 62.9 |
| NW | ASW | 6723 | 158.2 | 14.5 | 14.2 | 3 | 76.7 |
| NW | ASW | 6745 | 140.9 | −2.9 | 9.6 | 5 | 65.0 |
| NW | ASW | 6771 | 140.6 | −4.5 | 8 | 4 | 133.1 |
| BP | ES | 6882 | 140.7 | 15.2 | 21.5 | 8 | 7.6 |
| BP | ES | 7177 | 125.9 | 6.4 | 15.6 | 4 | 35.5 |
| NE2 | ASW | 6582 | 129.3 | 1.7 | 20.1 | 2 | 130.0 |
| BP | ASW | 6603 | 128.5 | −1.6 | 3.5 | 7 | 295.5 |
| NE2 | ASW | 6651 | 133 | 0.4 | 5.9 | 4 | 246.3 |
| BP | ASW | 6700a | 125.9 | −2.8 | 9.5 | 3 | 169.3 |
| NE2 | ASW | 6745 | 140.9 | −2.9 | 9.6 | 5 | 65.0 |
| NE2 | ASW | 6771 | 140.6 | −4.5 | 8 | 4 | 133.1 |
| NE2 | ASW | 6784 | 142.6 | −1.6 | 14.6 | 5 | 28.5 |
| NW | ASW | 6871 | 163.5 | 5.9 | 5.8 | 4 | 251.2 |
| NW | ES | 7166 | 173.6 | −0.7 | 13 | 10 | 14.8 |
| NE2 | ES | 6850 | 152.4 | 3.4 | 15.2 | 5 | 26.3 |
| NE1 | ES | 6727 | 168.3 | 13.3 | 19.4 | 4 | 17.8 |
| NW | ASW | 6522 | 179.2 | 10.5 | 8.2 | 6 | 68.0 |
| NE1 | ASW | 6700b | 162.1 | 11.9 | 9.6 | 3 | 165.4 |
| NE1 | ASW | 6723 | 158.2 | 14.5 | 14.2 | 3 | 76.7 |
| NE1 | ASW | 6744 | 156.5 | −1.9 | 8.3 | 4 | 124.0 |

**Table 4.** *Palaeomagnetic site means for fracture intervals, Wise County cores (N refers to sites in Table 3)*

| Site | Dec° | Inc° | $a_{95}$ | $k$ | $N$ | $P_{lat}$ | $P_{long}$ | dp | dm |
|---|---|---|---|---|---|---|---|---|---|
| 1. Bed parallel | 130.2 | 4.3 | 12.4 | 55.7 | 4 | 31.3 | 145.5 | 6.2 | 12.4 |
| 2. NE subvertical | 139.8 | −0.6 | 7.1 | 89.4 | 6 | 40 | 139.6 | 3.6 | 7.1 |
| 3. Matrix only | 151.7 | 11.2 | 28.9 | 19.2 | 3 | 42.9 | 122.3 | 14.9 | 29.3 |
| 4. NW vertical | 159.3 | 3.9 | 15.1 | 20.7 | 6 | 49.9 | 115.4 | 7.6 | 15.1 |
| 5. NE vertical | 161.2 | 9.5 | 10.5 | 77.5 | 4 | 48.1 | 111 | 5.4 | 10.6 |

Arch (Erlich & Coleman 2005). Structural interpretations based on seismic data (Sullivan *et al.* 2006) suggest a rotation of stress during development of the basin that may support the block rotation hypothesis. An alternative interpretation which does not involve regional block rotation is that the seismic data are the result of a linked system of coalesced, collapsed palaeocaves (McDonnell *et al.* 2007). Detailed structural analyses of the Fort Worth Basin, including more regional palaeomagnetic data, is needed to test the hypothesized rotation.

Although tectonic rotation around a horizontal axis may account for the east–SE-directed part of streak, the origin of the more south-directed ChRM directions is a crucial issue. Two or more components that are not resolved during demagnetization could cause a streak of directions. This is possible, although the ChRMs display linear decay on the orthogonal projections and this suggests the presence of only one ancient component. Another possibility is a mixing trend between two or more CRMs acquired between Pennsylvanian and Triassic time (Fig. 7).

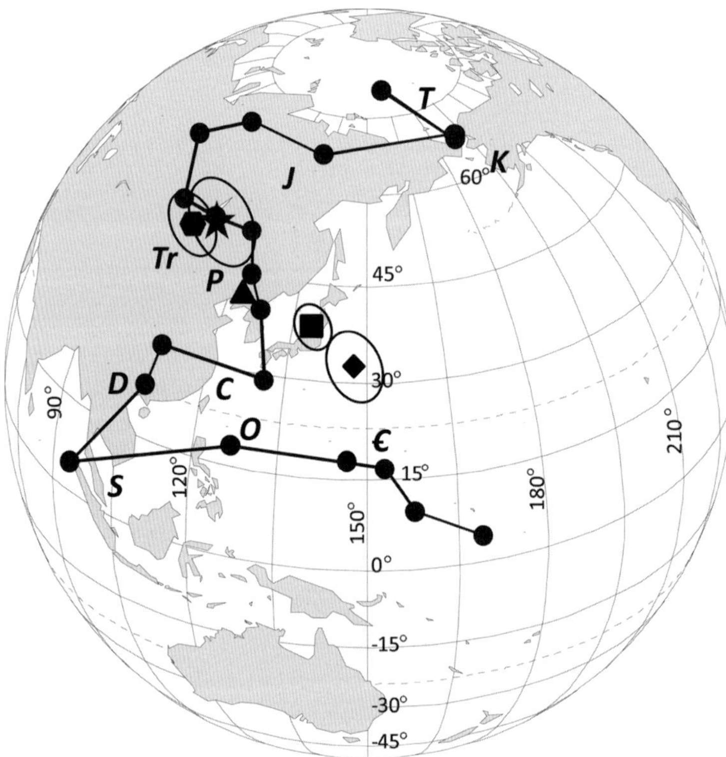


**Fig. 11.** Palaeopoles with 95% error ellipses for the sites around fractures plotted on the APWP. Diamond is bedding-parallel (stratiform) deposits; dark square is NE and subvertical fractures; triangle is matrix only; star is vertical NW tectonic fractures; octagon is vertical and NE fractures. The rock near vertical fractures is interpreted to carry the youngest diagenetic remagnetizations whereas the rocks containing stratiform and subvertical fractures appear to preserve older remagnetizations. The error ellipse for the matrix sites is not included because it is very large (Table 4).

There appears to be a connection between the directions of the CRMs and fracture orientation. Fluids in the fractures and matrix, either sourced internally or externally, are hypothesized to have been agents for chemical remagnetization. After correcting for the *c.* 15–20° anticlockwise rotation, the specimens close to the NE subvertical and bedding-parallel veins yield mid–late Pennsylvanian CRM directions. These NE subvertical and bedding-parallel veins contain sulphates, albite and sulphides such as sphalerite. Most cements from the ASW and ES cores contain slightly-to-moderately elevated $^{87}$Sr/$^{86}$Sr values (average or ave = 0.7082, $n = 7$, standard deviation or Stdev = 0.00012; Dennie 2010) compared to coeval values of 0.7079–0.7080 (MacArthur *et al.* 2001). The CRM is interpreted as having formed during burial diagenesis and remobilization of early syndepositional cements that formed on a sediment-starved stagnant maximum-flooding-surface seafloor. The timing of the CRM corresponds to the early burial window (Ewing 2006) and remagnetization as a result of clay diagenesis (Katz *et al.* 2000; Moreau *et al.* 2005; Tohver *et al.* 2008) or maturation of organic matter (Blumstein *et al.* 2004) is possible. The CRM in the matrix sites is also interpreted to have formed from burial diagenetic processes.

Specimens collected near NW and vertical fractures contain a Permian–early Triassic CRM that corresponds to deep burial (Ewing 2006). These fractures contain largely calcite, silica and liquid and solid hydrocarbons. The cements have slightly-to-moderately elevated $^{87}$Sr/$^{86}$ Sr values (ave = 0.7082, $n = 3$, Stdev = 0.00013; Dennie 2010) and are interpreted as having formed from either internally or externally derived fluids during deep burial. The presence of hydrocarbons in inclusions suggests that the cements were precipitated after the onset of hydrocarbon maturation. The fluids which caused the alteration could also have caused the CRM. It is interesting to note that the pole for this CRM (lat 49.9°N, long 115.4°E, dp = 7.6°; dm = 15.1°; Table 4) is close to the pole from the Barnett Shale in outcrop (lat 49.1°N, long 119.3°E, dp = 1.5°; dm = 3.0°; Kent & Opdyke 1979). This could mean that the Barnett Shale in outcrop was remagnetized by the same processes as the rock around the NW-oriented fractures.

The NE and vertical fractures also contain a late Permian–early Triassic CRM that is interpreted to have formed during deep burial. The veins have a mixture of remobilized cements and contain anhydrite (gypsum), late barite-celestine, late calcite, dolomite and pyrite. The barite and calcite in these fractures has moderate–elevated $^{87}$Sr/$^{86}$ Sr values (ave = 0.7089, $n = 3$, Stdev = 0.00059; Dennie 2010), suggesting that externally derived fluids altered the rocks and could have caused the CRM.

Gale *et al.* (2007) proposed that a NE fracture set (interpreted as equivalent to our slightly more NNE set) is older than a NW vertical fracture set. Fracture cements, morphology and overlapping ranges of fracture azimuths suggest the north–south fractures identified in Gale *et al.* (2007) in the eastern part of the basin are the same as NE fractures identified in the ASW and ES cores. The palaeomagnetic data suggest that the alteration in both NW and NE sets occurred at about the same time.

Several points should be mentioned regarding these interpretations. The magnetization data from near some fractures is limited. Because of this and the limitations of sampling near the veins, the hypothesized association between the orientation of the fractures and the magnetization directions should be tested by additional sampling around the different fracture sets and in the matrix. Regarding the inferred ages for the poles, Bilardello & Kodama (2010*a*, *b*) presented evidence that a correction for inclination shallowing could cause the late Palaeozoic APWP to shift to a lower latitude by 4–10°. This change would shift some of the pole positions reported in this study to a younger segment of the APWP.

The samples from the SC core, located closest to the Ouachita thrust zone, contain the most elevated $^{87}$Sr/$^{86}$Sr values (ave = 0.7090, $n = 7$, Stdev = 0.00067; Dennie 2010). This is consistent with the interpretation that externally derived fluids did enter and alter the shale, and probably resulted in the acquisition of one or more CRMs. At the base of the Barnett, the (Ordovician) Viola Limestone and lower Barnett also contain elevated Sr isotope values, suggesting the Ordovician unconformity acted as regional fluid conduit for externally derived fluids. The Barnett Shale, however, contains authigenic illite that formed from smectite. Feldspars, which have relatively high $^{87}$Sr, were probably dissolved to provide K for the illite. Some $^{87}$Sr was therefore probably released and contributed to some of the elevated $^{87}$Sr/$^{86}$Sr values in the cements. As a result, slightly or moderately elevated Sr isotope values may not indicate alteration by externally derived fluids.

Most $\delta^{34}$S values (e.g. $-7.12‰$) reported by Deng (2011) for some vein sulphate in the Barnett are depleted relative to the Phanerozoic seawater sulphate field (Kampschulte & Strauss 2004) and are consistent with alteration by externally derived fluids. The depleted sulphates probably represent a mixture between $^{34}$S-enriched and $^{34}$S-depleted components. The enriched component likely represents an externally derived evaporite sulphate, with some contributions from pore-water sulphate within the Barnett. Sulphide minerals related to microbial sulphate reduction are the dominant

[34]S-depleted source. A substantial fraction of sulphur in some vein sulphates would need to be derived from oxidized pyrite, which is a strong indication that the Barnett was altered by externally derived, oxygenated fluids (Deng 2011).

## Conclusions

We interpret the palaeomagnetic data to indicate that the Barnett Shale has undergone multiple remagnetization events beginning in the mid-Pennsylvanian Period and continuing into the early Triassic Period. The magnetizations reside in magnetite and are interpreted to be CRMs. The directions are streaked from a SE and shallow direction to a south and shallow direction. Both the scribe and VRM-oriented cores produce a similar distribution of specimen directions, which suggests that the data distribution is valid and not caused by core rotations. The palaeomagnetic data suggest the possibility of a local or regional anticlockwise rotation of the sampling area by up to 20° beginning in the late Pennsylvanian. This hypothesis needs to be tested, although it is consistent with the regional tectonic history of the basin.

Subdivision of the palaeomagnetic data does not reveal an obvious control on the CRMs by lithofacies or depth. One possibility is that one or more CRMs could have been acquired by multiple remagnetization processes between the Pennsylvanian and the Triassic and that the streak represents a mixing trend between discrete CRMs with overlapping data distributions that merge into a streak. There is a general correlation between fracture orientation and the CRM data. NE-oriented subvertical and stratiform fractures are found in rock that contains Pennsylvanian CRMs. NE- and NW-oriented vertical fractures are common to rocks that contain late Permian–Triassic CRMs and are interpreted to have formed during deep burial. The CRM associated with the NE vertical fractures is interpreted to have been caused by alteration associated with externally derived fluids, probably derived from the Ouachita thrust front. The depleted sulphur isotope and elevated Sr isotope results suggest that the Barnett Shale was altered by externally derived fluids from the Ouachita thrust front (Dennie 2010; Deng 2011).

The authors thank Devon Energy for providing support for this study and reviews by J. Gale, J. Urrutia-Fucugauchi and J. Geissman.

## References

AUDUNSSON, H. & LEVI, S. 1989. Drilling-induced remanent magnetization in basalt cores. *Geophysical Journal*, **98**, 613–622.

BEIER, J. A. & FELDMAN, H. R. 1991. Sulfur isotopes and paragenesis of sulfide minerals in the Silurian Waldron Shale, southern Indiana. *Geology*, **19**, 389–392.

BEST, M. E. & KATSUBE, T. J. 1995. Shale permeability and its significance in hydrocarbon exploration. *The Leading Edge*, **14**, 165–170.

BILARDELLO, D. & KODAMA, K. P. 2010a. Palaeomagnetism and magnetic anisotropy of Carboniferous redbeds from the Maritime Provinces of Canada: evidence for shallow palaeomagnetic inclinations and implications for North American apparent polar wander. *Geophysical Journal International*, **180**, 1013–1029.

BILARDELLO, D. & KODAMA, K. P. 2010b. A new inclination shallowing correction of the Mauch Chunk Formation of Pennsylvania, based on high-field AIR results: Implications for the Carboniferous North American APW path and Pangea reconstructions. *Earth and Planetary Science Letters*, **299**, 218–227.

BLUMSTEIN, A. M., ELMORE, R. D., ENGEL, M. H., ELLIOT, C. & BASU, A. 2004. Paleomagnetic dating of burial diagenesis in Mississippian carbonates, Utah. *Journal of Geophysical Research, B, Solid Earth and Planets*, **109**, 16.

BOWKER, K. A. 2003. Recent developments of the Barnett Shale play, Fort Worth Basin. *West Texas Geological Society Bulletin*, **42**, 4–11.

BOWKER, K. A. 2007. Barnett shale gas production, Fort Worth basin: issues and discussion. *American Association of Petroleum Geologist bulletin*, **91**, 523.

DE WALL, H. & WORM, H. 2001. Recognition of drilling–induced remanent magnetization by Q-factor analysis: a case study from the KTB-drillholes. *Journal of Applied Geophysics*, **46**, 55–64.

DEAN, R. S. & ROSS, G. J. 1976. Anomalous gypsum in clays and shales. *Clays and Clay Minerals*, **24**, 103–104.

DENG, D. 2011. *An integrated paleomagnetic, geochemical, and diagenetic study of the Mississippian Barnett Shale, Fort Worth Basin, Texas.* MSc thesis, University of Oklahoma, Norman, Oklahoma.

DENNIE, D. 2010. *An integrated paleomagnetic and diagenetic investigation of the Barnett Shale and underlying Ellenburger Group carbonates, Fort Worth Basin, Texas.* PhD thesis, University of Oklahoma, Norman, Oklahoma.

DUNLOP, D. J. & ARGYLE, K. S. 1991. Separating multidomain and single-domain-like remanences in pseudo-single-domain magnetites (215–540 nm) by low-temperature demagnetization. *Journal of Geophysical Research*, **96**, 2007–2017.

ELMORE, R. D. 2001. A review of palaeomagnetic data on the timing and origin of multiple fluid-flow events in the Arbuckle Mountains, Southern Oklahoma. *Petroleum Geoscience*, **7**, 223–229.

ELMORE, R. D., LONDON, D., BAGLEY, D. & GAO, G. 1993. Remagnetization by basinal fluids: testing the hypothesis in the Viola Limestone, southern Oklahoma. *Journal of Geophysical Research*, **98**, 6237–6254.

ELMORE, R. D., BANERJEE, S., CAMPBELL, T. & BIXLER, G. 1998. Palaeomagnetic dating of ancient fluid-flow events and paleoplumbing in the Arbuckle Mountains,

Southern Oklahoma. *In*: PARNELL, J. (ed.) *Dating and Duration of Fluid Flow Events and Rock-Fluid Interaction*. Geological Society, London, Special Publications, **144**, 9–25.

ERLICH, R. & COLEMAN, J. 2005. Drowning of the Upper Marble Falls carbonate platform (Pennsylvanian). *Sedimentary Geology*, **175**, 479–499.

EWING, T. E. 2006. Mississippian Barnett Shale, Fort Worth Basin: North-central Texas: Gas-shale play with multi-tcf potential: discussion. *American Association of Petroleum Geologist Bulletin*, **90**, 963–966.

FISHER, R. A. 1953. Dispersion on a sphere. *Geophysical Journal of the Royal Astronomical Society*, **217**, 295–305.

FLIPPIN, J. W. 1982. The stratigraphy, structure, and economic aspects of the Paleozoic strata in Erath County, north-central Texas. *In*: MARTIN, C. A. (ed.) *Petroleum Geology of the Fort Worth Basin and Bend Arch Area*. Dallas Geological Society, Dallas, 129–155.

GALE, J. F., REED, R. M. & HOLDER, J. 2007. Natural fractures in the Barnett Shale and their importance for hydraulic fracture treatments. *American Association of Petroleum Geologists Bulletin*, **91**, 603.

HALE-ERLICH, W. S. & COLEMAN, J. L. 1993. Ouachita–Appalachian juncture: a Paleozoic transpressional zone in the southeastern USA. *American Association of Petroleum Geologists Bulletin*, **77**, 552–568.

HARRINGTON, J. F. & HORSEMAN, S. T. 1999. Gas transport properties of clays and mudrocks. *In*: APLIN, A. C., FLEET, A. J. & MACQUAKER, J. H. S. (eds) *Muds and Mudrocks: Physical and Fuid-Flow Properties*. Geological Society, London, Special Publications, **158**, 107–124.

HAUBOLD, H. 1999. Alteration of magnetic properties of Palaeozoic platform carbonate rocks during burial diagenesis (Lower Ordovician sequence, Texas, USA). *In*: TARLING, D. H. & TARLING, P. (eds) *Paleomagnetism and Diagenesis of Sediments*. Geological Society, London, Special Publications, **151**, 181–203, doi: 10.1144/GSL.SP.1999.151.01.18

HENRY, J. D. 1982. Stratigraphy of the Barnett Shale (Mississippian) and associated reefs in the northern Fort Worth basin. *In*: MARTIN, C. A. (ed.) *Petroleum Geology of the Fort Worth Basin and Bend Arch Area*. Dallas Geological Society, Dallas, 157–178.

HESLOP, D., DEKKERS, M. J., KRUIVER, P. P. & VAN OORSCHOT, I. H. M. 2002. Analysis of isothermal remanent magnetisation acquisition curves using the expectation-maximisation algorithm. *Geophysical Journal International*, **148**, 58–64.

HESLOP, D., MCINTOSH, G. & DEKKERS, M. J. 2004. Using time- and temperature-dependent Preisach models to investigate the limitations of modelling isothermal remanent magnetization acquisition curves with cumulative log Gaussian functions. *Geophysical Journal International*, **157**, 55–63.

HOWELL, L. G. & MARTINEZ, J. D. 1957. Polar movement as indicated by rock magnetism. *Geophysics*, **22**, 384–397.

JARVIE, D. M., HILL, R. J., RUBLE, T. E. & POLLASTRO, R. M. 2007. Unconventional shale-gas systems: the Mississippian Barnett Shale of north-central Texas as one model for thermogenic shale-gas assessment.

American Association of Petroleum Geologists Bulletin, **91**, 475.

JOHNSON, K. S., AMSDEN, T. W., DENISON, R. E., DUTTON, S. P., GOLDSTEIN, A. G. & RASCOE, B. JR. 1988. Southern midcontinent region. *In*: SLOSS, L. L. (ed.) *Sedimentary Cover—North American Craton*. Geological Society of America, Boulder, CO, The Geology of North America, D-2, 307–359.

KAMPSCHULTE, A. & STRAUSS, H. 2004. The sulfur isotopic evolution of Phanerozoic seawater based on the analysis of structurally substituted sulfate in carbonates. *Chemical Geology*, **204**, 255–286.

KATZ, B., ELMORE, R. D., COGOINI, M., ENGEL, M. H. & FERRY, S. 2000. Associations between burial diagenesis of smectite, chemical remagnetization, and magnetite authigenesis in the Vocontian Trough, SE France. *Journal of Geophysical Research*, **105**, 851–868.

KENT, D. V. & OPDYKE, N. D. 1979. The Early Carboniferous palaeomagnetic field of North America and its bearing on tectonics of the northern Appalachians. *Earth and Planetary Science Letters*, **44**, 365–372.

KIRSCHVINK, J. L. 1980. The least-squares line and plane and the analysis of palaeomagnetic data. *Geophysical Journal of the Royal Astronomical Society*, **62**, 699–718.

KRUIVER, P. P., DEKKERS, M. J. & HESLOP, D. 2001. Quantification of magnetic coercivity components by the analysis of acquisition curves of isothermal remanent magnetization. *Earth and Planetary Science Letters*, **189**, 269–276.

LOUCKS, R. G. & RUPPEL, S. C. 2007. Mississippian Barnett Shale: Lithofacies and depositional setting of a deep-water shale-gas succession in the Fort Worth Basin, Texas. *American Association of Petroleum Geologists Bulletin*, **91**, 579.

LOUCKS, R. G., REED, R. M., RUPPEL, S. C. & JARVIE, D. M. 2009. Morphology, genesis, and distribution of nanometer-scale pores in siliceous mudstones of the Mississippian Barnett Shale. *Journal of Sedimentary Research*, **79**, 848–861.

LU, G., MCCABE, C., HANOR, J. S. & FERRELL, R. E. 1991. A genetic link between remagnetization and potassic metasomatism in the Devonian Onondaga Formation, northern Appalachian basin. *Geophysical Research Letters*, **18**, 2047–2050.

MACARTHUR, J. M., HOWARTH, R. J. & BAILEY, T. R. 2001. Strontium isotope stratigraphy: lOWESS version 3: best fit to the marine sr-isotope curve for 0–509 Ma and accompanying look-up table for deriving numerical age. *The Journal of Geology*, **109**, 155–170.

MCDONNELL, A., LOUCKS, R. G. & DOOLEY, T. 2007. Quantifying the origin and geometry of circular sag structures in northern Fort Worth Basin, Texas: Paleocave collapse, pull-apart fault systems, or hydrothermal alteration? *American Association of Petroleum Geologists Bulletin*, **91**, 1295–1318.

MECKEL, L. D. JR, SMITH, D. & WELLS, L. 1992. Ouachita foredeep basins: regional paleogeography and habitat of hydrocarbons. *In*: MACQUEEN, R. W. & LECKIE, D. A. (eds) *Foreland Basins and Fold Belts*. American Association of Petroleum Geologists, Tulsa, OK, Memoir, **55**, 427–444.

MILLIKEN, K., CHOH, S., PAPAZIS, P. & SCHIEBER, J. 2007. 'Cherty' stringers in the Barnett Shale are agglutinated foraminifera. *Sedimentary Geology*, **198**, 221–232.

MOREAU, M. G., ADER, A. & ENKIN, R. J. 2005. The magnetization of clay-rich rocks in sedimentary basins: low-temperature experimental formation of magnetic carriers in natural samples. *Earth and Planetary Science Letters*, **230**, 193–210.

NELSON, R. A., LENOX, L. C. & WARD, B. J. JR. 1987. Oriented core: its use, error, and uncertainty. *American Association of Petroleum Geologists Bulletin*, **71**, 357–367.

NEUZIL, C. E. 1994. How permeable are clays and shales? *Water Resources Research*, **30**, 145–150.

OLIVER, J. 1992. The spots and stains of plate tectonics. *Earth Science Review*, **32**, 77–106.

PAPAZIS, P. K. 2005. *Petrographic characterization of the Barnett Shale, Fort Worth Basin, Texas.* MSc thesis, The University of Texas at Austin.

POLLASTRO, R. M., JARVIE, D. M., HILL, R. J. & ADAMS, C. W. 2007. Geologic framework of the Mississippian Barnett Shale, Barnett-Paleozoic total petroleum system, Bend arch–Fort Worth Basin, Texas. *American Association of Petroleum Geologists Bulletin*, **91**, 405–436.

PULLAIAH, G., IRVING, E., BUCHAN, K. L. & DUNLOP, D. J. 1975. Magnetization changes caused by burial and uplift. *Earth and Planetary Science Letters*, **28**, 133–143.

SCHIEBER, J., KRINSLEY, D. & RICIPUTI, L. 2000. Diagenetic origin of quartz silt in mudstones and implications for silica cycling. *Nature*, **405**, 981–985.

SINGH, P. 2008. *Lithofacies and sequence stratigraphic framework of the Barnett shale, northeast Texas.* PhD thesis, University of Oklahoma.

SULLIVAN, E. C., MARFURT, K. J., LACAZETTE, A. & AMMERMAN, M. 2006. Application of new seismic attributes to collapse chimneys in the Fort Worth Basin. *Geophysics*, **71**, B111–B119.

SUK, D., VAN DER VOO, R., PEACOR, D. R. & LOHMANN, K. C. 1993. Late Paleozoic remagnetization and its carrier in the Trenton and Black River carbonates from the Michigan Basin. *Journal of Geology*, **101**, 795–808.

THOMAS, W. A. 1993. Low-angle detachment geometry of the late Precambrian-Cambrian Appalachian-Ouachita rifted margin of southeastern North America. *Geology*, **21**, 921–924.

TOHVER, E., WEIL, A. B., SOLUM, J. G. & HALL, C. M. 2008. Direct dating of carbonate remagnetization by 40Ar/ 39Ar analysis of the smectite–illite transformation. *Earth and Planetary Science Letters*, **274**, 524–530.

VAN ALSTINE, D. & BUTTERWORTH, J. 2002. Palaeomagnetic Core-Orientation Helps Determine the Sedimentological, Paleostress, and Fluid-Migration History in the Maracaibo Basin, Venezuela. Available at: http://www.appliedpaleomagnetics.com/Articles/Htm/ 2002Congress.htm

VAN DER VOO, R. 1993. *Palaeomagnetism of the Atlantic, Tethys and Iapetus Oceans.* Cambridge University Press, Cambridge, UK.

VAN DER VOO, R. & WATTS, D. 1978. Palaeomagnetic results from igneous and sedimentary rocks from the Michigan Basin borehole. *Journal of Geophysical Research*, **83**, 5844–5848.

WALPER, J. L. 1982. Plate tectonic evolution of the Fort Worth basin. *In*: MARTIN, C. A. (ed.) *Petroleum Geology of the Forth Worth Basin and Bend Arch Area.* Dallas Geological Society, Dallas, 237–241.

ZIJDERVELD, J. D. A. 1967. A.C. demagnetization of rocks: analysis of results. *In*: COLLINSON, D. E., CREER, K. M. & RUNCORN, S. K. (eds) *Methods in Palaeomagnetism.* NATO Advanced Study Institute on Palaeomagnetic Methods held in University of Newcastle Upon Tyne, New York, Elsevier Science, 254–286.

# Multiple magnetizations in Ordovician–Devonian carbonates in the Williston Basin (Manitoba, Canada)


ERIKA SZABÓ[1]* & MARIA T. CIOPPA[2]

[1]*Department of Earth Sciences, University of Western Ontario, 1151 Richmond Street, London, ON, Canada, N6A 5B7*

[2]*Department of Earth and Environmental Sciences, University of Windsor, 401 Sunset Avenue, Windsor, ON, Canada, N9B 3P4*

**Corresponding author (e-mail: eszabo@uwo.ca)*



**Abstract:** Palaeomagnetic and rock magnetic data collected from the Upper Ordovician Red River, Silurian Interlake and Devonian Winnipegosis, Souris River and Birdbear carbonates in one well from southwestern Manitoba (Canada) reveal a complex magnetization history for the north-eastern Williston Basin. Rock magnetic analysis (thermal demagnetization, anhysteretic remanent magnetization (ARM)/saturation isothermal remanent magnetization (SIRM), S-ratios, partial ARM (pARM), SIRM crossover curves and points and coercivity) show three magnetic carriers for different magnetizations seen in this well. An Early–Mid-Jurassic remagnetization observed in the lower Red River and Souris River formations is carried by single-domain to pseudo-single-domain (SD–PSD) magnetite and was probably produced by basement fluids circulating along fractures and faults created by the Hartney impact/volcanic structure and/or tectonic movements along the Superior Boundary Zone. In the Winnipegosis Formation a possible primary depositional or early depositional magnetization (Devonian age) is carried by PSD pyrrhotite. In the Birdbear Formation two different magnetizations of uncertain age are present: one carried in hematite in the upper strata, possibly originating from the younger Amaranth Formation, and magnetite dominates in the lower strata. The upper Red River and Interlake formations contain both magnetite and pyrrhotite; however, the weaker palaeomagnetic data reveal little in terms of a magnetization age.


Palaeomagnetic results obtained from core samples from the Upper Ordovician Red River carbonates of the Williston Basin, central North America, revealed a complex magnetization history that shows spatial and temporal variation (unpublished data; Szabó & Cioppa 2006). The two dominant palaeomagnetic components present in these strata could be either Silurian–Devonian or Late Carboniferous (Pennsylvanian)–Triassic and Late Triassic–Mid-Jurassic in age; the use of inclination-only analysis did not allow further constraint. In the absence of other reorientation tools, it was considered the examination of palaeomagnetic data from several formations in a single core (including the Red River carbonates) might be informative.

The current study aims to evaluate the palaeomagnetic data from the Red River carbonates through comparison of its palaeomagnetic and rock magnetic properties to those of younger formations (Silurian Interlake Group, Devonian Winnipegosis, Souris River and Birdbear formations) from the same core. Variations in magnetization and carriers within a single core may better constrain the original magnetizations seen in the Red River carbonates and, consequently, may

assist in identifying the appropriate mechanisms causing the magnetizations observed. In addition, the current study will test whether orogenic fluid flow was actively producing magnetization in this intracratonic basin, far from orogens such as the Appalachians or the Rocky Mountains.

## Geological framework

The samples for the present study were collected from the Calstan Hartney 16-33-5-24W1 well, located in south-western Manitoba (Canada) on the far north-eastern side of the Williston Basin, a *c.* 500 km wide intracratonic basin extending across Saskatchewan and Manitoba in Canada and North Dakota and Montana in the USA (Fig. 1). A protracted sedimentation and subsidence sequence influenced by the North American orogenies built up a stack of Palaeozoic, Mesozoic and Cenozoic carbonates and clastic sediments that hold major oil-producing structures. Two major tectonic episodes, the Mississippian–Permian Alleghenian orogeny and the Late Cretaceous–Palaeogene Laramide orogeny, had the most influence on hydrocarbon


*From*: ELMORE, R. D., MUXWORTHY, A. R., ALDANA, M. M. & MENA, M. (eds) 2012. *Remagnetization and Chemical Alteration of Sedimentary Rocks*. Geological Society, London, Special Publications, **371**, 107–122.
First published online June 26, 2012, http://dx.doi.org/10.1144/SP371.5

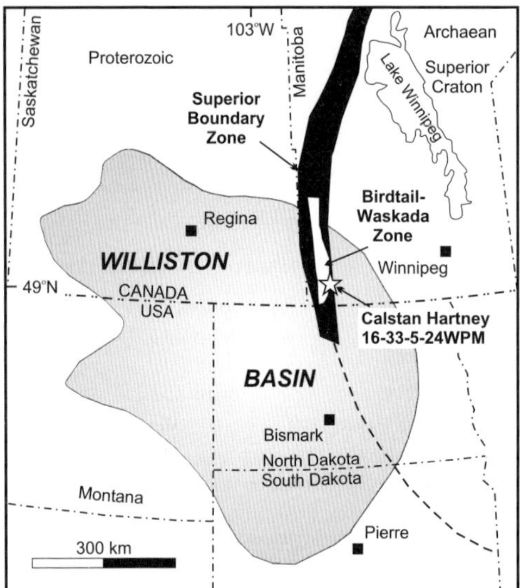

**Fig. 1.** Regional setting of the Williston Basin and location of the Calstan Hartney 16-33-5-24 WPM well relative to the Superior Boundary Zone (modified from Nicolas & Barchyn 2008).

maturation and migration in the Williston Basin (Burrus *et al.* 1996; Osadetz *et al.* 2002). Deformation of the basin due to these orogenies is not extensive or high grade (e.g. Gerhard *et al.* 1990; Rédly & Hajnal 1995). However, faulting occurred in strata as young as the Cretaceous, and is probably related to basement (Precambrian) faults that separate different blocks that moved independently as a result of the tectonic stress and the weight of sediments (e.g. Gerhard *et al.* 1990). One of the most active areas in this sense is the 40–50 km wide Superior Boundary Zone (SBZ) in south-western and central Manitoba (Fig. 1), which separates the Archaean Superior to the east and the Proterozoic cratons to the west and contains many basement-penetrating faults (Dietrich & Magnusson 1998).

The Red River carbonates were deposited in the Williston Basin during the Late Ordovician (Fig. 2). They are primarily dolostone interbedded with dolomitic limestone, minor limestone and anhydrite in the upper section (Martiniuk *et al.* 1998; Nicolas & Barchyn 2008). In the Manitoban subsurface these strata are divided into the lower Red River and upper Red River. However, as discussed by Nicolas & Barchyn (2008), the subdivisions do not seem to have been formally defined; the terminology of 'lower and upper Red River beds' will therefore

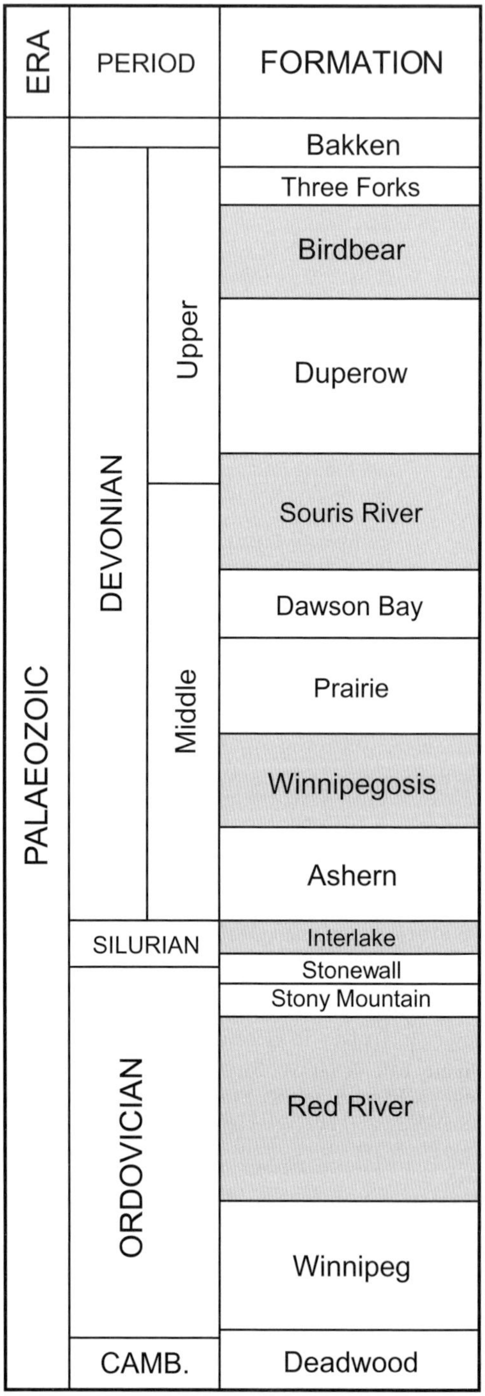

**Fig. 2.** Stratigraphy of the Lower Palaeozoic strata in the Williston Basin subsurface of Manitoba with current study strata highlighted in light grey. CAMB: Cambrian.

be used in this paper. Petrologically, the lower Red River carbonates are burrow-mottled fossiliferous dolostone (Bezys & Bamburak 2004), partly

| ERA | PERIOD | | FORMATION |
|---|---|---|---|
| PALAEOZOIC | DEVONIAN | Upper | Bakken |
| | | | Three Forks |
| | | | Birdbear |
| | | | Duperow |
| | | | Souris River |
| | | | Dawson Bay |
| | | Middle | Prairie |
| | | | Winnipegosis |
| | | | Ashern |
| | SILURIAN | | Interlake |
| | | | Stonewall |
| | | | Stony Mountain |
| | ORDOVICIAN | | Red River |
| | | | Winnipeg |
| | CAMB. | | Deadwood |

correlating to the well-known Tyndall stone of the Selkirk Member in Manitoba outcrops. The upper Red River carbonates are composed of cyclical beds of non-fossiliferous dolostone and anhydrite (Martiniuk *et al.* 1998).

The Silurian Interlake Group in Manitoba (Fig. 2) is divided into the lower and upper units (Nicolas & Barchyn 2008) and consists of crystalline dolostone interbedded with oolitic, fossiliferous, stromatolitic and biohermal interbeds (Glass 1990). Moving up the section, the Middle Devonian Winnipegosis Formation (Fig. 2) is divided into lower and upper members (Nicolas & Barchyn 2008) consisting of platform-facies dolostone or reefoid-facies dolostone, respectively (Baillie 1951; Norris *et al.* 1982; Bezys & Bamburak 2004). The Middle–Upper Devonian Souris River carbonates (Fig. 2) form several cycles of shale, limestone, dolostone and anhydrite (Bezys & Bamburak 2004; Nicolas & Barchyn 2008). From bottom to top, the subsurface subdivisions consist of the First Red Beds, Davidson, Harris and Hatfield members (Nicolas & Barchyn 2008). The Birdbear Formation is a carbonate-evaporite shelf layer of Late Devonian age (Fig. 2; Martiniuk *et al.* 1995) divided into lower and upper members and consisting of non-argillaceous limestone and dolostone and dolostone and interbedded evaporates, respectively (Martiniuk *et al.* 1995).

## Sampling and core description

Samples from the Red River consisted of three plugs from the upper Red River and 11 plugs from the lower Red River beds. The lower Red River samples were from medium grey to pale orange limestone, burrowed floatstones to rudstones with crinoidal wackestone intervals (Fig. 3a). The upper bed samples were from core of darker brown to grey, slightly calcitic and laminated dolomite, that had a wackestone to floatstone precursor with argillaceous layers (Fig. 3b). The samples (10 plugs) of the Interlake Group were from buff and grey-coloured dolostone, laminated, brecciated and with slightly calcitic porous zones (Fig. 3c). The Winnipegosis Formation was examined in 7 plugs and consisted of buff-grey fossiliferous and porous dolostone, with laminations and vug-filled anhydrite (Fig. 3d). The samples (6 plugs) from the Souris River Formation are from brecciated and fossiliferous brown-grey to brown-yellow dolostones, containing a replacement white and clear mineral, probably anhydrite, in vugs and fossils (Fig. 3e). The Birdbear samples (6 plugs) were from interbedded and laminated brown dolostones, calcareous mudstones and grey-blue anhydrite with silty and brecciated layers cemented by white calcite (Fig. 3f).

## Palaeomagnetic and rock magnetic methodology

Overall, a total of 43 plugs each 2.5 cm in diameter were drilled between 1465.1 and 896.6 m depth in the well, following the palaeomagnetic sampling and orientation method described by Lewchuk *et al.* (1998) and Cioppa *et al.* (2000). With the exception of the Birdbear Formation, the sections of core sampled were reassembled into segments from which at least three plugs were collected. This made possible a direct comparison between palaeomagnetic directions found in plugs of the same segment, providing information about consistency and validity. The plugs were cut into 2–3 specimens (126 in total) and stored for at least three months in a magnetically shielded room ($<1\%$ EMF) at the University of Windsor, in order to allow the least-stable viscous magnetization components to decay. Out of the 126 specimens, 105 were subject to demagnetization: 57 using the alternating field and 48 using the thermal technique. Calculations of palaeomagnetic directions for individual specimens were performed with principal component analysis on lines that contained at least four points and had a maximum angular deviation of $15°$ (Kirschvink 1980), as implemented in Enkin (1994). Statistical Fisher means were calculated on palaeomagnetic directions from specimens from the same plug (Fisher 1953), followed by inclination-only means (Enkin & Watson 1996) computed on all of the plugs showing a certain palaeomagnetic component.

Rock magnetic analysis was performed on one specimen from each plug but one (42 specimens); the only available specimen of the uppermost plug was reserved for a separate study. Rock magnetic experiments included: stepwise partial anhysteretic remanent magnetization (pARM) acquisition, ARM acquisition, saturation isothermal remanent magnetization (SIRM) acquisition and demagnetization, S-ratios, coercivity measurements ($H_{cr}$) and thermal demagnetization of the IRM. The pARM was measured in steps of 10 mT up to 100 mT using a DC field of 0.05 mT with an alternating field (AF) peak field of 100 mT (after Everitt 1961), followed by ARM acquisition and demagnetization and then by SIRM acquisition. The IRM was measured in 16 steps to 900 mT to reach the SIRM in magnetite and pyrrhotite (Symons & Cioppa 2000), and was then AF demagnetized in 11 steps to 140 mT. Subsequently, the IRM was measured at fields as high as 2000 mT (up to 21 steps) in order to detect any high coercivity minerals that might be present. The IRM S-ratios were obtained by inducing an IRM at 1200 mT field, followed by an opposite IRM at 100 mT and 300 mT. The $S_{100}$ and $S_{300}$ are the ratios of the IRM at 100

**Fig. 3.** Calstan Hartney 16-33-5-24 WPM core photographs: (**a**) Tyndall-type limey wackestone with dolomitized burrows in the lower Red River strata; (**b**) crenulated laminitic dolomite in the upper Red River strata; (**c**) dolomite breccia with limestone mud, with micrite filling the spaces between the clasts in Interlake Group; (**d**) very porous fossiliferous dolomite in the Winnipegosis Formation; (**e**) brecciated dolomite containing vugs filled with anhydrite in Souris River Formation; (**f**) transition from laminated mudstone/anhydrite to more massive and nodular anhydrite in the Birdbear Formation. Note the reddish colour resulting from the hematite content.

and 300 mT fields divided by the SIRM at 1200 mT. $H_{cr}$ was determined first by imposing a SIRM at 1200 mT field peak, followed by measurements of the IRM in increased reversed fields at 5 mT intervals until the SIRM was removed. IRM thermal demagnetization was performed on two samples (upper Red River and Birdbear) using a 1200 mT field and then thermally demagnetizing the specimens in 15 steps up to 700 °C.

The above palaeomagnetic and rock magnetic experiments were carried out using a Sapphire Instruments SI-4 AF demagnetizer, a Magnetic Measurements MMTD-1 (Magnetic Measurements Thermal Demagnetizer) thermal demagnetizer, a vertical-configuration 2 G 755R cryogenic magnetometer with a lower sensitivity limit of $c.\ 2 \times 10^{-6}$ A m$^{-1}$ and a Sapphire Instruments SI-6 pulse magnetizer in the Palaeomagnetics and Rock Magnetic Laboratory at the University of Windsor, Canada.

## Rock magnetic results and analysis

### Magnetic mineralogy

Thermal demagnetization data of the natural remanent magnetization (NRM) and isothermal remanent

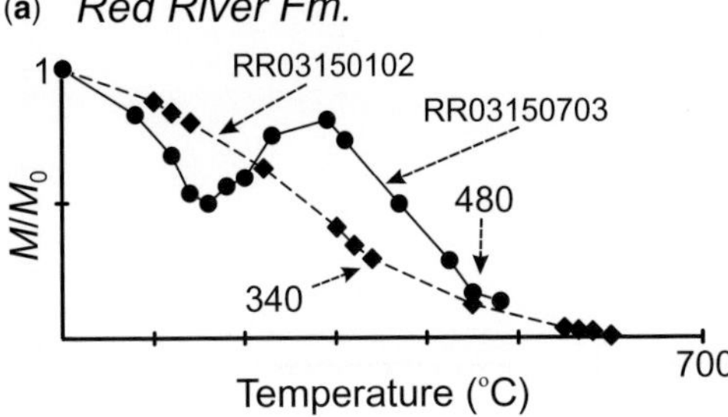


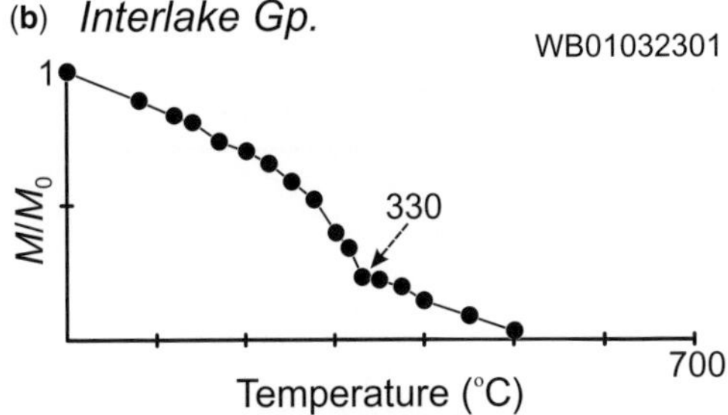


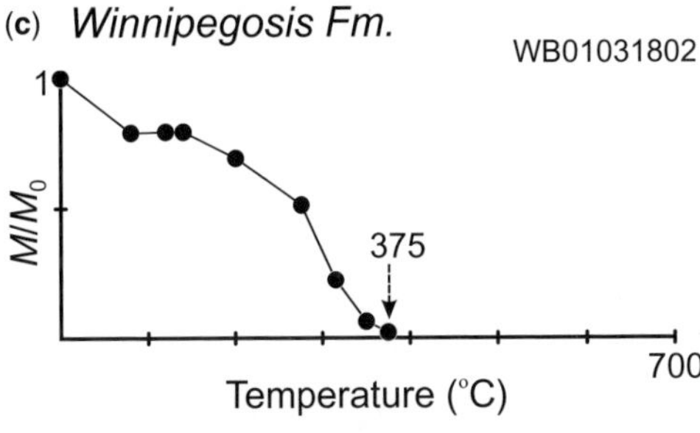


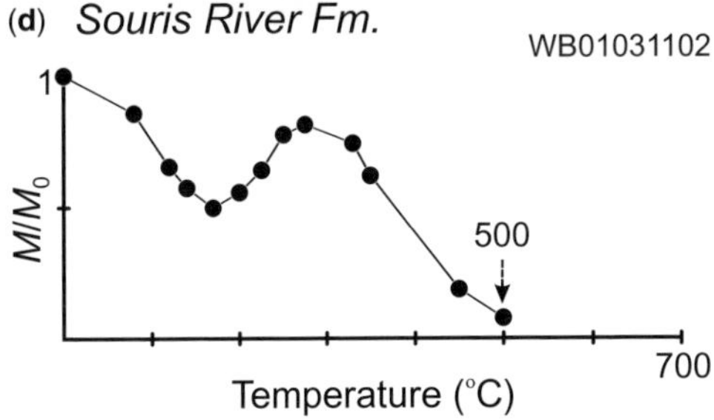


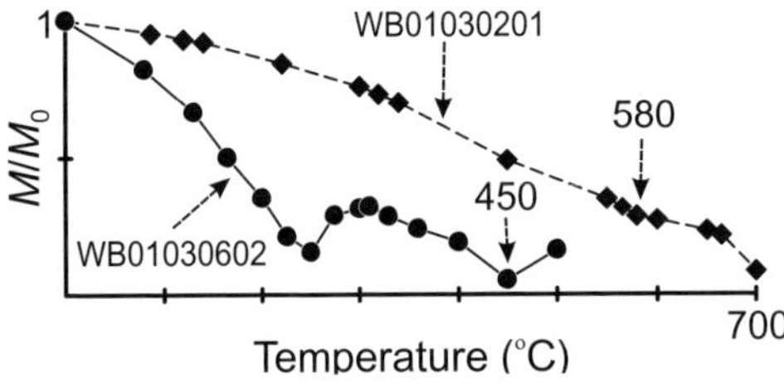


magnetization (IRM) revealed a mix of magnetic minerals (Fig. 4). NRM demagnetization of the lower Red River, Souris River and two plugs from the lower section of the Birdbear carbonates shows first a drop in intensity at *c.* 140–250 °C which is due to a reversal and then a second drop at between 450 and 500 °C (Fig. 4a (sample RR03150703), 4d, 4e (sample WB01030602)). This indicates that the magnetization is probably carried by magnetite. Thermal demagnetization of both the NRM and IRM in the upper Red River and Interlake Group carbonates revealed unblocking temperatures at *c.* 340 °C and *c.* 450–500 °C (Fig. 4a (sample RR03150102) and 4b), suggesting that both pyrrhotite and magnetite are present in this part of the core. The Winnipegosis samples display unblocking temperatures of *c.* 330–350 °C, close to the pyrrhotite unblocking temperature (*c.* 320 °C; Fig. 4c). An IRM demagnetized specimen from the upper part of the Birdbear carbonate section shows two unblocking temperatures at *c.* 580 °C (magnetite) and *c.* 700 °C (likely hematite, sample WB01030201; Fig. 4e).

This variation in magnetic mineralogy is also reflected in other magnetic parameters. Soft magnetic minerals such as magnetite and pyrrhotite occur from the lower Red River to the lower section of the Birdbear carbonates, and are indicated by $S_{300}$ ratios close to unity (Fig. 5), $H_{cr}$ values, pARM spectra and SIRM crossover plots. The pARM spectra in these parts of the core show well-defined peaks, mostly between 20 and 40 mT (Fig. 6a–e). The IRM in all of these samples could be reduced to zero, a $H_{cr}$ value was determined (Fig. 5) and saturation was reached at fields of *c.* 300 mT (Fig. 7). In contrast, the upper section of the Birdbear Formation contains a harder magnetic mineral (hematite) which is reflected by lower S-ratios (Fig. 5), relatively flat pARM spectra (Fig. 6e) and a hard IRM that could not reach saturation by 2000 mT (Fig. 7a).

## Magnetic particle domains and sizes

The magnetic domains and possible grain sizes were studied using pARM spectra and SIRM crossover plots and points. The IRM crossover point is the point where the IRM acquisition curve intersects the AF demagnetizing curve (Symons & Cioppa 2000).

**Fig. 4.** Representative examples of normalized intensity decay plots during thermal demagnetization of the natural remanent magnetization (NRM, solid lines) and isothermal remanent magnetization (IRM, dashed lines) of Ordovician–Devonian carbonates. *M*, intensity of magnetization; $M_0$, initial intensity of magnetization. See text for details.

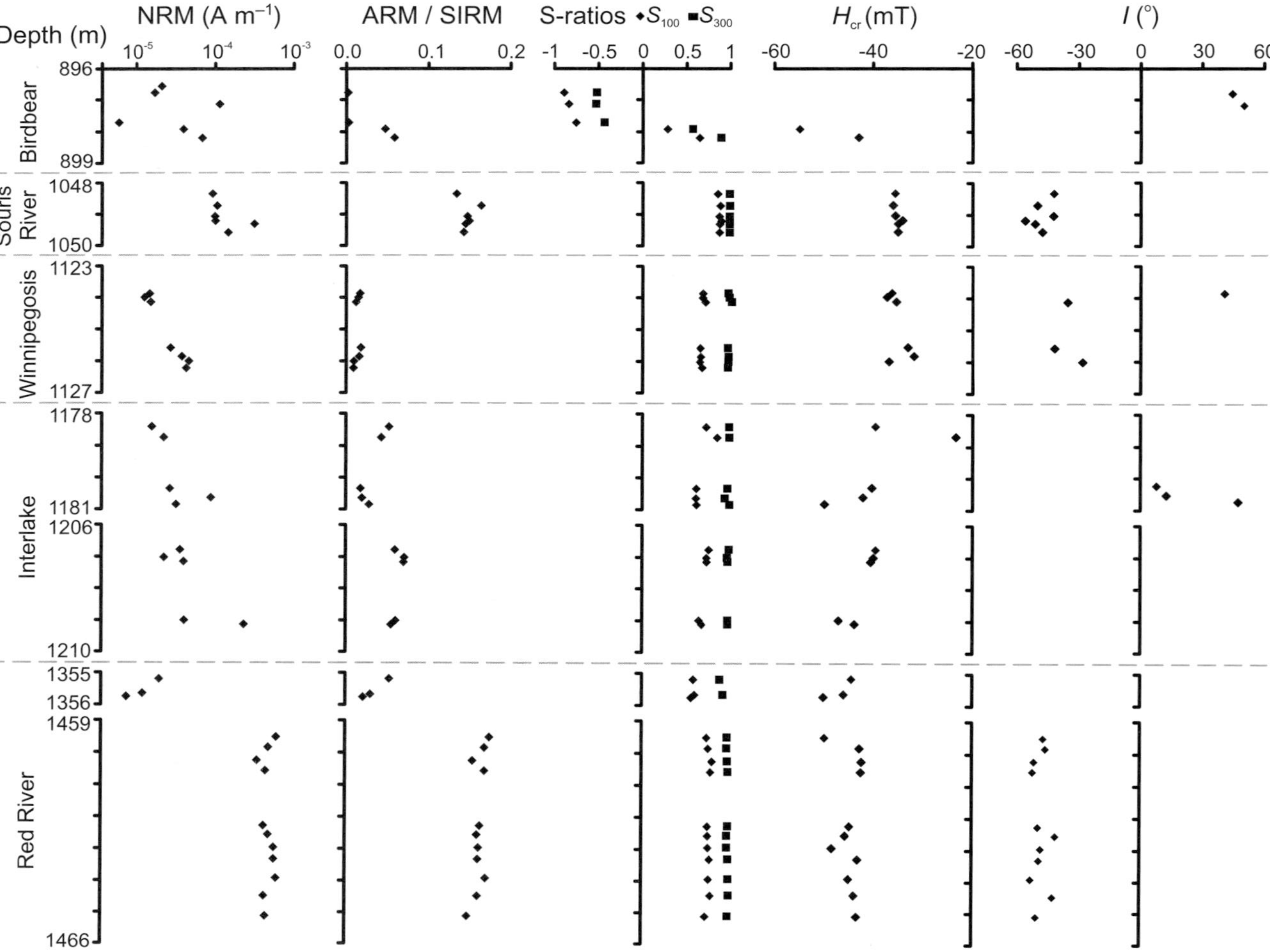

**Fig. 5.** Variation in magnetic and palaeomagnetic parameters with depth in Ordovician–Devonian strata. NRM, natural remanent magnetization; ARM, anhysteretic remanent magnetization; SIRM, saturation isothermal remanent magnetization; $H_{cr}$, coercivity of the remanence; $I$, inclination values of *B* palaeomagnetic components. See text for further details and discussion.

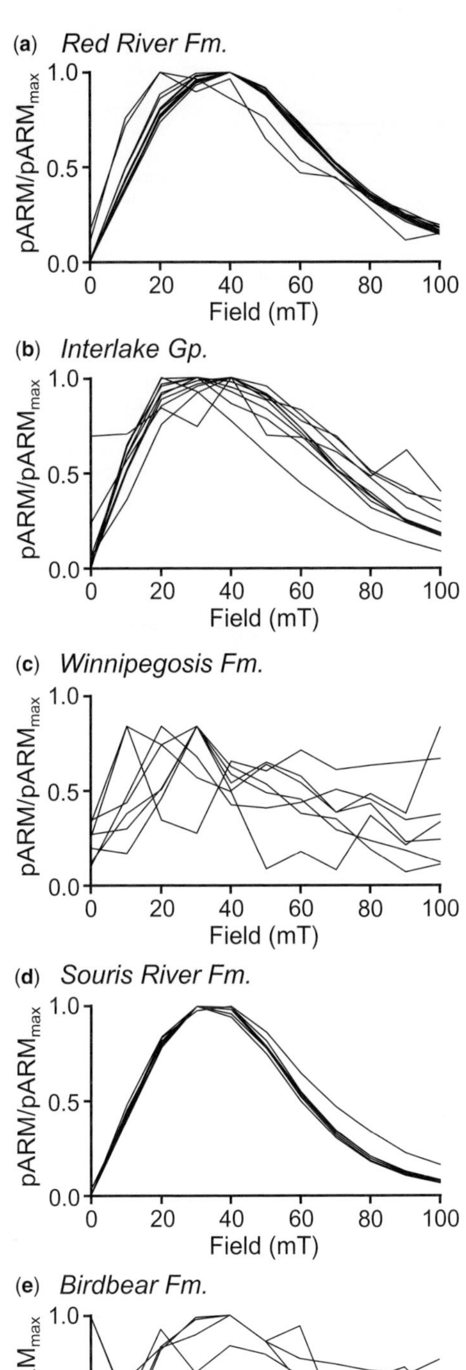



**Fig. 6.** Representative partial anhysteretic remanent magnetization (pARM) spectra. See text for discussion.

The magnetite present in the lower Red River, Interlake, Souris River and lower Birdbear section is a mixture of pseudo-single-domain (PSD) and single-domain (SD) grain size particles. This is illustrated by pARM peaks between 30 and 40 mT (Fig. 6a, b, d & e; Jackson *et al.* 1988), SIRM curves that overlap the SD–PSD magnetite SIRM crossover plots (Fig. 7a; Symons & Cioppa 2000) and SIRM crossover points close to 0.5, which is the theoretical value for pure SD magnetite (Fig. 8; Cisowski 1981; Symons & Cioppa 2000). The two lowermost samples from the Birdbear section have SIRM curves and points plotting slightly to the right, likely due to the presence of minor amounts of hematite that is notable in the upper Birdbear Formation (Fig. 7a). In contrast, the SIRM curves from the Souris River samples plot to the left with corresponding crossover points being slightly lower than 0.5, a indication of the presence of multidomain (MD) grains (Fig. 7a). This data suggests that the Souris River samples may contain somewhat larger magnetic grains than the lower Red River samples.

The pyrrhotite in the upper Red River, Interlake Group and Winnipegosis carbonates has pARM peaks between 10 and 30 mT (Figs 6a–c), and the instability of the Winnipegosis pARM spectra may reflect a low concentration of pyrrhotite. The SIRM curves and points for samples containing pyrrhotite are within the PSD pyrrhotite domain defined by the SIRM curves (Figs 7b & 8; Symons & Cioppa 2000), while a mixture of magnetite and pyrrhotite could have caused the difference between these samples and those from the upper Red River and Interlake (Fig. 8). The Winnipegosis samples fall into a well-defined group, perhaps due to contributions from MD grains, and suggest that a distinctive magnetic mineralogy may be present in this formation. The high coercivity of hematite caused the relatively flat curves of the pARM spectra in the upper part of the Birdbear Formation (Fig. 6e), as well as the lack of a SIRM crossover point. However, the corresponding SIRM crossover curves indicate a SD–PSD magnetic grain size for hematite (Fig. 7a).

## Rock magnetic profiles

The intensity of the NRM, the ARM/SIRM and the $S_{100}$ ratios echo the observed magnetic mineralogy variation. The highest NRM intensity values are seen in the sections where magnetite dominates: for example, Souris River ($7.1 \times 10^{-5} - 4.1 \times 10^{-4}$ A m$^{-1}$) and lower Red River ($2.3 \times 10^{-4} - 8.4 \times 10^{-4}$ A m$^{-1}$; Fig. 5), with the values suggesting that higher magnetite concentration

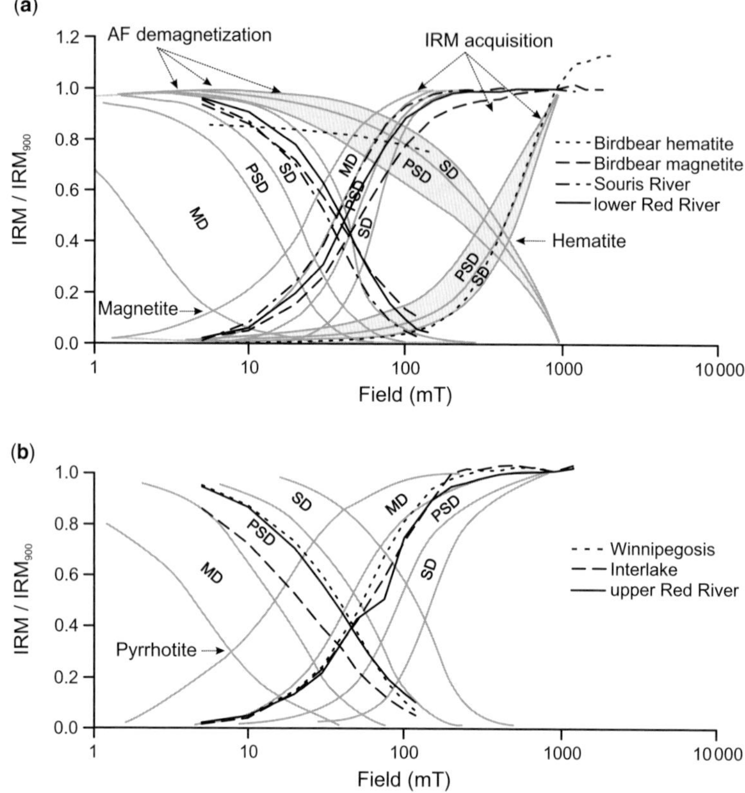


**Fig. 7.** (**a**) Representative saturation isothermal remanent magnetization (SIRM) acquisition and demagnetization crossover plots for the lower Red River, Souris River and Birdbear formations. Light-grey solid lines are the SIRM crossover template plots for magnetite; light-grey lines with grey-shaded areas represent SIRM crossover plots for hematite (Symons & Cioppa 2000). (**b**) Representative SIRM acquisition and demagnetization crossover plots from the upper Red River, Interlake and Winnipegosis carbonates. Light-grey solid lines are the SIRM crossover template plots for pyrrhotite. See text for details. MD, PSD, SD show the area for multidomain, pseudo-single-domain and single-domain magnetic phases, respectively. IRM$_{900}$ is the IRM measured at 900 mT.

could occur in the latter. The highest NRM values correlate with the highest ARM/SIRM ratios $(0.10 - 0.16)$ and slightly higher $S_{100}$ values (Fig. 5). In the sections of the core where pyrrhotite dominates (upper Red River, Interlake Group and Winnipegosis Formation), lower NRM values $(3.8 \times 10^{-6} - 9.2 \times 10^{-5} \text{ A m}^{-1})$, lower ARM/ SIRM ratios $(0.009 - 0.07)$ and lower $S_{100}$ values are present (Fig. 5). In the upper section of the Birdbear Formation, the presence of hematite is reflected by mostly lower NRM intensities $(4.9 \times 10^{-6} - 1.2 \times 10^{-4} \text{ A m}^{-1})$, lower ARM/SIRM values $(0.002 - 0.003)$ and lower and negative $S_{100}$ values.

## Palaeomagnetic results

In all of the plugs collected from the lower Red River carbonates, two different palaeomagnetic components were isolated (Fig. 9a). The first component **RRA** is removed at low fields (0–20 mT) and low temperatures (0–280 °C) and has positive and steep inclination values between 67.2 and 84.7°. The second component **RRB** is isolated at higher fields (30–100 mT) and temperatures (290–510 °C), has lower inclination values $(-38.7$ to $-56.4°)$, and is reversed compared to **RRA** (Figs 5 & 9a).

The upper Red River samples show different behaviour. The remanence in the upper samples becomes unstable above 80 mT or 310–340 °C. From an initial moderate to steeply positive inclination direction, demagnetization shows a trend towards negative shallow and moderate inclination directions.

The next strata up in the core, the Interlake, has poor palaeomagnetic data beyond an initial low coercivity component **ILA**. This is removed at low fields (0–20 mT) and low temperatures (80–280 °C)

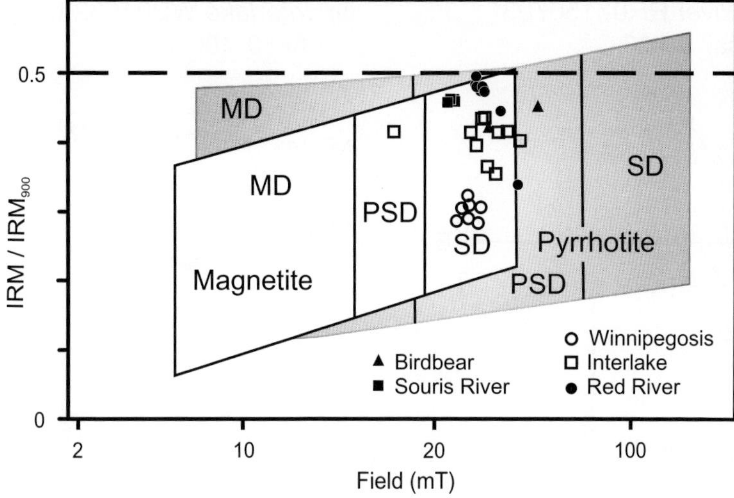


**Fig. 8.** Saturation isothermal remanent magnetization (SIRM) crossover points (open/closed symbols) in selected Ordovician to Devonian carbonates in the Williston Basin plotted against theoretical domains for magnetite and pyrrhotite of Symons & Cioppa (2000). See text for discussion. IRM$_{900}$, IRM intensity at 900 mT; MD, multidomain grain size; PSD, pseudo-single-domain grain size; SD, single-domain grain size.

and has high positive inclination values (62.0–79.9°). A second component *ILB* was isolated in a few specimens (3 of 18) during AF demagnetization at fields of 20–70 mT. Its inclination is also positive with significantly shallower values (7.8–47.4°; Figs 5 & 9b). However, thermal demagnetization shows a trend of the palaeomagnetic directions towards moderately negative directions and it is suspected that the shallow directions seen in the AF demagnetization data are still overprinted by the steep positive *ILA* components.

The Winnipegosis Formation displays somewhat better palaeomagnetic data than the Interlake and upper Red River formations. Here only one component *WPB* was isolated at high fields and low temperatures (16–115 mT; 0–330 °C; Figs 5 & 9c). With one exception, its inclination is reversed and has intermediate values (−23.5 to −41.7°).

The second section in the core with good palaeomagnetic data is the Souris River Formation. Similar to other strata in this well, two palaeomagnetic components were isolated in most of the samples. The first one, namely *SRA*, is characterized by low coercivities, is removed at low demagnetizing fields (0–15 mT) and low temperatures (80–330 °C) and carries positive inclination values of 58.6–74.9° (Fig. 9d). The second component *SRB* is isolated at higher fields (30–100 mT) and higher temperatures (250–550 °C) and has moderate reversed inclination compared to *SRA* (−40.1 to −58.3°; Figs 5 & 9d).

Analysis of the Birdbear Formation was complicated as the core could not be reconstructed. Some similarities were apparent between the plugs, however, and two types of behaviour were present. The two lowest of the six plugs were similar to the Souris River and lower Red River formations, carrying two components. *BBA* is a low temperature (<300 °C), low coercivity (0–20 mT) component, with steep inclination magnetization (absolute value of inclinations ranged from 62.6 to 74.7°; Fig. 9e). The second component *BBB* was only isolated during thermal demagnetizations at temperatures of 250–450°, and has medium to steep inclination values that are reversed compared to *BBA* (Fig. 9e). The uppermost three plugs in the Birdbear Formation showed evidence of a high coercivity mineral during AF demagnetization (see sample WB01030301, Fig. 9f). While there is little evidence of goethite, there were significant chemical changes above 500 °C thus preventing isolation of a characteristic remanent magnetization (ChRM) at temperatures characteristic of hematite. The low temperature/coercivity component *BBA* was observed in these plugs (47.5–75.3°); however, *BBB* was present to slightly higher temperatures (500 °C), was not reversed compared to *BBA* and could be defined in only two specimens (44.1° and 50.3°; Figs 5 & 9f).

## Interpretation of palaeomagnetic data

The palaeomagnetic directions discussed in the previous section lack azimuthal orientation. Trials to restore the core orientation by matching the low

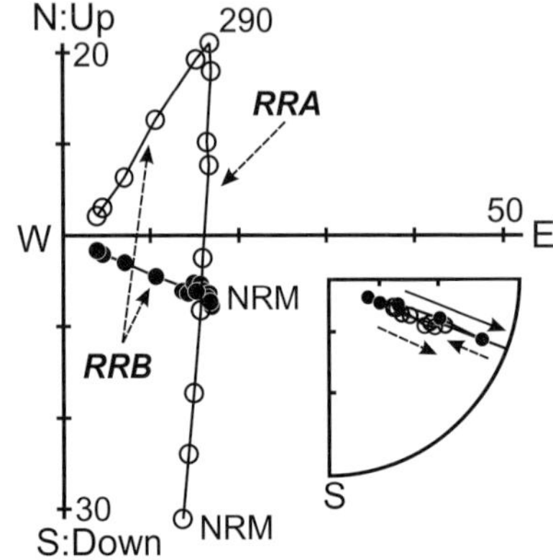

(a) Red River RR03150703
Thermal: 0–480°C
N:Up
20
290
RRA
W
50
E
NRM
RRB
30
S:Down
NRM
S

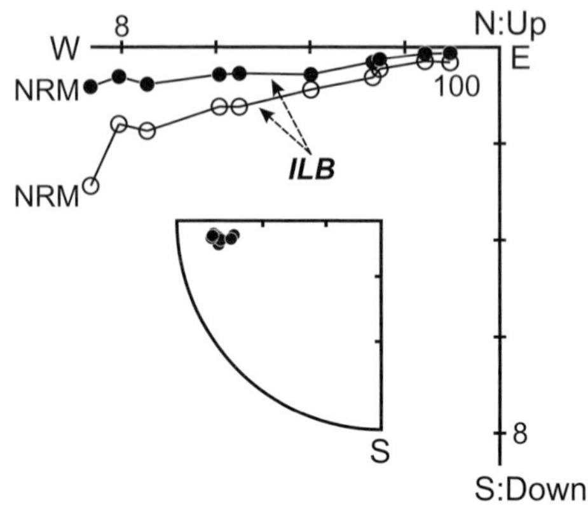

(b) Interlake WB01032302
AF: 0–100 mT
W
8
N:Up
E
NRM
100
NRM
ILB
S
8
S:Down

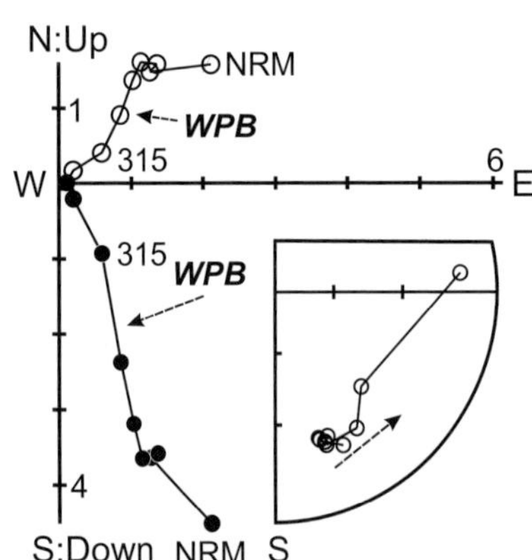

(c) Winnipegosis WB01031802
Thermal: 0–375°C
N:Up
NRM
1
WPB
315
W
6
E
315
WPB
4
S:Down
NRM
S

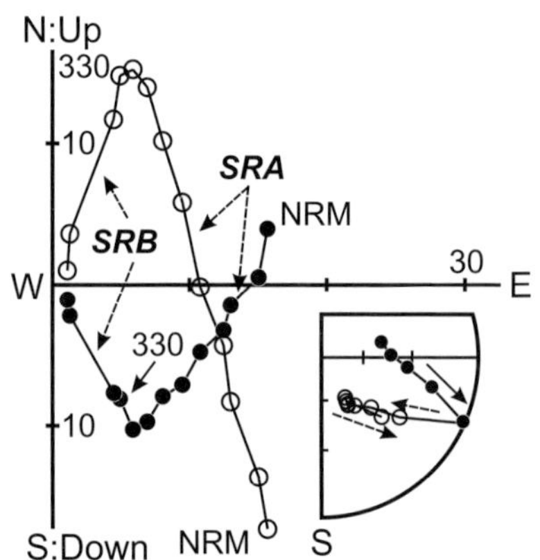

(d) Souris River WB01031102
Thermal: 0–500°C
N:Up
330
10
SRA
SRB
NRM
W
30
E
330
10
S:Down
NRM
S

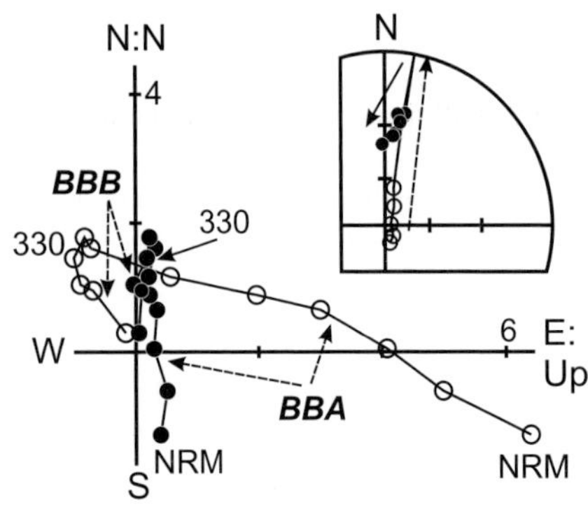

(e) Birdbear WB01030602
Thermal: 0–450°C
N:N
N
4
BBB
330
330
W
6
E:
Up
BBA
S
NRM
NRM

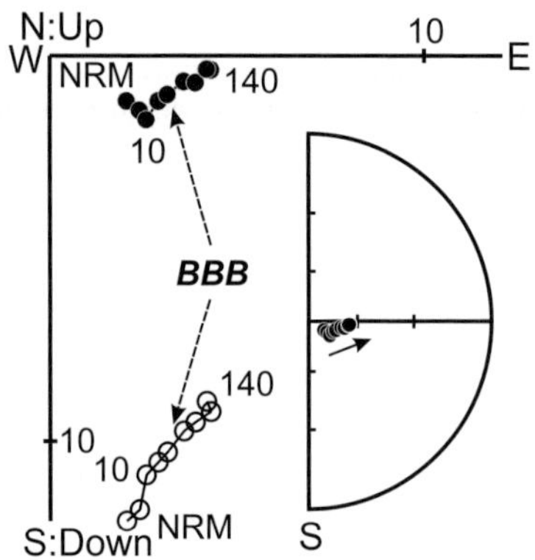

(f) Birdbear WB01030301
AF: 0–140 mT
N:Up
10
W
E
NRM
140
10
BBB
140
10
10
NRM
S:Down
NRM
S

field/low unblocking temperature components with the present-day magnetic field (Cioppa *et al.* 2000) failed to plot the other higher-temperature components on the North American apparent polar wander path (NAAPWP; Van der Voo 1990; Besse & Courtillot 2002). There are indications that the *A* component in both Red River and Souris River samples are a combination between the present Earth's magnetic field and a drilling-induced magnetization with some influence from the *B* magnetization as well, and therefore it cannot be used for orientation purposes. Because of the unknown declination values for the *B* components, the only possible way to examine the palaeomagnetic data in this study is inclination-only analysis. The inclination-only averages for the formations (excluding Birdbear and Interlake due to poor data) studied in this well are shown in Table 1. Figure 10 illustrates the inclination-only means that were calculated for the lower Red River, Winnipegosis and Souris River formations plotted on the expected inclinations defined from the NAAPWP (Van der Voo 1990; Besse & Courtillot 2002) at the well location. The inclination-only mean values for the Souris River Formation ($I = -49.1 \pm 4.5°$, $k = 104.6$, $N = 10$ specimens/ 6 plugs) and lower Red River carbonates ($I = -48.1 \pm 2.4°$, $k = 194.7$, $N = 21$ specimens/11 plugs) give the same possible magnetization ages: Late Silurian–Mid-Devonian or Early–Mid-Jurassic (Fig. 10). The inclination-only mean for the Winnipegosis Formation ($I = -36.6 \pm 7.1°$, $k = 62.2$, $N = 6$ specimens/4 plugs) gives a magnetization age that is Late Silurian–Late Devonian or Permian–Early Jurassic limit to Mid-Jurassic (Fig. 10).

## The age and nature of (re)magnetization events

While the Birdbear palaeomagnetic data from this well is poor and cannot be used for age determination, a similarity between the hematite palaeomagnetic inclination from the upper section of the Birdbear Formation and a magnetization found in the Amaranth Formation in Manitoba (Szabó *et al.* 2009) suggests that the hematite could have originated from the Amaranth Formation in the Hartney structure.

In three other sections of the core (upper Red River, Interlake and Winnipegosis strata), pyrrhotite is the main magnetic carrier; however, the first two also contain poor palaeomagnetic data. The inclination average of the characteristic palaeomagnetic components observed in the Winnipegosis Formation ($-36.6 \pm 7.1°$) is slightly higher (although within the error limits) than the inclination average value found in another well from Manitoba ($I = -27.3 \pm 10.4°$; Cioppa 2006). In that study, the inclination of the ChRM was depth-dependent, showing reversals and possibly indicating a primary or early post-depositional Devonian magnetization for this formation. The slight difference in the inclinations of the two wells might be due to the low number of plugs from which the magnetization in the Winnipegosis Formation was defined in this study, as well as to post-structure stabilization processes that induced movements along faults and fracturing of the Devonian strata in the Hartney structure (Anderson 1980). High-angle fracturing in the Winnipegosis section was noticed during the logging and palaeomagnetic sampling of this study well.

The lower Red River carbonates and Souris River Formation contain the best palaeomagnetic data and their rock magnetic characteristics suggest a common origin for the (re)magnetization. The oldest possible magnetization age is Late Silurian– early Mid-Devonian (Fig. 10); however, the Souris River Formation was deposited between the late Mid-Devonian and early Late Devonian (Fig. 2). This makes a Silurian–Devonian age for the *RRB*/*SRB* magnetization improbable. It follows that the most likely age for this magnetization is Early–Mid-Jurassic (Fig. 10). The most common suggestions for the origin of remagnetizations in basinal strata are burial, either thermoviscous remanent magnetization (TVRM) or chemical remanent magnetization (CRM), and fluid-flow (nominally orogenic). The maximum burial depth for the deepest sediments in the basin was *c.* 3.1 km (Osadetz *et al.* 2002, p. 238, Fig. 9) in the Late Cretaceous–Palaeogene. Neither the maximum temperature nor the timing of maximum burial are consistent with this component being a simple burial remagnetization (TVRM). A CRM carried by authigenic magnetite during either illitization or organic thermal maturation processes has been documented in other studies (e.g. Blumstein *et al.* 2004). Petrographic

---

**Fig. 9.** Characteristic demagnetization behaviours of the *A* and *B* palaeomagnetic components in the (**a**) lower Red River Formation; (**b**) Interlake Group; (**c**) Winnipegosis Formation; (**d**) Souris River Formation; (**e**) lower section of Birdbear Formation; and (**f**) upper section of Birdbear Formation. Each figure illustrates data based on either thermal or alternating field (AF) demagnetization of one specimen and includes a vector diagram and an equal-angle stereographic projection. Scale is in units of $10^{-5}$ A m$^{-1}$. Open (closed) circles in the stereographic projections represent negative (positive) inclinations. Closed (open) circles in the vector diagrams are projections on the horizontal (vertical) plane. NRM, natural remanent magnetization.

**Table 1.** *Summary of palaeomagnetic data*

| Formation | NM(p/s)* | NA(p/s)[†] | I[‡] (°) | Error ($\pm$) | K[§] | Remanence type |
|---|---|---|---|---|---|---|
| Birdbear | 6/16 | | | | | |
| Souris River | 6/14 | 6/10 | −49.1 | 4.5 | 104.6 | *SRB* |
| Winnipegosis | 7/16 | 4/6 | −36.6 | 7.1 | 62.2 | *WPB* |
| Interlake | 10/19 | | | | | |
| Red River | 14/41 | 11/21 | −48.1 | 2.4 | 194.7 | *RRB* |

*Numbers of samples measured (plugs/specimens).
[†]Numbers of samples used for calculating inclination-only means (plugs/specimens).
[‡]Value of palaeomagnetic direction inclination-only mean.
[§]Fisher precision parameter.

observations of this well, as well as other studies, have not documented any clay mineral content in the carbonates of the Red River Formation. In addition, the magnetization age recorded in the lower Red River and Souris River formations does not coincide with either the timing of early maturation of the Upper Ordovician–Middle Devonian strata (Early Carboniferous) nor with when they entered the petroleum generation stage (Late Cretaceous; Osadetz *et al.* 2002). Moreover, the vertical separation of Red River and Souris River formations suggests that the two could not have experienced illitization or organic thermal maturation at the same time, producing the same magnetization, while leaving the intervening strata (Winnipegosis Formation, Interlake Group) unaffected. Widespread fluid-flow-induced remagnetization has been suggested both in this basin (Koehler *et al.* 1997; Enkin *et al.* 2001) and elsewhere in the foreland basins around North America (e.g. McCabe & Elmore 1989; Symons *et al.* 1999; Cioppa *et al.* 2001). However, there are certain contra-indications for this hypothesis. First, palaeomagnetic data collected from the Red River carbonates from across the Williston

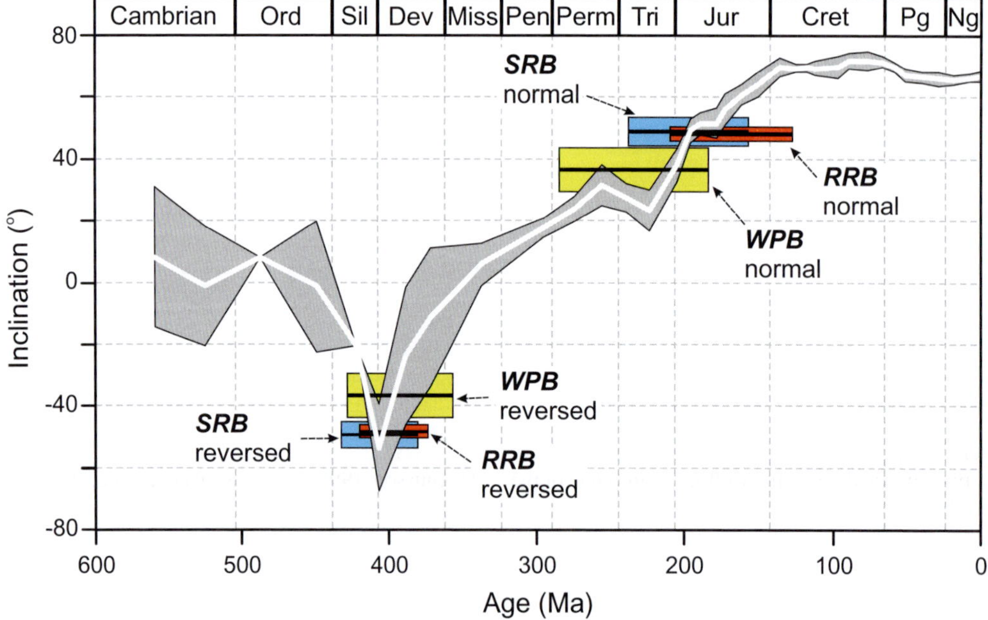


**Fig. 10.** Inclination-only mean values of the Red River (***RRB***), Souris River (***SRB***) and Winnipegosis (***WPB***) palaeomagnetic components (solid black bars) (see Table 1), plotted against expected inclinations (white line) at 49.4°/259° in the Williston Basin. The curve was calculated using the values for the North American apparent polar wander path of Van der Voo (1990) and Besse & Courtillot (2002). Modified after Enkin *et al.* (2001). Coloured and dark grey areas represent 95% confidence intervals. Ord, Ordovician; Sil, Silurian; Dev, Devonian; Miss, Mississippian; Pen, Pennsylvanian; Perm, Permian; Tri, Triassic; Jur, Jurassic; Cret, Cretaceous; Pg, Palaeogene; Ng, Neogene.

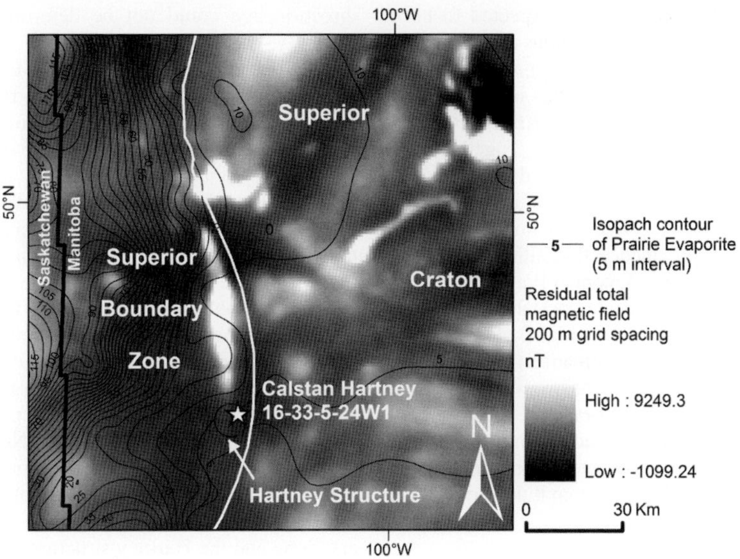


**Fig. 11.** Location of the Calstan Hartney 16-33-5-24 WPM well against the residual total magnetic field (Canadian Aeromagnetic Database 2011) and isopach contour lines for the Devonian Prairie Evaporite in the area (Li & Morozov 2007).

Basin (unpublished data) show that the highest NRM intensities and relevant magnetization carried in magnetite are present near the centre of the basin, associated with oil provinces and basement structures. However, at the margin of the basin, a hematite magnetization is present in the Red River carbonates. Second, there is some evidence for primary magnetizations in the Winnipegosis and lower Amaranth formations (Cioppa 2006; Szabó *et al.* 2009). This pattern is inconsistent with that associated with the widespread late Palaeozoic Alleghenian remagnetization which is carried prevalently by magnetite in carbonates and hematite in clastic sediments (McCabe & Elmore 1989). Neither pervasive large-scale fluid-flow remagnetization or burial remagnetization can therefore be invoked easily in the Williston Basin, and another explanation must be sought.

A closer look at the well location may assist in determining the causative factors for the remagnetization observed. First, the well sits at the edge of the Birdtail–Waskada Zone (BWZ) which overlaps the active Superior Boundary Zone (SBZ, Figs 1 & 11). Dissolution of salts has occurred in this north-trending zone (McCabe 1971) during the Devonian and Mississippian and at the Palaeozoic and Mesozoic boundaries (Nicolas & Barchyn 2008), as a result of basinal fluid circulation along reactivated basement-penetrating faults (Dietrich & Magnusson 1998; Gale & Conley 2000). Such reactivation has been documented elsewhere in the Williston Basin and is suggested to have resulted from post-

Precambrian movements of crustal blocks in the Early Palaeozoic and continuing until the Cretaceous (e.g. Bezys 1996), and to have created the 'Prairie-type' and upper Mississippi-Valley-type (MVT) mineralization within the Palaeozoic strata (Fedikow *et al.* 1996; Gale & Conley 2000; Bamburak & Klyne 2004). Volcanic activity may also have occurred as late as the Cretaceous in the SBZ (Bezys *et al.* 1996) and could have produced basinal fluids and mineralization. Coincidentally (or not), the presence of a strong magnetic anomaly just north of the well (Fig. 11) suggests that this anomalous area could have been the source of the magnetite carrying the *RRB/SRB* magnetization.

Second, the well was drilled on the interior margin of the Hartney structure (Fig. 11), which is considered to be either an impact-related crater or a volcanogenic-originated caldera (McCabe 1971; Sawatzky 1975; Anderson 1980; Dietrich & Magnusson 1998). Seismic data revealed that the top of the Palaeozoic strata in this well is about 180 m below the regional trend, and the Mississippian strata are missing (McCabe 1971; Anderson 1980). Magnetizations existing pre-impact/eruption are likely to have been affected by this event. Although the strata of the Souris River Formation and other Devonian strata in this well and in the Hartney structure area are brecciated and faulted due to the crater/caldera stabilization (Anderson 1980), similar fractures were not seen in the Red River Formation. Thus, if the magnetization was pre-impact, the Souris River strata would be expected to have scattered palaeomagnetic directions with higher error and lower Fisher precision parameter than the Red River results. However, this is not the case (Table 1), suggesting that the **SRB/RRB** magnetization is post-impact/eruption. Together, the basement-related faults of the SBZ and the fracturing and faulting due to the Hartney structure provide an excellent conduit for local basinal fluid flow. This is supported by the evidence that although present around the anomaly, the Devonian Prairie Evaporite is missing from the strata inside the Hartney structure (Fig. 11). The NRM intensity is higher at the bottom of the core in the Red River Formation (Fig. 5), which suggests a closer proximity to the origin of magnetizing fluids. For all of these reasons, the mostly likely explanation for the magnetization seen in the Souris River and Red River carbonates (and probably in the lower Birdbear) is a CRM from fluids originating in the basement.

### Constraints on the age of the Hartney structure?

The palaeomagnetic evidence from this and the Szabó *et al.* (2009) study can be used to provide constraints on the age of the Hartney structure. Anderson (1980) indicated that the first undisturbed strata in the Hartney anomaly are Lower Cretaceous sediments, which lie above disturbed Jurassic (?) red beds (McCabe 1971; Anderson 1980) probably belonging to the Lower Amaranth Formation. Anderson (1980) suggested an Early Cretaceous age for the formation of the Hartney structure (Anderson 1980), but recent palaeomagnetic data indicate a pre-Triassic deposition age for the Lower Amaranth red beds (Szabó *et al.* 2009). If the red beds discussed by Anderson (1980) were Amaranth, the data from this study and from Szabó *et al.* (2009) suggest that the formation of the Hartney structure occurred sometime between the Triassic and Mid-Jurassic, similar to the age proposed by Sawatzky (1975) and to the post-Mississippian–pre-Jurassic age suggested by McCabe (1971).

## Conclusions/Summary

The magnetic characteristics observed in the Calstan Hartney 16-33-5-24W1 well from the Williston Basin are complex and contain evidence of at least two episodes of remagnetization, as well as potentially evidence of a depositional magnetization. While the study was complicated by a lack of good palaeomagnetic data, detailed rock magnetic analysis of the magnetic mineral carriers associated with specific magnetizations showed several distinctive events, even when the magnetization

direction/age could not be determined with any certainty.

The oldest magnetization observed is carried by pyrrhotite in the Winnipegosis Formation, distinguished from the other magnetizations in the well by its magnetic characteristics. The oldest possible age suggested by inclination-only analysis (Late Silurian–Late Devonian) and similarities to one other study of the formation (Cioppa 2006) indicate that it is potentially a depositional remanent magnetization.

The next-oldest magnetization is the best defined and is a magnetite-carried remagnetization present in the lower Red River and Souris River formations. Its age is probably Early–Mid-Jurassic – a time when the Williston Basin was uplifted and no major orogenies were active on the North American craton. The magnetization was probably caused by basement-related fluid flow on conduits along fractures and faults related to both the Superior Boundary Zone and the Hartney structure.

The hematite-carried magnetization of the upper section of the Birdbear may have originated from the Lower Amaranth red beds present in the Hartney structure, while magnetite is present in the lower section. However, the palaeomagnetic data are insufficient to determine a precise magnetization age.

The palaeomagnetic data collected from the upper Red River and Interlake strata are weak and do not give a reliable magnetization age. Their magnetic mineralogy is a combination of pyrrhotite (similar to the Winnipegosis) and magnetite, but there is not enough evidence to prove a link to that formation's magnetization.

The magnetic data supports the idea that, when originally deposited, most carbonates have insufficient magnetic mineral content to record a depositional age and that the magnetization seen in the Souris River and lower Red River formations is likely due to external factors such as remagnetizing fluids. It also does not support the idea of far-travelled orogenic fluid flow as a mechanism for remagnetization of the intracratonic Williston Basin, but instead suggests that localized basement-originating flow may be the mechanism. As a side note, this data also constrains the formation age for the Hartney structure to sometime between Late Triassic and Mid-Jurassic, similar to the age suggested by McCabe (1971) and Sawatzky (1975).

The research for this paper was carried out by E. Szabò as part of her PhD thesis and was financially supported by a Natural Science and Engineering Research Council of Canada (NSERC) University Faculty Award and Discovery Grant to MTC. We thank A. Chertov who assisted with sample preparation and measurement. Thanks also to R. Bezys from the Manitoba Geological Survey and the people from the Core Laboratory at the Manitoba

Industry, Economic Development and Mines who provided assistance in core selection and retrieval. The paper was improved by suggestions from C. MacNiocaill and an anonymous reviewer.

# References


ANDERSON, C. E. 1980. A seismic reflection study of a probable astrobleme near Hartney, Manitoba. *Canadian Journal of Exploration Geophysics*, **16**, 7–18.

BAILLIE, A. D. 1951. *Devonian Geology of the Lake Manitoba Lake Winnipegosis Area*. Manitoba Mines and Natural Resources, Winnipeg, Manitoba, Mine Branch Publication 49–2.

BAMBURAK, J. D. & KLYNE, K. 2004. *A possible new Mississippi Valley-type mineral occurrence near Pemmican Island in the north basin of Lake Winnipegosis, Manitoba (NTS 63B12 and 13, 63C9 and 16)*. Manitoba Geological Survey, Manitoba Industry, Economic Development and Mines, Report of Activities, 2004, 266–278.

BESSE, J. & COURTILLOT, V. E. 2002. Apparent and true polar wander and geometry of the geomagnetic field over the last 200 Myr. *Journal of Geophysical Research*, **107**, 2300, doi: 10.1029/2000JB000050.

BEZYS, R. K. 1996. *Sub-Palaeozoic structure in Manitoba's northern Interlake along the Churchill Superior Boundary Zone: A Detailed Investigation of the Falconbridge William Lake Study Area*. Geological Services, Manitoba Energy and Mines, Manitoba, Open File OF94–3.

BEZYS, R. K. & BAMBURAK, J. D. 2004. *Lower to Middle Palaeozoic Stratigraphy of Southwestern Manitoba*. Manitoba Geological Survey, WCSB/TGI II Field Trip.

BEZYS, R. K., FEDIKOW, M. A. F. & KJARSGAARD, B. A. 1996. *Evidence of Cretaceous (?) volcanism along the Churchill-Superior Boundary Zone, Manitoba (NTS 63G/4)*. Manitoba Energy and Mines, Minerals Division, Report of Activities 1996, 122–126.

BLUMSTEIN, A. M., ELMORE, R. D. & ENGEL, M. H. 2004. Palaeomagnetic dating of burial diagenesis in Mississippian carbonates, Utah. *Journal of Geophysical Research*, **109**, B04101, doi: 10.1029/2003JB002698.

BURRUS, J., OSADETZ, K., WOLF, S., DOLIGEZ, B., VISSER, K. & DEARBORN, D. 1996. A two-dimensional regional basin model of Williston Basin hydrocarbon systems. *AAPG Bulletin*, **80**, 265–291.

CANADIAN AEROMAGNETIC DATA BASE 2011. *Geoscience Data Repository*. Geological Survey of Canada, Earth Sciences Sector, Natural Resources Canada Government of Canada. http://gdr.nrcan.gc.ca/aeromag/index_e.php.

CIOPPA, M. T. 2006. Fluid flow in the Duperow and Winnipegosis formations, Williston Basin, Canada: evidence from preliminary magnetic measurements. *Journal of Geochemical Exploration*, **89**, 65–68.

CIOPPA, M. T., GILLEN, K. P., LEWCHUK, M. T. & SYMONS, D. T. A. 2000. Palaeomagnetism provides alternative core orientation methods. *Oil & Gas Journal*, **98**, 46–53.

CIOPPA, M. T., LONNEE, J. S., SYMONS, D. T. A., AL-AASM, I. S. & GILLEN, K. P. 2001. Facies and lithological controls on palaeomagnetism: an example from the Rainbow South field, Alberta, Canada. *Bulletin of Canadian Petroleum Geology*, **49**, 393–407.

CISOWSKI, S. 1981. Interacting v. non-interacting single domain behaviour in natural and synthetic samples. *Physics of the Earth and Planetary Interiors*, **26**, 56–62.

DIETRICH, J. R. & MAGNUSSON, D. H. 1998. Basement controls on Phanerozoic development of the Birdtail-Waskada salt dissolution zone, Williston Basin, Southwestern Manitoba. *In*: CHRISTOPHER, J. E., GILBOY, C. F., PATERSON, D. F. & BEND, S. L. (eds) *Eighth International Williston Basin Symposium*. Saskatchewan Geological Society, Saskatchewan, Special Publication, **13**, 166–174.

ENKIN, R. J. 1994. *A computer program package for analysis and presentation of palaeomagnetic data, Version 4*. http://gsc.nrcan.gc.ca/sw/paleoe.php.

ENKIN, R. J. & WATSON, G. S. 1996. Statistical analysis of palaeomagnetic inclination data. *Geophysical Journal International*, **126**, 495–504.

ENKIN, R. J., BAKER, J. & OSADETZ, K. G. 2001. *Palaeomagnetic indications for a Late Palaeozoic age for part of the Watrous Formation, Williston Basin, Southern Saskatchewan*. Summary of Investigations, Saskatchewan Geological Survey, Saskatchewan Energy and Mines, Miscellaneous Reports, 2001–4.1, 72–76.

EVERITT, C. F. W. 1961. Thermoremanent magnetization I: Experiments on single-domain grains. *Philosophical Magazine*, **6**, 713–726.

FEDIKOW, M. A. F., BEZYS, R. K., BAMBURAK, J. D. & ABERCROMBIE, H. J. 1996. *Prairie-type microdisseminated Au mineralization – a new deposit type in Manitoba's Phanerozoic rocks (NTS 63C/14)*. Manitoba Energy and Mines, Minerals Division, Report of Activities 1996, 108–121.

FISHER, R. A. 1953. Dispersion on a sphere. *Proceedings of the Royal Society, London, Series A*, **217**, 295–305.

GALE, G. H. & CONLEY, G. G. 2000. *Metal contents of selected Phanerozoic drill cores and the potential for carbonate-hosted Mississippi Valley-type deposits in Manitoba*. Geological Survey, Manitoba Industry, Trade and Mines, Open File Report OF2000–3.

GERHARD, L. C., ANDERSON, S. B. & FISCHER, D. W. 1990. Petroleum Geology of the Williston Basin. *In*: LEIGHTON, M. W. (ed.) *Interior Cratonic Basins*. American Association of Petroleum Geologists, Tulsa, Bulletin, **51**, 507–559.

GLASS, D. J. 1990. *Lexicon of Canadian Stratigraphy*. Canadian Society of Petroleum Geologists, Western Canada, 4.

JACKSON, M., GRUBER, W., MARVIN, J. & SUBIR, K. B. 1988. Partial Anhysteretic Remanence and its anisotropy: applications and grain-size dependence. *Geophysical Research Letters*, **15**, 440–443.

KIRSCHVINK, J. L. 1980. The least-squares line and plane and the analysis of palaeomagnetic data. *Geophysical Journal of the Royal Astronomical Society*, **62**, 699–718.

KOEHLER, G., KYSER, T. K., ENKIN, R. & IRVING, E. 1997. Palaeomagnetic and isotopic evidence for the diagenesis and alteration of evaporates in the Palaeozoic Elk Point Basin, Saskatchewan,

Canada. *Canadian Journal of Earth Sciences*, **34**, 1619–1629.

LEWCHUK, M. T., AL-AASM, I. S., SYMONS, D. T. A. & GILLEN, K. P. 1998. Dolomitization of Mississippian carbonates in the Shell Waterton gas field, southwestern Alberta: insights from palaeomagnetism, petrology and geochemistry. *Bulletin of Canadian Petroleum Geology*, **46**, 387–410.

LI, J. & MOROZOV, I . 2007. *Geophysical Investigations of the Precambrian Basement of the Williston Basin in south-eastern Saskatchewan and south-western Manitoba.* Williston Basin Targeted Geoscience Initiative (TGI) Report, University of Saskatchewan.

MARTINIUK, C. D., YOUNG, H. R. & LEFEVER, J. A. 1995. Lithofacies and petroleum potential of the Birdbear Formation (Upper Devonian), southwestern Manitoba and north-central North Dakota. *In*: VERN HUNTER, L. D. & SCHALLA, R. A. (eds) *Seventh International Williston Basin Symposium.* Saskatchewan Geological Society, Saskatchewan, Special Publication, **12**, 89–102.

MARTINIUK, C. D., YOUNG, H. R. & KLASSEN, H. J. 1998. *Regional Overview of the Geology and Petroleum Potential, Red River Formation, Southwestern Manitoba.* Manitoba Energy and Mines, Manitoba, Petroleum Open File POF17-98.

MCCABE, H. R. 1971. Stratigraphy of Manitoba, an introduction and review. *In:* TURNOCK, A. C. (ed.) *Geoscience studies in Manitoba.* Geological Association of Canada, Special Paper, **9**, 167–187.

MCCABE, C. & ELMORE, D. R. 1989. The occurrence and origin of Late Palaeozoic remagnetization in the sedimentary rocks of North America. *Reviews of Geophysics*, **27**, 471–494.

NICOLAS, M. P. B. & BARCHYN, D. 2008. *Summary Report on Palaeozoic stratigraphy, Mapping and Hydrocarbon Assessment, Southwestern Manitoba.* Williston Basin project (targeted geoscience initiative TGI II), Manitoba Geological Survey, Manitoba Science, Technology, Energy and Mines, Geoscientific Paper GP2008-2.

NORRIS, A. W., UYENO, T. T. & MCCABE, H. R. 1982. *Devonian rocks of the Lake Winnipegosis-Lake Manitoba outcrop belt, Manitoba.* Geological Survey of Canada, Canada, Memoir, **392**.

OSADETZ, K. G., KOHN, B. P., FEINSTEIN, S. & O'SULLIVAN, P. B. 2002. Thermal history of Canadian Williston Basin from apatite fission-track thermochronology-implications for petroleum systems and geodynamic history. *Tectonophysics*, **349**, 221–249.

RÉDLY, P. & HAJNAL, Z. 1995. Tectono-stratigraphic evolution of the Williston Basin – a regional seismic stratigraphic study. *In*: VERN HUNTER, L. D. & SCHALLA, R. A. (eds) *Seventh International Williston Basin Symposium.* Saskatchewan Geological Society, Saskatchewan, Special Publication, **12**, 341–350.

SAWATZKY, H. B. 1975. Astroblemes in the Williston Basin. *AAPG Bulletin*, **59**, 694–710.

SYMONS, D. T. A. & CIOPPA, M. T. 2000. Crossover plots: a useful method for displaying SIRM data in palaeomagnetism. *Geophysical Research Letters*, **27**, 1779–1782.

SYMONS, D. T. A., ENKIN, R. J. & CIOPPA, M. T. 1999. Palaeomagnetism in the Western Canada Sedimentary Basin: dating fluid flow and deformation events. *Bulletin of Canadian Petroleum Geology*, **47**, 534–547.

SZABÓ, E. & CIOPPA, M. T. 2006. Multiple magnetization events in the Red River carbonates, Williston Basin: evidence for fluid flow? *Journal of Geochemical Exploration*, **89**, 384–388.

SZABÓ, E., CIOPPA, M. T. & AL-AASM, I. S. 2009. Palaeomagnetic and geochemical evidence for a pre-Triassic age for the Lower Amaranth Member of the Williston Basin of Manitoba (Canada). *Canadian Journal of Earth Sciences*, **46**, 855–873, doi: 10.1139/E09-058.

VAN DER VOO, R. 1990. Phanerozoic palaeomagnetic poles from Europe and North America and comparison with continental reconstruction. *Reviews of Geophysics*, **28**, 167–206.

# A multidisciplinary investigation of multiple remagnetizations within the Southern Canadian Cordillera, SW Alberta and SE British Columbia


M. S. ZECHMEISTER[1,2]*, S. PANNALAL[1] & R. D. ELMORE[1]

[1]*School of Geology and Geophysics, University of Oklahoma, Norman, OK 73019, USA*

[2]*Present Address: Shell Exploration and Production Company, Houston, TX 77079, USA*

**Corresponding author (e-mail: Matthew.Zechmeister@shell.com)*



**Abstract:** An integrated palaeomagnetic, geochemical and petrographic study was conducted on two folds in the Front Range of the Southern Canadian Cordillera in order to better understand the timing and origin of chemical remanent magnetizations (CRMs) relative to orogenesis. The folds are contained within Mississippian carbonates (330–335 Ma) which contain a pervasive pre-tilting to early syn-tilting Early Cretaceous, high-temperature CRM residing in magnetite. An intermediate-temperature CRM is a late syn-tilting to post-tilting, possibly Tertiary remagnetization, residing in pyrrhotite. A fluid conduit test (FCT) conducted on bedding-parallel veins shows that they are associated with the magnetite CRM, whereas late-stage tensile veins show a relationship to a pyrrhotite CRM. Elevated $^{87}Sr/^{86}Sr$ data indicate alteration by fluids with a radiogenic signature; along with the FCT results, these data are consistent with the interpretation that the magnetite CRM formed as a result of hydrocarbons and/or evolved basinal fluids that migrated ahead of the deformation front. Based on the presence of sulphur-enriched bitumen, barite and sphalerite, common by-products of thermal sulphate reduction (TSR), the pyrrhotite CRM is interpreted to be the result of late-stage TSR caused by warm basement fluids which moved along faults and fractures.


In terms of fold and thrust belts, diagenesis and tectonism processes are inherently linked. Palaeomagnetic studies in fold and thrust belts can provide useful information on the timing of diagenetic events relative to orogenesis (e.g. McCabe & Elmore 1989; Stamatakos *et al.* 1996; Enkin *et al.* 2000; Katz *et al.* 2000; Elmore *et al.* 2001; O'Brien *et al.* 2007). During orogenesis, tectonic stacking of thrust sheets along with uplift and erosion increases burial in foreland basins and leads to elevated burial temperatures. Such elevated temperatures can cause the acquisition of chemical remanent magnetizations (CRMs) through clay diagenesis (Katz *et al.* 2000) or maturation of organic matter (Banerjee *et al.* 1997; Blumstein *et al.* 2004), as well as thermoviscous remagnetizations (TVRMs). Orogenic processes also generate regional fluid migration either through gravitational recharge (e.g. Ge & Garven 1994) or tectonically induced fluid flow (Oliver 1986; Machel & Cavell 1999). This can produce diagenetic alteration including precipitation of magnetic minerals and acquisition of CRMs in fold and thrust belts (e.g. McCabe & Elmore 1989; O'Brien *et al.* 2007). Stress reorganization of domain walls (e.g. Hudson *et al.* 1989; Borradaile 1997) or strain alteration of pre-existing magnetizations during deformation (e.g. Kodama 1988; Elmore *et al.* 2006; Zechmeister 2010) have also been proposed as remagnetization mechanisms in folded rocks.

This study has utilized an integrated palaeomagnetic, rock magnetic, geochemical and petrographic approach to investigate Mississippian carbonates (330–335 Ma) in the Canadian Cordillera. The objective of the study is to determine the origin and timing of a regional multi-component remagnetization in the rocks. Previous studies in the Canadian Cordillera and Western Canada sedimentary basin (WCSB) reported an early syn-tilting to pre-tilting high-temperature magnetic component contained in magnetite that displayed a wide range in ages from the Tertiary to the Late Jurassic (e.g. Enkin *et al.* 2000). The Enkin *et al.* (2000) study was based on samples from several east–west transects through the Canadian Cordillera and did not investigate individual folds in detail (Enkin *et al.* 2000). Other studies utilized drill core from only one limb of a fold and were unable to perform complete tilt tests (e.g. Lewchuk *et al.* 1998; Cioppa & Symons 2000). These studies proposed multiple hypotheses for the origin of the magnetite component, each of which will be evaluated in the discussion. The previous studies also found an intermediate-temperature (IT) component which was commonly interpreted as a TVRM in magnetite (e.g. Robion *et al.* 2004).


*From*: ELMORE, R. D., MUXWORTHY, A. R., ALDANA, M. M. & MENA, M. (eds) 2012. *Remagnetization and Chemical Alteration of Sedimentary Rocks*. Geological Society, London, Special Publications, **371**, 123–144.
First published online August 22, 2012, http://dx.doi.org/10.1144/SP371.11
© The Geological Society of London 2012. Publishing disclaimer: www.geolsoc.org.uk/pub_ethics

Cioppa *et al.* (2003) suggested the IT component may reside in both magnetite and pyrrhotite.

The objective of this integrated study was to clarify the origin of this complex remagnetization by conducting a detailed study of two well-exposed folds in the Southern Canadian Cordillera. Palaeomagnetic analysis utilized tilt tests to determine the age of the magnetization relative to folding. High-temperature, room-temperature and low-temperature rock magnetic experiments were conducted to determine magnetic carriers. Fluid conduit tests (FCTs) which compare the magnetic intensity and geochemistry of veined material to that of the unveined host rock (e.g. Elmore *et al.* 1993) were used to investigate the relationship between fluids and each remagnetization event. Geochemical analysis and fluid inclusion microscopy were then used to investigate the nature of these vein- and vug-forming fluids. Transmitted light and electron microprobe microscopy were used to determine lithology and identify diagenetic features and opaque minerals.

## Geological and tectonic setting

The foreland fold and thrust belt in the Canadian Cordillera is the result of a west–east-directed horizontal compression due to terrane accretion, accommodating up to *c.* 200 km of horizontal shortening

(e.g. Price 1981). Initiation of thrusting in the westernmost portion of the foreland belt began in the Early Cretaceous (Carr & Simony 2006) and is expressed as thin-skinned detachment thrusting and folding of Mesoproterozoic–Cretaceous strata (e.g. Monger & Price 1979; Price 1994). The majority of shortening in the Canadian Cordillera occurred during the Late Cretaceous–Eocene (Price 1981; van der Pluijm *et al.* 2006).

The folds sampled in this study are located within the eastern Front Ranges of the Southern Canadian Cordillera (Fig. 1) and are contained within the Mississippian Rundle Group. In SW Alberta, the Mt Kidd syncline was sampled within Kananaskis Country Provincial Park along Kananaskis Road (HWY-40; Fig. 1a). This structure originates in the Rundle Thrust Sheet and is a plunging asymmetrical syncline (plunge, trend of fold axis: S17°, 178). The structure (Fig. 2a) is contained within the Banff, Livingstone, Mount Head and Etherington formations (Fig. 1a; McMechan 1995). In SE British Columbia the Line Creek anticline was sampled within the Lewis Thrust Sheet on the property of Tek Coal (Fig. 1b). The fold is an asymmetrical upright fold (fold axis: S2°, 169; Fig. 2b) contained in the Mount Head and Etherington formations (Fig. 1b; Price *et al.* 1992).

The Mississippian Rundle Group is considered to be predominantly platform–basin carbonates

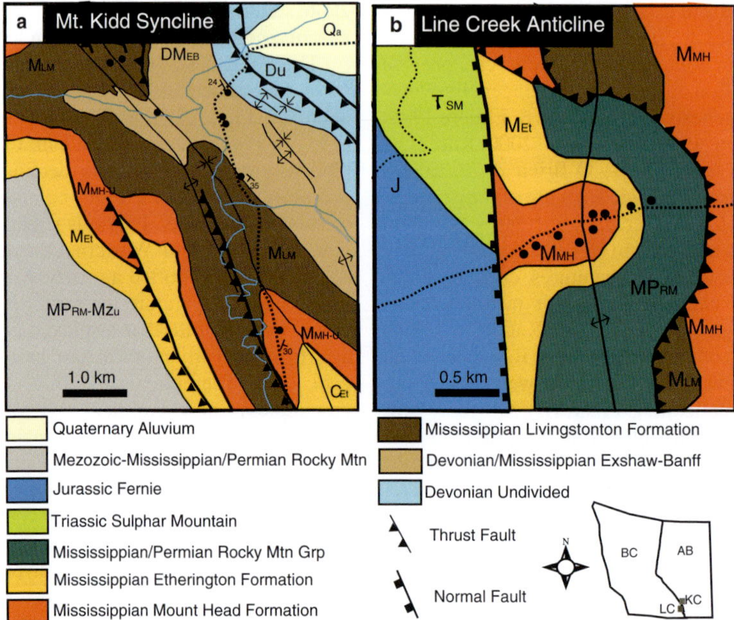



**Fig. 1.** Geological maps of the folds sampled. Black dots indicate major outcrops sampled. (**a**) Mt Kidd syncline, in Kananaskis Country Provincial park in SW Alberta (modified from McMechan 1995). (**b**) Line Creek Anticline in SE British Columbia (modified from Price *et al.* 1992). Inset map on lower left shows the study areas (squares).

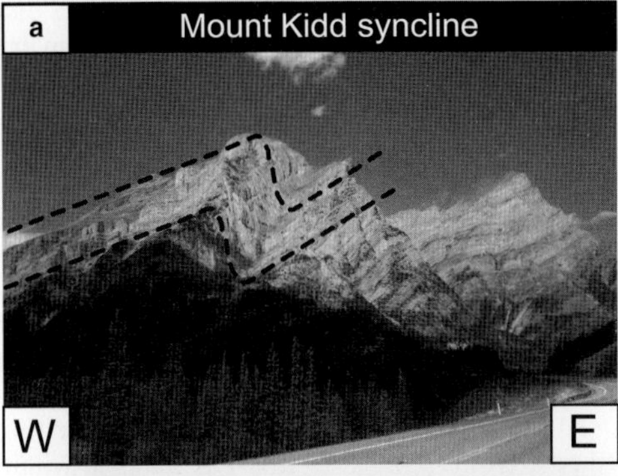

**Fig. 2.** Field photos of structures sampled. (**a**) Mt Kidd syncline with the road for scale. (**b**) Line Creek anticline. Trees are *c.* 3 m for scale.

deposited within a passive-margin setting (e.g. Hardebol *et al.* 2007) or a foreland basin during the Antler Orogeny (e.g. Root 2001). The paragenetic sequence for the carbonates is complex including early matrix dolomitization, pervasive dolomite recrystallization and late formation of vug-rimming saddle dolomites and sulphide mineralization (e.g. Al-Aasm 2000; Cooley *et al.* 2011). Hydrocarbon generation and migration in the Front Ranges is believed to have occurred within Carboniferous strata during the Late Jurassic–Early Cretaceous prior to deep Laramide burial (Kalkreuth & McMechan 1988). Hydrocarbon emplacement was followed by thermochemical sulphate reduction (TSR) of the hydrocarbons, creating extensive sour gas reservoirs throughout the Front Ranges, Foothills and Western Canada Sedimentary Basin (Orr 1974; Machel *et al.* 1995; Roure *et al.* 2005; Vandeginste *et al.* 2009).

## Methodology

Sites for palaeomagnetic and rock magnetic analysis (*c.* 8 specimens/site) were sampled with a portable gasoline drill and each core was oriented with an inclinometer and Brunton compass in the field. Thirty-six sites were collected from the Mt Kidd syncline within the Mississippian carbonates (Fig. 1a). Fifty-six sites were collected within the asymmetrical Line Creek anticline from an almost continuous exposure of outcrops along Line Creek (Fig. 1b). An FCT was conducted at Mt Kidd along a 95 cm long traverse across three *c.* 2.0 cm thick bedding-parallel veins. At Line Creek no definitive large-scale isolated veins were observed so the FCT compared results from a bed devoid of veins and a bed containing intense tensile veining a few metres away.

Palaeomagnetic samples were cut to standardized lengths (*c.* 2.2 cm) and natural remanent magnetizations (NRMs) were measured using a 2G-Enterprises cryogenic magnetometer with DC squids in the shielded palaeomagnetic lab at the University of Oklahoma. Prior to thermal demagnetization, the samples were subjected to two low-temperature demagnetization (LTD) treatments to remove the unstable remanence in multi-domain (MD) grains (Dunlop & Argyle 1991). This was accomplished by submerging the samples in liquid nitrogen for two hours and warming them to room temperature in a zero field. The NRM was then re-measured after each treatment. Cores were then subjected to stepwise thermal demagnetization in 22 steps from 100 to 580 °C using an ASC scientific Thermal Specimen Demagnetizer.

Demagnetization data was plotted on Zijderveld (1967) diagrams using the Super Interactive Analysis of Paleomagnetic Data (IAPD) program, which was also used for principal component analysis (PCA; Kirschvink 1980) to determine the direction of the different remanence components. Only specimens with a maximum angle of deviation (MAD) value $<10°$ were considered as coherent and used to create site statistics. Site means were calculated using Fisher (1953) statistics in Super IAPD. Only sites with suitable statistics ($k > 10$, $\alpha 95 < 15°$) were used in the tilt tests. The optimal clustering (OC) tilt test (Watson & Enkin 1993) and the direction-correction (DC) tilt test (Enkin 2003) were performed using the Geological Survey of Canada Paleomagnetism Analysis program package (PMGSC) version 4.2 software program (Enkin, pers. comm., 2004). Prior to conducting the tilt tests, the plunge of the Mt Kidd Syncline was removed because unfolding around a horizontal axis is needed to accurately apply the incremental un-tilting tests. This was accomplished by considering the site mean vector data as a line on the surface of a particular bed. The plunge was corrected to horizontal by following a modified version of the method of tracking a line during the unfolding of a plunging fold as outlined in Marshak & Mitra (1988; method 6–14, p. 119–120). Palaeopoles as well as their errors were calculated using the PMGSC version 4.2 software program (Enkin 2004). The poles were plotted on the Mesozoic portion of the apparent polar wander path (APWP) for North America (Besse & Courtillot 2002) in order to determine the age of the magnetizations.

Isothermal remanent magnetization (IRM) acquisition/decay experiments were performed after the NRM had been subjected to alternating field (AF) demagnetization at 120 mT using a 2 G Enterprises Automated Degaussing System. Acquisition of the room temperature (RT) IRM was conducted in 25 steps from 10 to 2500 mT using an ASC Scientific Impulse Magnetizer. Cumulative log-Gaussian (CLG) analysis was then performed on the data using the IRM-CLG 1.0 software of Kruiver *et al.* (2001) to investigate the room-temperature coercivity spectrum (Heslop *et al.* 2004). Finally, these specimens were again demagnetized using AF demagnetization and had an RTIRM imparted along three mutually perpendicular directions within the cores using peak fields of 120, 500 and 2500 mT. This was followed by thermal demagnetization for 25 steps from 20 to 680 °C in order to produce tri-axial thermal decay curves which are useful to investigate unblocking temperatures (Lowrie 1990).

Eight samples also underwent a series of low-temperature (LT) experiments using a Quantum Design Magnetic Property Measurement System at the Institute for Rock Magnetism. These experiments involved an RT saturation IRM (SIRM) which was then subjected to progressive cooling to 10 K followed in a few experiments by warming to room temperature, all in a zero field. This experiment is useful to determine LT magnetic transitions (i.e. 120 K Verwey and 34 K Besnus transitions; Rochette *et al.* 2011) which are indicative of particular magnetic phases (Dekkers *et al.* 1989; Rochette *et al.* 1990).

Petrographic analysis using transmitted and reflected light was conducted on thin sections (minimum of one per site) produced from oriented cores as well as hand samples in order to determine lithology and to investigate diagenetic- and deformation-related features. An automated Cameca SX50 electron probe micro-analyser equipped with an integrated energy-dispersive x-ray analyser at the University of Oklahoma was used to investigate the nature of common opaque minerals in the carbonates. Backscatter electron imaging (BSEI) was used to image the opaques while energy-dispersive x-ray analysis (EDXA) was used for qualitative analysis of the composition of the opaque minerals.

Doubly polished thin sections from calcite- and dolomite-filled veins and vugs were prepared for preliminary fluid inclusion analysis. Petrographic analysis was performed on a Zeiss research microscope using a Zeiss 40× long working distance objective. Fluid inclusion microthermometry was carried out using a Linkam TH-600 heating/freezing stage connected to a Linkam TMS 90 control unit. The temperature at which the vapour bubble disappeared in an individual inclusion was recorded as the homogenization temp ($T_h$). These temperatures were recorded in increasing order so that lower-temperature inclusions were not damaged. Careful observations were made to look for stretching of the inclusions. No inclusions studied exhibited stretching. The small size of the inclusions ($>10$ µm) made it difficult to obtain ice melting temperatures ($T_m$).

Strontium isotope analysis was performed at the University of Texas (Austin) in two batches according to the methods described by Gao *et al.* (1992). Four samples were ran from Mississippian carbonate matrix and vein material from Line Creek and seven samples from Mississippian carbonate matrix and vein material from Mt Kidd. Samples consisted of dolomitized limestones; after removal of the exchangeable and water-soluble Sr the samples were dissolved in 4% acetic acid for 20 min to remove as much calcite as possible before dissolving them in 8% acetic acid to dissolve the dolomite. The NIST SRM 987 standard mean value for the 2008 samples was $0.710264 \pm 0.000007$ ($2\sigma = 0.000015, n = 33$). For the 2009 batch, the NIST SRM 987 standard mean was $0.71026 \pm 0.000005$ ($2\sigma = 0.000012$, $n = 35$). The strontium values were normalized relative to the National Bureau of Standards NBS $987 = 0.71014$. The $^{87}Sr/^{86}Sr$ values were then plotted and compared to the coeval seawater values for Carboniferous seawater (McArthur *et al.* 2001).

## Results and interpretation

### Palaeomagnetism

Based on the Zijderveld diagrams the Mississippian carbonates at Mt Kidd and Line Creek contain two stable components after removal of a modern viscous remanent magnetization (VRM) by 200 °C (Fig. 3a, b). An intermediate-temperature (IT) component with southerly declinations and steep up inclinations was removed between 220 and 340 °C (Fig. 3a, b). The maximum unblocking temperature suggests pyrrhotite carries the component although, according to the palaeomagnetic data alone, greigite and magnetite cannot be completely ruled out. The characteristic remanent magnetization (ChRM) has northerly directions and steep down inclinations (Fig. 3a, b) and is removed from 340 to 540 °C, which suggests that magnetite carries the component. The temperature range for both components can be clearly seen on graphs of percent NRM versus temperature (Fig. 3a, b). Low-temperature

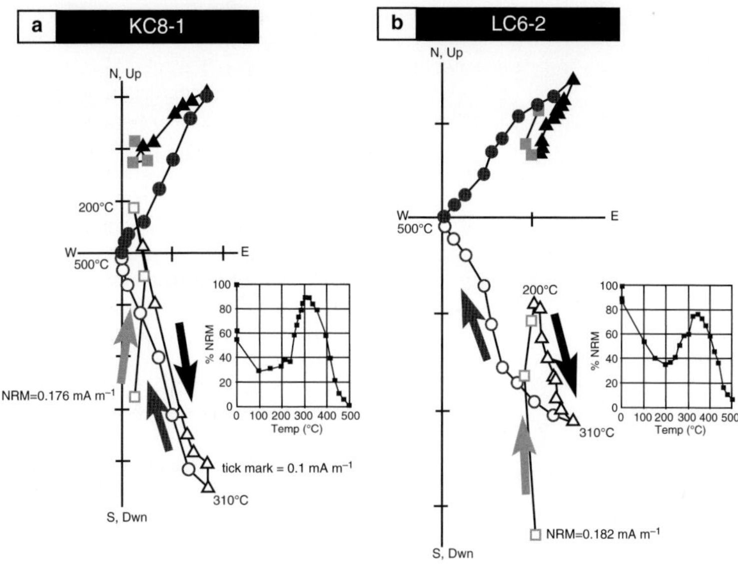



**Fig. 3.** Representative orthogonal projections (Zijderveld diagrams) in geographical coordinates for stepwise thermal demagnetization of (**a**) KC and (**b**) LC specimens. Open symbols represent the vertical component and the closed symbols indicate the horizontal component. The normal-polarity ChRM is indicated by circles, the reversed-polarity intermediate-component is indicated by triangles and the modern VRM is indicated by squares. Arrows indicate the direction of thermal decay. The small insert map shows the percentage of magnetization removed by a particular temperature during stepwise demagnetization.

treatment prior to thermal demagnetization shows an overall decrease in the NRM intensity for the specimens at all localities but the decrease varies from site to site (5–41% for KC; 10–30% for LC), suggesting a highly variable concentration of unstable MD grains.

## Rock magnetism

Cumulative log Gaussian analysis (Kruiver *et al.* 2001) was performed on the IRM acquisition data to quantitatively examine the coercivity spectrum. The LC and KC samples are similar and require three components to fit the raw data (Fig. 4a, b; Table 1). The first component is a very low-coercivity component which contributed less than 10% of the total IRM and had a coercivity of *c.* 17 mT which is in the range of MD magnetite. The second component is a low-coercivity component which contributed between 50 and 70% of the IRM and has a coercivity range of 40–52 mT which is in the range for pseudo-single-domain (PSD) to single-domain (SD) magnetite (Hunt *et al.* 1995). The third component is a moderate-coercivity component which contributed 20–40% of the total IRM and had a coercivity of 158–251 mT, which overlaps the range for pyrrhotite (Hunt *et al.* 1995).

Thermal decay of a tri-axial IRM is dominated by the low-coercivity component in the majority of the samples, with all the curves completely decayed by 580 °C suggesting the presence of magnetite (Fig. 5a, b). In many samples, a minor inflection in the low-coercivity tri-axial decay curve is observed around 320 °C, which is consistent with the presence of pyrrhotite. In some sites demagnetization was not completed until 680 °C, suggesting the presence of non-remanence-carrying late-stage hematite likely related to post-exhumation oxidation of iron-bearing minerals.

Two out of the eight samples cooled to 10 K after having been imparted with a RTSIRM showed a decrease in magnetic intensity at *c.* 34 K (Fig. 6a), indicative of pyrrhotite (Dekkers *et al.* 1989; Besnus Transition). The remaining six samples did not show the 34 K Besnus transition although all of the samples examined contained a $M_{max}$ between 200 and 250 K (Fig. 6b), which is also diagnostic of pyrrhotite (Dekkers *et al.* 1989; Rochette *et al.* 1990). Although all of the samples examined contained the intermediate component, the samples which displayed the Besnus transition had the most pronounced IT component. The RTSIRM samples do not show a well-defined Verwey transition (Fig. 6), perhaps because it is suppressed due to unblocking of a range of magnetite grain sizes with variable

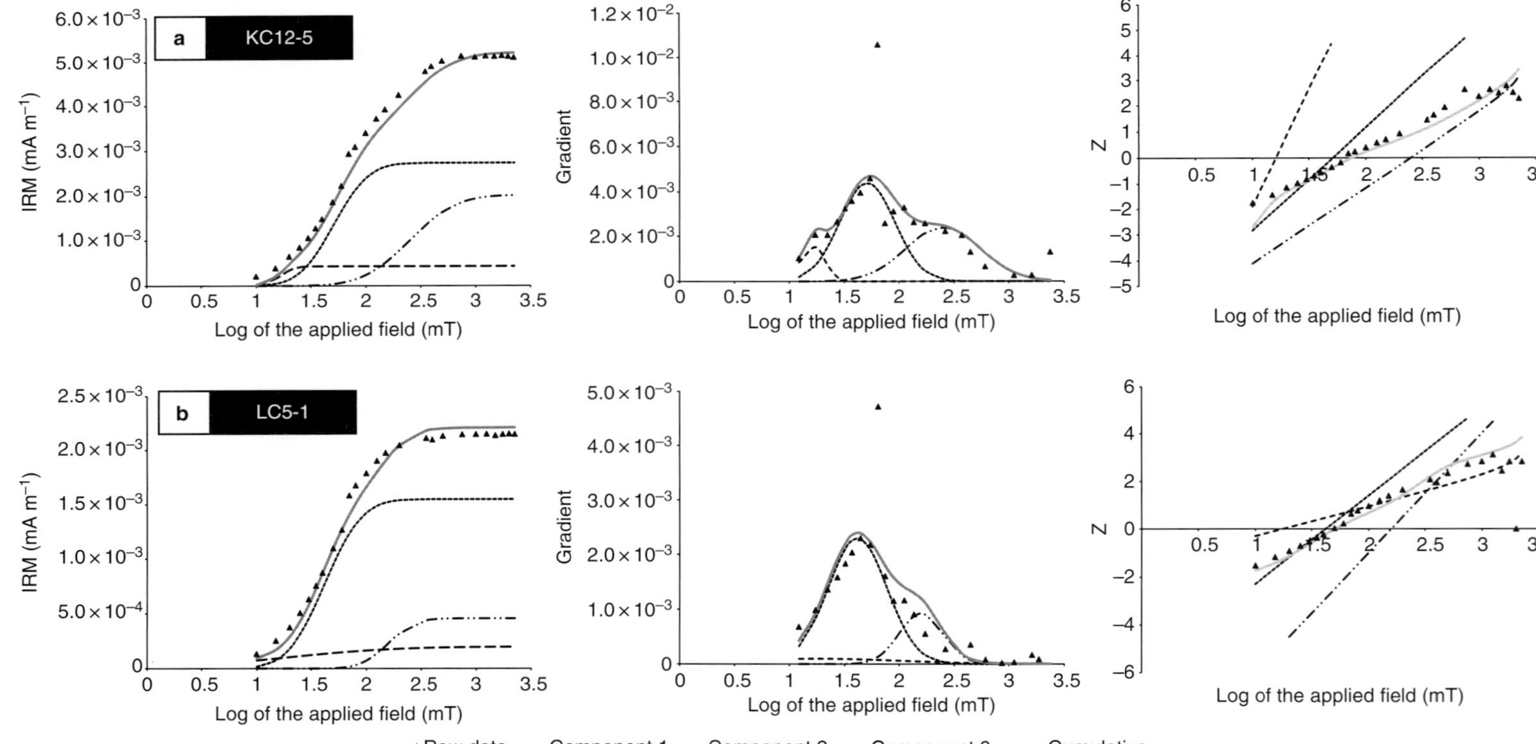



**Fig. 4.** Representative results from cumulative log-Gaussian analysis of a 25-step IRM acquisition for Mt Kidd (KC) and Line Creek (LC). (**a**) KC12-5 from the west-dipping shallow limb from Mt Kidd syncline showing 3 components. (**b**) LC5-6 from the west-dipping shallow back limb of Line Creek anticline showing 3 components.

**Table 1.** *Component results from cumulative log-Gaussian analysis*

| Sample | Contribution (%) | SIRM (mA m$^{-1}$) | Log ($B_{1/2}$) (mT) | $B_{1/2}$ (mT) | DP (mT) | Mineralogy |
|---|---|---|---|---|---|---|
| KC12-5 | | | | | | |
| 1 | 8.3 | $4.33 \times 10^{-4}$ | 1.21 | 16.2 | 0.11 | MD magnetite |
| 2 | 52.8 | $2.75 \times 10^{-3}$ | 1.71 | 51.3 | 0.25 | PSD-SD magnetite |
| 3 | 38.9 | $2.03 \times 10^{-3}$ | 2.4 | 251.2 | 0.34 | Pyrrhotite |
| LC5-1 | | | | | | |
| 1 | 9 | $2.00 \times 10^{-4}$ | 1.25 | 17.8 | 0.8 | MD magnetite |
| 2 | 70.1 | $1.55 \times 10^{-3}$ | 1.62 | 41.7 | 0.27 | PSD-SD magnetite |
| 3 | 20.9 | $4.61 \times 10^{-4}$ | 2.2 | 158.5 | 0.2 | Pyrrhotite |

coercivites (pers. com. Mike Jackson, 2008) or oxidation of magnetite (Özdemir *et al.* 1993).

## Tilt test results

*Line Creek Anticline, BC.* Out of the 56 sites collected from Line Creek, 51 of the sites (91%) contained specimens with the ChRM (Table 2). Out of these, 35 of the sites (60%) contained specimens with coherent (stable) magnetizations with MAD angles <10° needed to generate Fisher statistics. Out of these, 4 sites did not have suitable site statistics

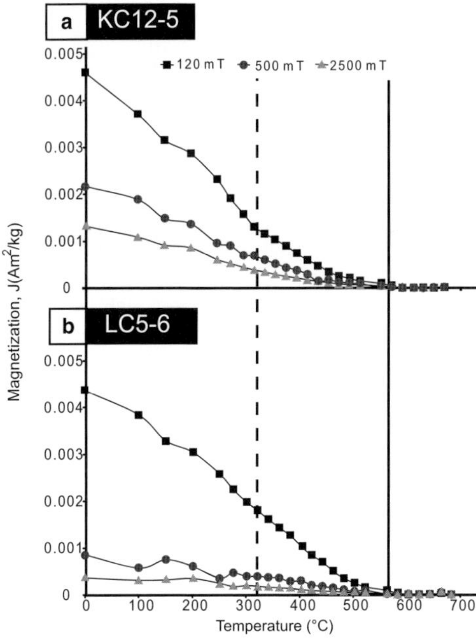



**Fig. 5.** Representative tri-axial thermal decay curves of isothermal remanent magnetizations (IRM) for Mt Kidd and Line Creek. The dashed line highlights the pyrrhotite unblocking temperature (325 °C) and the solid line represents the magnetite unblocking temperature (580 °C) (**a**) KC12-5, Mt Kidd syncline. (**b**) LC5-6, Line Creek Anticline.

($k > 10$, $\alpha 95 < 15°$) and were excluded from the tilt tests. The specimens in the 16 remaining sites contained incoherent (unstable) magnetizations with very weak NRM values (<0.010 mA m$^{-1}$) and/or displayed poorly defined decay which did not allow for the components to be sufficiently resolved. For the Line Creek anticline 100% un-tilting of the ChRM site mean data appears to generate the best grouping with no crossover of data, suggesting a pre-tilting magnetization (Fig. 7a). This is confirmed by both tilt tests, with the OC tilt test showing optimal grouping at 97.8 ± 3.1% (Fig. 8; Table 3) and the DC tilt test showing the best grouping at 96.6 ± 6.0% (Table 3).

Fifty-three of the sites (95%) contained specimens with the IT component, but only 34 contained coherent magnetizations suitable for site statistics. Thirty of these sites (56%) had suitable site statistics needed to perform the tilt tests (Table 4). The site mean data for the IT component have two distinct groupings prior to tilt correction, and display a crossover geometry during un-tilting (Fig. 7c). The OC tilt test shows the best grouping is at 69.5 ± 2.5% (Fig. 8; Table 3) and the DC tilt test shows the optimal grouping is achieved at 69.8 ± 7.3% (Table 3). These results suggest that the IT component was acquired during an intermediate stage of folding.

*Mt Kidd Syncline, AB.* Out of the 36 sites sampled at Mt Kidd, 31 of these sites (86%) had specimens which displayed the ChRM. Out of these, 23 of the sites (74%) contained specimens with coherent magnetizations and 22 of these had suitable site statistics to perform the tilt tests (Table 5). Upon un-tilting of the ChRM site mean data for the Mt Kidd syncline the remanence vectors display a crossover geometry and do not fully group, suggesting a syn-tilting result (Fig. 7b). This observation is confirmed by both the OC and DC tilt tests. The OC tilt test shows that optimal un-tilting is achieved at 84.9 ± 4% un-tilting (Fig. 8) and the DC tilt test shows similar results with optimal un-tilting at 85.6 ± 10.3% (Table 3). These results

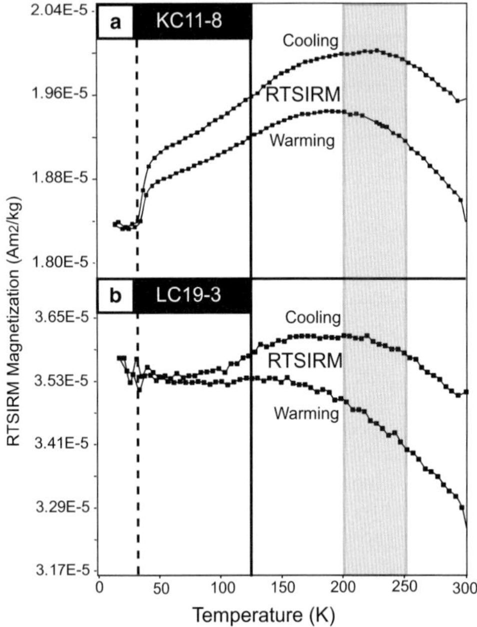



**Fig. 6.** Representative curves generated during LT SIRM experiments from Mt Kidd (KC) and Line Creek (LC). The dashed line highlights the 34 K Besnus transition, the solid line highlights the Verwey transition (120 K) and the shaded region is highlighting the Ms peak at 200–250 K. (**a**) KC11-8, Mt Kidd Syncline. (**b**) LC19-3, Line Creek Anticline.

suggest that the magnetization was acquired early in the folding process.

Thirty-two of the sites (88%) from the KC syncline contained the IT component, with 22 sites (56%) containing coherent magnetizations and 21 of these having suitable site statistics to conduct tilt tests (Table 6). The best grouping of the IT site mean data for the KC syncline occurs at *c.* 0% un-tilting with the remanence vectors moving away from each other upon 100% un-tilting (Fig. 7d). This suggests that the IT component is post-tilting. The OC tilt test shows that optimal grouping occurs at $-3.6 \pm 8.0\%$ (Fig. 8) and the DC tilt test shows that optimal grouping occurs at $-3.5 \pm 10.1\%$ (Table 3). These results suggest that the IT component was acquired after the units had been folded.

## Palaeopoles and the apparent polar wander path

The palaeopole for the pre-tilting ChRM (74.7°N, 194.0°E; dp/dm = 6.2°/5.5°) at Line Creek falls along the Mesozoic portion of the APWP in the Early Cretaceous (Aptian; *c.* 120–110 Ma; Fig. 9).

The palaeopole for the ChRM from the syn-tilting Mt Kidd syncline (70.6°N, 202.1°E; dp/dm = 7.5°/6.9°) falls to the east of the Early Cretaceous portion of the APWP (Fig. 9). The syn-tilting IT pole at Line Creek falls off the path (62.5N, 170.9E; dp/dm = 7.3/6.0; Fig. 9) and the post-tilting IT pole for Mt Kidd also plots off the path (74.5°N, 325.8°E; dp/dm = 5.7/4.7) on the other side of the North Pole from the APWP (Fig. 9).

## Fluid conduit test

The FCT at Mt Kidd was conducted using 12 specimens taken along a profile which crossed three bedding-parallel veins (Fig 10). Petrographic examination of the veins showed that they contain complex twinning as well as some bent twins (Fig. 11a). This twin morphology is observed in the host rocks and results from the high strain and elevated temperatures associated with folding (i.e. Ferrill *et al.* 2004). This suggests that these veins formed prior to deformation. Orthogonal projections created from samples collected for the FCT at Mt Kidd show that both components exist in all specimens (Fig. 10a). A large spike in total magnetic intensity of the ChRM and not the IT component is observed at each of the three veins, suggesting a relationship between bedding-parallel veins and the ChRM (Fig. 10b). The mean for the ChRM could be calculated (Table 5) but not for the IT component; this is because of erratic decay of the IT component in the majority of the specimens which resulted in poor site statistics. The site (KC24) mean for the ChRM has a similar direction to other sites from the west-dipping limb at Mt Kidd (Table 5), suggesting a related origin.

At Line Creek, the site from a bed devoid of veins (LC23) contained both components; unlike the other sites however, at Line Creek the ChRM is reversed (Fig. 12a; Table 2). The presence of reversed directions is not unusual and has been reported in other studies (e.g. Enkin *et al.* 2000). The bed which contains a high amount of calcite and dolomite tensile veins (LC55; Fig. 11b) is dominated by the IT component (Fig. 12b) with only a few specimens containing an incoherent ChRM. The intensity of the IT component appears to overwhelm the ChRM. Tri-axial decay curves generated for both sites are unblocked by 580 °C. In the veined samples, 85–90% of the IRM for the low- and moderate-coercivity curves is removed by 325 °C, whereas only 60–70% is removed in the unveined sample (Fig. 13a, b). This also suggests a larger concentration of pyrrhotite compared to magnetite in the veined sample. These results suggest that the fluids which precipitated these tensile veins may have exerted a strong control on the IT component.

**Table 2.** *Line creek anticline high-temperature data*

| Site | Lat/long | Lithology | Strike/dip | Statistics | | | Tilted | | Tilt-corrected | |
| --- | --- | --- | --- | --- | --- | --- | --- | --- | --- | --- |
| | | | | $N/N_0$ | $K$ | $\alpha_{95}$ | Dec | Inc | Dec | Inc |
| LC01 | N49.892/W114.831 | OFGS | 185.8/20W | 5/5 | 88.3 | 8.2 | 50 | 60.7 | 9.4 | 69.4 |
| LC02 | N49.892/W114.831 | FPS | 185.8/20W | 4/7 | 55.9 | 12.4 | 40.7 | 53.7 | 10.8 | 60.8 |
| LC04 | N49.892/W114.831 | FGS | 153.5/31W | 6/8 | 20.5 | 15.1 | 28.7 | 55.4 | 329.2 | 71 |
| LC05 | N49.892/W114.830 | DFWS | 158.8/32W | 7/8 | 151.6 | 4.9 | 29.3 | 56.8 | 325.1 | 69 |
| LC06 | N49.892/W114.829 | DWS | 179.8/34W | 8/8 | 48.3 | 8.1 | 42.3 | 51.8 | 348.3 | 62.3 |
| LC07 | N49.892/W114.829 | PGS | 153.8/25W | 9/9 | 26.0 | 10.3 | 353.4 | 64.8 | 301.6 | 61.7 |
| LC10 | N49.892/W114.829 | DWS | 169.5/11W | 7/7 | 208.5 | 4.2 | 353.8 | 74.7 | 317.5 | 71.9 |
| LC11 | N49.893/W114.826 | ChCD | 157.8/10W | 6/7 | 115.8 | 6.3 | 42.1 | 73.6 | 12.2 | 81.5 |
| LC13 | N49.893/W114.826 | DPS | 157.5/10W | 6/6 | 62.5 | 8.5 | 1.9 | 72.9 | 327.7 | 74.2 |
| LC15 | N49.893/W114.823 | PGS | 345.5/63E | 8/9 | 18.8 | 13.1 | 268.8 | 27.8 | 352.4 | 78.2 |
| LC16 | N49.893/W114.823 | DMS | 328.5/55E | 6/8 | 12.3 | 19.9 | 281.3 | 41.3 | 352.2 | 56.1 |
| LC19 | N49.894/W114.821 | PS-GS | 345.5/70E | 7/8 | 53.4 | 8.3 | 272.6 | 11.2 | 319.9 | 71.3 |
| LC23 | N49.894/W114.820 | DWS | 350.8/84E | 3/8 | 282.9 | 7.3 | 294.8 | −83.5 | 264.4 | -0.6 |
| LC24 | N49.892/W114.831 | PS-GS | 350.8/84E | 7/8 | 14.1 | 25.3 | 270.6 | −20.6 | 280.4 | 61.7 |
| LC26 | N49.892/W114.831 | DWS | 175.8/46W | 7/7 | 62.8 | 7.7 | 58.2 | 53.9 | 317.1 | 69.5 |
| LC27 | N49.892/W114.831 | PS | 155.8/32W | 7/8 | 27.5 | 11.7 | 45.3 | 66.3 | 286.1 | 77.4 |
| LC28 | N49.892/W114.830 | DFPS | 159.8/36W | 8/8 | 24.8 | 11.3 | 32.8 | 59.7 | 308.9 | 69.3 |
| LC30 | N49.892/W114.830 | DWS-PS | 155.8/32W | 8/8 | 221.7 | 3.7 | 59.7 | 51.1 | 36.1 | 82.3 |
| LC31 | N49.892/W114.829 | FGS | 166.8/25W | 8/16 | 114.1 | 5.2 | 31.6 | 66.3 | 322.2 | 71.7 |
| LC32 | N49.892/W114.829 | OPS | 168.8/25W | 6/9 | 117.8 | 6.2 | 54.7 | 57.5 | 10.7 | 76.3 |
| LC33 | N49.894/W114.822 | PFPS | 351.8/79E | 6/8 | 131.1 | 5.9 | 283.1 | 8.3 | 346.4 | 68.8 |
| LC34 | N49.894/W114.822 | PFPS | 352.8/72E | 6/8 | 45.7 | 10 | 280.1 | 25.7 | 26.7 | 71.2 |
| LC37 | N49.894/W114.822 | DFPS | 339.8/79E | 7/8 | 66.1 | 7.5 | 282.8 | 12.1 | 340.3 | 57.8 |
| LC38 | N49.894/W114.822 | – | 356.8/70E | 6/6 | 97.7 | 6.8 | 280.4 | 14 | 334.1 | 75.7 |
| LC39 | N49.894/W114.822 | DFPS | 350.8/76E | 8/8 | 132.5 | 4.8 | 279.3 | 10.6 | 328.2 | 70.3 |
| LC40 | N49.893/W114.822 | FGS-PS | 348.8/73E | 6/8 | 52.2 | 9.4 | 276.2 | 15.6 | 353.1 | 73.2 |
| LC41 | N49.893/W114.822 | PPS | 345.8/69E | 3/3 | 16.0 | 31.9 | 266.7 | 26 | 14.5 | 78.8 |
| LC42 | N49.893/W114.822 | DPS | 331.8/71E | 9/9 | 211.6 | 3.5 | 273.8 | 17 | 272.8 | 10.2 |
| LC44 | N49.893/W114.822 | DPS | 345.8/70E | 8/8 | 130.4 | 4.9 | 275.1 | 18.1 | 343.1 | 71.7 |
| LC46 | N49.892/W114.823 | PS | 335.8/37E | 5/7 | 12.4 | 22.6 | 306.1 | 46.3 | 350.9 | 51.6 |
| LC47 | N49.892/W114.823 | DWS | 330.8/30E | 8/8 | 22.7 | 11.9 | 288.6 | 46.6 | 326.7 | 59.3 |
| LC48 | N49.893/W114.827 | DWS | 330.8/39E | 9/9 | 59.8 | 6.7 | 273.1 | 37.1 | 312.1 | 63.3 |
| LC50 | N49.893/W114.827 | CD | 330.8/39E | 8/8 | 57.0 | 7.4 | 26.8 | 63.1 | 351.8 | 64.2 |
| LC52 | N49.892/W114.829 | DFPS | 160.8/22W | 8/10 | 101.3 | 5.5 | 48.4 | 62.1 | 2.2 | 79 |
| LC54 | N49.892/W114.830 | – | 155.8/25W | 7/7 | 82.4 | 6.7 | 74.5 | 62 | 122.1 | 85.1 |

*Notes*: Sites that were weak and incoherent as well as sites which contain specimens with MAD angles $>15°$ were excluded. $N/N_0$ is the number of specimens with well-defined component versus the number of samples demagnetized; $K$ is a measure of grouping, $\alpha_{95}$ is the 95% cone of confidence. Dec is declination; Inc is inclination in both tilted and tilt-corrected coordinates. Lithology: CD = crystalline dolomite, FPS = Fossiliferous packstone, DFP = dolomitized fossiliferous packstone, DFG = dolomitized fossiliferous grainstone, FGS = fossiliferous grainstone, DFWS = dolomitized fossiliferous wackestone, DWS = dolomitized wackestone, OFGS = oolitic fossiliferous grainstone, PFGS = pellodial fossiliferous grainstone, PS = packstone, WS = wackestone, GS = grainstone, DMS = dolomitized mudstone, DP = dolomitized packstone, DPGS = dolomitized pellodial grainstone, DGS = dolomitized grainstone, ChCD = cherty crystalline dolomite (Dunham 1962).

## Petrology

*Transmitted/reflected light microscopy and fluid inclusion analysis.* At Line Creek, 22 representative lithologies were sampled (Table 2) and classified according to Dunham (1962). These samples range from pure limestone with no dolomitization to samples that have been completely dolomitized. At KC the carbonates are similar to Line Creek and also show a large variation in lithologies and dolomitization, with 16 different lithologies sampled (Table 4). The samples from both locations show various concentrations of detrital siliciclastic material (quartz and feldspar), opaque minerals (pyrite, hematite) and degraded hydrocarbons (i.e. bitumen). It is worth noting that at both locations there is no observable relationship between sites that have coherent or incoherent magnetizations and the petrographic characteristics. In addition, there is no relationship between the ChRM and type or amount of dolomitization.

Calcite veins sampled from both structures were milky and the only observable inclusions were too small for analysis. A late-stage dolomite

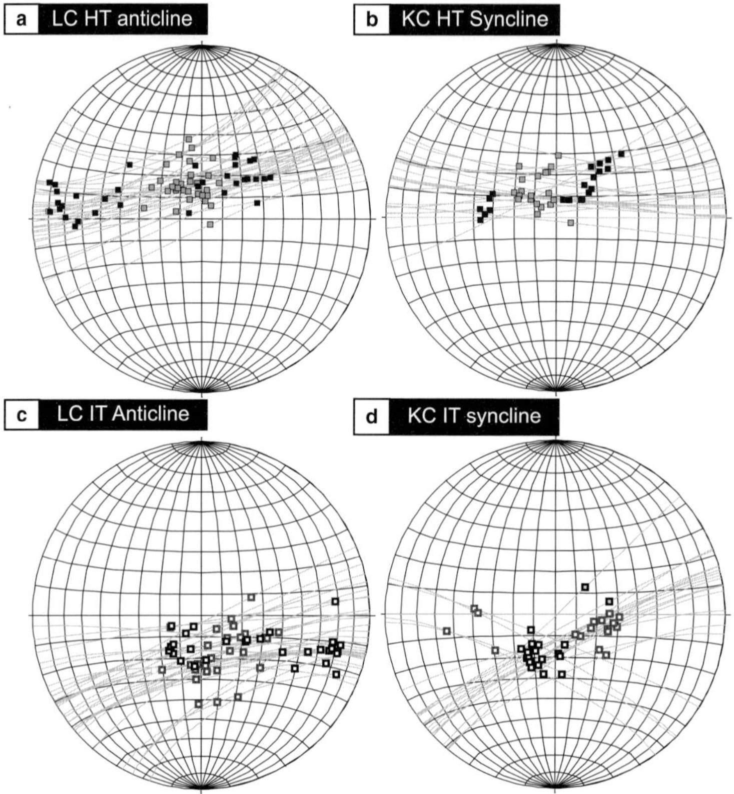



**Fig. 7.** Equal-area projections of site mean data. Normal polarity is indicated by solid black (*in situ*), and solid grey squares (tilt corrected) and reversed polarities are indicated by open black (*in situ*) and open grey (tilt corrected) squares. Grey lines denote small circles which track the path during un-tilting and highlight the concentric nature of the fold axis. (**a**) High-temperature ChRM from Line Creek (LC) Anticline, which shows greatest grouping is in the tilt-corrected position. (**b**) High-temperature ChRM from the Mt Kidd (KC) Syncline, which shows a cross-over geometry. (**c**) IT component from LC, which shows a cross-over geometry. (**d**) IT component from KC which shows the greatest grouping in the tilted coordinates.

vug within one of the thick veins collected at Mt Kidd contained large enough inclusions for analysis. These inclusions were two phase aqueous brine inclusions along growth planes. Preliminary analysis of these inclusions suggests fluids with homogenization temperatures between 140 and 212 °C.

*Electron microprobe results.* Pyrite grains were the most abundant and were commonly partially replaced by hematite (Fig. 14a). Degraded hydrocarbons are abundant throughout the samples examined and opaque minerals are common in the hydrocarbons. The EDXA showed that the solid bitumen (Fig. 14b) is enriched in sulphur with the weight percent sulphur ranging from 12.25 to 29.12%. Sphalerite (Fig. 14c) and barite (Fig. 14a) were also observed. Based on EDXA, frambodial pyrrhotite (Fig. 14d) was identified; however, due

to the small crystal size compared to the beam size, it is possible that the EDXA signal was a representation of pyrite and hematite, which has a spectrum that looks similar to pyrrhotite.

*Geochemistry.* Strontium isotope data were collected from 15 samples to investigate possible fluid alteration. Seven were host composed of calcite, 4 were host composed of dolomite, 2 were calcite veins and 2 were dolomite veins. All samples examined have elevated $^{87}Sr/^{86}Sr$ values when compared to coeval seawater values for Mississippian carbonates (Fig. 15; McArthur *et al.* 2001). The calcite hosts have a mean value of $0.708128 \pm 0.000006$ and the dolomite hosts have a slightly lower mean value of $0.708040 \pm 0.000006$. The vein material shows similar results (calcite veins mean value of $0.707970 \pm 0.000005$ and the dolomite veins mean value of $0.708015 \pm 0.000006$).

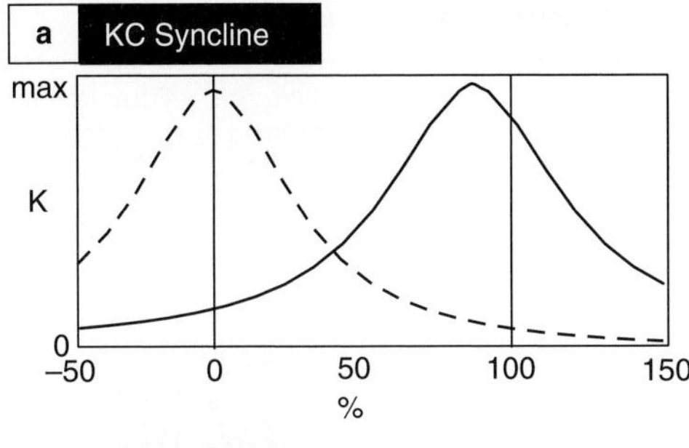


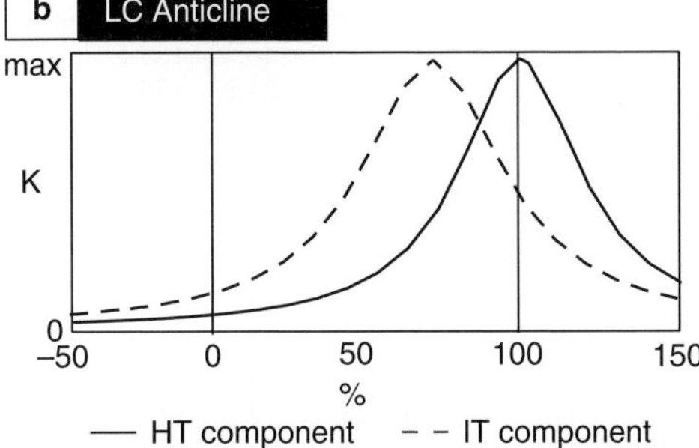


**Fig. 8.** Tilt test (Watson & Enkin 1993) results for the ChRM and IT component at Line Creek and Mt Kidd. ChRM, solid; IT, dashed.

## Discussion

The Mississippian carbonates at Line Creek and Mt Kidd contain a multi-component remagnetization similar to what has been observed in other areas of the Canadian Cordillera (e.g. Lewchuk *et al.* 1998; Enkin *et al.* 2000; Robion *et al.* 2004). The ChRM in both folds is predominantly normal polarity with a maximum unblocking temperature of 540 °C, which is in the range of magnetite. Magnetite is confirmed in tri-axial thermal decay experiments and cumulative log-Gaussian analysis of a RTIRM. LTSIRM experiments do not show a well-developed Verwey transition, which may be attributed to unblocking of a wide range of magnetite grain sizes (pers. comm. M. Jackson, 2008) or oxidation (Özdemir *et al.* 1993). The carbonate units only experienced moderate burial temperatures (180–250 °C) based on vitrinite reflectance data from the overlying Jurassic coals at Line Creek (per. comm. Jason Halko, Tek coal, 2007), calcite twin morphology (Zechmeister 2010) and the burial history reconstruction of the Lewis Thrust (Feinstein *et al.* 2007). The magnetite component is interpreted as a CRM as opposed to a TVRM based on the unblocking temperature, moderate burial temperatures and the magnetite relaxation-time–unblocking-temperature curves for SD grains (e.g. Dunlop *et al.* 2000), which is applicable since the MD contribution has been removed by LTD. This CRM interpretation contrasts with some studies in the southern Canadian Cordillera which reported magnetizations with lower maximum unblocking temperatures (<450 °C) that were suggested to be burial remagnetizations (Symons & Cioppa 2002).

The intermediate component has a reversed polarity and a maximum unblocking temperature of 340 °C, with most samples unblocked around 320 °C. This suggests either a TVRM in magnetite or a CRM in pyrrhotite. A TVRM in magnetite is

**Table 3.** *Summary of tilt tests*

| | $N$ | $D$ (°) | $I$ (°) | $k$ | $\alpha_{95}$ (°) | %Unfolding | Pole | dp (°) | dm (°) |
|---|---|---|---|---|---|---|---|---|---|
| *HT component* | | | | | | | | | |
| **KC syncline** | | | | | | | | | |
| Optimal correction | 22 | 330.9 | 75.3 | 63.3 | 4.1 | 84.9 ± 4.0 | N70.6/E202.1 | 7.5 | 6.9 |
| Direction correction | 22 | – | – | – | – | 85.63 ± 10.3 | – | – | – |
| **LC anticline** | | | | | | | | | |
| Optimal correction | 31 | 338.6 | 72.6 | 55.02 | 3.5 | 97.8 ± 3.1 | N71.8/E192.0 | 6.4 | 5.7 |
| Direction correction | 31 | – | – | – | – | 96.58 ± 6.0 | – | – | – |
| *IT component* | | | | | | | | | |
| **KC syncline** | | | | | | | | | |
| Optimal correction | 21 | 202.4 | −68.8 | 102.7 | 3.3 | −1.8 ± 5.3 | N74.5/325.8 | 7.3 | 6.0 |
| Direction correction | 21 | – | – | – | – | −2.13 ± 10.8 | – | – | – |
| **LC anticline** | | | | | | | | | |
| Optimal correction | 30 | 137.0 | −66.8 | 37.4 | 4.4 | 69.4 ± 2.5 | N62.5/E170.9 | 5.7 | 4.7 |
| Direction correction | 30 | – | – | – | – | 69.79 ± 7.3 | – | – | – |

$N$, number of sites used in the tilt test; $I$, inclination; $D$, declination; $k$, statistical measure of grouping; $\alpha_{95}$, 95% cone of confidence. Percent untilting and confidence interval from the optimal clustering (OC; Watson & Enkin 1993) and direction correction tilt tests (DC, Enkin 2003). The pole represents the inferred location at the time of acquisition. dp, dm: semi-minor and semi-major axis respectively of the 95% error ellipse.

**Table 4.** *Line creek anticline intermediate-temperature data (notation and lithology as for Table 2)*

| Site | Lat/long | Lithology | Strike/dip | Statistics | | | Tilted | | Tilt-corrected | |
|---|---|---|---|---|---|---|---|---|---|---|
| | | | | $N/N_0$ | $K$ | $\alpha_{95}$ | Dec | Inc | Dec | Inc |
| LC01 | N49.892/W114.831 | OFGS | 185.8/20W | 7/7 | 56.0 | 8.1 | 220.7 | −79 | 129.7 | −73.7 |
| LC04 | N49.892/W114.831 | FGS | 153.5/31W | 6/8 | 35.0 | 11.5 | 214.1 | −83.8 | 70.5 | −64.2 |
| LC05 | N49.892/W114.830 | DFWS | 158.8/32W | 8/8 | 62.4 | 7.1 | 173.4 | −69 | 108.5 | −57.1 |
| LC06 | N49.892/W114.829 | DWS | 179.8/34W | 7/8 | 48.6 | 8.7 | 187.3 | −66 | 130.6 | −51.9 |
| LC07 | N49.892/W114.829 | PGS | 153.8/25W | 4/4 | 13.9 | 25.5 | 126.7 | −58.7 | 102.6 | −42.4 |
| LC10 | N49.892/W114.829 | DWS | 169.5/11W | 7/7 | 31.7 | 10.9 | 119.5 | −65.2 | 108.4 | −56.1 |
| LC11 | N49.893/W114.826 | ChCD | 157.8/10W | 6/7 | 53.3 | 9.3 | 134.7 | −72.9 | 111.9 | −67.1 |
| LC13 | N49.893/W114.826 | DPS | 157.5/10W | 5/6 | 72.1 | 14.6 | 110.8 | −59.2 | 101.7 | −51.3 |
| LC15 | N49.893/W114.823 | PGS | 345.5/63E | 5/9 | 77.8 | 8.7 | 104.9 | −22.9 | 163.3 | −63.1 |
| LC16 | N49.893/W114.823 | DMS | 328.5/55E | 6/8 | 24.7 | 13.8 | 113.6 | −46.4 | 181.5 | −47.6 |
| LC19 | N49.894/W114.821 | PS-GS | 345.5/70E | 7/8 | 130.9 | 5.3 | 103.8 | −21.6 | 174 | −63.5 |
| LC22 | N49.894/W114.820 | OGS | 351.8/80E | 7/8 | 107.3 | 5.9 | 108.3 | −34 | 221.3 | −55.4 |
| LC23 | N49.894/W114.820 | DWS | 350.8/84E | 7/8 | 51.5 | 8.5 | 84.1 | −21.4 | 249.4 | −74.3 |
| LC24 | N49.892/W114.831 | PS-GS | 350.8/84E | 3/8 | 117.2 | 11.4 | 105.5 | −27.7 | 216.4 | −58.1 |
| LC26 | N49.892/W114.831 | DWS | 175.8/46W | 6/7 | 29.0 | 12.7 | 250.9 | −75.1 | 107.3 | −73.8 |
| LC27 | N49.892/W114.831 | PS | 155.8/32W | 6/8 | 42.5 | 10.4 | 216.6 | −67.7 | 107.3 | −73.6 |
| LC28 | N49.892/W114.830 | DFPS | 159.8/36W | 6/8 | 6.5 | 28.4 | 195.3 | −68.2 | 109 | −61.4 |
| LC29 | N49.892/W114.830 | OFPS | 180.8/37W | 3/3 | 232.0 | 8.1 | 223.1 | −67.4 | 130 | −63.2 |
| LC30 | N49.892/W114.830 | DWS-PS | 155.8/32W | 5/8 | 144.2 | 6.4 | 224.8 | −69.4 | 95.9 | −75.4 |
| LC31 | N49.892/W114.829 | FGS | 166.8/25W | 9/16 | 67.7 | 6.3 | 197.3 | −73.6 | 118.2 | −68.4 |
| LC32 | N49.892/W114.829 | OPS | 168.8/25W | 5/9 | 725.0 | 2.8 | 204.5 | −66.6 | 139.9 | −68.4 |
| LC33 | N49.894/W114.822 | PFPS | 351.8/79E | 5/8 | 134.0 | 6.6 | 105.1 | −18 | 191.8 | −66.4 |
| LC37 | N49.894/W114.822 | DFPS | 339.8/79E | 3/8 | 125.0 | 11.1 | 112.9 | −14.4 | 169.2 | −47.9 |
| LC38 | N49.894/W114.822 | – | 356.8/70E | 6/6 | 38.6 | 10.9 | 101.6 | −17 | 167.2 | −75.7 |
| LC39 | N49.894/W114.822 | DFPS | 350.8/76E | 6/8 | 70.4 | 8 | 104.2 | −17.9 | 183.8 | −67.2 |
| LC40 | N49.893/W114.822 | FGS-PS | 348.8/73E | 5/8 | 56.7 | 10.3 | 110.4 | −22.4 | 184.3 | −59.8 |
| LC41 | N49.893/W114.822 | PPS | 345.8/69E | 3/3 | 35.7 | 20.9 | 100.8 | −19.7 | 167.1 | −66.5 |
| LC42 | N49.893/W114.822 | DPS | 331.8/71E | 8/9 | 66.5 | 6.8 | 103.1 | −23.7 | 166.3 | −51.4 |
| LC44 | N49.893/W114.822 | DPS | 345.8/70E | 8/8 | 72.4 | 6.6 | 101 | −21.9 | 174.9 | −66.4 |
| LC47 | N49.892/W114.823 | DWS | 330.8/30E | 5/8 | 29.3 | 14.4 | 117 | −64.9 | 186 | −64.5 |
| LC48 | N49.893/W114.827 | DWS | 330.8/39E | 6/9 | 33.4 | 11.8 | 118.8 | −37 | 155 | −47.2 |
| LC50 | N49.893/W114.827 | CD | | 5/8 | 54.4 | 10.5 | 195 | −60.7 | 165.3 | −59 |
| LC52 | N49.892/W114.829 | DFPS | 160.8/22W | 7/10 | 75.0 | 7 | 224.3 | −71.1 | 130.2 | −80.3 |
| LC54 | N49.892/W114.830 | – | 155.8/25W | 7/7 | 7.5 | 23.5 | 267.4 | −61.6 | 326.4 | −79.8 |

discounted because the units only experienced moderate burial temperatures. These moderate burial temperatures would likely only have affected the unstable MD contribution, which was removed by LTD prior to thermal demagnetization. The effect of the moderate burial temperatures on the MD contribution may explain the low unblocking temperatures of the ChRM reported in other studies conducted in the Canadian Cordillera (e.g. Symons & Cioppa 2002). Pyrrhotite is generally considered to have a maximum unblocking temperature of 325 °C, although temperatures as high as 350 °C have been reported (Rochette *et al.* 1990). The presence of pyrrhotite is highlighted by the presence of the 34 K Besnus transition as well as cumulative log-Gaussian analysis. Siderite also has a magnetic transition near 40 K (Housen *et al.* 1996); however, no siderite has been observed in any of the petrographic analysis. Also, siderite would not display the peak

$M_{max}$ between 200 and 250 K which is indicative of pyrrhotite and is observed in all LTSIRM experiments conducted in this study (Fig. 6). For KC11–8, the cycling of cooling and warming allow for estimation of the pyrrhotite grain size (Rochette *et al.* 1990). The ratio of heating over cooling at the 34 K Besnus transition (h/c ratio) (Fig. 6a) was 0.98 and the ratio of IRM after and before cycling (R) was 0.95. This suggests that this sample is dominated by *c.* 2 μm pyrrhotite grains which are in the SD range. This pyrrhotite component is also considered to be a CRM based on the SD pyrrhotite relaxation-time–unblocking-temperature curves (Dunlop *et al.* 2000). This is in contrast with previous studies where the IT component has been interpreted to be a TVRM in magnetite (e.g. Enkin *et al.* 2000; Robion *et al.* 2004).

The Line Creek anticline contains an Early Cretaceous pre-tilting CRM and the Mt Kidd syncline

**Table 5.** *Mt Kidd syncline high-temperature site statistics (notation and lithology as for Table 2)*

| Site | Lat/long | Lithology | Strike/dip | Statistics | | | Tilted | | Tilt-corrected | |
|---|---|---|---|---|---|---|---|---|---|---|
| | | | | $N/N_0$ | $K$ | $\alpha 95$ | Dec | Inc | Dec | Inc |
| KC1 | N50.826/W115.165 | OFGS | 026.3/28E | 7/8 | 63.2 | 7.7 | 276.1 | 58 | 304.2 | 81.5 |
| KC2 | N50.826/W115.165 | FGS | 021.3/30E | 8/8 | 91.4 | 5.8 | 268.1 | 53.5 | 280 | 81.1 |
| KC3 | N50.826/W115.165 | FPS | 021.3/30E | 7/8 | 59.7 | 7.9 | 276.1 | 53.1 | 305 | 78.9 |
| KC4 | N50.826/W115.165 | OFPS | 021.3/45E | 6/8 | 67.4 | 8.2 | 272.5 | 55.7 | 109.2 | 82.6 |
| KC6 | N50.861/W115.175 | DGS | 136.3/42W | 7/8 | 77.1 | 6.9 | 34.9 | 60.7 | 305.2 | 75.4 |
| KC7 | N50.861/W115.175 | DGS | 166.3/42W | 8/8 | 345.7 | 3 | 43.5 | 59.9 | 323.9 | 60.1 |
| KC8 | N50.862/W115.175 | DGS | 151.3/42W | 8/8 | 178.9 | 4.2 | 43.4 | 63.9 | 306.8 | 69.5 |
| KC9 | N50.862/W115.175 | ChCD | 155.3/35W | 7/7 | 138.0 | 5.2 | 39.3 | 56.9 | 2.1 | 60.8 |
| KC10 | N50.868/W115.173 | WS-PS | 155.3/26W | 8/8 | 80.3 | 6.2 | 22.3 | 81.1 | 300.5 | 66.9 |
| KC11 | N50.868/W115.173 | WS-PS | 153.3/27W | 8/8 | 79.5 | 6.2 | 37.6 | 79.7 | 300.4 | 69.8 |
| KC12 | N50.868/W115.173 | DWS | 153.3/27W | 7/7 | 338.9 | 3.3 | 55.4 | 75.5 | 310.9 | 74.4 |
| KC22 | N50.826/W115.165 | OFGS | 028.3/34E | 7/7 | 101.8 | 6 | 285.5 | 56.8 | 336.6 | 83.4 |
| KC24 | N50.826/W115.165 | FPS | 028.3/34E | 6/12 | 88.2 | 7.2 | 290.1 | 57.6 | 351.9 | 81.7 |
| KC26 | N50.861/W115.175 | CD | 151.3/34W | 7/7 | 264.5 | 3.7 | 42.5 | 52.9 | 354.2 | 66 |
| KC27 | N50.861/W115.175 | DGS | 152.3/44W | 7/7 | 53.0 | 8.4 | 46.1 | 46.1 | 349 | 68.3 |
| KC29 | N50.862/W115.175 | ChCD | 155.3/35W | 6/7 | 118.0 | 6.2 | 47.1 | 56.1 | 346.7 | 68.3 |
| KC37 | N50.861/W115.175 | DGS | 151.3/38W | 3/4 | 270.3 | 7.5 | 30.9 | 37.1 | 0.2 | 51.5 |
| KC38 | N50.862/W115.175 | DGS | 151.3/38W | 4/4 | 103.2 | 9.1 | 30.8 | 39.5 | 357.4 | 53.2 |
| KC39 | N50.868/W115.173 | DWS | 165.3/25W | 6/7 | 124.0 | 6 | 51.3 | 68.4 | 318 | 66 |
| KC40 | N50.868/W115.173 | DWS | 154.3/28W | 8/9 | 1019.4 | 1.7 | 52.8 | 70.4 | 295.6 | 70.2 |
| KC41 | N50.868/W115.173 | WS-PS | 146.3/25W | 4/4 | 5.7 | 42.4 | | | | |
| KC42 | N50.868/W115.173 | DMS | 144.3/20W | 7/7 | 140.1 | 5.1 | 55.5 | 74.9 | 342.9 | 78.1 |
| KC43 | N50.868/W115.173 | DPGS | 156.3/25W | 9/9 | 229.3 | 3.4 | 41.4 | 69.6 | 336 | 68.8 |

contains an early syn-tilting, apparently Cretaceous CRM. The fact that the poles for each do not coincide could be because the syn-tilting result is not unique or because of structural complications (Zechmeister 2010). The pyrrhotite component is syn-tilting at the Line Creek Anticline and post-tilting

**Table 6.** *Mt Kidd syncline intermediate-temperature data (notation and lithology as for Table 2)*

| Site | Lat/long | Lithology | Strike/dip | Statistics | | | Tilted | | Tilt-corrected | |
|---|---|---|---|---|---|---|---|---|---|---|
| | | | | $N/N_0$ | $k$ | $\alpha_{95}$ | Dec | Inc | Dec | Inc |
| KC1 | N50.826/W115.165 | OFGS | 026.3/28E | 6/8 | 24.3 | 13.9 | 190.9 | −61.4 | 239.3 | −56.6 |
| KC2 | N50.826/W115.165 | FGS | 021.3/30E | 6/8 | 37.9 | 11.0 | 237.2 | −76.5 | 273.9 | −50.7 |
| KC4 | N50.826/W115.165 | OFPS | 021.3/45E | 5/8 | 240.6 | 4.9 | 212.9 | −65.5 | 261.3 | −35.8 |
| KC6 | N50.861/W115.175 | DGS | 136.3/42W | 6/8 | 90.7 | 7.1 | 224.7 | −67.2 | 48.2 | −70.8 |
| KC7 | N50.861/W115.175 | DGS | 166.3/42W | 8/8 | 288.4 | 3.3 | 214.5 | −71.5 | 101.1 | −59.7 |
| KC8 | N50.862/W115.175 | DGS | 151.3/42W | 8/8 | 656.0 | 2.2 | 218.1 | −72.4 | 76.7 | −63.4 |
| KC9 | N50.862/W115.175 | ChCD | 155.3/35W | 6/7 | 226.5 | 4.5 | 208.3 | −70.6 | 97 | −67.6 |
| KC10 | N50.868/W115.173 | WS-PS | 155.3/26W | 7/8 | 23.0 | 12.9 | 161.9 | −75.1 | 98.1 | −61.8 |
| KC11 | N50.868/W115.173 | WS-PS | 153.3/27W | 7/8 | 22.4 | 13 | 175.1 | −71.6 | 105.2 | −64 |
| KC12 | N50.868/W115.173 | DWS | 153.3/27W | 7/7 | 89.0 | 7.1 | 207 | −65 | 130 | −74.2 |
| KC22 | N50.826/W115.165 | OFGS | 028.3/34E | 4/7 | 68.5 | 11.2 | 208.4 | −73.8 | 270.9 | −52.7 |
| KC26 | N50.861/W115.175 | CD | 151.3/34W | 7/7 | 122.3 | 5.5 | 207.9 | −64.5 | 110.2 | −71.7 |
| KC27 | N50.861/W115.175 | DGS | 152.3/44W | 7/7 | 124.3 | 5.4 | 212.3 | −66.7 | 89.1 | −64 |
| KC29 | N50.862/W115.175 | ChCD | 155.3/35W | 6/7 | 69.4 | 8.1 | 216 | −68 | 100.1 | −71.2 |
| KC34 | N50.871/W115.192 | DGS | 336.3/35E | 4/7 | 11.1 | 28.9 | 196.6 | −52.9 | 215.7 | −25.4 |
| KC37 | N50.861/W115.175 | DGS | 151.3/38W | 4/4 | 191.5 | 6.7 | 237.2 | −64.1 | 69.7 | −77.7 |
| KC38 | N50.862/W115.175 | DGS | 151.3/38W | 4/4 | 294.6 | 5.4 | 211.8 | −68.5 | 90.6 | −68.4 |
| KC39 | N50.868/W115.173 | DWS | 165.3/25W | 6/7 | 35.3 | 11.4 | 204.1 | −62.6 | 128 | −63.2 |
| KC40 | N50.868/W115.173 | DWS | 154.3/28W | 7/9 | 381.6 | 3.1 | 198.3 | −64.9 | 107.5 | −63.5 |
| KC41 | N50.868/W115.173 | WS-PS | 146.3/25W | 4/4 | 268.4 | 5.6 | 173 | −70.3 | 92.2 | −59.1 |
| KC42 | N50.868/W115.173 | DMS | 144.3/20W | 7/7 | 37.2 | 10 | 196.2 | −68.1 | 134.1 | −76.5 |
| KC43 | N50.868/W115.173 | DPGS | 156.3/25W | 8/9 | 97.9 | 5.6 | 173.3 | −61.6 | 127.9 | −58.9 |

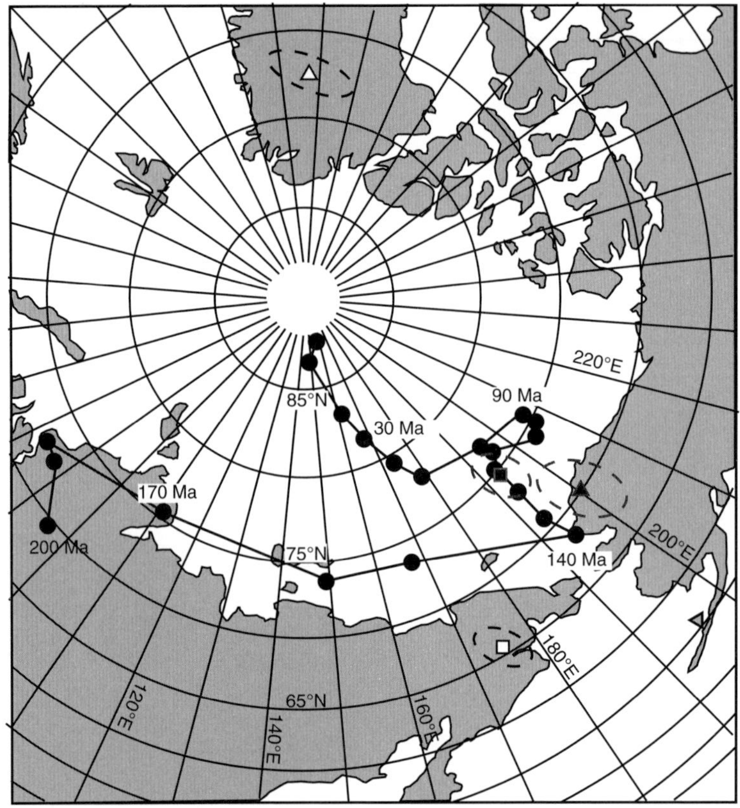


**Fig. 9.** Mesozoic section of the apparent polar wander path for North America (modified from Besse & Courtillot 2002). Triangles, Mt Kidd syncline; squares, Line Creek anticline; closed symbols, ChRM; open symbols, IT component. dp/dm errors (Table 3) are indicated by the dashed ovals.

at Mt Kidd. The pole for post-tilting CRM at Mt Kidd plots off of the APWP. This may be because the process forming the remagnetization occurred too rapidly and did not average out secular variation. Although the timing of remanence acquisition is uncertain, it may be Late Tertiary in age. The syn-tilting IT Line Creek pole plots off the path, possibly because the syn-tilting result is not unique.

## Origin of magnetite CRM

The FCT at Mt Kidd was conducted on bedding-parallel veins. Similar bedding-parallel veins in the southern Canadian front ranges are interpreted as pre-deformational based on cross-cutting relationships and the presence of folded veins (Cooley *et al.* 2011). The FCT results suggest a connection between the CRM in magnetite and the fluids that formed the veins. The carbonates from both field locations have elevated strontium isotope values

compared to Mississippian coeval seawater which suggest alteration by externally derived fluids. The CRM in magnetite is interpreted to have formed as a result of a regional pervasive fluid migration of evolved fluids. This contrasts with Machel & Cavell (1999) who proposed that regional fluid flow was not sufficient to cause regional remagnetization in the Canadian Cordillera, opting for burial diagenesis as a likely mechanism.

Hydrocarbon maturation (Cioppa & Symons 2000) and clay diagenesis (Machel & Cavell 1999) have also been invoked as a remagnetization mechanism in the Canadian Cordillera. Burial clay diagenesis is not considered to be a likely remagnetization mechanism due to a relatively negligible clay content observed in the Mississippian carbonates. Dolomitization has also been invoked as the origin of the magnetite component in the Canadian Cordillera (e.g. Lewchuk *et al.* 1998; Symons *et al.* 1998; Roure *et al.* 2005); this can however be discounted due to the presence of the magnetite

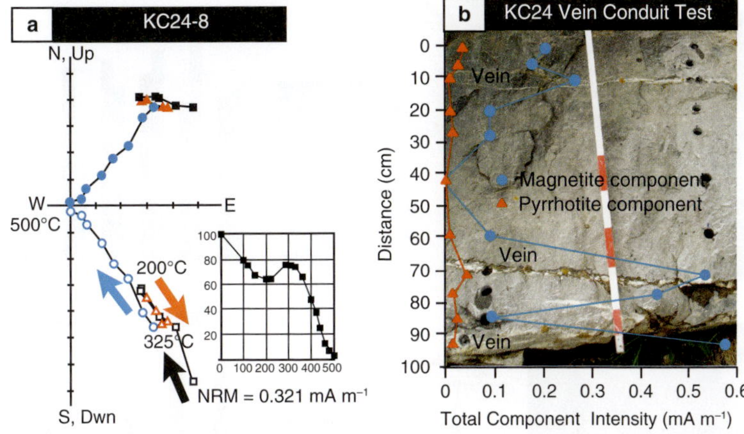



**Fig. 10.** Bedding-parallel fluid conduit test from Mt Kidd. (**a**) Orthogonal projection in geographical coordinates for stepwise thermal demagnetization of a KC sample from the vein test with a thermal demagnetization plot (same conventions as Fig. 3). (**b**) Field photo showing veins, core (1 inch diameter) locations and component intensities versus distance from the veins. Blue circles, ChRM (magnetite) component intensity; orange triangles, IT component (pyrrhotite) intensity. The ChRM is interpreted to reside in magnetite so the component intensity is a combination of the VRM contained in MD magnetite, which is calculated from the intensity drop caused by the liquid nitrogen treatment, and the total intensity of the ChRM (PSD–SD grains). The MD intensity was added to the PSD/SD intensities because they are concluded to form during the same event.

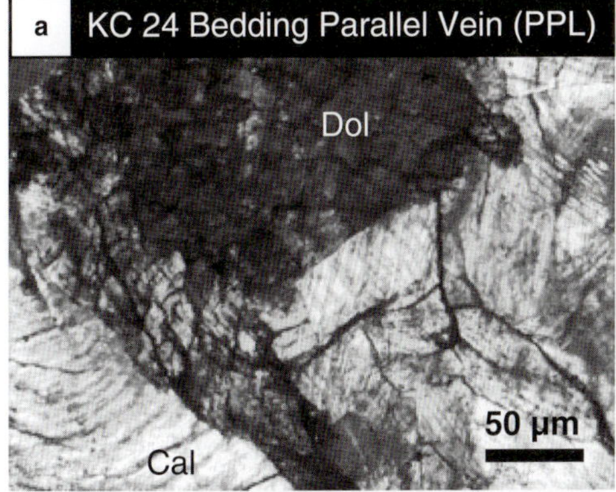



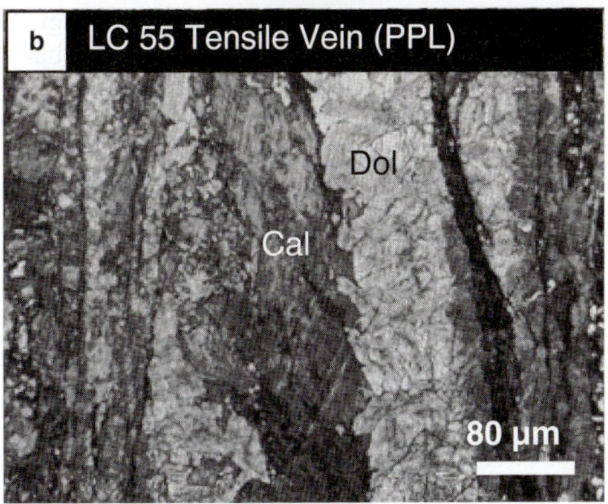



CRM in both limestones and dolostones at each locality. Dolomitization however played an important role by creating porosity for later fluid migration events. This idea is supported by the fact that dolomitized mudstones and wackestones contain the magnetite component, which was probably aided by dolomitization, which increased porosity and permeability.

Since the timing of the ChRM relative to folding is considered pre-tilting at Line Creek, the fluids would have needed to migrate through these units prior to or just at the onset of deformation. This is supported by the FCT at KC which shows a strong correlation between pre-deformational bedding-parallel veins and the magnetite component. Laramide deformation is believed to have initiated in the westernmost portion of the Alberta foreland Basin by around 135 Ma (Evenchick *et al.* 2007). This suggests that the observed Early Cretaceous (*c.* 120 Ma; Aptian) remagnetization event could have been sourced from the highlands to the west

**Fig. 11.** Transmitted light photographs of vein features associated with each fluid conduit test. (**a**) KC 24 bedding-parallel calcite vein with a late-stage dolomite vug. Bent twins are observed in the lower left corner and suggest high strain alteration. (**b**) LC 55 tensile vein with interfingering calcite and dolomite.

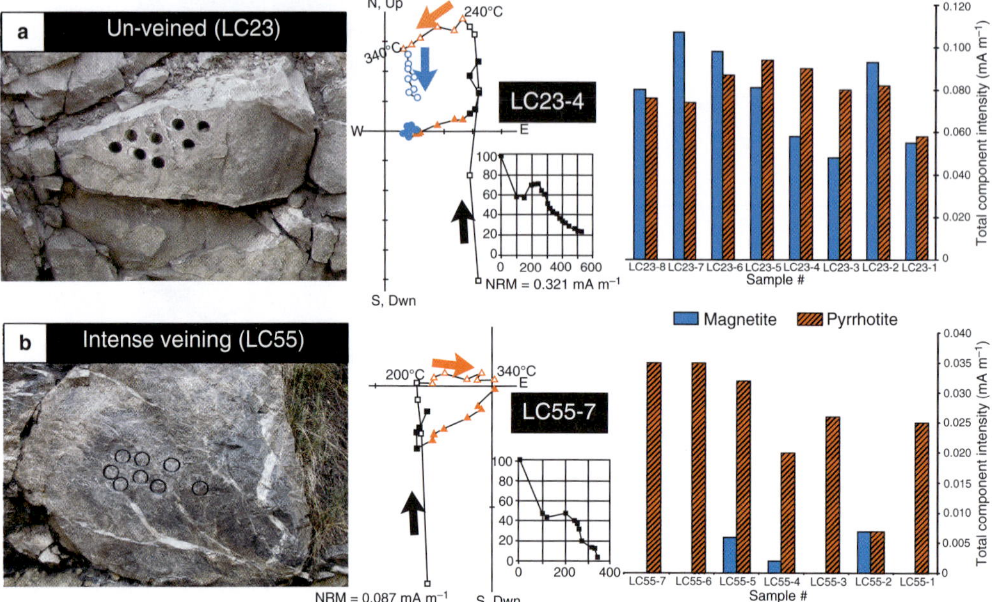

**Fig. 12.** Tectonic fluid conduit test from Line Creek. (**a**) Field photo of unveined rock (left) with core holes (1 inch diameter). Middle diagram is an orthogonal projection (LC23-7) in geographical coordinates for stepwise thermal demagnetization of an unveined sample. The ChRM is indicated by circles, the IT component is indicated by triangles and the modern VRM is indicated by squares. Arrows indicate the direction of thermal decay. The small insert graph shows the percentage of magnetization removed by a particular temperature during stepwise demagnetization. Graph on right is component intensity of both components for individual specimens. (**b**) Field photo of intensely veined rock (left), orthogonal projection (LC55-8) in geographical coordinates for stepwise thermal demagnetization of a veined sample (middle) and graph of component intensity for both components.

through gravitational processes and/or by the initiation of tectonically induced fluid fluxes during Laramide shortening in the foreland fold and thrust belt.

## Origin of the pyrrhotite CRM

The presence of pyrrhotite as a stable remanence carrier has important implications on what diagenetic processes may have affected these rocks. Pyrrhotite is now recognized as an important carrier of a palaeomagnetic signal (e.g. Dekkers *et al.* 1989; Rochette *et al.* 1990; Jackson *et al.* 1993; Xu *et al.* 1998; Crouzet *et al.* 2003; Weaver *et al.* 2002; Gillett & Karlin 2004; Font *et al.* 2006; Preeden *et al.* 2008). Pyrrhotite is considered an important redox indicator that can form as a result of diagenesis (Hall 1986), and has been shown to be useful in geothermometry (Dunlop *et al.* 2000; Schill *et al.* 2002). Jackson *et al.* (1993) proposed that the Cenozoic TVRM widely reported throughout the Appalachian fold and thrust belt may in fact be a CRM contained in pyrrhotite. Authigenic pyrrhotite capable of carrying a CRM can form by a

number of chemical mechanisms. These include pre-existing magnetite reacting with pyrite and organic matter as a result of burial metamorphism (Gillett 2003), the breakdown of pyrite around 400 °C (Rochette 1987), oxidation of pre-existing pyrite (Salmon *et al.* 1988), thermochemical sulphate reduction (Pierce *et al.* 1998), hydrocarbon migration (Machel & Burton 1991), diagenesis associated with gas hydrates (Housen & Musgrave 1996; Larrasoaña *et al.* 2007; van Dongen *et al.* 2007; Rowan & Roberts 2008) and release of pore fluids (Urbat *et al.* 2000).

Sulphate reduction is known to have occurred throughout the Canadian Cordillera due to warm sulphate-rich formational waters moving up along faults and fractures (e.g. Goldhaber & Reynolds 1991; Pierce *et al.* 1998; Mountjoy *et al.* 1999) and could be a viable remagnetization mechanism. Bacterial sulphate reduction (BSR) occurs in low-temperature diagenetic environments due to sulphate-reducing bacteria (0–80 °C) which, above 80 °C, cease to metabolize (e.g. Machel & Foght 2000). On the other hand, TSR is the result of a redox reaction that occurs during deep

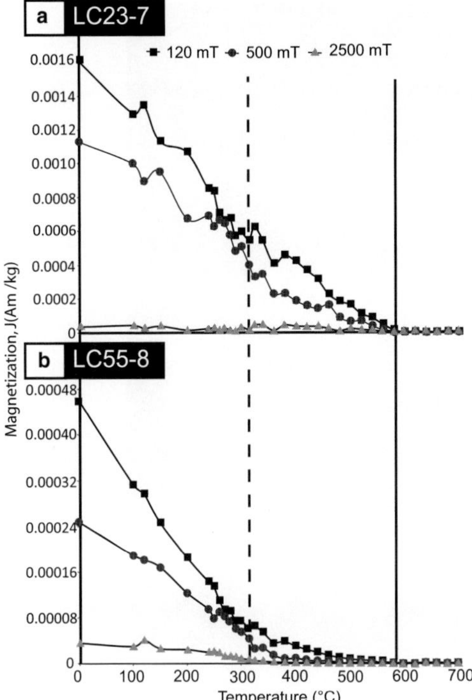


**Fig. 13.** Tri-axial IRM thermal decay curves for the LC vein conduit test. The dashed line highlights the pyrrhotite unblocking temperature (325 °C) and the solid line represents the magnetite unblocking temperature (580 °C). (**a**) Unveined sample (LC23-7) and (**b**) intensely veined sample (LC55-8).

burial diagenesis within a temperature range of $c.\,100-200$ °C through sulphate reduction of hydrocarbons with concomitant oxidation of the organic compounds (e.g. Machel 2001). Evidence that TSR/BSR has occurred in the Mississippian carbonates comes from the observation of multiple common TSR/BSR by-products including authigenic pyrite, frambodial pyrrhotite, sphalerite and sulphur-enriched bitumen (29.12 Wt% sulphur). The preliminary fluid inclusions data from late-stage vug-filling dolomite (140–212 °C) are consistent with the temperature range for TSR. The fluids which caused TSR may also have caused alteration resulting in the elevated Sr isotope values. The elevated Sr values indicate diagenetic alteration and cannot be directly related to the pre-deformational fluids hypothesized to have caused the magnetite CRM or fluids which caused the pyrrhotite CRM.

Based on the palaeomagnetic, rock magnetic, petrographic and geochemical data, one possible remagnetization mechanism for the pyrrhotite component is TSR. The pyrrhotite CRM could have formed when warm fluids moved upwards along faults and fractures. This interpretation is supported by the LC FCT. The veins examined in the LC FCT are considered to have formed during brittle thinning associated with steepening of the front limb of the LC Anticline. This type of dialant veining has also been observed within the Livingstone Anticlinorium to the east of LC and is associated with the late stages of folding, with mixtures of calcite and dolomite precipitated from warm basement-derived fluids (Cooley *et al.* 2011). This late-stage veining may have resulted in a large enough volume of warm sulphate-rich water to be in contact with magnetite, iron-rich organic material and/or pyrite to use as a source of the Fe needed to form pyrrhotite. Alteration of magnetite may explain the lack of a magnetite CRM in the veined samples in the LC FCT. Since the pyrrhotite is a pervasive remagnetization, minor amounts of the warm fluids must have migrated out into the un-veined sections leading to pervasive TSR and formation of pyrrhotite at the expense of magnetite, iron-rich organics and/or pyrite, but leaving enough magnetite to preserve the ChRM.

The nature of sulphate reduction and associated precipitation of pyrrhotite could lead to problems when trying to date TSR using palaeomagnetism. One issue is the time-span over which TSR occurs. TSR formation of sour gas reservoirs typically occurs over time periods of tens–hundreds of thousands of years under the right conditions and, at most, may require a few million years (Goldhaber & Orr 1995). If TSR occurs at the lower end of the spectrum, then it is possible that the formation of pyrrhotite may not average out secular variation. This may be reflected at Mt Kidd by the fact that the palaeopole plots off of the Late Tertiary portion of the path. Another issue relating to the nature of TSR is the limited stratigraphic extent of TSR ($c.\,10-100$ m). This may result in multiple events occurring at different time intervals within a single structure. At Mt Kidd, the exposure limits sampling to a relatively small vertical section within the structure. The results from the tilt tests at Mt Kidd gave a post-tilting remagnetization, which is what would be expected for a Late Tertiary TSR. At Line Creek, however, the exposure was much more expansive, allowing sampling across a stratigraphic section comprising several hundred metres. The results from the tilt test gave a syn-tilting result ($c.\,60\%$ un-tilting). Another unresolved issue is the overprint by a later pyrrhotite-forming remagnetization event due to BSR during exhumation to shallow levels, which would further complicate the palaeomagnetic signal. More work needs to be done to resolve these issues before palaeomagnetic dating of TSR/BSR can be considered reliable.

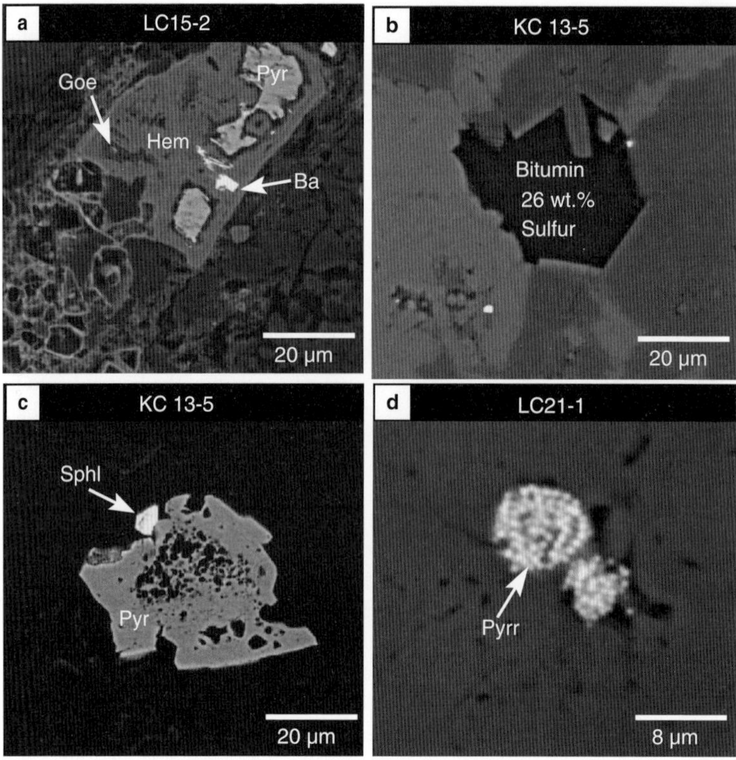



**Fig. 14.** Electron backscatter images from host rock. (**a**) Pyrite being replaced by hematite and goethite; (**b**) sulphur-rich solid bitumen; (**c**) spahlerite on pyrite; and (**d**) possible frambodial pyrrhotite.

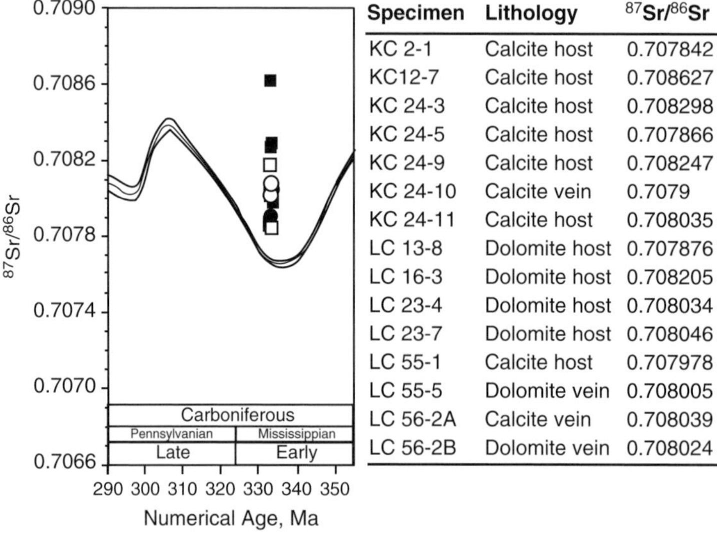



| Specimen | Lithology | $^{87}$Sr/$^{86}$Sr |
|---|---|---|
| KC 2-1 | Calcite host | 0.707842 |
| KC12-7 | Calcite host | 0.708627 |
| KC 24-3 | Calcite host | 0.708298 |
| KC 24-5 | Calcite host | 0.707866 |
| KC 24-9 | Calcite host | 0.708247 |
| KC 24-10 | Calcite vein | 0.7079 |
| KC 24-11 | Calcite host | 0.708035 |
| LC 13-8 | Dolomite host | 0.707876 |
| LC 16-3 | Dolomite host | 0.708205 |
| LC 23-4 | Dolomite host | 0.708034 |
| LC 23-7 | Dolomite host | 0.708046 |
| LC 55-1 | Calcite host | 0.707978 |
| LC 55-5 | Dolomite vein | 0.708005 |
| LC 56-2A | Calcite vein | 0.708039 |
| LC 56-2B | Dolomite vein | 0.708024 |

**Fig. 15.** Plot and table of $^{87}$Sr/$^{86}$Sr results for both host rock and vein material. Squares, host carbonates; circles, vein material; solid symbols, calcite; open symbols, dolomite. Data plotted on the Carboniferous coeval seawater curve (modified from McArthur *et al.* 2001).

## Conclusions

Detailed palaeomagnetic analysis of two folds in the Front Ranges of the Canadian Cordillera has uncovered a multi-component remagnetization in Mississippian carbonates at the Mt Kidd syncline and at the Line Creek Anticline. Based on unblocking temperatures, and supported by rock magnetic experiments, the ChRM is a CRM residing in magnetite and the IT reversed component is a CRM residing in pyrrhotite. The magnetite CRM is Early Cretaceous and formed as a result of regional fluid migration ahead of the deformation front. The pyrrhotite CRM is apparently syn-tilting at Line Creek and post-tilting at Mt Kidd.

Based on the positive correlation between bedding-parallel veins and the magnetite CRM, as well as the elevated Sr-isotope results, the most probable origin of the magnetite CRM is a regional fluid-flow event related to the initiation of thrusting and folding in the Main Ranges of the foreland belt prior to the onset of Laramide deformation in the Front Ranges. The origin of the pyrrhotite component observed in the Mississippian carbonates is consistent with warm Fe-rich fluids migrating along faults and fractures causing TSR of hydrocarbons. This model is based on the petrographic observation of multiple common by-products of TSR which include authigenic pyrite, frambodial pyrrhotite, sphalerite and sulphur-enriched bitumen. The exact timing of pyrrhotite remagnetization is unclear due to the general nature of TSR.

This research was supported by an ExxonMobil Student Research Grant, a R. E. McAdams Memorial AAPG student research grant, a Geological Society of America Allan V. Cox Student Research Award and an Institute for Rock Magnetism visiting research fellowship. Thanks to the Alberta Parks and Recreations and J. Halko at Tek Coal for access to the field areas. Thanks also to R. A. Price for suggesting the Line Creek Area and K. Cole for sparking my interest in Kananaskis Country. Thanks to G. Morgan for his help using the electron microprobe and E. Manning for the preliminary fluid inclusion work. Thanks to V. O'Brien, A. Theil and B. Helmke for their assistance in the field over four summer field seasons.

## References

AL-AASM, I. S. 2000. Chemical and isotopic constraints for recrystallization of sedimentary dolomites from the Western Canada Sedimentary Basin. *Aquatic Geochemistry*, **6**, 227–248.

BANERJEE, S., ELMORE, R. D. & ENGEL, M. H. 1997. Chemical remagnetization and burial diagenesis; testing the hypothesis in the Pennsylvanian Belden Formation, Colorado. *Journal of Geophysical Research*, **102**, 24825–24842.

BESSE, J. & COURTILLOT, V. 2002. Apparent and true polar wander and the geometry of the geomagnetic field over the last 200 Myr. *Journal of Geophysical Research*, **107**, 2300, doi: 10.1029/2000JB000050.

BLUMSTEIN, A. M., ELMORE, R. D., ENGEL, M. H., ELLIOT, C. & BASU, A. 2004. Paleomagnetic dating of burial diagenesis in Mississippian carbonates, Utah. *Journal of Geophysical Research*, **109**, 16, BO4101, doi: 10.1029/2003JB002698.

BORRADAILE, G. J. 1997. Deformation and Paleomagnetism. *Surveys in Geophysics*, **18**, 405–435.

CARR, S. D. & SIMONY, P. S. 2006. Ductile thrusting versus channel flow in the southeastern Canadian Cordillera; evolution of a coherent crystalline thrust sheet. *Geological Society Special Publications*, **1**, 561–587.

COOLEY, M. A., PRICE, R. A., KYSER, T. K. & DIXON, J. M. 2011. Stable-isotope geochemistry of syntectonic veins in Paleozoic carbonate rocks in the Livingstone Range anticlinorium and their significance to the thermal and fluid evolution of the southern Canadian foreland thrust and fold belt. *AAPG Bulletin*, **95**, 1851–1882.

CIOPPA, M. T. & SYMONS, D. T. A. 2000. Timing of hydrocarbon generation and migration; Paleomagnetic and rock magnetic analysis of the Devonian Duvernay Formation, Alberta, Canada. *Journal of Geochemical Exploration*, **69–70**, 387–390.

CIOPPA, M. T., AL-AASM, I. S., SYMONS, D. T. A. & GILLEN, K. P. 2003. Dating penecontemporaneous dolomitization in carbonate reservoirs; paleomagnetic, petrographic, and geochemical constraints. *American Association of Petroleum Geologists Bulletin*, **87**, 71–88.

CROUZET, C., GAUTAM, P., SCHILL, E. & APPEL, E. 2003. Multicomponent magnetization in western Dolpo (Tethyan Himalaya, Nepal); tectonic implications. *Tectonophysics*, **377**, 179–196.

DEKKERS, M. J., MATTEI, J. L., FILLION, G. & ROCHETTE, P. 1989. Grain-size dependence of the magnetic behavior of pyrrhotite during its low-temperature transition at 34 K. *Geophysical Research Letters*, **16**, 855–858.

DUNHAM, R. J. 1962. Classification of carbonate rocks according to depositional texture. *In*: HAM, W. E. (ed.) *Classification of Carbonate Rocks. A Symposium. American Association of Petroleum Geologists, Memoir*, **1**, 108–171.

DUNLOP, D. J. & ARGYLE, K. S. 1991. Separating multi domain and single-domain-like remanences in pseudo-single-domain magnetites (215–540 nm) by low-temperature demagnetization. *Journal of Geophysical Research*, **96**, 2007–2017.

DUNLOP, D. J., ÖZDEMIR, O., CLARK, D. A. & SCHMIDT, P. W. 2000. Time-temperature relations for the remagnetization of pyrrhotite (Fe (sub 7) S (sub 8)) and their use in estimating paleotemperatures. *Earth and Planetary Science Letters*, **176**, 107–116.

ELMORE, R. D., LONDON, D., BAGLEY, D. & GAO, G. 1993. Remagnetization by basinal fluids: testing the hypothesis in the Viola Limestone, southern Oklahoma. *Journal of Geophysical Research*, **98**, 6237–6254.

ELMORE, R. D., KELLEY, J., EVANS, M. & LEWCHUK, M. T. 2001. Remagnetization and orogenic fluids; testing the hypothesis in the Central Appalachians. *Geophysical Journal International*, **144**, 568–576.

ELMORE, R. D., FOUCHER, J. L.-E., EVANS, M., LEWCHUK, M. & COX, E. 2006. Remagnetization of the Tonoloway

Formation and the Helderberg Group in the Central Appalachians; testing the origin of syn-tilting magnetizations. *Geophysical Journal International*, **166**, 1062–1076.

ENKIN, R. J. 2003. The direction-correction tilt test; an all-purpose tilt/fold test for paleomagnetic studies. *Earth and Planetary Science Letters*, **212**, 1–2.

ENKIN, R. J. 2004. Paleomagnetism Data Analysis: Version 4.2. Geological Survey of Canada. On http://gsc.nrcan.gc.ca/sw/paleo_e.php.

ENKIN, R. J., OSADETZ, K. G., BAKER, J. & KISILEVSKY, D. 2000. Orogenic remagnetizations in the front ranges and inner foothills of the southern Canadian Cordillera; chemical harbinger and thermal handmaiden of Cordilleran deformation. *Geological Society of America Bulletin*, **112**, 929–942.

EVENCHICK, C. A., MCMECHAN, M. E., MCNICOLL, V. J. & CARR, S. D. 2007. A synthesis of the Jurassic–Cretaceous tectonic evolution of the central and southeastern Canadian Cordillera: exploring links across the orogen. *Geological Society of America Special Paper*, **433**, 117–145.

FEINSTEIN, S., KOHN, B., OSADETZ, K. & PRICE, R. A. 2007. Thermochronometric reconstruction of the pre-thrust paleogeothermal gradient and initial thickness of the Lewis thrust sheet, southeastern Canadian Cordillera foreland belt. *Geological Society of America Special Paper*, **433**, 167–182.

FERRILL, D. A., MORRIS, A. P., EVANS, M. A., BURKHARD, M., GROSHONG, R. H. JR. & ONASCH, C. M. 2004. Calcite twin morphology; a low temperature deformation geothermometer. *Journal of Structural Geology*, **26**, 1521–1529.

FISHER, R. A. 1953. Dispersion on a sphere. *Geophysical Journal of the Royal Astronomical Society*, **217**, 295–305.

FONT, E., TRINDADE, R. I. F. & NEDELEC, A. 2006. Remagnetization in bituminous limestones of the Neoproterozoic Araras Group (Amazon craton); hydrocarbon maturation, burial diagenesis, or both? *Journal of Geophysical Research*, **111**, 17, B06204, doi: 10.1029/2005BJ004106.

GAO, G., ELMORE, R. D. & LAND, L. S. 1992. Geochemical constraints on the origin of calcite veins and associated limestone alteration, Ordovician Viola Group, Arbuckle Mountains, Oklahoma, U.S.A. *Chemical Geology*, **98**, 3–4.

GE, S. & GARVEN, G. 1994. A theoretical model for thrust-induced deep groundwater expulsion with application to the Canadian Rocky Mountains. *Journal of Geophysical Research*, **99**, 851–813.

GILLETT, S. L. 2003. Paleomagnetism of the Notch Peak contact metamorphic aureole, revisited; pyrrhotite from magnetite + pyrite under submetamorphic conditions. *Journal of Geophysical Research*, **108**, 11, 2446, doi: 10.1029/2002JB002386.

GILLETT, S. L. & KARLIN, R. E. 2004. Pervasive late Paleozoic-Triassic remagnetization of miogeoclinal carbonate rocks in the Basin and Range and vicinity, SW USA; regional results and possible tectonic implications. *Physics of the Earth and Planetary Interiors*, **141**, 95–120.

GOLDHABER, M. B. & ORR, W. L. 1995. Kinetic controls on thermochemical sulfate reduction as a source of sedimentary H2S. *American Chemical Society Symposium Series*, **612**, 412–425.

GOLDHABER, M. B. & REYNOLDS, R. L. 1991. Relations among hydrocarbon reservoirs, epigenetic sulfidization, and rock magnetization; examples from the South Texas coastal plain. *Geophysics*, **56**, 748–757.

HALL, A. J. 1986. Pyrite-pyrrhotite redox reactions in nature. *Mineralogical Magazine*, **50**, 223–229.

HARDEBOL, N. J., CALLOT, J.-P., FAURE, J.-L., BERTOTTI, G. & ROURE, F. 2007. Kinematics of the SE Canadian fold-and-thrust belt; implications for the thermal and organic maturation history. *In*: LACOMBE, O., CALLOT, J. P. & FAURE, R. L. (eds) *Thrust Belts and Foreland Basins*. New Frontiers in Earth Sciences. Springer-Verlag, Berlin, Germany, 179–202.

HESLOP, D., MCINTOSH, G. & DEKKERS, M. J. 2004. Using time- and temperature-dependent Preisach models to investigate the limitations of modeling isothermal remanent magnetization acquisition curves with cumulative log Gaussian functions. *Geophysical Journal International*, **157**, 55–63.

HOUSEN, B. A. & MUSGRAVE, R. J. 1996. Rock-magnetic signature of gas hydrates in accretionary prism sediments. *Earth and Planetary Science Letters*, **139**, 509–519.

HOUSEN, B. A., BANERJEE, S. K., MOSKOWITZ, B. M. & COURTILLOT, V. 1996. Low-temperature magnetic properties of siderite and magnetite in marine sediments. *Geophysical Research Letters*, **23**, 2843–2846.

HUDSON, M. R., REYNOLDS, R. L. & FISHMAN, N. S. 1989. Synfolding magnetization in the Jurassic Preuss Sandstone, Wyoming-Idaho-Utah thrust belt. *Journal of Geophysical Research*, **94**, 13 681–13 705.

HUNT, C. P., BANERJEE, S. K., HAN, J., SOLHEID, P. A., OCHES, E., SUN, W. & LIU, T. 1995. Rock-magnetic proxies of climate change in the loess-Palaeosol sequences of the western Loess Plateau of China. *Geophysical Journal International*, **123**, 232–244.

JACKSON, M., ROCHETTE, P., FILLION, G., BANERJEE, S. & MARVIN, J. 1993. Rock magnetism of remagnetized Paleozoic carbonates; low-temperature behavior and susceptibility characteristics. *Journal of Geophysical Research*, **98**, 6217–6225.

KALKREUTH, W. & MCMECHAN, M. E. 1988. Burial history and thermal maturity, Rocky Mountain Front Ranges, foothills, and foreland, east-central British Columbia and adjacent Alberta, Canada. *American Association of Petroleum Geologist Bulletin*, **72**, 1395–1410.

KATZ, B., ELMORE, R. D., COGOINI, M., ENGEL, M. H. & FERRY, S. 2000. Associations between burial diagenesis of smectite, chemical remagnetization, and magnetite authigenesis in the Vocontian Trough, SE France. *Journal of Geophysical Research*, **105**, 851–868.

KIRSCHVINK, J. L. 1980. The least-squares line and plane and the analysis of palaeomagnetic data. *Geophysical Journal of the Royal Astronomical Society*, **62**, 699–718.

KODAMA, K. P. 1988. Remanence rotation due to rock strain during folding and the stepwise application of the fold test. *Journal of Geophysical Research*, **93**, 3357–3371.

KRUIVER, P. P., DEKKERS, M. J. & HESLOP, D. 2001. Quantification of magnetic coercivity components by the analysis of acquisition curves of isothermal remanent magnetization. *Earth and Planetary Science Letters*, **189**, 269–276.

LARRASOAÑA, J. C., ROBERTS, A. P., MUSGRAVE, R. J., GRACIA, E., PINERO, E., VEGA, M. & MARTINEZ-RUIZ, F. 2007. Diagenetic formation of greigite and pyrrhotite in gas hydrate marine sedimentary systems. *Earth and Planetary Science Letters*, **261**, 350–366.

LEWCHUK, M. T., AL-AASM, I. S., SYMONS, D. T. A. & GILLEN, K. P. 1998. Dolomitization of Mississippian carbonates in the Shell Waterton gas field, southwestern Alberta; insights from paleomagnetism, petrology and geochemistry. *Bulletin of Canadian Petroleum Geology*, **46**, 387–410.

LOWRIE, W. 1990. Identification of ferromagnetic minerals in a rock by coercivity and unblocking temperature properties. *Geophysical Research Letters*, **17**, 159–162.

MACHEL, H. G. 2001. Bacterial and thermochemical sulfate reduction in diagenetic settings; old and new insights. *Sedimentary Geology*, **58**, 2295–2318.

MACHEL, H. G. & BURTON, E. A. 1991. Chemical and microbial processes causing anomalous magnetization in environments affected by hydrocarbon seepage. *Geophysics*, **56**, 598–605.

MACHEL, H. G. & CAVELL, P. A. 1999. Low-flux, tectonically-induced squeegee fluid flow ('hot flash') into the Rocky Mountain foreland basin. *Bulletin of Canadian Petroleum Geology*, **47**, 510–533.

MACHEL, H. G. & FOGHT, J. 2000. Products and depth limits of microbial activity in petroliferous subsurface settings. *In*: RIDING, R. E. & AWRAMIK, S. M. (eds) *Microbial Sediments*. Springer-Verlag, Berlin, Germany, 106–120.

MACHEL, H. G., KROUSE, H. R. & SASSEN, R. 1995. Products and distinguishing criteria of bacterial and thermochemical sulfate reduction. *Applied Geochemistry*, **108**, 1108–1119.

MARSHAK, S. & MITRA, G. 1988. *Basic methods of structural geology; Part 1, Elementary techniques; Part 2, Special topics*. Prentice Hall, NJ.

MCARTHUR, J. M., HOWARTH, R. J. & BAILEY, T. R. 2001. Strontium isotope stratigraphy; LOWESS Version 3; best fit to the marine Sr-isotope curve for 0–509 Ma and accompanying look-up table for deriving numerical age. *Journal of Geology*, **109**, 155–170.

MCCABE, C. & ELMORE, R. D. 1989. The occurrence and origin of late Paleozoic remagnetization in the sedimentary rocks of North America. *Reviews of Geophysics*, **27**, 471–494.

MCMECHAN, M. E. 1995. *Geology rocky mountain foothills and front ranges in Kananaskis country, Alberta*. Geological Survey of Canada, map, **1865A**.

MONGER, J. W. H. & PRICE, R. A. 1979. Geodynamic evolution of the Canadian Cordillera; progress and problems. *Canadian Journal of Earth Sciences*, **16**, 771–791.

MOUNTJOY, E. W., GREEN, D., MACHEL, H. G., DUGGAN, J. & WILLIAMS-JONES, A. E. 1999. Devonian matrix dolomites and deep burial carbonate cements; a comparison between the Rimbey-Meadowbrook reef trend and the deep basin of west-central Alberta. *Bulletin of Canadian Petroleum Geology*, **47**, 487–509.

O'BRIEN, V. J., MORELAND, K. M., ELMORE, R. D., ENGEL, M. H. & EVANS, M. A. 2007. Origin of orogenic remagnetizations in Mississippian carbonates, Sawtooth Range, Montana. *Journal of Geophysical Research*, **112**, B06103, doi: 10.1029/2006JB004699.

OLIVER, J. 1986. Fluids expelled tectonically from orogenic belts; their role in hydrocarbon migration and other geologic phenomena. *Geology*, **14**, 99–102.

ORR, W. L. 1974. Changes in sulfur content and isotopic ratios of sulfur during petroleum maturation; study of big horn basin paleozoic oils. *American Association of Petroleum Geologist Bulletin*, **68**, 713–743.

ÖZDEMIR, O., DUNLOP, D. J. & MOSKOWITZ, B. M. 1993. The effect of oxidation on the Verwey transition in magnetite. *Geophysical Research Letters*, **20**, 1671–1674.

PIERCE, J. W., GOUSSEV, S. A., CHARTERS, R. A., AMBERCROMBIE, H. J. & DEPAOLI, G. R. 1998. Intrasedimentary magnetization by vertical fluid flow and exotic geochemistry. *The Leading Edge*, **17**, 89.

PREEDEN, U., PLADO, J., MERTANEN, S. & PUURA, V. 2008. Multiply remagnetized Silurian carbonate sequence in Estonia. *Estonian Journal of Earth Sciences*, **57**, 170–180.

PRICE, R. A. (ed.) 1981. *The Cordilleran Foreland Thrust and Fold Belt in the Southern Canadian Rocky Mountains*. Geological Society, London, Special Publications, **9**, 427–448.

PRICE, R. A. 1994. Cordilleran tectonics and the evolution of the Western Canada sedimentary basin. *In*: MOSSOP, G. & SHELTSIN, I. (eds) *Geological Atlas of the Western Canada Sedimentary Basin*. Alberta Research Council and Canadian Society of Petroleum Geologists, Calgary, 13–24.

PRICE, R. A., GRIEVE, D. A. & PATENAUDE, C. 1992. *Geology, Tornado Mountain British Columbia-Alberta*. Geological Survey of Canada, map, **1823A**.

ROBION, PH., FAURE, J. L. & SWENNEN, R. 2004. Late cretaceous chemical remagnetization of the paleozoic carbonates from the undeformed foreland of the Western Canadian Cordillera. *In*: SWENNEN, R., ROURE, F. & GRANATH, J. (eds) *Fluid Flow, Deformation History and Reservoir Appraisal in Foreland Fold-and-Thrust Belts*. American Association of Petroleum Geologists, Tulsa, OK, Hedberg Series 1.

ROCHETTE, P. 1987. Metamorphic control of the magnetic mineralogy of black shales in the Swiss Alps; toward the use of 'magnetic isogrades'. *Earth and Planetary Science Letters*, **84**, 446–456.

ROCHETTE, P., FILLION, G., MITTEI, J.-L. & DEKKERS, M. J. 1990. Magnetic transition at 30–34 Kelvin in pyrrhotite; insight into a widespread occurrence of this mineral in rocks. *Earth and Planetary Science Letters*, **98**, 319–328.

ROCHETTE, P., FILLION, G. & DEKKERS, M. J. 2011. Interpretation of low-temperature data part 4: the low-temperature magnetic transition of monoclinic pyrrhotite. *The IRM Quarterly*, **21/1**, 7–10.

ROOT, K. G. 2001. Devonian Antler fold and thrust belt and foreland basin development in the southern Canadian Cordillera; implications for the Western Canada Sedimentary Basin. *Bulletin of Canadian Petroleum Geology*, **268**, 561–587.

ROURE, F., SWENNEN, R. *ET AL.* 2005. Incidence and importance of tectonics and natural fluid migration on reservoir evolution in foreland fold-and-thrust belts. *Oil & Gas Science and Technology*, **60**, 67–106.

ROWAN, C. J. & ROBERTS, A. P. 2008. Widespread remagnetizations and a new view of Neogene tectonic rotations within the Australia-Pacific plate boundary zone, New Zealand. *Journal of Geophysical Research*, **113**, B03103, doi: 10.1029/2006JB004594.

SALMON, E., EDEL, J. B., PIQUE, A. & WESTPHAL, M. 1988. Possible origins of Permian remagnetizations in Devonian and Carboniferous limestones from the Moroccan Anti-Atlas (Tafilalet) and Meseta. *Physics of the Earth and Planetary Interiors*, **52**, 339–351.

SCHILL, E., APPEL, E. & GAUTAM, P. 2002. Towards pyrrhotite/magnetite geothermometry in low-grade metamorphic carbonates of the Tethyan Himalayas (Shiar Khola, central Nepal). *Journal of Asian Earth Sciences*, **20**, 195–201.

STAMATAKOS, J., HIRT, A. M. & LOWRIE, W. 1996. The age and timing of folding in the Central Appalachians from paleomagnetic results. *Geological Society of America Bulletin*, **108**, 815–829.

SYMONS, D. T. A. & CIOPPA, M. T. 2002. Conodont CAI and magnetic mineral unblocking temperatures: implications for the Western Canada Sedimentary Basin. *Physics and Chemistry of the Earth*, **27**, 1189–1193.

SYMONS, D. T. A., LEWCHUK, M. T. & SANGSTER, D. F. 1998. Laramide orogenic fluid flow into the Western Canada sedimentary basin; evidence from paleomagnetic dating of the Kicking Horse Mississippi valley-type ore deposit. *Economic Geology and the Bulletin of the Society of Economic Geologists*, **93**, 68–83.

URBAT, M., DEKKERS, M. J. & KRUMSIEK, K. 2000. Discharge of hydrothermal fluids through sediment at the Escanaba Trough, Gorda Ridge (ODP Leg 169): assessing the effects on the rock magnetic signal. *Earth and Planetary Science Letters*, **176**, 481–494, doi: 10.1016/S0012-821X(00)00024-8.

VAN DER PLUIJM, B. A., VROLIJK, P. J., PEVEAR, D. R., HALL, C. M. & SOLUM, J. 2006. Fault dating in the Canadian Rocky Mountains. Evidence for late Cretaceous and early Eocene orogenic pulses. *Geology*, **34**, 837–840.

VAN DONGEN, B. E., ROBERTS, A. P., SCHOUTEN, S., JIANG, W.-T., FLORINDO, F. & PANCOST, R. D. 2007. Formation of iron sulfide nodules during anaerobic oxidation of methane. *Geochimica et Cosmochimica Acta*, **71**, 5155–5167.

VANDEGINSTE, V., SWENNEN, R., GLEESON, S. A., ELLAM, R. M., OSADETZ, K. & ROURE, F. 2009. Thermochemical sulphate reduction in the Upper Devonian Cairn Formation of the Fairholme carbonate complex (South-West Alberta, Canadian Rockies): evidence from fluid inclusions and isotopic data. *Sedimentology*, **56**, 439–460.

WATSON, G. S. & ENKIN, R. J. 1993. The fold test in paleomagnetism as a parameter estimation problem. *Geophysical Research Letters*, **20**, 2135–2137.

WEAVER, R., ROBERTS, A. P. & BARKER, A. J. 2002. A late diagenetic (syn-folding) magnetization carried by pyrrhotite; implications for paleomagnetic studies from magnetic iron sulphide-bearing sediments. *Earth and Planetary Science Letters*, **200**, 371–386.

XU, W., VAN DER VOO, R. & PEACOR, D. R. 1998. Electron microscopic and rock magnetic study of remagnetized Leadville carbonates, central Colorado. *Tectonophysics*, **296**, 333–362.

ZECHMEISTER, M. S. 2010. *Integrated diagenetic, structural and paleomagnetic study of remagnetized folded Mississippian –Triassic units from the Southern Canadian Cordillera, Alberta & British Columbia.* Dissertation, University of Oklahoma.

ZIJDERVELD, J. D. A. 1967. A.C. demagnetization of rocks: analysis of results. *In*: COLLINSON, D. E., CREER, K. M. & RUNCORN, S. K. (eds) *Methods in Paleomagnetism.* Elsevier, Amsterdam, 254–286.

# Remagnetization of the Alamo Breccia, Nevada

STACEY C. EVANS, R. DOUGLAS ELMORE*, DEVIN DENNIE
& SHANNON A. DULIN

*School of Geology and Geophysics, University of Oklahoma, Norman, OK 73019, USA*

**Corresponding author (e-mail: delmore@ou.edu)*

**Abstract:** The Devonian Alamo Breccia is a thick (<30–130 m) unit, interpreted as a bolide impact deposit, which is bracketed by marine carbonates. Samples were collected within the breccia and above/below the breccia for a contact test to determine if the breccia acted as a conduit for fluids that could have caused the widespread chemical remanent magnetizations (CRMs) present in Palaeozoic Era rocks in Nevada. The carbonates above, below and in the breccia contain a Cretaceous Period syn-tilting CRM that resides in pyrrhotite and a pre-tilting late Palaeozoic Era CRM that resides in magnetite. The contact test is negative. Despite these results, diagenetic alteration by externally derived fluids is interpreted as the most likely mechanism of remagnetization. This hypothesis is supported by $^{87}Sr/^{86}Sr$ values in the breccia and surrounding rocks that suggest alteration by fluids with a radiogenic signature. The fluids were not localized in the breccia but are interpreted to have moved pervasively through the rocks. The results differ from some other studies that found that fluids caused localized CRMs around fluid conduits.

Chemical remanent magnetizations (CRMs) can form by many diagenetic mechanisms, including alteration associated with migration of fluids (e.g. Elmore *et al.* 1993), maturation of organic matter (e.g. Banerjee *et al.* 1997; Blumstein *et al.* 2004) and clay diagenesis (e.g. Katz *et al.* 2000; Woods *et al.* 2002; Tohver *et al.* 2008). Palaeomagnetic analysis can be useful in deciphering the timing of such events by determining the direction of the CRM and plotting the corresponding pole position on the apparent polar wander path (APWP). This can be combined with geochemical and other data in an integrated study to understand the origin of diagenetic events.

Many studies have inferred that fluids were responsible for CRMs (e.g. McCabe & Elmore 1989; McCabe *et al.* 1989; Lu *et al.* 1991; Saffer & McCabe 1992), but relatively few studies have focused on testing for a connection between the fluids and CRMs by conducting contact tests on the fluid conduits (e.g. Elmore *et al.* 1993, 1998; Elmore 2001; Blumstein *et al.* 2005). These studies have found that remagnetizing fluids have moved through conduits and remagnetized only the conduits and surrounding rocks. Other studies (e.g. Gillett & Karlin 2004) have found that CRMs can be pervasive in many units but questions remain as to how these rocks were remagnetized. Burial remagnetization mechanisms may be possible in some cases (Fruit *et al.* 1995; Banerjee *et al.* 1997; Katz *et al.* 2000; Blumstein *et al.* 2004; Moreau *et al.* 2005) but cannot explain all pervasive remagnetizations (Gillett & Karlin 2004).

The overall objective of this study is to investigate how fluids move through rocks and cause remagnetizations. This will be accomplished by testing if the Alamo Breccia in Nevada, a potential palaeoaquifer, was in fact a conduit for remagnetizing fluids that caused the widespread remagnetization (Gillett & Karlin 2004) in the Basin and Range area of the southwestern United States.

The Devonian Period Alamo Breccia is a thick (up to *c.* 100 m; Warme & Sandberg 1996) breccia unit in southern Nevada that is bracketed above and below by shallow marine carbonates. Because of presumed higher permeability, the breccia is likely to have acted as a conduit for fluids which may have caused remagnetization compared to the rocks above and below. It is interpreted as a breccia formed from a bolide impact onto shallow marine carbonate shelf. The brecciated rocks were reworked by tsunamis resulting from the impact (Warme & Kuehner 1998). At some localities, the breccia contains extensive thin calcite veins that are probably related to post-impact deformation events.

To test if the Alamo Breccia was a conduit we conducted an integrated palaeomagnetic and geochemical study to identify, date and determine the origin of remagnetizations present in the unit. Samples from within the breccia and from the carbonates above and below the breccia were collected to conduct a contact test. The test was conducted at two different locations in south central Nevada (Hancock Summit and Mt Irish; Fig. 1). At Hancock Summit, samples were also collected

*From*: ELMORE, R. D., MUXWORTHY, A. R., ALDANA, M. M. & MENA, M. (eds) 2012. *Remagnetization and Chemical Alteration of Sedimentary Rocks*. Geological Society, London, Special Publications, **371**, 145–162.
First published online August 22, 2012, http://dx.doi.org/10.1144/SP371.8

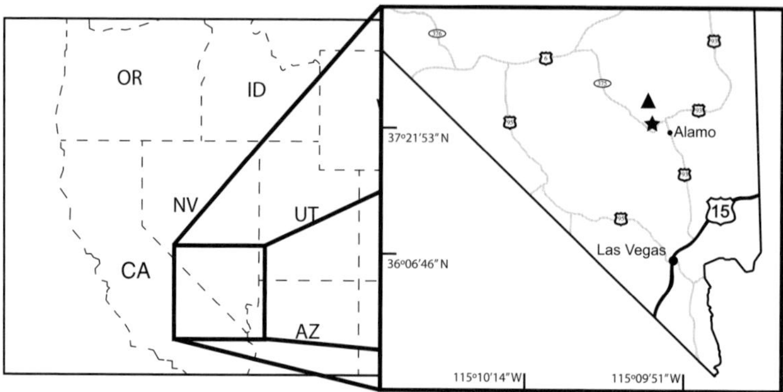



**Fig. 1.** Map of the western United States with inset of study area in Nevada. The star represents the Hancock Summit location, and the triangle represents the Mt Irish location. Major cities and roads are also shown.

from around *c.* 1–2 cm thick veins within the breccia to perform vein contact tests. Samples were subjected to a low-temperature treatment and detailed thermal demagnetization to isolate the magnetizations present. Rock magnetic analysis was conducted to investigate the mineralogy of the magnetic minerals. Petrographic and geochemical studies were performed to determine the nature of the diagenetic alteration and help determine the origin(s) of the remagnetizations.

## Geological background

The Alamo Breccia is found in at least 25 mountain ranges throughout south and central Nevada and western Utah (Pinto & Warme 2008). The area contains Palaeoproterozoic Era metamorphic rocks and igneous intrusions that were overlain by a thick wedge of mainly carbonate rocks deposited during the Neoproterozoic Era through the Mid-Devonian Period. The wedge of carbonate and siliciclastic rocks is up to almost 9000 m thick with *c.* 1800 m deposited during the Devonian Period (Stewart 1980; Gillett & Karlin 2004). From the Late Devonian Period to the Early Mississippian Epoch there was compressional tectonic activity due to the Antler Orogeny, and then the Sonoma Orogeny in the Late Permian Period to the Early Triassic Period. In the Mid-Jurassic Period to the end of the Mesozoic Era, the Sevier Orogeny caused additional compressional tectonic activity. In the middle of the Cenozoic Era, the area began undergoing extensional tectonics that created the present-day basin and range topography (Stewart 1980; Gillett & Karlin 2004). Due to this younger tectonic deformation, the precise centre of the impact crater is not known. Miocene Epoch and post-Miocene Epoch anticlockwise vertical axis rotations of

around 10° have been inferred for the study area (Hudson *et al.* 1998) and particularly the Hancock Summit location (Gillett & Karlin 2004).

The Alamo Breccia is one of the largest known carbonate megabreccias and is calculated to be *c.* 1300 km$^3$ in volume (Pinto & Warme 2008). The breccia is interpreted as a megasedimentary deposit formed by the ejecta curtain, tsunamis and other processes resulting from a bolide impact event on or near a shallow marine carbonate platform (Warme 2004). Conodonts have been used to date the Alamo Breccia as being deposited during early Frasnian punctata Zone (Sandberg & Warme 1993; Warme & Sandberg 1995, 1996). Taking into consideration radiometric dating of the Late Devonian Period (Kaufmann *et al.* 2004; Trapp *et al.* 2004), this converts to an age of *c.* 382 Ma (Morrow & Sandberg 2005) for the breccia.

Evidence that the Alamo Breccia is an impact deposit includes hematite-studded shocked quartz (Leroux *et al.* 1995; Morrow *et al.* 2005), accretionary lapilli (Warme *et al.* 2002) and a minor iridium anomaly. While the iridium anomaly is fairly weak (133–139 ppt), it is still noteworthy when compared to the background iridium levels of ≤25 ppt (Morrow *et al.* 2005).

The Alamo Breccia is the middle member of the Guilmette Formation. The Alamo Breccia Member is underlain by the Yellow Slope-forming Member and the Carbonate Platform Facies (member), which contain cyclic limestones and dolostones. The Alamo Breccia is overlain by the Upper Guilmette Member, which consists of the Slope Facies Member topped with quartzose sandstones (Warme *et al.* 2008).

The Alamo Breccia is divided into four stratigraphic divisions based on changes in lithology and the processes that formed each division (Fig. 2). The lowest section is Unit D, which consists

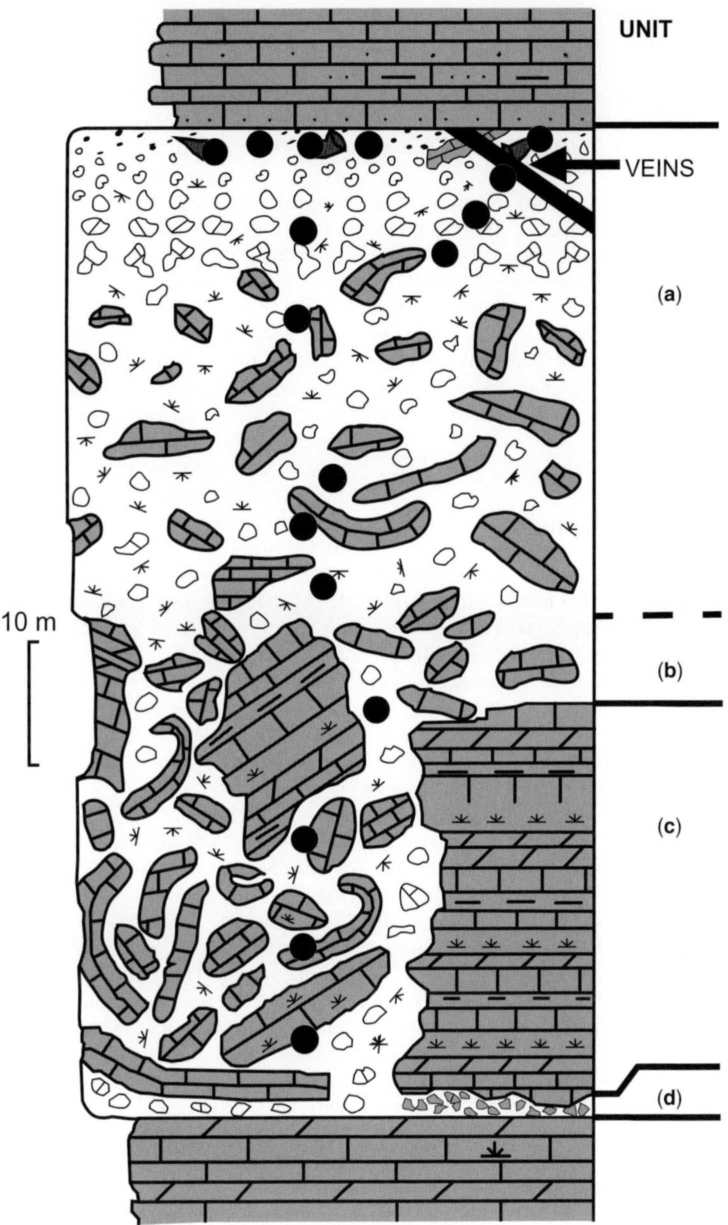



**Fig. 2.** Stratigraphic column showing different units of the Alamo Breccia. Unit A is a coarse–fine-grained graded polymict breccia. Unit B is a chaotic polymict breccia. Unit C consists of large megaclasts. Unit D is a thin monomict breccia. Dots show sample locations within the breccia at Hancock summit. Modified from Warme & Sandberg (1996).

of a thin monomict limestone breccia that separates the megaclasts of Unit C from the lower members of the Guilmette Formation. This unit is interpreted as fluidized rock that was preserved along a detachment surface only in locations where the large megaclasts of Unit C are maintained in nearly their original positions over it. Unit C consists of large (up to *c.* 500 m long by *c.* 80 m thick) megaclasts that are parallel–subparallel to bedding. Unit B is a chaotic, polymict, matrix-supported calcite-cemented breccia with clasts up to hundreds of metres long that resembles a debris-flow deposit. Unit B overlies Unit C and, where Units C and D are missing, extends to the underlying Carbonate

Platform Facies. The uppermost division is Unit A, which is a coarse–fine-grained, graded, polymict breccia. This unit is less chaotic than Unit B. Unit A usually exhibits at least two intervals of normally graded beds at the top of the unit (Warme & Sandberg 1996; Warme & Kuehner 1998; Pinto & Warme 2008).

At Hancock Summit, Unit A contains numerous veins. The veins range in size from *c.* 100 μm up to a few centimetres wide. The veins are generally widely spaced but are very closely grouped in some locations. There were two sets of vein orientations: south-southeasterly (161°) and westerly (270°). The westerly veins are generally smaller in size than the southeasterly veins. Some of the larger veins are over 1 m in length, and are oriented in an en echelon pattern.

## Previous palaeomagnetic studies

The Alamo Breccia has not been the subject of palaeomagnetic study. Gillett & Karlin (2004) conducted a regional palaeomagnetic study of Palaeozoic Era carbonate rocks in the Basin and Range area of the southwestern United States including the Guilmette Formation, and found that the rocks contained two ancient components. An intermediate-temperature component with relatively steep positive inclinations and generally north declinations was found between *c.* 230 °C and 350 °C and was interpreted to reside in magnetite and possibly pyrrhotite at some sites. At higher temperatures (320–500 °C), a low-inclination (LI) component was identified; it was interpreted to reside in magnetite and has south–SE declinations (in stratigraphic coordinates).

The timing and origin of both components were difficult to interpret. The intermediate component was interpreted to have been acquired during the Cretaceous Period normal superchron, which overlaps with timing of the Sevier Orogeny. At several sites however, this Cretaceous Period origin was difficult to justify and the component was interpreted as more likely being Cenozoic Era in age despite the normal polarity. Gillett & Karlin (2004) interpreted the intermediate component to be polygenic in origin and probably reflected local Cenozoic Era or Mesozoic Era events.

The timing and origin of the LI component was also problematic. The negative inclinations and southeasterly declinations suggested the magnetization had been acquired during the Palaeozoic Era Kiaman reversed superchron but some sites, based on the directions, suggest the remagnetization was as young as the Early Triassic Period. Gillett & Karlin (2004) suggested three possible origins for this component: brine mobilization during the

Sonoma Orogeny (Late Permian Period–Early Triassic Period); brine mobilization during the Ancestral Rockies uplift (Late Palaeozoic Era); and smectite destruction/magnetite authigenesis. Both of the mechanisms involving brine mobilization are unable to fully account for the geographic extent of remagnetization, while the destruction of smectite does not fully account for the stratigraphic extent unless coupled with thermal resetting of stratigraphically lower units. This led to the conclusion that the LI component was also polygenetic in origin.

Several palaeomagnetic studies have used contact tests to investigate units acting as palaeoaquifers for fluid migration events. For example, Elmore *et al.* (1998) used palaeomagnetic and petrographic/geochemical methods on Palaeozoic Era rocks in the Arbuckle Mountains to determine timing of fluid-flow events through the Reagan Sandstone, a basal palaeoaquifer. The base of the Reagan Sandstone and the upper section of the underlying Colbert Rhyolite were found to be chemically altered and contained a Late Palaeozoic Era CRM that was interpreted to have been caused by alteration associated with a fluid migration event. The underlying Colbert Rhyolite was also interpreted to carry a primary magnetization, whereas the upper portion of the Reagan Sandstone and the overlying Honey Creek Formation contained a Cambro-Ordovician Period magnetization (Elmore *et al.* 1998).

A similar study conducted in the Appalachian Mountains showed the presence of fluid-related CRMs located in a Devonian Period palaeoaquifer, the Oriskany Formation (Elmore *et al.* 2001). The Oriskany and the underlying aquitard, the Helderberg Group, both contain CRMs that reside in magnetite. The Oriskany Formation contained geochemical evidence that orogenic fluids had migrated through it, and the CRM was interpreted as related to the fluids. The Helderberg Group contained no geochemical evidence for externally derived fluids, and the CRM was interpreted as being related to burial diagenetic processes.

Other studies have used contact tests around faults and fractures to date fluid-flow events. For example, Blumstein *et al.* (2005) studied the rocks in and around the Moine Thrust Zone in northern Scotland; the thrust zone was interpreted to have been a conduit for multiple fluid-flow events. This method can also be applied to the smaller scale of veins. Elmore *et al.* (1993) found a pervasive Carboniferous Period CRM residing in magnetite in the Viola Limestone in the Arbuckle Mountains, but also found a Permian Period CRM residing in hematite that was localized around veins within the Viola. The Permian Period CRM was found in rocks near the veins that had elevated $^{87}Sr/^{86}Sr$ values. The $^{87}Sr/^{86}Sr$ values decrease with distance from the vein to coeval values, and the Permian Period CRM decreased until only the pervasive Carboniferous Period CRM was present. This suggests that the fluids that altered the rocks also caused the CRM in magnetite and that the CRM dates the fluid-flow event.

## Methods

Cores were collected from the breccia for palaeomagnetic, rock magnetic, geochemical and petrographic analyses using a portable gasoline-powered drill and oriented using an inclinometer and a Brunton compass. Bedding and vein orientation were also measured using a Brunton compass. Cores were also collected from inside and outside of the breccia in order to perform a breccia contact test. A contact test is performed by sampling within the hypothesized fluid conduit, and then sampling the overlying and underlying units. If the breccia was a conduit for remagnetizing fluids, a magnetization would be present within the breccia but would decrease away from the breccia. The intensity of the magnetizations might also be higher in the breccia as opposed to the overlying and underlying units.

During initial sampling cores were collected from throughout the breccia at Hancock Summit (Fig. 1) including individual breccia clasts, matrix and coarse–fine-grained polymict breccia at the top of the unit, as well as sites above and below the breccia. The breccia was also sampled at Mt Irish (Fig. 1), including accretionary lapilli clasts in the upper section of the Alamo Breccia (Warme *et al.* 2002). The rocks were resampled to confirm the results from the initial sampling and to perform additional demagnetization procedures. To resample, cores were collected from 9 sites (6–28 cores per site) within the Alamo Breccia at the Hancock Summit. Eight sites (2–8 cores per site) were also sampled from up to *c.* 11 m above the breccia and 11 sites (2–3 cores per site) were sampled from down to *c.* 50 m below the breccia. At Mt Irish, cores were collected from 7 sites (2–8 cores per site) in rocks overlying the Alamo Breccia up to *c.* 12 m and 5 sites (1–4 cores per site) from within the breccia. The underlying beds were not exposed.

Contact tests were also performed on veins within the breccia at Hancock Summit. Cores were collected from the veins and with increasing distance from the veins and analysed to determine if there were any connections between the veins and the magnetic characteristics (i.e. magnetic intensities, directions). As for the breccia contact test, if this test was positive there would be a component present at/near the vein and it would decrease with distance from the vein. Eight separate vein contact tests were performed with 97 cores.

In the palaeomagnetic laboratory at the University of Oklahoma, the cores were cut to standard 2.2 cm lengths using a double-blade core saw. The natural remanent magnetizations (NRMs) were measured using a three-axis 2 G Enterprises cryogenic magnetometer with DC SQUIDs (superconducting quantum interference device) in the magnetically shielded lab. Most cores collected during the initial sampling were only subject to stepwise thermal demagnetization. Some specimens were subjected to low-temperature demagnetization (LTD) in order to test its effectiveness and, as a result, all the specimens from the second sampling were subjected to two LTD steps by immersion in liquid nitrogen for a minimum of 30 min. This process is used to remove the magnetization held in multi-domain (MD) magnetite (Dunlop & Argyle 1991). Specimens were then subjected to thermal demagnetization at increasing temperature steps between 100 °C and 600 °C using an ASC Scientific Thermal Specimen Demagnetizer (Model TD-48 SC). The specimens did not exhibit stable decay above 500–540 °C.

The demagnetization data were analysed using the Super IAPD (Interactive Analysis of Paleomagnetic Data) software program (www.geodynamics.no/software.htm) and plotted on Zijderveld diagrams (Zijderveld 1967). Principal component analysis (Kirschvink 1980) was used to determine the magnetic components in specimens which had mean angular deviations (MADs) of less than 15°, although most had MADs of less than 10°. Mean directions of the magnetic components were determined using Fisher (1953) statistics. Because of different bedding attitudes at Hancock Summit (N020/40W) and Mt Irish (N168/22W), a regional tilt test was performed using the directional correction (Enkin 2003) and the Watson & Enkin (1993) tilt tests. Both tilt tests were performed using the Geological Survey of Canada Paleomagnetism Analysis program package (Enkin, version 4.2).

Isothermal remanent magnetization (IRM) acquisition and thermal decay measurements were collected to aid in determining the magnetic mineralogy present in the specimens. An IRM was induced in 14 representative specimens with an impulse magnetizer using a stepwise progression in 25 steps from 0 to 2500 mT. The IRM acquisition curves were analysed using the IRMUNMIX 2.2 (Isothermal Remanent Magnetization unmix) software program (Heslop *et al.* 2002) which was followed by cumulative log-Gaussian (CLG) analysis (Kruiver *et al.* 2001) to model the different coercivity contributions (Heslop *et al.* 2004). The specimens were subjected to alternating field (AF) demagnetization up to 120 mT and were given IRMs

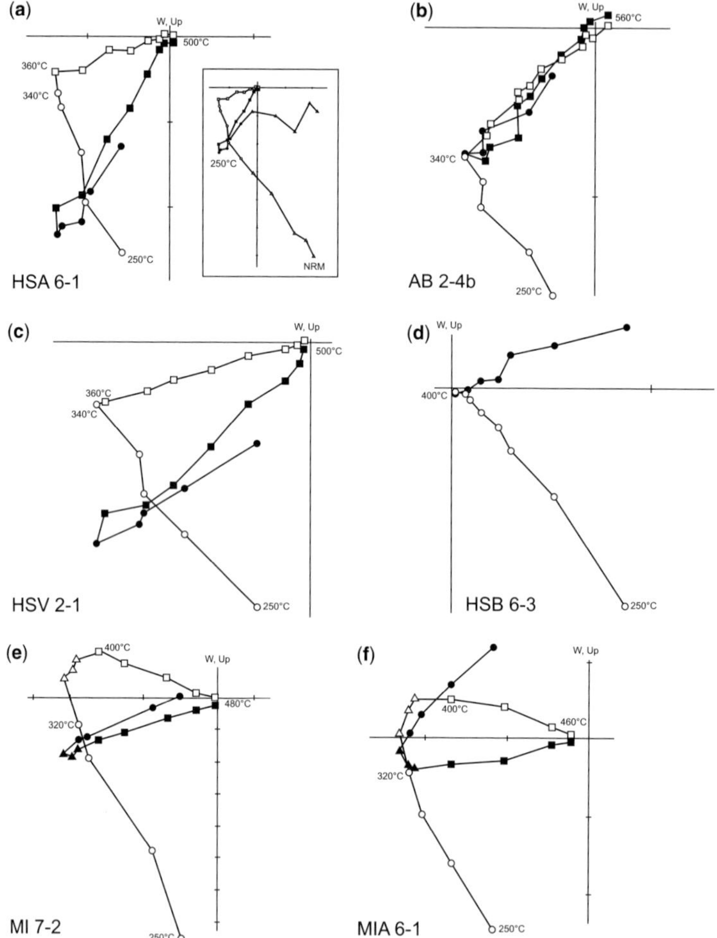

**Fig. 3.** Representative Zijderveld diagrams showing the decay of magnetization within specimens from the Alamo Breccia and the surrounding rocks. In all diagrams except (a), the VRM (up to 200 °C) has been removed in order to better illustrate the decay at higher temperatures. Closed symbols represent the horizontal component; open symbols represent the vertical component. Circles represent IT component; squares represent HT component. Triangles in (e) and (f) represent the curved portion of the demagnetization path; these points were not used in calculating components. (**a**) Hancock Summit specimen from above the breccia. Inset shows decay with VRM. (**b**) Hancock Summit breccia specimen that did not undergo LTD. This specimen is the same rock type as shown in (c) [note the steeper inclination of the HT component compared to (c)]. (**c**) Hancock Summit specimen from within the breccia. (**d**) Hancock Summit specimen from below the breccia. (**e**) Mt Irish specimen from within the breccia. (**f**) Mt Irish specimen from above the breccia. The NRM intensities are as follows: (a) 0.523, (b) 0.240, (c) 0.142, (d) 0.309, (e) 0.827 and (f) 0.346 mA m$^{-1}$. All tick marks equal 0.1 mA m$^{-1}$.

(2500 mT, 500 mT and 120 mT) in three perpendicular directions prior to nine of the specimens being thermally demagnetized to obtain the tri-axial decay curves (Lowrie 1990).

Petrographic analysis was conducted to identify magnetic minerals as well as diagenetic features. Fifty polished thin sections from inside and outside the breccia were examined using reflected and transmitted light.

Strontium isotope analysis was conducted at the University of Texas at Austin using the methodology described by Gao *et al.* (1992) to determine if the rocks had undergone chemical alteration. Analysis was performed on seven specimens from Hancock Summit: two limestone specimens from outside the breccia, four specimens from within the breccia and one calcite vein from within the breccia. Each specimen consisted of *c.* 200 mg aliquots. The

NIST SRM 987 Sr (National Institute of Standards and Technology Standard Reference Materials) standard was measured to a mean value of 0.710248 $\pm$ 0.000008. The measured strontium isotope values were normalized to the National Bureau of Standards' NBS 987 = 0.71014 and then compared with the coeval seawater values for Late Devonian-age limestones (McArthur *et al.* 2001).

**Table 1.** *Palaeomagnetic data. $N/N_0$: number of specimens with direction versus number of specimens demagnetized; k: precision parameter; $\alpha95$: cone of 95% confidence. MI refers to Mt Irish sites; all others are from Hancock Summit*

| Site | $N/N_0$ | Geographic | | | | Stratigraphic (tilt corrected) | | | |
|---|---|---|---|---|---|---|---|---|---|
| | | Declination | Inclination | $k$ | $\alpha95$ | Declination | Inclination | $k$ | $\alpha95$ |
| *Intermediate-temperature component* | | | | | | | | | |
| AB | 5/8 | 311.6 | 49.3 | 148 | 6.3 | 77.1 | 76.7 | 148 | 6.3 |
| HSA | 21/22 | 328 | 51.9 | 10.3 | 10.4 | 36.9 | 66.4 | 10.3 | 10.4 |
| HSB | 21/23 | 317.5 | 47.3 | 46 | 4.7 | 21.7 | 71 | 46 | 4.7 |
| HSV1 | 5/5 | 314.9 | 40.7 | 95.6 | 7.9 | 358.8 | 69.8 | 95.6 | 7.9 |
| HSV2 | 6/6 | 314.2 | 51.8 | 131.1 | 5.9 | 32.2 | 74.8 | 131.1 | 5.9 |
| HSV3 | 11/12 | 334.8 | 47 | 16.5 | 11.6 | 29.9 | 60.7 | 16.5 | 11.6 |
| HSV7 | 5/10 | 322.3 | 62.1 | 15.5 | 20 | 64.6 | 69.2 | 15.5 | 20 |
| MI1 | 8/9 | 325.7 | 59.8 | 193.9 | 4 | 299 | 35.5 | 193.9 | 4 |
| MI2 | 3/4 | 333.4 | 67.2 | 112.2 | 11.7 | 313.5 | 40.3 | 112.2 | 11.7 |
| MIA | 12/12 | 318.9 | 61.9 | 22.9 | 9.3 | 295.1 | 47.1 | 22.9 | 9.3 |
| *Non-LTD sites** | | | | | | | | | |
| HSBrclasts | 16/29 | 327.6 | 55.7 | 23.7 | 7.7 | 48.7 | 66.9 | 23.7 | 7.7 |
| AB2 | 7/9 | 294.4 | 56.8 | 19.4 | 14 | 86.9 | 83.6 | 19.4 | 14 |
| AB3 | 4/10 | 301.1 | 58 | 554.9 | 3.9 | 71.4 | 80.4 | 554.9 | 3.9 |
| AB1 | 3/6 | 314.1 | 57.5 | 60.5 | 16 | 57.3 | 74 | 60.5 | 16 |
| AB8 | 4/10 | 321.9 | 48.4 | 69 | 11.1 | 27.3 | 68.4 | 69 | 11.1 |
| MI1i | 5/7 | 328.5 | 56.1 | 252.9 | 4.8 | 305.4 | 44.4 | 252.9 | 4.8 |
| MI2i[†] | 13/16 | 2.6 | 61 | 53.8 | 5.7 | 323.5 | 58.9 | 53.8 | 5.7 |
| MI3i | 6/6 | 73.5 | 67.2 | 71.9 | 8 | 11 | 88.1 | 72.1 | 8 |
| *High-temperature component* | | | | | | | | | |
| AB | 7/8 | 126 | 29.5 | 83.1 | 6.7 | 125.4 | −20.1 | 83.1 | 6.7 |
| HSA | 19/22 | 122 | 36.2 | 15.1 | 8.9 | 119.6 | −2.1 | 15.1 | 8.9 |
| HSB | 3/23 | 119.3 | 26.6 | 73.6 | 14.5 | 118.4 | −12.8 | 73.6 | 14.5 |
| HSV1 | 5/5 | 134.6 | 29.9 | 113.7 | 7.2 | 131.5 | −7.2 | 113.7 | 7.2 |
| HSV2 | 6/6 | 128.2 | 20.1 | 74.7 | 7.8 | 127.8 | −17.1 | 74.7 | 7.8 |
| HSV3 | 9/12 | 132 | 27.2 | 60.8 | 6.7 | 129.7 | −9.4 | 60.8 | 6.7 |
| HSV7 | 6/10 | 136.8 | 23.5 | 66.1 | 8.3 | 134.9 | −11.7 | 66.1 | 8.3 |
| MI1 | 4/9 | 162.4 | −25 | 277.3 | 5.5 | 151.3 | −13.2 | 277.3 | 5.5 |
| MI2 | 4/4 | 156.2 | −32.5 | 208.7 | 6.4 | 149.2 | −8.5 | 208.7 | 6.4 |
| MIA | 10/12 | 164.4 | −20.7 | 102.1 | 4.8 | 156.7 | −17.9 | 102.1 | 4.8 |
| *Non-LTD sites** | | | | | | | | | |
| HSBrclasts | 21/29 | 124.5 | 32 | 20.2 | 7.3 | 122.4 | −7 | 20.2 | 7.3 |
| AB2 | 9/9 | 116.2 | 31.6 | 57.3 | 6.9 | 115.3 | −7.2 | 57.3 | 6.9 |
| AB3 | 10/10 | 118.8 | 35.2 | 63.7 | 6.1 | 117.1 | −3.5 | 63.7 | 6.1 |
| AB1 | No component | | | | | | | | |
| AB8 | No component | | | | | | | | |
| MI1i | 8/8 | 158.9 | −27.8 | 277.3 | 3.3 | 148.8 | −22.3 | 277.3 | 3.3 |
| MI2i | No component | | | | | | | | |
| MI3i | 5/6 | 143.5 | −23.2 | 391.2 | 3.9 | 137.1 | −12.9 | 391.2 | 3.9 |
| *Other components* | | | | | | | | | |
| HSA9 | 8/8 | 172.5 | −52.5 | 14.1 | 15.3 | 227.3 | −52.3 | 14.1 | 15.3 |
| MIA7 | 5/20 | 205.7 | −25.7 | 37.6 | 12.6 | 194 | −37.6 | 37.6 | 12.6 |

*Specimens not subjected to LTD and not used in statistical calculations (HSBrclasts are breccia clasts (18) from the bottom and middle of the breccia; AB2 and AB3 are from the coarse–fine-grained polymict breccia at the top of the unit; AB1 is from 1 m above the breccia; AB8 is from 40 m below the breccia; MI1i is from a lapilli clast; MI2i is from breccia clasts and the specimens contain an apparent Modern overprint; MI3 is from breccia matrix.
[†]MI2i has a high-temperature component removed by 500 °C.

## Results and interpretations

### *Palaeomagnetism*

*Magnetic components.* Examination of Zijderveld diagrams of thermal demagnetization results for specimens allowed for the identification of three magnetic components based on unblocking temperatures (Fig. 3a) in most specimens. A low temperature component with northerly declinations and steep down inclinations is found below 250 °C. Based on the direction and the unblocking temperatures, this component is interpreted as a modern viscous remanent magnetization (VRM) acquired in the current geomagnetic field. Since this component is considered to be noise from a palaeomagnetic point of view (Butler 1992), it is ignored for the remainder of this paper. An intermediate-temperature (IT) magnetization has northwesterly declinations and positive inclinations. In most specimens it is removed between 250 °C and 340 °C (Fig. 3a–f) although, in some specimens (*c.* 30%), it is not removed until up to 500 °C (e.g. site MI2i, Table 1). The IT component is found in sites

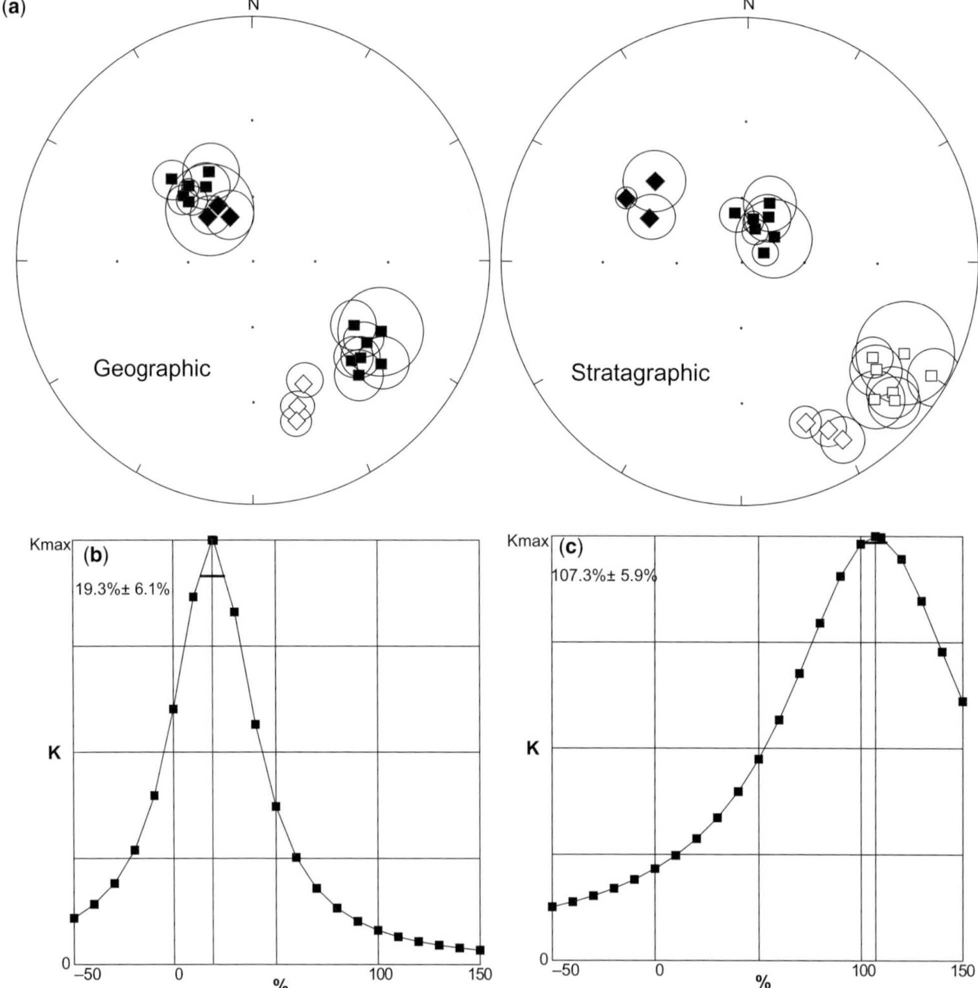



**Fig. 4.** Tilt test results. (**a**) Equal area plots showing the site means of the IT (N–NW declinations) and HT (SE declinations) components in geographic and stratigraphic coordinates. The $\alpha$95 values are shown for each site. Closed symbols represent positive inclinations; open symbols represent negative inclinations. Hancock Summit sites are represented by squares; Mt Irish is represented by diamonds. (**b**) Watson and Enkin tilt test (2003) results for the IT component. The best grouping is at 19.3% with an error of 6.1% (shown by the bar). (**c**) Watson & Enkin tilt test (2003) for the HT component shows the best grouping is at 107.0% with an error of 5.9%.

within, above and below the breccia. Most specimens from sites below the breccias at Hancock Summit only contain this component (Fig. 3d). A high-temperature (HT) component with southeasterly declinations and shallow inclinations is removed between 340 °C and 540 °C in most specimens from within the breccia as well as above and below the breccia (Fig. 3a, c, e, & f). It is found in only three specimens below the breccia at Hancock Summit (site HSB, Table 1). NRM intensities are also similar for the specimens within, above and below the breccia.

Most (c. 73%) specimens from breccia clasts and matrix (HSBrclasts, MI3i; Table 1), the coarse–fine-grained polymict breccia at the top of the unit (AB1, AB3; Table 1) and a lapilli clast (MI1i; Table 1) from the first sample set contain both the IT and HT components. The breccia and surrounding rock were resampled to further test the distribution of the HT component. In addition, some specimens from the second sampling were initially subjected to LTD; they showed a notable inclination shallowing in the HT component compared to the components in specimens from the first sampling, which only underwent thermal demagnetization (Fig. 3b, c). This is consistent with removal of a modern overprint due to the LTD treatment. Consequently, all the specimens from the second sampling were subjected to LTD. Because of the possibility of contamination, only the specimens subjected to LTD were used to calculate the mean directions and for the tilt test.

Analysis after LTD showed that the NRM intensity decreased c. 17% on average. Whereas the sites for non-LTD specimens were not used in the statistical calculations, the mean directions are similar to some of the LTD sites (Table 1) and lend support to the presence of both components throughout the breccia. It is also interesting to note that the LTD treatment did not appear to affect the IT component in most specimens (Fig. 3b, c).

Using specimen directions for the IT and HT components, group mean directions (Table 1) at Hancock Summit were calculated for the breccia (AB, HSV1, HSV2, HSV3 and HSV7), above the breccia (HSA) and below the breccia (HSB). At Mt Irish, group means (Table 1) were calculated for the breccia (MI1, MI2) and above the breccia (MIA).

One site (MIA7) at the top of the section at Mt Irish contains a high-intensity magnetization ($>15$ mA m$^{-1}$) with southwesterly declinations and negative inclinations (Table 1). The site is interpreted as having been hit by a lightning strike (Butler 1992) and is not considered further. One site above the breccia at Hancock Summit (HSA9; Table 1) had a HT component with a direction that is distinct from the other sites and was not considered in the calculation of the mean directions.

*Tilt test results.* Using the site means from Hancock Summit and Mt Irish, a regional tilt test was performed to determine when the magnetization was acquired relative to the deformation, which is interpreted to be related to Basin and Range extension. For the IT component the site means group best between geographic and stratigraphic coordinates (Fig. 4a), suggesting this component was acquired during tilting. The Watson & Enkin (1993) tilt test shows the best grouping at 19% ($\pm6\%$) untilting (Fig. 4b; Table 2) with declination 324° and inclination 58° ($k = 119.4$, $N = 10$, $\alpha95 = 4.4$). Because error calculations for this test may produce unrealistically small errors (Enkin 2003), the directional correction (DC) test (Enkin 2003) was also performed. The DC test also yielded a syndeformational result (Table 2) with the best grouping for the site means at 18% ($\pm14\%$) unfolding. The pole position (61°N, 168°E; semi-axes of the 95% ellipse of confidence around the pole: dp = 5°, dm = 7°) for the best grouping direction plots off the apparent polar wander path (APWP) for North

**Table 2.** *Tilt test results*

| | $N/N_0$ | Dec (°) | Inc (°) | $k$ | $\alpha95$ | Percent untilting | Pole | dp (°) | dm (°) |
|---|---|---|---|---|---|---|---|---|---|
| *Intermediate-temperature component* | | | | | | | | | |
| Direction-correction | 10/12 | – | – | – | – | 18.5 ± 14.2 | – | – | – |
| Best grouping | 10/12 | 323.6 | 57.6 | 119.4 | 4.4 | 19.3 ± 6.1 | 61.1°N, 167.6°E | 5 | 6.8 |
| *High-temperature component* | | | | | | | | | |
| Direction-correction | 10/12 | – | – | – | – | 96.6 ± 19.6 | – | – | – |
| Best grouping | 10/12 | 134.5 | − 13.9 | 33.2 | 8.5 | 107.0 ± 5.9 | 38.8°N, 129.9°E | 4.4 | 8.7 |
| Untilted | 10/12 | 134.4 | − 12.2 | 32.6 | 8.6 | 100.0 | 38.2°N, 129.2°E | 4.4 | 8.7 |
| Inclination only | 10/12 | – | − 12.0 | 91.6 | 3.5 | 100.0 ± 5.5 | 45.2°N, 118.9°E | 1.8 | 3.9 |

*Note*: Excluded other components (HSA 9, MIA 7). Same conventions as Table 1. Best grouping is the Watson & Enkin (1993) tilt test result.

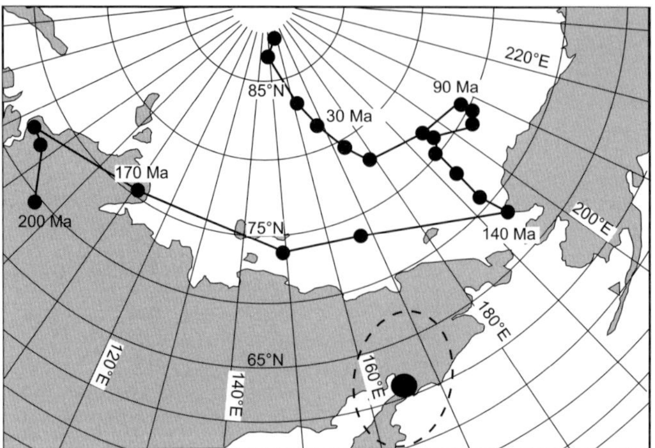


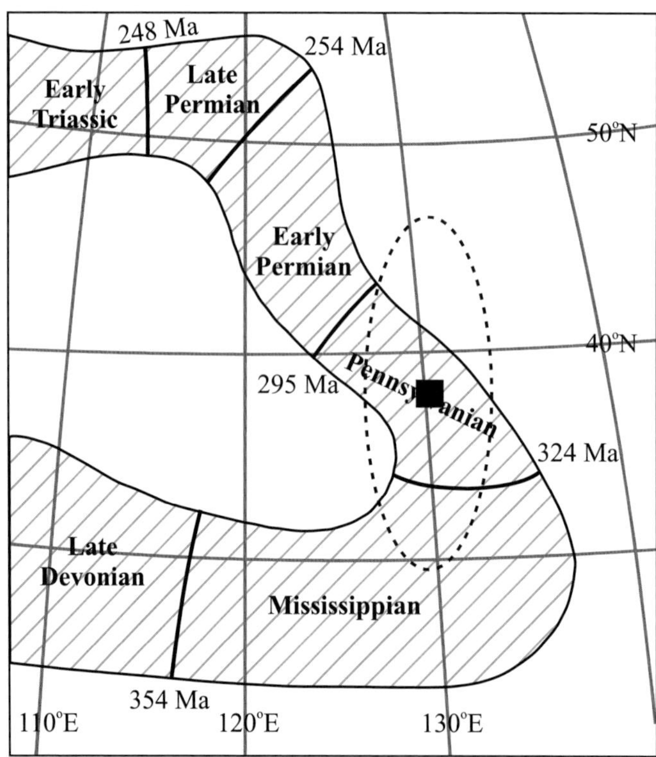


**Fig. 5.** (**a**) The apparent polar wander path (APWP) for North America for the middle Mesozoic–Cenozoic eras (modified from Besse & Courtillot 2002). The circle represents the pole position of the IT component and the ellipse represents the error (Table 2). (**b**) The APWP for North America for the Mid-Palaeozoic–Mid-Triassic (modified from Van der Voo 1993; Symons *et al.* 2005). The square represents the pole position for the HT component found with the Watson & Enkin (1993) tilt test (Table 2). The triangle represents the pole position from the inclination-only tilt test (see text), based on correcting for a 10° anticlockwise block rotation.

America (Fig. 5a). Correcting for a possible anti-clockwise block rotation (*c*. 10°) in the study area (Hudson *et al.* 1998) would move the pole (69°N, 165°E) closer to the Early Cretaceous Period part of the APWP.

For the HT component or characteristic remanent magnetization (ChRM) the site means group best in stratigraphic coordinates (Fig. 4a), suggesting this component was acquired before deformation. The Watson & Enkin (1993) tilt test shows the best grouping at 107% ($\pm$6%) unfolding (Fig. 4c; Table 2) with declination 134° and inclination $-14°$ ($k = 33.2$, $N = 10$, $\alpha 95 = 8.5$). The DC test (Enkin 2003) also yielded a pre-deformational result (Table 2), with the best grouping at 97% ($\pm$10) unfolding. When examining the site means for the ChRM in stratigraphic coordinates (Fig. 4a), the sites do not group well even at the best grouping. Because of the possibility of block rotation around a vertical axis at the study area (Hudson *et al.* 1998) which would affect the declination, an inclination-only test was performed. This test yielded a best grouping at 100% ($\pm$5%) unfolding (Table 2).

The age of the HT component was found by plotting the pole position for the 100% untilting direction on the APWP for North America (Fig. 5b). The pole plots at 38°N, 129°E (dp = 4°, dm = 9°), which lies in the Mid-Pennsylvanian Epoch. Due to the aforementioned possibility of block rotation, a 'corrected' pole was also calculated using the inclination from the inclination-only tilt test results and the declination corrected for the 10° inferred rotation (Hudson *et al.* 1998). This pole plots at 45°N, 119°E on the Early Permian Period part of the APWP (Fig. 5b). This is the preferred pole for the HT component because of the possibility of block rotation at Hancock Summit.

There is some concern that the poles used to construct the APWP did not take into consideration the effects of inclination shallowing. Bilardello & Kodama (2010*a, b*) have shown that inclination correction could cause the late Palaeozoic Era path to shift down in latitude up to 10°, but more commonly 4–5°. This would shift the pole towards the Late Permian Period.

*Breccia contact test.* One of the major objectives of this research was to determine if the breccia acted as a conduit for remagnetizing fluids. At Hancock Summit, the overlying rocks fail the contact test (Fig. 6). The same IT component and ChRM are found in specimens from within the breccia and in specimens from above the breccia. Most of the

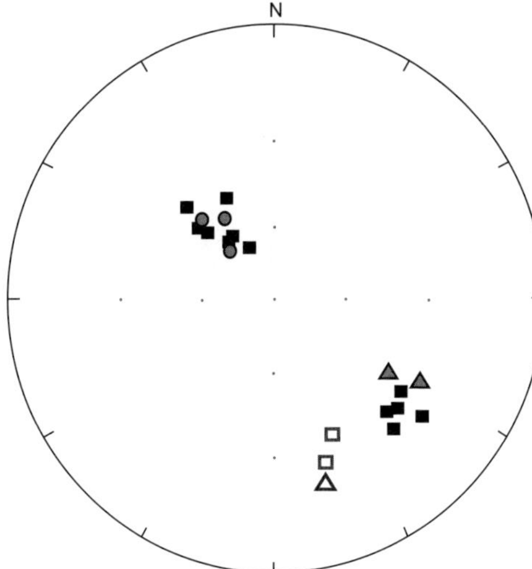

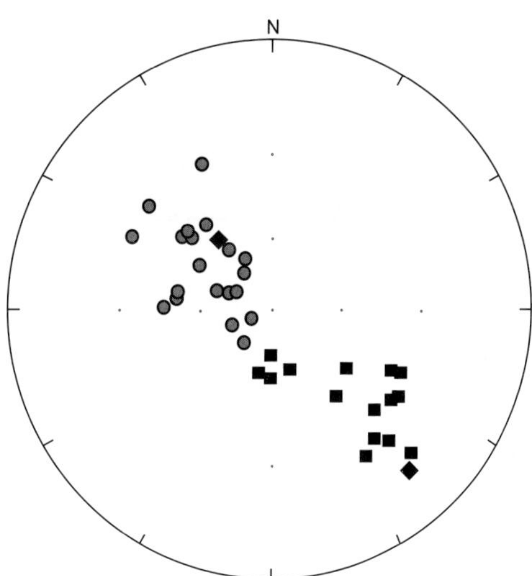

**Fig. 6.** Breccia contact test results. Squares represent site means of breccia specimens at both locations. Circles represent IT component site means of non-breccia specimens; triangles represent HT component site means of non-breccia specimens. The most southerly means with negative inclinations are from Mt Irish. Note that for both components at both locations, site means for breccia and non-breccia specimens have similar directions. Closed symbols represent positive inclinations; open symbols represent negative inclinations. All sites means are in geographic coordinates.

**Fig. 7.** Vein contact test from HSV8. Circles represent the IT component; squares represent the HT component. The black diamonds represents the mean directions of the other sites from Hancock Summit, including unveined breccia as well as sites from above and below the breccia. Closed symbols represent positive inclinations. All specimens are in geographic coordinates. Note the streak of directions between the IT and HT components.

specimens (sites HSB and AB8; Table 1) from below the breccia contain the IT component, with only three specimens containing the ChRM (site HSB; Table 1). At Mt Irish, the IT and HT components are both present in specimens from within the breccia as well as above the breccia (Fig. 6). These results suggest that the breccia was not a focused conduit, although it is possible that the sampling did not extend far enough above or below the breccia to be out of the alteration zone.

*Vein contact tests.* Eight vein contact tests were performed at Hancock Summit in order to test if the thin veins were conduits for fluids that caused re-magnetization. Of the eight vein sites at Hancock Summit, four sites (HSV1, HSV2, HSV3 and HSV7; Table 1) have specimen directions similar to the IT and HT components (Table 1). There is no relationship between distance from the veins and the magnetic intensities or directions. One site, HSV8 (Fig. 7), has streaked directions between the

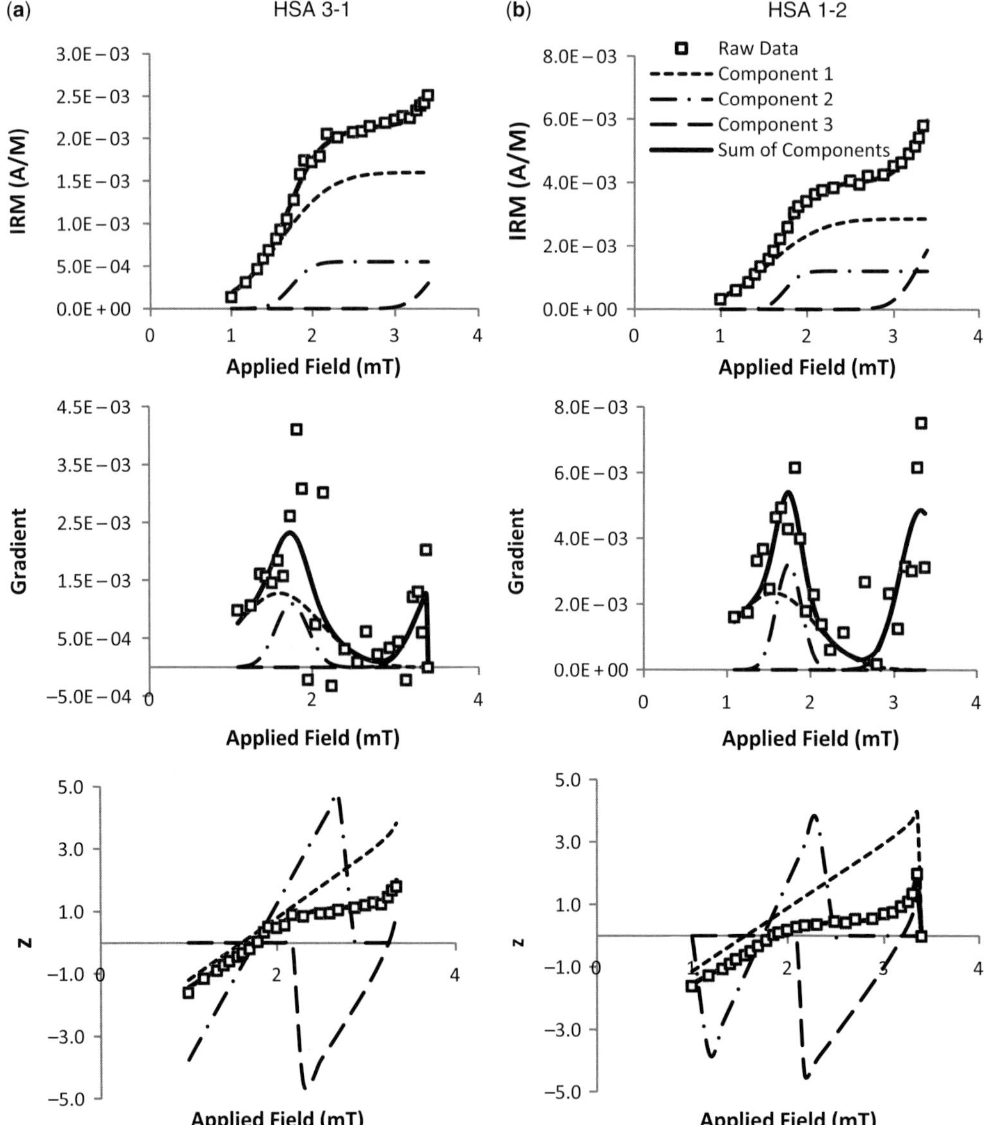



**Fig. 8.** Cumulative log-Gaussian (CLG) analysis of IRM acquisition data for a representative sample from (**a**) Hancock Summit and (**b**) Mt Irish. Both samples were modelled with three components that are interpreted as magnetite, pyrrhotite and goethite. See Table 3 for the CLG analysis results for all samples.

IT component direction (northwesterly) and HT component direction (southeasterly). This is interpreted to be because of vector addition between the two components. The results from remaining vein sites contain scattered specimen directions that are not well grouped and mean directions were not calculated.

## Rock magnetism

The CLG analysis (Kruiver *et al.* 2001; Heslop *et al.* 2004) of the IRM acquisition data indicates that the specimens contain three components (Fig. 8; Table 3). A preliminary interpretation based on the coercivities is that they are magnetite, pyrrhotite and goethite (Table 3). The tri-axial decay of the high-coercivity IRM component shows a large decrease at 100 °C which suggests the presence of goethite, although this phase is not interpreted as carrying a stable remanence based on the thermal demagnetization results. The decay of the low-coercivity IRM component shows a slight drop in intensity at 325–350 °C, suggesting the presence of pyrrhotite (Fig. 9). This low-coercivity component is removed by 580 °C, suggesting the presence of magnetite. Tri-axial decay also shows that a higher coercivity phase, namely hematite, is present until *c.* 680–700 °C. This phase is not interpreted as carrying a stable remanence. The CLG analysis and the thermal decay of the IRMs are consistent with the interpretation that the IT component and ChRM reside in pyrrhotite and magnetite, respectively.

## Petrography and geochemistry

The specimens of the Alamo Breccia are classified as crystalline dolomites, fossiliferous grainstones and packstones, and mudstones. The veins mostly contained calcite, but one vein contained calcite and quartz. Using reflected light some iron oxides

**Table 3.** *Coercivity spectrum analysis results*

| Specimen | Component | Coercivity (mT) | Contribution (%) | Mineral |
|---|---|---|---|---|
| AB 25-2 | 1 | 52.5 | 11.3 | Magnetite |
| | 2 | 131.8 | 19.7 | Pyrrhotite |
| | 3 | 2041.7 | 69.0 | Goethite |
| HSA 1-2 | 1 | 38.0 | 70.1 | Magnetite |
| | 2 | 60.3 | 10.9 | Pyrrhotite |
| | 3 | 1905.5 | 19.0 | Goethite |
| HSA 3-1 | 1 | 39.8 | 52.2 | Magnetite |
| | 2 | 56.2 | 18.0 | Pyrrhotite |
| | 3 | 3162.3 | 29.5 | Goethite |
| HSB 8-1 | 1 | 7.9 | 10.0 | Multi-Domain Magnetite |
| | 2 | 52.5 | 78.4 | Pyrrhotite |
| | 3 | 501.2 | 11.6 | Hematite |
| HSV 1-3 | 1 | 63.1 | 25.5 | Pyrrhotite |
| | 2 | 41.7 | 32.6 | Magnetite |
| | 3 | 2167.7 | 41.9 | Goethite |
| HSV 2-2 | 1 | 42.7 | 36.1 | Magnetite |
| | 2 | 58.9 | 18.4 | Pyrrhotite |
| | 3 | 2884.0 | 45.5 | Goethite |
| HSV 3-12 | 1 | 29.5 | 24.6 | Magnetite |
| | 2 | 66.1 | 25.5 | Pyrrhotite |
| | 3 | 2290.9 | 49.9 | Goethite |
| HSV 5-6 | 1 | 47.9 | 44.9 | Magnetite |
| | 2 | 1995.3 | 55.1 | Goethite |
| HSV 8-22 | 1 | 42.7 | 30.2 | Magnetite |
| | 2 | 53.2 | 12.0 | Pyrrhotite |
| | 3 | 2041.7 | 57.8 | Goethite |
| HSV 8-26 | 1 | 56.2 | 10.4 | Magnetite |
| | 2 | 70.8 | 27.0 | Pyrrhotite |
| | 3 | 2089.3 | 62.6 | Goethite |
| MIA 1-2 | 1 | 36.3 | 40.3 | Magnetite |
| | 2 | 55.0 | 17.0 | Pyrrhotite |
| | 3 | 2089.3 | 42.8 | Goethite |
| MIA 2-1 | 1 | 39.8 | 51.3 | Magnetite |
| | 2 | 57.5 | 24.9 | Pyrrhotite |
| | 3 | 2290.9 | 23.7 | Goethite |

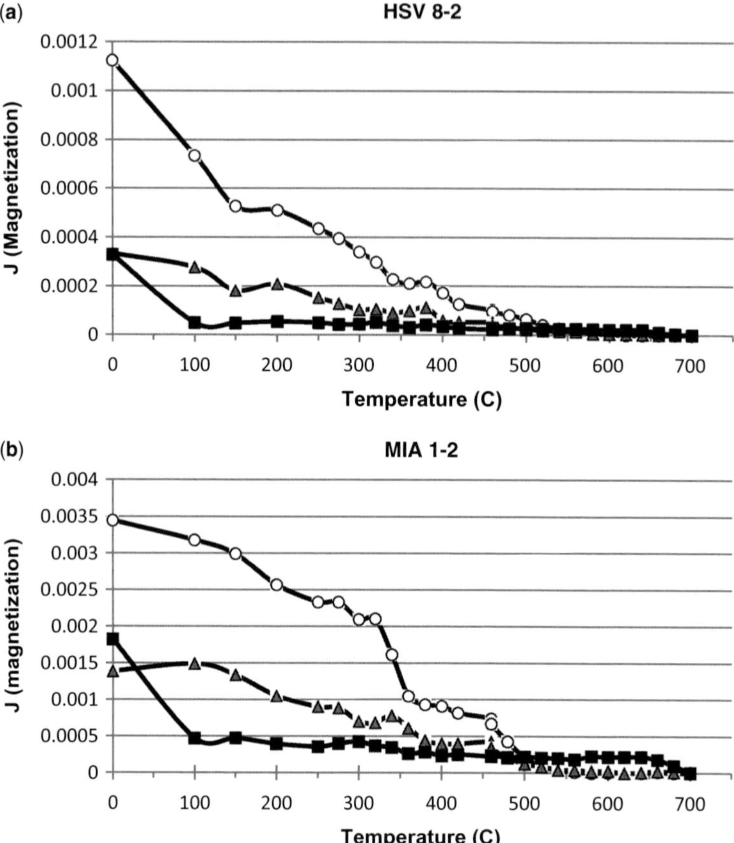



**Fig. 9.** Tri-axial thermal decay of an imparted IRM for representative samples (**a**) Hancock Summit and (**b**) Mt Irish. The coercivities for the curves are 120 mT (circles), 500 mT (triangles) and 2500 mT (squares). The drop in intensity in the low-coercivity curve between 320 °C and 350 °C indicates the presence of pyrrhotite. The low-coercivity curve reaches zero intensity below 580 °C, suggesting the presence of magnetite.

(e.g. hematite, goethite) were identified; magnetite and pyrrhotite were not identified however, possibly to due to their small size. Pyrite was also observed using reflected light.

Strontium isotope analysis results were used as an alteration indicator. The $^{87}Sr/^{86}Sr$ values were plotted on a coeval seawater curve (Fig. 10) for the Late Devonian Period (McArthur *et al.* 2001). The carbonate breccia samples are elevated in strontium, indicating that they were altered by externally derived fluids. The value from one calcite vein from within the breccia is highly elevated, indicating that externally derived fluids also precipitated the vein.

## Discussion

The Alamo Breccia and the surrounding carbonate rocks at Hancock Summit and Mt Irish contain two ancient magnetic components: a syn-tilting IT component with northwesterly declinations and positive inclinations is present between 250 °C and *c.* 340 °C; and a pre-tilting ChRM with southeasterly declinations and negative inclinations is found between 340 °C and 540 °C. These components are interpreted to reside in pyrrhotite and magnetite, respectively, based on maximum unblocking temperatures and rock magnetic analysis. Although the maximum unblocking temperatures for the IT component are high for pyrrhotite, Rochette *et al.* (1990) have shown that pyrrhotite can have an unblocking temperature up to 350 °C. The lack of effect that LTD had on the IT component is consistent with this interpretation. In a few specimens, the IT component is interpreted to reside in magnetite because the unblocking temperatures extend up to 500 °C.

Although the ChRM is pre-tilting, it is a secondary magnetization because the pole position

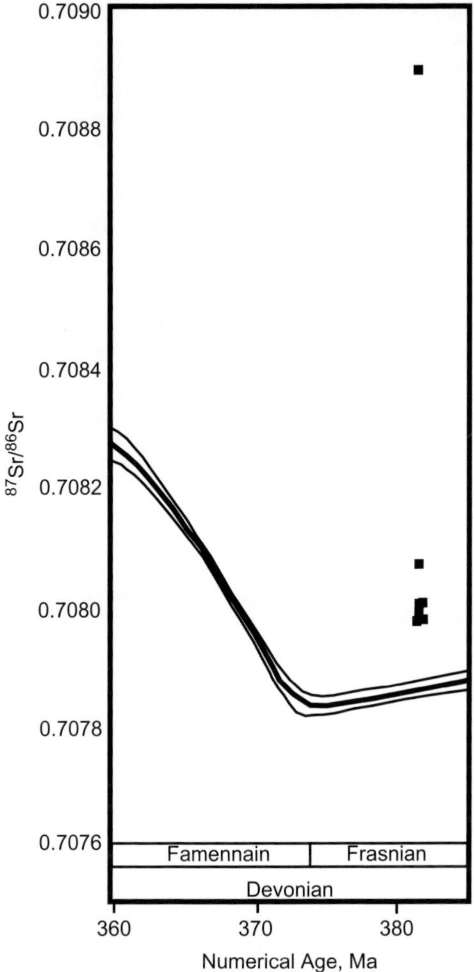



**Fig. 10.** Specimens from Hancock Summit plotted with the coeval strontium isotope values for the age of the Alamo Breccia in the Devonian Period (thick solid line with errors shown by thin lines; McArthur *et al.* 2001). The samples from within the breccia and from the rocks above/below the breccia have a radiogenic signature. The most radiogenic signature is from a vein within the Alamo Breccia. This enrichment suggests alteration or precipitation from externally derived fluids.

suggests an age younger than the Devonian Period impact that formed the breccia (Fig. 5b). The Alamo Breccia has been completely remagnetized. These results are consistent with those found in Gillett & Karlin's (2004) regional palaeomagnetic study of Palaeozoic Era rocks in Nevada.

Several different origins of these components are possible. The conodonts in the Alamo Breccia have a CAI (Conodont Alteration Index) of 3.0 (Jared Marrow, pers. comm. 2008) which corresponds to a burial temperature of 110–200 °C (Epstein *et al.* 1977). The IT component, which resides predominantly in pyrrhotite, has a maximum unblocking temperature of 340 °C. Based on this low burial temperature, the component's maximum unblocking temperature and the pyrrhotite relaxation time-unblocking temperature curves for single-domain (SD) grains (Dunlop *et al.* 2000), the pyrrhotite component cannot be a thermoviscous remanent magnetization (TVRM). The magnetization is instead interpreted as a CRM. Many previous studies have interpreted similar components as TVRMs in magnetite (e.g. Hudson *et al.* 1998; Enkin *et al.* 2000). Because of the unblocking temperature, the rock magnetic results, and the lack of effect that LTD had on this component, this magnetic mineral is interpreted as pyrrhotite.

The ChRM is also considered to be a CRM based on the maximum unblocking temperature of 540 °C and the SD magnetite relaxation-time–unblocking-temperature curves (Dunlop *et al.* 2000). The relaxation-time–unblocking-temperature curves for SD grains are applicable for this study since the MD contribution has been removed by liquid nitrogen treatment.

The ages for the two CRMs correlate fairly well with the ages proposed by Gillett & Karlin (2004) for the two components found in their study. The pole position for the pyrrhotite CRM in this study plots off the APWP (Fig. 5a), although accounting for the block rotation moves the pole close to the Early Cretaceous Period portion of the path. Other factors such as secular variation not being averaged out or tectonic complications due to the remagnetization being acquired during deformation could also account for the pole being off the path. The age of the IT CRM is interpreted to be Cretaceous Period or perhaps Cenozoic Era. The HT CRM is interpreted to be Permian Period in age (Fig. 5b).

There are a number of diagenetic processes that could have caused the CRM in magnetite. Alteration by fluids would not appear to be a likely mechanism because the results of the contact tests suggest that the breccia was not a focused conduit for fluids. But burial remagnetization mechanisms are also not likely mechanisms. For example, maturation of organic matter (e.g. Fruit *et al.* 1995; Banerjee *et al.* 1997; Blumstein *et al.* 2004) is not likely because the carbonates are not source rocks. The carbonates do contain some clay, and clay diagenesis (Katz *et al.* 2000; Moreau *et al.* 2005) is a possibility but is not considered likely because many of rocks with the CRM are not clay rich. The strontium isotope analysis indicates that the breccias, as well as the rocks around the breccias, were altered by externally derived fluids (Fig. 10). One hypothesis supported by the data is that the fluids were not constrained by the breccias and moved pervasively through the rocks and altered larger intervals of the section. Alteration by fluids was deemed very likely by Gillett & Karlin (2004) who found the CRM in many rock types. It should be noted, however, that they did not conduct geochemical studies. The interpretation that pervasive flow of fluids caused the CRM differs from other studies (e.g. Elmore *et al.* 1993, 1998; Elmore 2001; Blumstein *et al.* 2005) where remagnetizing fluids were only found to have migrated through conduits and locally remagnetized a contact zone around the conduits.

There are several chemical mechanisms that could explain the pyrrhotite CRM. Pyrrhotite authigenesis could be due to pre-existing magnetite reacting with pyrite as a result of burial metamorphism under reducing conditions (Gillett 2003). Salmon *et al.* (1988) suggested the formation of diagenetic pyrrhotite due to the oxidation of pre-existing pyrite. Thermochemical sulphate reduction (TSR) can also produce pyrrhotite (Pierce *et al.* 1998) but there is no evidence (e.g. sulphur-rich hydrocarbons, sphalerite, barite) that it occurred in these rocks. Hydrocarbon migration (Machel & Burton 1991) and gas hydrates (Housen & Musgrave 1996; Larrasoaña *et al.* 2007) are also ruled out as a chemical remagnetization mechanism due to being mostly absent from the rocks. The release of trapped pore fluid (Urbat *et al.* 2000) is a possibility but, due to lack of evidence, cannot be proven or rejected. As for the magnetite CRM, externally derived fluids can also be a possibility for pyrrhotite CRM.

The results of the vein contact tests at Hancock Summit do not suggest a connection between the veins and the CRMs. It is however unclear why some of the sites around the veins contain scattered directions.

## Conclusions

The Alamo Breccia has been completely remagnetized by two CRMs: a syn-tilting IT CRM residing in pyrrhotite acquired in the Cretaceous Period; and a pre-tilting HT Late Palaeozoic Era CRM residing in magnetite. The breccia is not interpreted as a conduit for focused fluids. Both CRMs are found within the breccia and beyond the inferred contact zone in the surrounding rocks. Results from the vein contact tests at Hancock Summit indicate that the veins did not influence the CRMs in the surrounding rocks.

There are a number of possibilities for the origin of the CRMs. The most likely mechanism for imparting the CRMs is interpreted as external fluids based on the elevated $^{87}Sr/^{86}Sr$ values, which suggest that the rocks have been altered by fluids with a radiogenic signature. Other origins of CRM impartment cannot be ruled out however,

such as clay diagenesis causing the CRM in magnetite or the pyrrhotite CRM forming due to the reaction of pre-existing magnetite with pyrite.

The results of this study suggest that the Alamo breccia was not a conduit for focused flow of remagnetizing fluids, but that the fluids moved pervasively through the rocks. This is surprising considering other studies that indicate that such fluids were localized in conduits (e.g. Elmore *et al.* 1993, 1998; Elmore 2001; Blumstein *et al.* 2005).

We thank M. Elwood Madden, K. Keranen and two anonymous reviewers for comments on the manuscript. We also thank the AAPG Grants-in-Aid for awarding the Jon R. Withrow grant to SCE. We are also grateful to J. Warme, J. Pannalal and P. Kincaid for their help in the field, and D. Dennie and S. Dulin for their previous work on the Alamo Breccia.

## References

BANERJEE, S., ENGEL, M. H. & ELMORE, R. D. 1997. Chemical remagnetization and Burial Diagenesis: testing the hypothesis in the Pennsylvanian Belden Formation, Colorado. *Journal of Geophysical Research*, **102**, 24825–24842, doi: 10.1029/97JB01893.

BESSE, J. & COURTILLOT, V. 2002. Apparent and true polar wander and the geometry of the geomagnetic field over the last 200 Myr. *Journal of Geophysical Research*, **107**, 2300, doi: 10.1029/2000JB000050.

BILARDELLO, D. & KODAMA, K. P. 2010a. Palaeomagnetism and magnetic anisotropy of Carboniferous redbeds from the Maritime Provinces of Canada: evidence for shallow palaeomagnetic inclinations and implications for North American apparent polar wander. *Geophysical Journal International*, **180**, 1013–1029.

BILARDELLO, D. & KODAMA, K. P. 2010b. A new inclination shallowing correction of the Mauch Chunk Formation of Pennsylvania, based on high-field AIR results: Implications for the Carboniferous North American APW path and Pangea reconstructions. *Earth and Planetary Science Letters*, **299**, 218–227.

BLUMSTEIN, A., ELMORE, R. D. & ENGEL, M. 2004. Paleomagnetic dating of Burial Diagenesis in Mississippian Carbonates, Utah. *Journal of Geophysical Research*, **109**, B04101, doi: 10.1029/2003JB002698.

BLUMSTEIN, R. D., ELMORE, R. D., ENGEL, M. H., PARNELL, J. & BARON, M. 2005. Multiple fluid migration events along the Moine Thrust Zone, Scotland. *Journal of the Geological Society, London*, **162**, 1031–1045.

BUTLER, R. F. 1992. *Paleomagnetism: Magnetic Domains to Geologic Terranes*. Blackwell Science Inc., Boston, MA.

DUNLOP, D. J. & ARGYLE, K. S. 1991. Separating multidomain and single-domain-like remanences in pseudo-single-domain magnetites (215–540 nm) by low-temperature demagnetization. *Journal of Geophysical Research*, **96**, 2007–2017.

DUNLOP, D. J., ÖZDEMIR, Ö., CLARK, D. A. & SCHMIDT, P. W. 2000. Time-temperature relations for the remagnetization of pyrrhotite (Fe7S8) and their use in estimating paleotemperatures. *Earth and Planetary Science Letters*, **176**, 107–116.

ELMORE, R. D. 2001. A review of palaeomagnetic data on the timing and origin of multiple fluid-flow events in the Arbuckle Mountains, southern Oklahoma. *Petroleum Geoscience*, **7**, 223–229.

ELMORE, R. D., LONDON, D., BAGLEY, D. & GAO, G. 1993. Remagnetization by basinal fluids: testing the hypothesis in the Viola Limestone, southern Oklahoma. *Journal of Geophysical Research*, **98**, 6237–6254.

ELMORE, R. D., CAMPBELL, T., BANERJEE, S. & BIXLER, W. G. 1998. Palaeomagnetic dating of ancient fluid-flow events in the Arbuckle Mountains, southern Oklahoma. *In*: PARNELL, J. (ed.) *Dating and Duration of Fluid Flow and Fluid-Rock Interaction*. Geological Society, London, Special Publications, **144**, 9–25.

ELMORE, R. D., KELLEY, J., EVANS, M. & LEWCHUK, M. T. 2001. Remagnetization and orogenic fluids: testing the hypothesis in the central Appalachians. *Geological Journal International*, **144**, 568–576.

ENKIN, R. J. 2003. The direction-correction tilt test: an all-purpose tilt/fold test for paleomagnetic studies. *Earth and Planetary Science Letters*, **212**, 151–166.

ENKIN, R. J., OSADETZ, K. G., BAKER, J. & KISILEVSKY, D. 2000. Orogenic remagnetizations in the front ranges and inner foothills of the southern Canadian Cordillera: chemical harbinger and thermal handmaiden of Cordilleran deformation. *Geological Society of America Bulletin*, **112**, 929–942.

EPSTEIN, A. G., EPSTEIN, J. B. & HARRIS, L. D. 1977. Conodont color alteration – an index to organic metamorphism. *U. S. Geological Survey Professional Paper*, **995**, 27.

FISHER, R. A. 1953. Dispersion on a sphere. *Proceedings of the Royal Society of London Series A*, **217**, 295–305.

FRUIT, D., ELMORE, R. D. & HALGEDAHL, S. 1995. Remagnetization of the folded Belden Formation, NW Colorado. *Journal of Geophysical Research*, **100**, 15009–15024.

GAO, G., ELMORE, R. D. & LAND, L. S. 1992. Geochemical constraints on the origin of calcite veins and associated limestone alteration, Ordovician Viola Group, Arbuckle Mountains, Oklahoma, U.S.A. *Chemical Geology*, **98**, 257–269.

GILLETT, S. L. 2003. Paleomagnetism of the Notch Peak contact-metamorphic aureole, revisited: pyrrhotite from magnetite + pyrite under submetamorphic conditions. *Journal of Geophysical Research*, **108**, 2446, doi: 10.1029/2002JB002386.

GILLETT, S. L. & KARLIN, R. E. 2004. Pervasive late Paleozoic–Triassic remagnetization of the miogeoclinal carbonate racks in the Basin and Range and vicinity, SW USA: regional results and possible tectonic implications. *Physics of the Earth and Planetary Interiors*, **141**, 95–120.

HESLOP, D., DEKKERS, M. J., KRUIVER, P. P. & VAN OORSCHOT, I. H. M. 2002. Analysis of isothermal remanent magnetisation acquisition urves using the expectation-maximisation algorithm. *Geophysical Journal International*, **148**, 58–64.

HESLOP, D., MCINTOSH, G. & DEKKERS, M. J. 2004. Using time- and temperature-dependent Preisach models to investigate the limitations of modeling isothermal remanent magnetization acquisition curves with cumulative log Gaussian functions. *Geophysical Journal International*, **157**, 55–63.

HOUSEN, B. A. & MUSGRAVE, R. J. 1996. Rock-magnetic signature of gas hydrates in accretionary prism sediments. *Earth and Planetary Science Letters*, **139**, 509–519.

HUDSON, M. R., ROSENBAUM, J. G., GROMMÉ, C. S., SCOTT, R. B. & ROWLEY, P. D. 1998. Paleomagnetic evidence for counterclockwise rotation in a broad sinistral shear zone, Basin and Range province, southeastern Nevada and southwestern Utah. *In*: FAULDS, J. E. & STEWART, J. H. (eds) *Accommodation Zones and Transfer Zones: The Regional Segmentation of the Basin and Range Province*. Geological Society of America, Boulder, CO, Special Paper, **323**, 149–180.

KAUFMANN, B., TRAPP, E. & MEZGER, K. 2004. The numerical age of the Upper Frasnian (Upper Devonian) Kellwasser Horizons: a new U-Pb Zircon date from Steinbruch Schmidt (Kellerwald, Germany). *Journal of Geology*, **112**, 495–501.

KATZ, B., ELMORE, R. D., ENGEL, M. H., COGOINI, M. & FERRY, S. 2000. Associations between burial diagenesis of smectite, chemical remagnetization and magnetite authigenesis in the Vocontian Trough of SE-France. *Journal of Geophysical Research*, **105**, 851–868.

KIRSCHVINK, J. L. 1980. The least-squares line and plane and the analysis of paleomagnetic data. *Geophysical Journal International*, **62**, 699–718.

KRUIVER, P. P., DEKKERS, M. J. & HESLOP, D. 2001. Quantification of magnetic coercivity components by the analysis of acquisition curves of isothermal remanent magnetization. *Earth and Planetary Science Letters*, **189**, 269–276.

LARRASOAÑA, J. C., ROBERTS, A. P., MUSGRAVE, R. J., GRÀCIA, E., PIÑERO, E., VEGA, M. & MARTÍNEZ RUIZ, F. 2007. Diagenetic formation of greigite and pyrrhotite in has hydrate marine sedimentary systems. *Earth and Planetary Science Letters*, **261**, 350–366.

LEROUX, H., WARME, J. E. & DOUKHAN, J.-C. 1995. Shocked quartz in the Alamo Breccia, southern Nevada: Evidence for a Devonian impact event. *Geology*, **23**, 1003–1006, doi: 10.1130/0091–7613.

LOWRIE, W. 1990. Identification of ferromagnetic minerals in a rock by coercivity and unblocking temperature properties. *Geophysical Research Letters*, **17**, 159–162.

LU, G., MCCABE, C., HANER, J. S. & FERRELL, R. E. 1991. A genetic link between remagnetization and potassic metasomatism in the Devonian Onondaga formation, northern Appalachian basin. *Geophysical Research Letters*, **18**, 2047–2050.

MCARTHUR, J. M., HOWARTH, R. J. & BAILEY, T. R. 2001. Strontium isotope stratigraphy: Lowess version 3: best fit to the marine Sr-Isotope curve for 0–509 Ma and accompanying look-up table for deriving numerical age. *The Journal of Geology*, **109**, 155–170.

MACHEL, H. G. & BURTON, E. A. 1991. Causes and spatial distribution of anomalous magnetization in hydrocarbon seepage environments. *American Association of Petroleum Geologist Bulletin*, **75**, 1864–1876.

MCCABE, C. & ELMORE, R. D. 1989. The occurrence and origin of Late Paleozoic remagnetization in the sedimentary rocks of North America. *Reviews of Geophysics*, **27**, 471–494.

McCabe, C., Jackson, M. & Saffer, B. 1989. Regional patterns of magnetite authigenesis in the Appalachian Basin: implications for the mechanism of Late Paleozoic remagnetization. *Journal of Geophysical Research*, **94**, 10 429–10 443.

Moreau, M. G., Ader, A. & Enkin, R. J. 2005. The magnetization of clay-rich rocks in sedimentary basins: Low-temperature experimental formation of magnetic carriers in natural samples. *Earth and Planetary Science Letters*, **230**, 193–210.

Morrow, J. R. & Sandberg, C. A. 2005. Revised dating of Alamo and some other late Devonian impacts in relation to resulting mass extinction. *Meteoritics & Planetary Science Supplement*, **40**, 5148.

Morrow, J. R., Sandberg, C. A. & Harris, A. G. 2005. Late Devonian Alamo Impact, southern Nevada, USA: Evidence of size, marine site, and widespread effects. *In*: Kenkmann, T., Hörz, F. & Deutsch, A. (eds) *Large Meteorite Impacts III*. Geological Society of America, Boulder, CO, Special Paper, **384**, 259–280.

Pinto, J. A. & Warme, J. E. 2008. Alamo Event, Nevada: Crater stratigraphy and impact breccia Realms. *In*: Evans, K. R., Horton, J. W. Jr, King, D. T. Jr. & Morrow, J. R. (eds) *The Sedimentary Record of Meteorite Impacts*. Geological Society of America, Boulder, CO, Special Paper, **437**, 99–137.

Pierce, J. W., Goussev, S. A., Charters, R. A., Ambercrombie, H. J. & DePaoli, G. R. 1998. Intrasedimentary magnetization by vertical fluid flow and exotic geochemistry. *The Leading Edge*, **17**, 89–92.

Rochette, P., Fillion, G., Mattéi, J.-L. & Dekkers, M. J. 1990. Magnetic transition at 30–24 Kelvin in pyrrhotite: insight into a widespread occurrence of this mineral in rocks. *Earth and Planetary Science Letters*, **98**, 319–328.

Saffer, B. & McCabe, C. 1992. Further studies of carbonate remagnetization in the northern Appalachian basin. *Journal of Geophysical Research*, **97**, 4231–4248.

Salmon, E., Edel, J. B., Pique, A. & Westphal, M. 1988. Possible origins of Permian remagnetizations in Devonian and Carboniferous limestones from the Moroccan Anti-Atlas (Tafilalet) and Meseta. *Physics of the Earth and Planetary Interiors*, **52**, 339–351.

Sandberg, C. A. & Warme, J. E. 1993. Conodont dating, biofacies, and catastrophic origin of Late Devonian (early Frasnian) Alamo Breccia, southern Nevada. *Geological Society of America Abstracts with Programs*. Geological Society of America, Denver, CO, **25**, 77.

Stewart, J. H. 1980. *Geology of Nevada*. Nevada Bureau of Mines and Geology, Nevada, Special Publication **4**, 136.

Symons, D. T. A., Panalal, S. J., Coveney, R. M. Jr. & Sangster, D. F. 2005. Paleomagnetism of the Late Paleozoic strata and mineralization in the Tri-State lead-zinc ore district. *Society of Economic Geologists*, **100**, 295–309.

Tohver, E., Weil, A. B., Solum, J. G. & Hall, C. M. 2008. Direct dating of carbonate remagnetization by 40Ar/39Ar analysis of the smectite–illite transformation. *Earth and Planetary Science Letters*, **274**, 524–530.

Trapp, E., Kaufmann, B., Mezger, K., Korn, D. & Weyer, D. 2004. Numerical calibration of the Devonian-Carboniferous boundary: two new U-Pb isotope dilution–thermal ionization mass spectrometry single-zircon ages from Hasselbachtal (Sauerland, Germany). *Geology*, **32**, 857–860.

Urbat, M., Dekkers, M. J. & Krumsiek, K. 2000. Discharge of hydrothermal fluids through sediment at the Escanaba Trough, Gorda Ridge (ODP Leg 169): assessing the effects on the rock magnetic signal. *Earth and Planetary Science Letters*, **176**, 481–494, doi: 10.1016/S0012-821X(00)00024-8.

Warme, J. E. 2004. The many faces of the Alamo impact breccia. *Geotimes*, **49**, 26–29.

Warme, J. E. & Kuehner, H.-C. 1998. Anatomy of an anomaly: the Devonian catastrophic Alamo impact breccia of southern Nevada. *International Geology Review*, **40**, 189–216.

Warme, J. E. & Sandberg, C. A. 1995. The catastrophic Alamo breccia of southern Nevada—record of a Late Devonian extraterrestrial impact. *Courier Forschungsinstitut Senckenberg*, **188**, 31–57.

Warme, J. E. & Sandberg, C. A. 1996. Alamo Megabreccia: record of a Late Devonian Impact in Southern Nevada. *GSA Today*, **6**, 1–7.

Warme, J. E., Morgan, M. & Kuehner, H.-C. 2002. Impact-generated carbonate accretionary lapilli in the Late Devonian Alamo Breccia. *In*: Koeberl, C. & MacLeod, K. G. (eds) *Catastrophic Events and Mass Extinctions: Impacts and Beyond*. Geological Society of America, Boulder, CO, Special Paper, **356**, 489–504.

Warme, J. E., Morrow, J. R. & Sandberg, C. A. 2008. Devonian carbonate platform of eastern Nevada: Facies, surfaces, cycles, sequences, reefs, and cataclysmic Alamo Impact Breccia. *In*: Duebendorfer, E. M. & Smith, E. I. (eds) *Field Guide to Plutons, Volcanoes, Faults, Reefs, Dinosaurs, and Possible Glaciation in the Selected Areas of Arizona, California, and Nevada*. Geological Society of America, Boulder, CO, Field Guide, **11**, 215–247.

Watson, G. S. & Enkin, R. J. 1993. The fold test in paleomagnetism as a parameter estimation problem. *Geophysical Research Letters*, **20**, 2135–2137.

Woods, S. D., Elmore, R. D. & Engel, M. H. 2002. Paleomagnetic dating of the smectite-to-illite conversion: testing the hypothesis in Jurassic sedimentary rocks, Skye, Scotland. *Journal of Geophysical Research*, **107**, 10.1029/2000JB000053, EPM 2-1-2-12.

Van der Voo, R. 1993. *Paleomagnetism of the Atlantic, Tethys, and Iapetus Oceans*. Cambridge University Press, Cambridge.

Zijderveld, J. D. A. 1967. A.C. demagnetization of rocks: Analysis of results. *In*: Collinson, D. E., Creer, K. M. & Runcorn, S. K. (eds) *Methods in Paleomagnetism*. Elsevier Science, Amsterdam, 254–286.

# Pyrrhotite remagnetizations in the Himalaya: a review

E. APPEL[1]*, C. CROUZET[2] & E. SCHILL[3]

[1]*Fachbereich Geowissenschaften, Universität Tübingen, Sigwartstrasse 10,
72076 Tübingen, Germany*

[2]*Institut des Sciences de la Terre, Université de Savoie, CNRS, 73376 Le Bourget du Lac, France*

[3]*Institut de Géologie et d'Hydrogéologie, Université de Neuchâtel, Rue Emile-Argand 11,
2009 Neuchâtel, Switzerland*

*Corresponding author (e-mail: erwin.appel@uni-tuebingen.de)*

**Abstract:** This paper reviews results on the nature and thermo-tectonic interpretation of widespread pyrrhotite remagnetizations in the Himalaya. Throughout the last two decades a large dataset has been acquired, in particular from low-grade metamorphic rocks of the Tethyan Sedimentary Series (TSS). The nature of this magnetization is a thermoremanence when the peak metamorphic temperature $T_{max}$ exceeded the Curie temperature ($T_c$ c. 325 °C) of pyrrhotite; in this case the remanence age can be related to last metamorphic cooling. For $T_{max} < T_c$, the remanence can be of chemical, thermochemical or thermoremanent origin. Cooling ages show a systematic trend of c. 50–20 Ma from the western to the eastern Himalaya. The pyrrhotite remagnetizations post-date main Himalayan folding and record late orogenic long-wavelength rotations and tiltings around vertical and horizontal axes. Remanence directions in the western Himalaya are well matching with large-scale deformations of rotational shortening and oroclinal bending, while in the central and eastern Himalaya they are predominantly controlled by meso-scale effects due to crustal doming. Stable pyrrhotite remanences are especially typical for low-grade marly limestones in the TSS, but were also found in medium-grade rocks of the Lesser Himalaya, highly metamorphic rocks of the Higher Himalayan Crystalline and diorite dykes intruding into the TSS.

Remagnetization is a common phenomenon in orogenic belts, due to regional or contact metamorphism as well as tectonic stress and strain in conjunction with fluid migration (e.g. D'Agrella-Filho *et al.* 2000; Elmore *et al.* 2006; Kim *et al.* 2009). It has long been regarded as disappointing when results of palaeomagnetic work reveal a magnetic overprint of the studied rocks. Later it was realized that the restriction of research to primary magnetizations is an unnecessary cut-down of useful data and their potential constraints. Strategies and methodologies have been developed to also use secondary magnetic remanences for interpreting tectonic features and processes. A typical case is the Himalayan range, where preservation of remanences with primary origin is only common in the foreland molasse basin of the Siwaliks (e.g. Tauxe & Opdyke 1982; Gautam & Fujiwara 2000), but is rarely found in the tectonic units north of it (Lesser Himalaya, Higher Himalayan Crystalline, Tethyan Himalaya Sequence). In order to decipher the kinematic evolution of the passive northern Indian margin before and during the India–Asia collision, palaeomagnetic work mainly focused on the Tethyan Sedimentary Series (TSS). Primary remanences were extracted from the youngest units

(Upper Cretaceous–Lower Tertiary), delivering direct constraints about the extent of 'Greater India' (Besse *et al.* 1984; Patzelt *et al.* 1996). These youngest units, however, are preserved in only a few places. In older sequences of Mesozoic and Palaeozoic rocks, the success rate was poor and very limited data are available (Klootwijk & Bingham 1980; Klootwijk *et al.* 1983; Appel *et al.* 1991; Schill *et al.* 2002a; Crouzet *et al.* 2003; Torsvik *et al.* 2009). A large portion of our present knowledge on the tectonic development of the northern Indian margin comes from remagnetized rocks, however. In this paper we review palaeomagnetic works in the Himalaya, mainly from the TSS, from which a large dataset of secondary magnetizations residing in pyrrhotite has been obtained (Fig. 1) and interpreted in terms of tectonic deformation and thermo-tectonic processes. These works were mainly carried out by the Tübingen group within the past two decades. With the understanding of the origin and age of pyrrhotite remanences and their occurrence in certain rock types and formations, major interest was addressed to sample suitable recorder 'materials' systematically along the arc. This paper summarizes the present knowledge of the nature and origin of secondary

*From*: ELMORE, R. D., MUXWORTHY, A. R., ALDANA, M. M. & MENA, M. (eds) 2012. *Remagnetization and Chemical Alteration of Sedimentary Rocks*. Geological Society, London, Special Publications, **371**, 163–180.
First published online August 22, 2012, http://dx.doi.org/10.1144/SP371.1

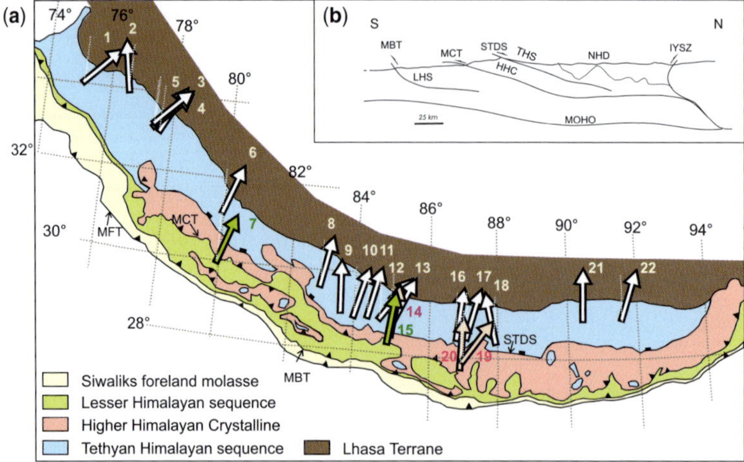

**Fig. 1.** (a) Palaeomagnetic rotations (arrows) resulting from secondary pyrrhotite remanences along the Himalayan arc (simplified geological map after Yin 2006). Locations are marked with numbers (for data see Table 1): (1) Kashmir, (2) Zanskar, (3) Sarchu, (4) Losar, (5) Spiti, (6) Malari, (7) LH Kumaon, (8) west Dolpo, (9) Hidden Valley, (10) Manang, (11) Nar Phu, (12) Larkya, (13) Shiar, (14) HHC Buri, (15) LH Buri, (16) Rongbuk, (17) Kharta Valley, (18) Dinggye, (19) HHC Khumbu N, (20) HHC Khumbu S, (21) Dykes Nakarze, (22) Dykes Qonggai (the north direction at each sampling area is shown by a white dashed line). For data see Table 1. The small inserted figure (b) shows a schematic north–south cross-section through the Himalayan range (MFT, Main Frontal Thrust; MBT, Main Boundary Thrust; LHS, Lesser Himalayan Sequence; MCT, Main Central Thrust; HHC, Higher Himalayan Crystalline; STDS, South Tibetan Detachment System; THS, Tethyan Himalayan Sequence; NHD, North Himalayan Domes; IYSZ, Indus–Yarlung Suture Zone).

pyrrhotite remanences in low-grade metamorphic rocks of the Tethyan Himalaya (and additionally some results from the Lesser Himalaya and the Higher Himalayan Crystalline) and it's tectonic and tectono-metamorphic constraints, and also highlights further possibilities of future studies.

## History of research on pyrrhotite remagnetization in the Himalaya

When the first palaeomagnetic results from the Tethyan Sedimentary Series (TSS) were published (Klootwijk & Bingham 1980; Klootwijk *et al.* 1983), the authors reported a component with unblocking temperatures of maximum 330 °C. They interpreted it as a secondary remanence, based on an increased scatter of remanence directions after bedding correction and a clear discrepancy with the expected direction for a primary remanence. In the TSS of Thakkola area (western Nepal), Klootwijk & Bingham (1980) found this type of remagnetization in Devonian–Carboniferous thin-bedded to massive meta-limestones, and attributed it to metamorphism associated with the Main Central Thrust (MCT). Mean directions of normal and reverse polarities were demagnetized at 200–270 °C and 270–330 °C, respectively, differing

within 95% confidence limits. This discrepancy was explained by remanence acquisition at different stages of ongoing deformation. In the Kashmir Basin (NW Himalaya) a similar component was isolated in Lower Palaeozoic–Triassic low-grade meta-sediments, dominating the natural remanence in these rocks. From its mean direction, a clockwise rotation of 45° of the Panjal Nappe v. peninsular India was concluded (Klootwijk *et al.* 1983). The authors did not relate this secondary component to pyrrhotite in any of these studies, although from the present knowledge the observed demagnetization behaviour was clearly indicating it. The time to consider iron sulphides as an important remanence carrier became ripe with fundamental studies on the behaviour of pyrrhotite remanences at high and low temperatures (Dekkers 1989), its 30–34 K low-temperature transition (Rochette *et al.* 1990), domain structures (Soffel 1977, 1981) and the finding of pyrrhotite as a carrier of successive polarities in low-grade metamorphic sediments in the western Alps (Ménard & Rochette 1992; Rochette *et al.* 1992; Crouzet *et al.* 2001*a*).

Appel *et al.* (1991) intensively studied meta-sediments of the TSS in Manang area (western Nepal), only *c.* 20–30 km east of the working area of Klootwijk & Bingham (1980). For the first time, pyrrhotite was recognized as the carrier of a dominant secondary remanence in the Himalaya, based on thermal unblocking at the Curie temperature of pyrrhotite (*c.* 320–330 °C), the 32 K low-temperature transition and typical domain structures observed by Bitter patterns. A 99% negative fold test confirmed a secondary origin that is post-dating main Himalayan folding. Comparing the overall observed mean direction to that expected from the apparent polar wander path (APWP) of India for Cenozoic times revealed a 19° steeper inclination and a 16° clockwise deviation; the authors attributed this to steepening along the MCT and to uniform rotational shortening between India and Asia by 520–1100 km at 84°E, respectively.

Patzelt *et al.* (1996) obtained a first palaeomagnetic constraint for the extent of 'Greater India' through primary magnetizations in Upper Cretaceous and Lower Tertiary sequences of Gamba and Duela (southern Tibet). Stimulated by these results, comparable stratigraphic sequences were studied in the westernmost Himalaya of NW Zanskar by Appel *et al.* (1995). Unfortunately, no primary component could be found; instead, a new extensive dataset of well-defined secondary pyrrhotite remanences was obtained from mainly Middle Cretaceous–Early Paleocene limestones. A local trend of declinations and inclinations was observed and interpreted by smaller-scale tectonic modulation within the large-scale deformation regime. The overall results showed an counterclockwise deviation of declinations seemingly not in agreement with the expected clockwise trend in the region at the eastern side of the western Himalayan syntaxis. The authors noted that the strike of the main tectonic features in the sampling area shows an anomalous orientation, and they argued that the observed counterclockwise rotation might be explained by meso-scale tectonic block rotation processes post-dating remanence acquisition.

Meanwhile remanences carried by pyrrhotite were also found in one site of leucogranites (25–15 Ma ages) in the Everest region (Rochette *et al.* 1994), as well as in Lesser Himalayan medium-grade meta-carbonates of Alaknanda area (India) and south of the Manaslu peak (western Nepal), close to MCT (Schill *et al.* 1998). Klootwijk *et al.* (1982) had already detected such a component in the Krol belt (near Alaknanda valley); however, at that time they did not recognize pyrrhotite as the remanence carrier.

Until the mid-1990s, results based on pyrrhotite remanences were more a by-product of studies aiming to reveal primary remanence components. Based on the high stability of secondary pyrrhotite remanences and the increasing understanding of the remanence origin, a systematic investigation of suitable rock formations along the Himalayan arc was started during the ensuing period. Schill *et al.*

(2001) published a large dataset from the Spiti-Losar-Sarchu region in the western Himalaya, attempting to separate the effects of block rotations on different scales, that is, large-scale (rotational shortening, oroclinal bending), meso-scale (*c.* 100 km) and local effects. They also recognized small-circle distributions of site mean directions and applied small-circle methods (Waldhör *et al.* 2001; Waldhör & Appel 2006) to analyse them. Furthermore, for the first time, illite crystallinity and K–Ar ages were used to derive crucial information about peak metamorphic temperatures and the age of remanence acquisition in pyrrhotite (Schill *et al.* 2001). The pyrrhotite remanence directions, corrected for meso-scale tectonic strike attitudes, delivered a consistent 0–20° clockwise rotation of the Spiti-Losar-Sarchu region v. stable India since Oligocene.

The 'hunting' of secondary pyrrhotite remanences in the TSS progressed further eastwards with a steadily high success rate. Schill *et al.* (2002*b*) studied Triassic carbonates at Malari (*c.* 80°E), followed by a very dense network of data from mainly Carboniferous and Triassic meta-limestones between 83 and 85°E. These data included (from west to east): western Dolpo (Crouzet *et al.* 2003); Hidden Valley (Crouzet *et al.* 2001*b*); Larkya (Schill *et al.* 2004); Nar Phu (Schill *et al.* 2003); and Shiar Khola (Schill *et al.* 2002*a*). In Hidden Valley Crouzet *et al.* (2001*b*) found a systematic variation of normal and reverse polarities (R1-N1-R2-N2) with altitude, and discussed the remanence acquisition mechanism in detail. Schill *et al.* (2004) summarized the data from western Nepal, and noted a trend from 'no rotation' in the west to 'clockwise rotation' in the east. From this observation, they proposed a hypothetical model of a large-scale dextral shear zone eastward of *c.* 84–85°, and related it to the extrusion of the Tibetan Plateau. The absence of palaeomagnetic data in the eastern Himalaya prevented this hypothesis from being tested, however. For this reason, a new series of studies in the central to the eastern Himalayan sector was initiated after the year 2005.

Of these regions in the east, results from the Everest/Makalu area and from SE Tibet are already available. Unfortunately, the excellent recorder 'materials' of the TSS, predominantly studied so far, die out eastwards of the Yadong graben (*c.* 89°E); other rock formations in this region (mainly Triassic flysch meta-sediments) turned out to be rather unsuitable for palaeomagnetic studies. This was a major disappointment for the large amount of work done in SE Tibet. Surprisingly, secondary pyrrhotite remanences were identified in several Cretaceous dioritic dykes of SE Tibet by Antolin *et al.* (2010), indicating a *c.* 20° clockwise rotation in Qonggyai area (eastern part) and no rotation in Nagarze area (western part). A totally different interpretation was suggested from results of the TSS in the Everest/Makalu region. Departures of declinations were explained by the Ama Drime massif development, producing apparent rotations through tilting around a north–south axis (Crouzet *et al.* in press). In the Higher Himalayan Crystalline (HHC) south of Everest (Khumbu area), such apparent rotations were also observed and related to crustal doming (El Bay *et al.* 2011). The work in the HHC was at the same time the first comprehensive study revealing secondary pyrrhotite as a predominant and stable remanence carrier in high metamorphic rocks. A first indication that this type of remagnetization can deliver meaningful results had already been mentioned by Schill *et al.* (2004) however, when studying such rocks on a north–south traverse at 84.5°E.

## Origin of pyrrhotite

In nature, pyrrhotite is a complex pattern of different minerals with the general formula $Fe_{1-x}S$. As only monoclinic pyrrhotite ($Fe_7S_8$) is able to carry a significant remanent magnetization, palaeomagnetic studies usually focus on it. Occurrence of (monoclinic) pyrrhotite in low-grade metamorphic sediments, especially marly limestones, is a general phenomenon. Examples have been reported worldwide, especially in mountain belts: in the Appalachian belt (Carpenter 1974); the Dalradian schists of Scotland (Hall 1982); the marly limestones of the Dauphinoise zone from the western Alps (Rochette & Lamarche 1986; Rochette 1987; Crouzet *et al.* 1996); the metapelites from Ardennes (Robion *et al.* 1995, 1997); and in the Himalaya (Appel *et al.* 1991, 1995). Pyrrhotite was also discovered in metamorphic slates close to several granitic bodies (Skye island, Woods *et al.* 2000; Elba island,

Wehland *et al.* 2005*a*, *b*, *c*). The understanding of pyrrhotite formation and its remanence acquisition during regional low-grade metamorphism, as well as the relation with granite emplacement, is of general importance.

Pyrrhotite can be formed at elevated temperature by simple desulphidation (Carpenter 1974; Ferry 1981), or the breakdown of magnetite and pyrite (Ferry 1981; Tracy & Robinson 1988) under reductive conditions above *c.* 200 °C (Lambert 1973; Hall 1986). Pyrrhotite is usually destroyed at higher temperatures above *c.* 450–500 °C, often associated with new formation of magnetite (Bina *et al.* 1991). This breakdown of pyrrhotite at *c.* 500 °C has been outlined by Beugnies (1986) and Robion *et al.* (1995) using field petrography and rock magnetic evidences, respectively. Paradoxically, pyrrhotite can also be stable up to 700 °C (Ferry 1981) as long as carbon is present to ensure reducing conditions (Poulson & Ohmoto 1989). Aubourg & Pozzi (2010) also demonstrate the formation of pyrrhotite at very low temperature ($<100$ °C) in several types of claystones. According to Horng & Roberts (2006), pyrrhotite formation will be extremely slow below *c.* 180 °C. The relatively low-temperature limit where pyrrhotite starts to form explains its frequent occurrence in low-grade metamorphic sequences of the TSS. As the upper limit of its stability range is often *c.* 450–500 °C, it can be understood why in higher metamorphic TSS sediments magnetite is dominating. The surprisingly rather frequent occurrence of pyrrhotite in the highly variable rock types of the HHC could be explained by formation of pyrrhotite during cooling, rather than by specific conditions keeping pyrrhotite stable up to 700 °C. In the diorite dykes of SE Tibet it is very likely that pyrrhotite did not exist primarily, but formed after sulphur migration from the host rock (Triassic flysch) during metamorphism (Antolin *et al.* 2010).

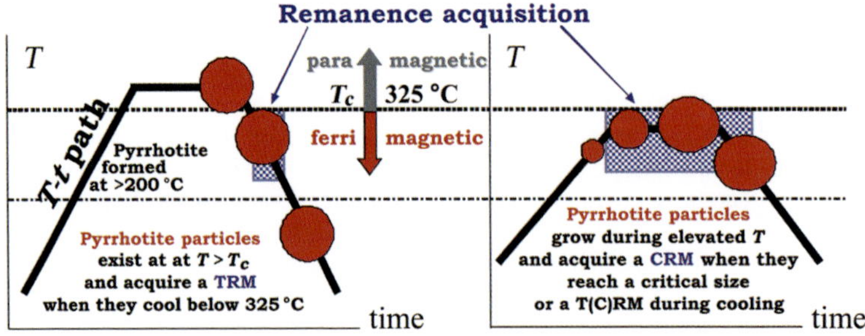



**Fig. 2.** Scheme of remanence acquisition in pyrrhotite during a metamorphic temperature–time (*T–t*) path, for $T_{max}$ higher (left) or lower (right) than the Curie temperature ($T_c$) of pyrrhotite, leading to a thermoremanent magnetization (TRM) or a chemical remanent magnetization (CRM), respectively. (The blue field shows the possible time and temperature range of remanence acquisition.) In the case of $T_{max} < T_c$, the nature of the remanence can be also a thermochemical remanence (TCRM) or even a partial (p)TRM.

## Mechanism and age of remanence acquisition in pyrrhotite

The remanence mechanism has been discussed in several papers listed in the previous section; for a most comprehensive presentation we refer to Crouzet *et al.* (2001*b*). The nature of the remanence process (see Fig. 2) depends on the peak metamorphic temperature ($T_{max}$). If $T_{max}$ exceeded the Curie temperature of pyrrhotite ($T_c$ c. 325 °C), a thermoremanent magnetization (TRM) will be acquired during cooling when grains are crossing through the blocking temperature during the last metamorphic cooling event (as in igneous rocks); the age of this TRM can be related to the

metamorphic cooling age through $T_c$. In the case of $T_{max} < T_c$ a chemical remanence (crossing through the blocking volume by grain growth) or a thermochemical remanence (grain growth and cooling) are possible. In the high-grade metamorphic HHC and in the medium-grade carbonates of the Lesser Himalaya, the precondition of $T_{max} > T_c$ was certainly fulfilled. In the low-grade metamorphic rocks of the TSS, both $T_{max} > T_c$ and $T_{max} < T_c$ may have been the case. The occurrence of antipodal components during thermal demagnetization, observed in several localities (Hidden Valley, Kharta), is very difficult to explain if we consider that the pyrrhotite remanence is a CRM. It would imply growth of pyrrhotite

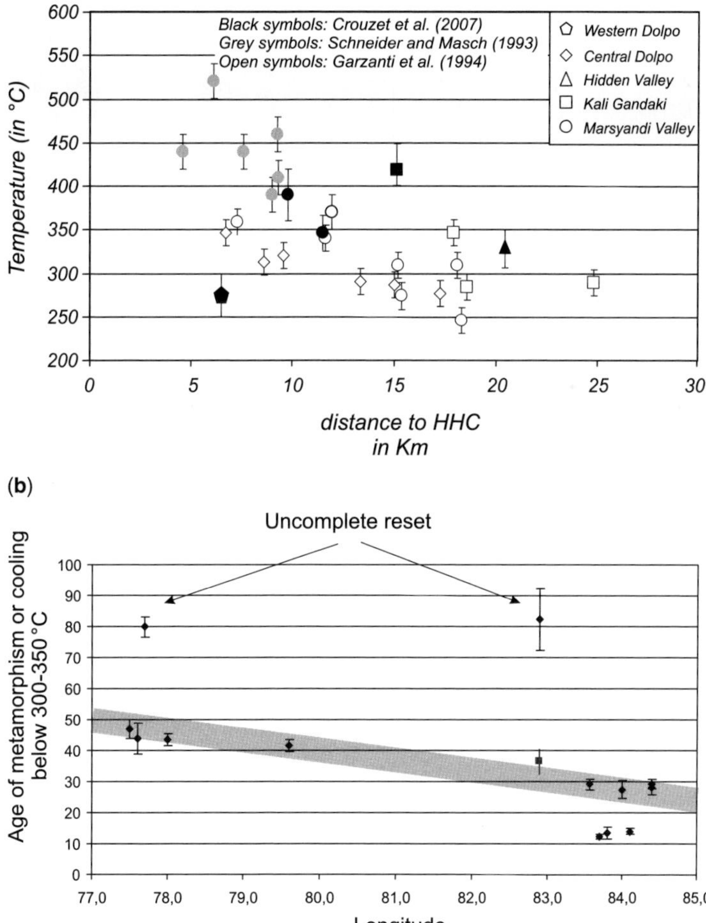



**Fig. 3.** (**a**) Peak metamorphic temperatures in the Tethyan Sedimentary Series (TSS) v. distance to the Higher Himalayan Crystalline; (**b**) K–Ar and Ar–Ar ages from the TSS (away from STDS and granite outcrops) v. longitude, showing a decreasing trend towards the east (figures from Crouzet *et al.* 2007).

**Table 1.** *Summary of secondary pyrrhotite characteristic remanent magnetization ChRM (Pyr), predominantly from the Tethyan Sedimentary Series (others, i.e. Higher Himalayan Crystalline, Lesser Himalaya and dykes are indicated by HHC, LH, Dykes, respectively). References are (1) Klootwijk* et al. *(1983), (2) Appel* et al. *(1995), (3) Schill* et al. *(2001), (4) Schill* et al. *(2002b) including data from Klootwijk* et al. *(1982) for Kumaon, (5) Crouzet* et al. *(2003), (6) Crouzet* et al. *(2001b), (7) Klootwijk & Bingham (1980), (8) Appel* et al. *(1991), (9) Schill* et al. *(2003), (10) Schill* et al. *(2004), (11) Schill* et al. *(2002a), (12) Crouzet* et al. *(in press), (13) El Bay* et al. *(2011), (14) Rochette* et al. *(1994), (15) Antolin* et al. *(2010)*

| | Kashmir (1) | Zanskar (2) | Sarchu (3) | Losar (3) | Spiti (3) | Malari (4) | LH Kumaon (4) | W-Dolpo (5) | Hidden Valley (6) | Thakkhola (7) | Manang (8) | Nar Phu (9) |
|---|---|---|---|---|---|---|---|---|---|---|---|---|
| Geog. position | 34.0°N, 75.0°E | 33.9°N, 76.5°E | 33.0°N, 77.6°E | 32.5°N, 77.7°E | 32.1°N, 78.0°E | 30.8°N, 80.0°E | 30°N, 79°E | 29.4°N, 83.0°E | 28.8°N, 83.6°E | 28.7°N, 83.7°E | 28.7°N, 84.0°E | 28.8°N, 84.3°E |
| [1]Sampled rocks | Cambrian to Triassic | Cretaceous to Eocene | Triassic | Triassic | Carbon. and Triassic | Triassic | Carbonates not datable | Triassic | Carbon. and Triassic | Devonian | Carbon. and Triassic | Permian and Triassic |
| [2]ChRM (Pyr) rem. age (Ma) | *c.* 50? | 50–44 | 49–39 | *c.* 44 (i) | 46.5–45 | *c.* 40 (i) | 39–11 | *c.* 35 (i) | 31–29 | *c.* 29 (i) | 30–25 | 32–25 |
| No. sites | 7 + 4 + 5 | 27 | 7 | 14 | 5 + 18 + 8 + 7 | 10 | 26 | 23 | 23 | 1 | 12 | 13 |
| [3]*In situ* mean direction, D/I (°) | 12.5/38, 24/35.5, 27.5/27 | 336.6/39.6 | 28.8/50.2 | 32.5/19.7 | 27.4/6.8, 22.7/36.9, 6.5/49.6, 16.9/20.1 | 7.6/34.0 | 11.0/26.3 | 191.7/−30.9 | 357.2/59.2 | 296/69 | 16.4/65.9 | 355.5/61.4 |
| [4]Alpha 95 (°) | 7, 7.5, 9 | 5.9 | 14.0 | 15.8 | 9.8, 7.9, 13.7, 7.1 | 13.4 | 10.0 | 5.7 | 3.6 | n.a. | 3.2 | 16.5 |
| [4]$k$ | 75, 149, 73.5 | 22.9 | 19.5 | 7.5 | 62.1, 20.2, 17.4, 73.9 | 13.9 | 9.0 | 29.5 | 85.7 | n.a. | 183 | 7.3 |
| [5]Polarities | Norm & **rev** | Norm & rev | Norm & **rev** | Norm & **rev** | Norm & **rev** | **Norm** & rev | Norm & rev | Rev | Norm & rev | Norm & rev | Rev | Norm & rev |
| [6]Fold test | None | negative, 99% (E) | negative, 99% (E,F) | Insignificant | Negative, 99% (E,F) | Insignificant | Negative, 95% (F) | Negative, 99% (F) | Negative, 99% (E) | n.a. | Negative, 99% (E) | Negative, 99% (F) |
| [7]Magnetite, component | Primary? | None | None | Secondary | Secondary | Secondary | None | Primary | None | Primary? | Primary | None |
| [8]Angle (°) D v. exp. D | cw 45 (KR81) | ccw 10 (K91) | cw 35–55 (BC91) | Cw 40–50 (BC91) | cw 20–35 (BC91) | cw 21 (BC91) | cw 20 (BC91) | cw 10–15 (A99) | ccw 2* (A99, BC02) | n.a. | cw 17* (A99, BC02) | cw 16* (A99, BC02) |
| [9]Inclination, anomaly (°) | Steeper (+15) | Steeper (+17) | Steeper (+26) | Shallower (−4) | Steeper (+8) | Steeper (+8) | Shallower (−16) | Shallower (−4) | Steeper (+24) | Steeper (+33) | Steeper (+28) | Steeper (+24) |

| | Larkya (10) | Shiar (11) | HHC, Buri (10) | LH, Buri (10) | Rongbuk (12) | Kharta, Valley (12) | Dinggye (12) | HHC, Khumbu N (13) | HHC-LG, Khumbu (14) | HHC, Khumbu S (13) | Dykes, Nagarze (15) | Dykes, Qonggyai (15) |
|---|---|---|---|---|---|---|---|---|---|---|---|---|
| Geog. position | 28.7°N, 84.6°E | 28.6°N, 85.1°E | 28.4 °N, 84.9 °E | 28.2 °N, 84.8 °E | 28.3 °N, 86.8 °E | 28.4 °N, 87.2 °E | 28.2 °N, 87.8 °E | 27.8 °N, 86.8 °E | 27.8 °N, 86.8 °E | 27.5 °N, 86.7 °E | 29.0 °N, 90.4 °E | 29.0 °N, 91.6 °E |
| [1]Sampled rocks | Triassic | Permian and Triassic | Gneisses | Carbonates not datable | Ordovician to Triassic | Ordovician to Triassic | Ordovician to Triassic | Crystalline | Leucogranite | Gneisses | Cretaceous dykes | Cretaceous dykes |
| [2]ChRM (Pyr) rem. age (Ma) | 30–17 | 25–17 | 17–5 | 9–3 | 22–18 | 19–13 | 19–13 | 16 | 25–15 | 16 | 25–15 | 25–15 |
| No. sites | 4 + 10 | 31 | 7 | 5 | 4 | 8 | 20 | 4 | 1 | 8 | 3 | 7 |
| [3]In situ mean direction, D/I (°) | 32.3/21.2, 353.6/61.2 | 26.9/27.2 | 14.8/29.2 | 358.8/24.6 | 191.4/48.3 | 20.6/37.5 | 351.7/25.1 | 39.7/31.9 | 6.1/52.5 | 6.8/43.6 | 1/20 | 19/28 |
| [4]Alpha 95 (°) | 20.1, 10.2 | 6.8 | 8.1 | 24.1 | 7.8 | 6.9 | 9.4 | 22.0 | n.a. | 11.1 | Not det., cyl. best fit | Not det.cyl., best fit |
| [4]k | 21.8, 23.2 | 15.2 | 56.1 | 11.0 | 97.6 | 51.9 | 13.1 | 18.5 | n.a. | 25.8 | not det., cyl. best fit | Not det., cyl. best fit |
| [5]Polarities | Norm | Rev | Norm & rev | Norm & **rev** | rev | Norm & Rev | Norm & Rev | norm & rev | rev | Norm & rev | Norm & rev | Norm & rev |
| [6]Fold test | Negative, 95% (E) | Insignificant | n.a. | n.a. | insignificant | negative, 99% (F) | Insignificant | n.a. | n.a. | n.a. | n.a. | Negative, 99% (F) |
| [7]Magnetite, component | None | Primary? | None | None | none | Primary? | None | Secondary | ? | Secondary | None | None |
| [8]Angle (°) D v. exp. D | cw 30–47 (A99,BC02) | cw 27* (A99,BC02) | cw 10 (A99,BC02) | cw 10 (A99,BC02) | cw 8 (BC02) | cw 17 (BC02) | ccw 12 (BC02) | cw 38 (A99) | n.a. | cw 5 (A99) | cw 2 (A99,BC02) | cw 20 (A99,BC02) |
| [9]Inclination, anomaly (°) | Shallower (−1) | Shallower (−18) | Shallower (−17_) | Shallower (−22) | Steeper (+3) | Shallower (−9) | Shallower (−21) | Shallower (−13) | Steeper (+8) | Shallower (−1) | Shallower (−27) | Shallower (−19) |

[1]Stratigraphic positions of rocks yielding significant ChRM (Pyr) results are given.
[2]Remanence acquisition ages of ChRM (Pyr) are derived from thermochronological data of Bonhomme & Garzanti (1991), Oliver *et al.* (1995), Wiesmayr & Grasemann (1999), Godin *et al.* (2006), Crouzet *et al.* (2007), Kellet & Godin (2009), Kali *et al.* (2010) and Dunkl *et al.* (2011) or are interpolated (i) from the linear west–east trend as shown by Crouzet *et al.* (2007).
[3]D, I: declination, inclination (normal polarity direction)
[4]Alpha95: 95% and k: 95% confidence limit and precision parameter from Fisher statistics
[5]Polarities: (norm) normal, (rev) reverse polarities (dominant polarity in bold)
[6]According to E: McElhinny (1964) or F: McFadden (1990)
[7]The presence of a statistically significant magnetite component is indicated by its nature (primary or secondary)
[8]Angles (as given by the respective authors; data with asterisk recalculated by Schill *et al.* 2004) between the measured mean declination (D) and the expected declination (exp. D); cw: clockwise, ccw: counterclockwise. Expected values were determined using APWPs of KR81: Klootwijk & Radhakrishnamurty (1981); K91: Klootwijk *et al.* (1991); BC91: Besse & Courtillot (1991); A99: Acton (1999); and BC02: Besse & Courtillot (2002) for a remanence age as given in this table and based the present shape of the Indian Plate.
[9]The inclination anomaly is the angle between the measured inclination and the expected inclination; given angles in brackets for steeper (+) and shallower (−). For multiple results only the average is listed. All expected values are calculated using the APWP of Acton (1999).

grains at different times during cooling. Moreover, the grain size should be very homogenous and become smaller with decreasing temperature; such a scenario is clearly unrealistic. The antipodal components are more easily explained by the record of successive reversals of the Earth's magnetic field during slow cooling of the rocks (Crouzet *et al.* 2001*b*). Using several geothermometers, Crouzet *et al.* (2007) estimated $T_{max}$ in Hidden Valley approximately equal to $T_c$; a TRM is therefore likely and in accordance with the observed systematic sequence of different polarities with altitude in this area. In some other study areas, such as at Manang (Appel *et al.* 1991), pyrrhotite remanences were obtained from very different stratigraphic levels representing different $T_{max}$ because of different depths of burial and metamorphic overprint; consequently (p)TRMs and CRMs could be expected. As no significant differences in pyrrhotite remanence directions from different levels were observed, we conclude that even in the case of $T_{max} < T_c$ magnetic viscosity of the pyrrhotite particles is high enough at elevated temperature and the magnetization is only blocked once the temperature starts to decrease. Schill *et al.* (2002*c*) performed a Thellier–Thellier test on samples from Shiar Khola where $T_{max}$ very likely did not reach $T_c$; the results showed a clear linear correlation of the laboratory-induced pTRMs with the NRM intensities demagnetized in the same intervals, demonstrating a pTRM nature of the latter.

In order to determine whether the pyrrhotite remanence is a TRM $(T_{max} > T_c)$ or whether a CRM is possible $(T_{max} < T_c)$, the peak metamorphic temperature $(T_{max})$ must be known. To establish when a TRM was formed, the time when the rocks cooled through $c.$ 325 °C has to be determined. Different low-temperature geothermometers can be used, depending on the available rock types and minerals in the study areas. The most common methods are illite or chlorite crystallinity, vitrinite reflectivity, chlorite-chloritoid and calcite-dolomite geothermometry and calcite twin lamellae. Crouzet *et al.* (2007) studied several geothermometers in different areas of the TSS and compiled data from other authors. The results show that the distance to the Higher Himalayan Crystalline plays an important role, with lower $T_{max}$ at more distant locations (Fig. 3a). Relevant cooling ages for passing through $T_c$ of pyrrhotite can be obtained by K–Ar or Ar–Ar on illite or muscovite and by fission track analysis (both methods having closure temperatures around $T_c$ of pyrrhotite). In Table 1 the results from different areas are summarized for the palaeomagnetic sampling areas. Figure 3b displays results for a profile from the western to the central Himalaya, revealing a systematic decrease of cooling ages from west to east.

## Analytical evidence for pyrrhotite and remanence properties

A simple north–south cross-section through the Himalaya is shown in Figure 1b. The TSS of the Tethyan Himalaya represents the former passive margin of India and comprises a $c.$ 6–7 km thick sequence from Cambro-Ordovician–Eocene rocks (e.g. Garzanti 1999). It has undergone regional metamorphism (Schneider & Masch 1993; Garzanti *et al.* 1994) and also possible contact metamorphism near frequently occurring Early Miocene leucogranite intrusions (e.g. Guillot *et al.* 1995). Due to burial and/or tectonic thickening, most of the sequence has been thermally overprinted with medium- to low-grade metamorphism, with only a few locations showing no visible overprint at their topmost part of the sequence. The grade of metamorphic overprint varies considerably at different locations. For example, at Gamba and Tingri (southern Tibet between $c.$ 87 and 89 °E) the Cretaceous and Tertiary units are non-metamorphosed while the comparable stratigraphic units in Zanskar (western Himalaya) are clearly low-grade overprinted. On the other hand, Torsvik *et al.* (2009) very surprisingly reported a hematite-based primary remanence even in Ordovician red beds from Spiti area (western Himalaya). This indicates that even in the lowest parts of the TSS sequence metamorphism did not erase primary remanences at this location, at least for the hematite component.

Rocks which represent the most suitable conditions for secondary pyrrhotite remanences consist of marly carbonate rocks (favourable conditions for pyrrhotite formation because of reductive conditions and sulphur contents), with peak metamorphic temperatures $(T_{max})$ above the Curie temperature $(T_c \ c.$ 325 °C) of pyrrhotite (i.e. secondary remanences being very likely a TRM). The temperature $T_{max}$ of suitable rocks should not exceed $T_c$ by too much in order to avoid: (1) growth of over-large multidomain pyrrhotite grains with relatively lower coercivity, as evidenced in some contact metamorphic aureoles (e.g. Skye Island and Elba Island; Wehland *et al.* 2005*a, b*) or in the HHC (El Bay *et al.* 2011); or (2) oxidation of pyrrhotite as observed in thermomagnetic runs (at 450 °C or even lower $T$ depending of the grain size and the availability of oxygen; Dekkers 1990; Bina & Daly 1994; Wang *et al.* 2008) and in the Ardennes massifs (Robion *et al.* 1995). Such ideal recorder 'materials' have been found in the TSS at different stratigraphic levels, depending on the variable degree of regional metamorphism along the Himalayan arc. In particular, the Middle–Upper Triassic (marly limestones) and the Lower Triassic (condensed limestones) sequences have turned out to be the best choice in most of the sampling areas. Only in NW

Zanskar, where metamorphic overprint significantly affected higher stratigraphic levels, Middle–Upper Cretaceous (pelagic limestones) and Paleocene (open-shelf carbonates) units turned out to be the best choice. Another suitable rock type throughout the TSS was found in Carboniferous rocks (dark limestones with argillaceous layers) and partly also in Devonian and Permian strata (carbonates).

The demagnetization behaviour of suitable rock types of the TSS is in principle similar at different areas and mostly straightforward to analyse. Alternating field demagnetization (AfD) in fields up to 150 mT is usually not sufficient to remove a significant portion of the remanence, indicating that single-domain (SD) or pseudo-single domain (PSD) properties dominate. The occurrence of pyrrhotite in such a small grain-size range is supported by reflected light microscopy (particles often near or below the limit of optical resolution) and scanning electron microscopy (Fig. 4a, b), as well as first-order reversal curves (FORC) analysis (Wehland *et al.* 2005*a*). During thermal demagnetization (ThD) pyrrhotite remanences are unblocked at temperatures up to a maximum of *c.* 325 °C, often restricted to a very narrow temperature interval of *c.* 10–50 °C (Fig. 5a) near the Curie temperature. In some rocks the unblocking temperature ($T_{ub}$) range may be larger, even up to about 100 °C or more, which indicates the presence of SD and/or PSD particles with a larger variation of grain size and therefore a wider spectrum of relaxation times. Often, normal and reverse polarity components were observed during ThD in single specimens (Fig. 5b), especially in the case of a wider unblocking range (Crouzet *et al.* 2001*b*). In part of the rocks a pyrrhotite component exists exclusively, while in others it coexists with a magnetite remanence (Fig. 5c). Separation of the pyrrhotite component is mostly easy to perform because of its thermal unblocking in a limited temperature window. Additionally, most of the magnetite remanence can be removed by AfD before starting the ThD treatment.

Following the procedure of Crouzet *et al.* (1996, 2001*a*) for samples from the Western Alps, Wehland *et al.* (2005*c*) imparted laboratory TRMs while changing the field direction during cooling of the samples. Their results demonstrate that pyrrhotite bearing low-grade meta-carbonates can indeed record different polarity directions as pTRMs in single specimens. The fact that different polarities occur in the studied samples is also a strong indication that the period of remanence acquisition was sufficiently long to average out palaeosecular variations in the records of single specimens.

The studied metamorphic rocks are often highly deformed. An important issue is therefore whether the observed remanence directions of pyrrhotite are strongly influenced by its high magnetocrystalline anisotropy. It is difficult to answer this question by applying measurements of the anisotropy of magnetic susceptibility (AMS), as pyrrhotite has a low susceptibility and therefore the AMS is usually dominated by magnetite or paramagnetic minerals. El Bay *et al.* (2011) performed a more direct experiment, imparting anhysteretic remanences (ARM) with a known constant magnetic field direction and measuring the angle between the direction of the applied field and the resulting ARM direction; their results did not show a significant angular deviation.

In certain rock types, especially the Triassic flysch of the northern Tethyan Himalaya and the Higher Himalayan Crystalline, a high scatter of specimen directions is observed within sites despite reasonable good demagnetization behaviour of single specimens. This can be explained by remanence acquisition during metamorphic cooling while ductile (plastic) deformation was still prevailing below the upper limit of blocking temperatures (*c.* 300–325 °C), which will result in a large scatter of remanence directions at site-scale or

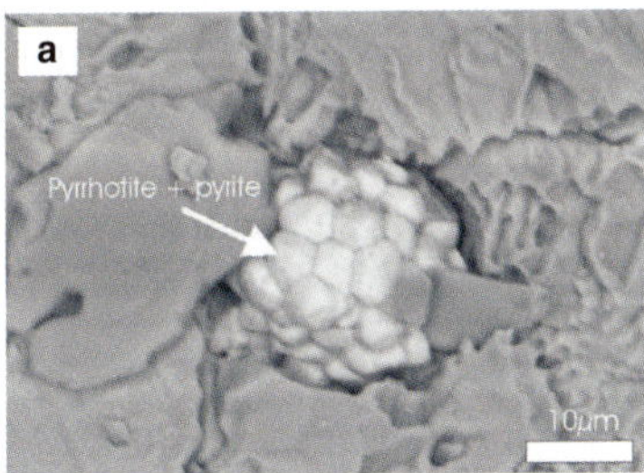


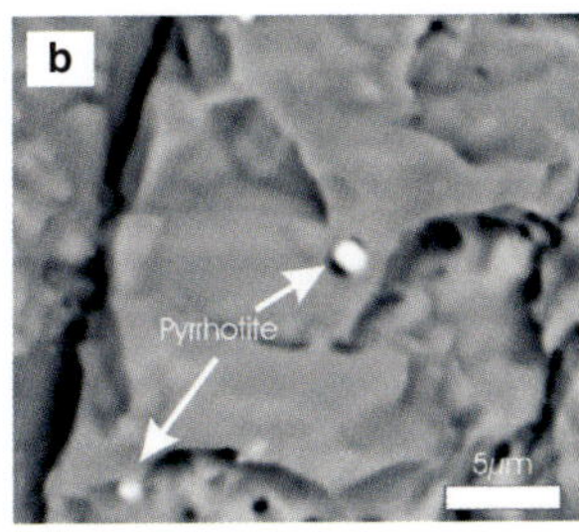

**Fig. 4.** (**a**) Cluster of iron sulphides (pyrite and pyrrhotite) and (**b**) small pyrrhotite particles in the single-domain grain-size range, both from Larkya area (Wehland *et al.* 2005*c*). (**c**) Bitter patterns of magnetic domain structures on a large multidomain pyrrhotite grain (reflected light microscopy) from the TSS at Yadong area (southern Tibet, unpublished).

E. APPEL *ET AL.*

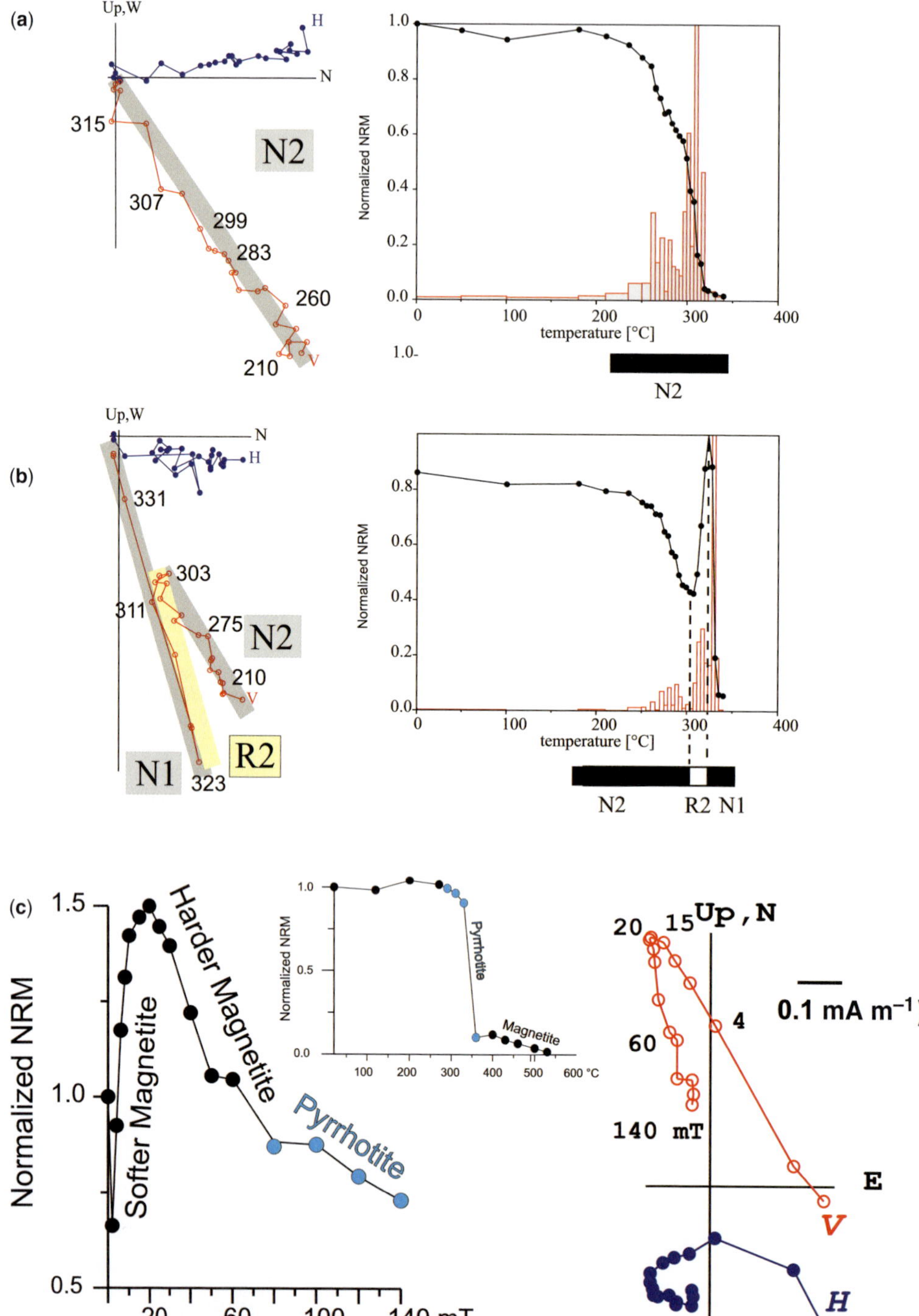



**Fig. 5.** Thermal demagnetization behaviour of pyrrhotite in low-grade meta-carbonates from Hidden Valley with (**a**) one component; (**b**) three antiparallel components (both from Crouzet *et al.* 2001*b*; R1-N1-R2-N2 are subsequent reverse- and normal-polarity phases recorded in the Hidden Valley samples – R1 does not appear in the shown examples); and (**c**) alternating field demagnetization (small figure: thermal demagnetization) of a sample from W-Dolpo, containing pyrrhotite and magnetite (Crouzet *et al.* 2003).

even microscopic scale. The critical temperature for the transition from ductile to brittle deformation depends on the mineralogy of the rock and can lie within the range of *c.* 200–400 °C (e.g. Carter & Tsenn 1987).

The presence of pyrrhotite in the samples is already well indicated by the unblocking behaviour during ThD and its stability during AfD (Fig. 5a). Acquisition of isothermal remanences (IRM) and ThD of saturation isothermal remanence magnetization (SIRM) are further supportive. Direct identification of the Curie temperature by thermal runs of susceptibility is often hindered by the high coercivity of pyrrhotite; the magnetite content or even paramagnetic minerals are hiding the characteristic Hopkinson Peak and subsequent decay (Fig. 6a). More often, the 32 K low-temperature transition can be detected (Fig. 6b); however, sometimes it is also found to be absent in samples where the presence of pyrrhotite is clearly supported by ThD or other results. Furthermore, magnetic domain structures can be observed (most easily by Bittern patterns) on larger particles, revealing typical patterns of uniaxial anisotropy (Fig. 4c). The fact that volume domains are visible even without removing the irregularly stressed surface layer (produced by mechanical polishing) is a further clear evidence of pyrrhotite as no other magnetic mineral shows this phenomenon. Unfortunately, domain structure observations are restricted to MD grains with grain sizes above a few μm (SD–MD transition at 1–2 μm; Soffel 1977).

At this stage it should be noted that in higher metamorphic rocks (e.g. deeper stratigraphic levels of the TSS, Lesser Himalaya, Higher Himalayan Crystalline, dykes in SE Tibet) larger pyrrhotite grains (Fig. 4c) occur frequently, but mostly show a lower coercivity (even comparable to soft MD magnetite) and often an unstable ThD behaviour. It has been shown that it is still possible to retrieve meaningful results from such rock types (Schill *et al.* 1998, 2002*b*, 2004; Antolin *et al.* 2010; El Bay *et al.* 2011), but with a clearly lower success rate.

## Tectono-metamorphic implications

Klootwijk *et al.* (1985) interpreted the first (primary) palaeomagnetic data in terms of oroclinal bending of the Himalayan arc and rotational underthrusting of India beneath Asia around a pivot in the western syntaxis area (see principle in Fig. 7c, d). From the angle between declinations of the measured and expected remanence vectors, they determined this magnitude to 650 km at *c.* 84°E. Appel *et al.* (1991) adopted the concept and used the secondary pyrrhotite component from Manang area to estimate the magnitude of rotational shortening (Fig. 7c) at the same longitude, resulting in 520–1100 km depending on different tectonic models. When Patzelt *et al.* (1996) got the first reliable palaeolatitudes (Fig. 7b) for Tertiary (Paleocene) units at Gamba/Duela (southern Tibet *c.* 89°E), they used the results of rotational shortening to estimate the extent of 'Greater India' (Fig. 7a) along the entire Himalayan arc. Comparison of declinations from the TSS at Malari (*c.* 80°E) and the Lesser Himalaya (Alaknanda valley, Kumaon Himalaya) south of it indicated that oroclinal bending and rotational shortening has been

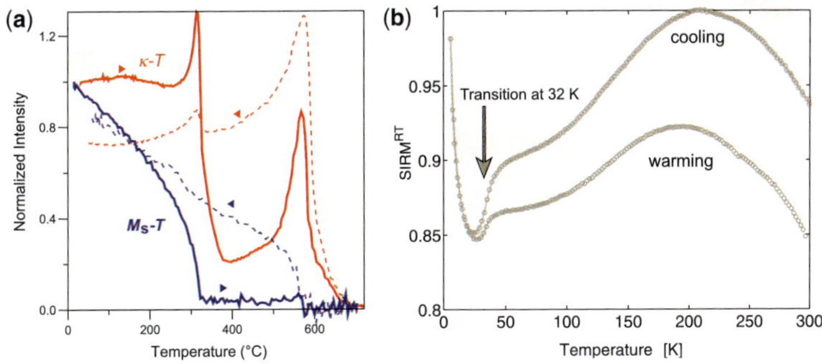

**Fig. 6.** (a) Heating and cooling curves of susceptibility $\kappa$ (red) and saturation magnetization (blue) for a sample from Kumaon Lesser Himalaya (Schill *et al.* 1998) showing the Curie temperature (at *c.* 320–330 °C in the $M_s$–$T$ and $\kappa$–$T$ curves) of pyrrhotite and a Hopkinson Peak corresponding to pyrrhotite (at *c.* 300 °C in the $\kappa$–$T$ curve). Magnetite is produced during heating (note: the enhancement in the $\kappa$–$T$ curve above 500 °C is not a Hopkinson peak) while most of the pyrrhotite is destroyed (evidenced by the decreased pyrrhotite Hopkinson Peak in the $\kappa$–$T$ cooling curve and the smaller bump at *c.* 320 °C in the $M_s$–$T$ cooling curve) during heating at higher temperatures. (b) Low-temperature cycling (zero-field cooling and warming) on a sample from Tethyan Sedimentary Series (Kharta Valley) showing the 32 K transition of pyrrhotite (Crouzet *et al.* in press).

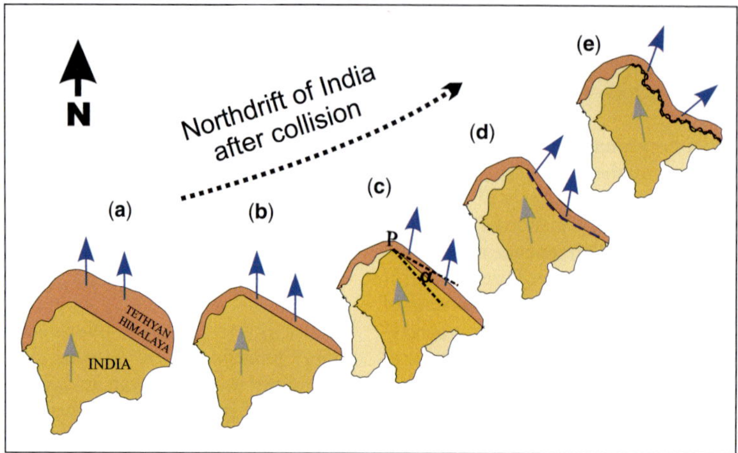



**Fig. 7.** Principle of magnetic rotations at the northern Indian Plate margin (blue arrows) in respect to the stable Indian Plate (grey arrows) subsequent to the India–Asia collision. (**a**) Remanence acquisition, (**b**) north–south shortening (not detectable by rotations), (**c**) uniform rotational underthrusting of India beneath Asia around a pivot P in the western syntaxis, (**d**) oroclinal bending and (**e**) modulation by meso-scale effects.

accommodated along the Late Miocene Main Boundary Thrust rather than along the older Main Central Thrust (Schill *et al.* 2002*b*).

The increasing database turned out several important constraints on tectonic features and processes. First, the control of palaeomagnetic remanences by large-scale processes of oroclinal bending and rotational shortening in the western part of the Himalaya (between *c.* 75 and 83°E) could be confirmed. Second, modulation of the large-scale effects by meso-scale rotations (*c.* 100 km scale; Fig. 7e) is evident. These meso-scale effects do not violate the interpretation of the general pattern as they can be related to variation in the general strike directions appearing in geological maps, such as for the results from NW Zanskar (Appel *et al.* 1995; Schill *et al.* 2001; Fig. 1). Inclinations of pyrrhotite remanences were often found to be significantly steeper than expected for the position of the northern Indian margin (Appel *et al.* 1991; Rochette *et al.* 1994; Crouzet *et al.* 2001*b*), which was initially interpreted as an effect of ramping of the Tethyan Himalya on the Main Central Thust (MCT). Later it became evident that inclinations vary within a large range for secondary pyrrhotite directions of similar remanence ages at different areas (Table 1). It was Schill *et al.* (2003) who first related the variation of inclinations to a medium-scale effect of crustal doming, observed for the Chako antiform in the Nar Phu valley (Gleeson & Godin 2006; Godin *et al.* 2006).

At this point, we have to recall what the pyrrhotite remagnetizations show us. Numerous negative fold tests demonstrate that they post-date the main Himalayan folding, that is small-scale to mega-scale folds, which can be unambiguously recognized and measured in the field. Subsequently to main Himalayan folding, which may be restricted to the Eohimalayan (pre-Miocene) phase, late-orogenic long-wavelength deformation with possible rotation around vertical and horizontal axes occurred. Such long-wavelength deformations can hardly be recognized from field evidence. At the time when the secondary pyrrhotite remanence formed (during cooling, that is predominantly in Oligocene and Miocene times), long-wavelength deformation was already occurring or the very last phase of the main Himalayan folding may still have been in progress. This is demonstrated by a systematic variation of inclinations in the subsequent polarity components recorded at Hidden Valley (Crouzet *et al.* 2001*b*) and at Rongbuk area (Crouzet *et al.* in press), as well as by small-circle distributions observed in Larkya (Schill *et al.* 2004), Shiar (Schill *et al.* 2002*a*) and Dinggye (Crouzet *et al.* in press) areas. A surprising result from Larkya is the east–west alignment of the small circle (Fig. 8a), which indicates tilting around a north–south axis; this is clearly incompatible with steepening along the east–west-striking MCT. In Shiar area the small circle does not include the expected Earth magnetic field direction during remanence acquisition, which means that a vertical-axis rotation must also have occurred (Fig. 8b). Tilting most probably post-dates the block rotation, as otherwise a widespread NW–SE-directed tilting (folding) must have happened (which is very unlikely in the general north–south-oriented compression of the India–Asia convergence).

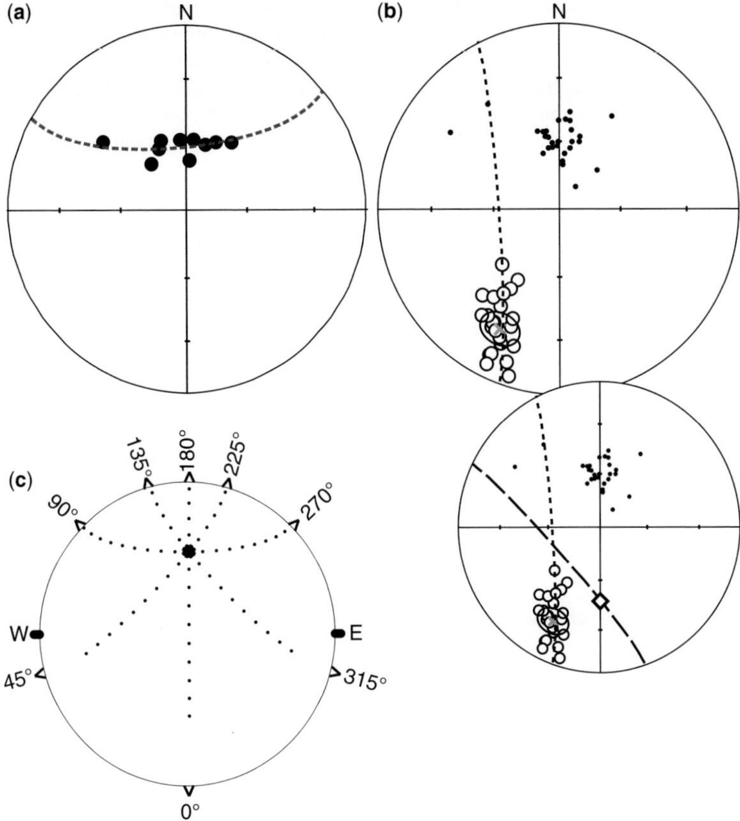



**Fig. 8.** (**a, b**) Examples of small circles fitted through site mean directions (closed/open symbols denote lower/upper hemisphere), which record rotations around sub-horizontal axes during or after remanence acquisition. (a) East–west-directed tilting indicated by site mean directions distributed on an east–west oriented small circle (Larkya south area; Schill *et al.* 2004); the expected Earth magnetic field during remanence acquisition (not shown) lies on the small circle. (b) North–south directed tilting indicated by site mean directions distributed on a north–south-oriented small circle (Shiar Khola area; Schill *et al.* 2002*a*). Small dots represent (upper hemishere) bedding poles. Additionally, a clockwise block rotation of *c.* 30° must have occurred, which can be determined by rotating the small circle around a vertical axis until it intersects the expected (reverse) Earth magnetic field direction (diamond) at the time of remanence acquisition. (**c**) Principle of remanence vector rotation by tilting in different directions (e.g. 0° and 90° denote N and E tilting directions, respectively). The chosen remanence before rotation is D/I = 0°/45° (large full symbol). Remanence directions resulting by tilting are distributed on small circles (small dots; calculated in 10° increments for eight different tilting directions; lower hemisphere projection); these apparent rotations may be misinterpreted as block rotations.

Oroclinal bending and rotational shortening, as well as the influence of a large-scale dextral shear zone (Schill *et al.* 2004), may also influence the resulting remanence directions of pyrrhotite in the eastern part of the Himalaya. However, meso-scale deformation is obviously severely imprinting in this region and often dominates palaeomagnetic remanence directions. The observed rotations may represent apparent rotations (Fig. 8c) due to crustal doming, which could produce tilting in different directions and with different angles depending on the relative sampling position with respect to the dome structure. Miocene North Himalayan doming

structures are a very common tectonic feature in the Tethyan Himalaya, and wherever they appear at the surface as gneiss domes they can be well mapped (Watts *et al.* 2005). We also find crustal doming in the HHC, for example in the Ama Drime Massif (ADM) in the Everest–Makalu region with a huge magnitude of exhumation from a 14–16 kbar level, corresponding to 50–60 km depth (Liu *et al.* 2007; Kali *et al.* 2010; Leloup *et al.* 2010). In this region, the pyrrhotite remanences show a clockwise deviation of declinations to the west of the ADM and an counterclockwise deviation to the east of it. This feature is probably caused by westward and

eastward tilting at the western and eastern flanks of
the ADM, respectively (Crouzet *et al.* in press). In
the Khumbu region south of Everest, the observed
clockwise deviation is also interpreted as an apparent rotation caused by westward tilting at the
western side of the leucogranite body (El Bay
*et al.* 2011). The effect of Late Orogenic doming was
also observed in the easternmost sampling areas of
SE Tibet (Antolin *et al.* 2010), revealed by systematic variation of inclinations along a north–south
direction (Fig. 9).

## Conclusions and future perspectives

More than 300 sites, by far most of them from low-
grade metamorphic carbonates of the TSS, have
delivered significant results using remagnetizations
residing in pyrrhotite. This component has been
acquired during the last metamorphic cooling during
Eocene–Oligocene (in the west) to Miocene (in the
east) times. This type of remanence records net
movements of the sampling units, that is rotations
around vertical and horizontal axes, and provides
far more than only 'second choice' results when

primary remanences are erased. The observed directions can be explained by large-scale tectonic processes as oroclinal bending and rotational crustal
shortening (in the western and central region of
the arc) and meso-scale tectonic processes mostly
related to crustal doming (in the central to eastern
Himalaya).

It would be desirable to fill data gaps along the
arc in order to make interpretations more stable
and to detect further unrecognized tectonic features.
There is no method other than palaeomagnetism which can reveal long-wavelength deformation
structures. In particular, around doming structures
it would be interesting to acquire denser grids of
palaeomagnetic data. Such structures can be related
to channel flow in the crust and exhumation of the
channel (Grujic *et al.* 2002; Jamieson *et al.* 2006).
Doming structures may be widely present in the
crust without cropping out at surface. Palaeomagnetic 'mapping' could reveal the spatial distribution
and shape of such crustal features, and thus could
support modelling of crustal kinematics and
crustal evolution at the southern margin of the
India–Asia collision zone.

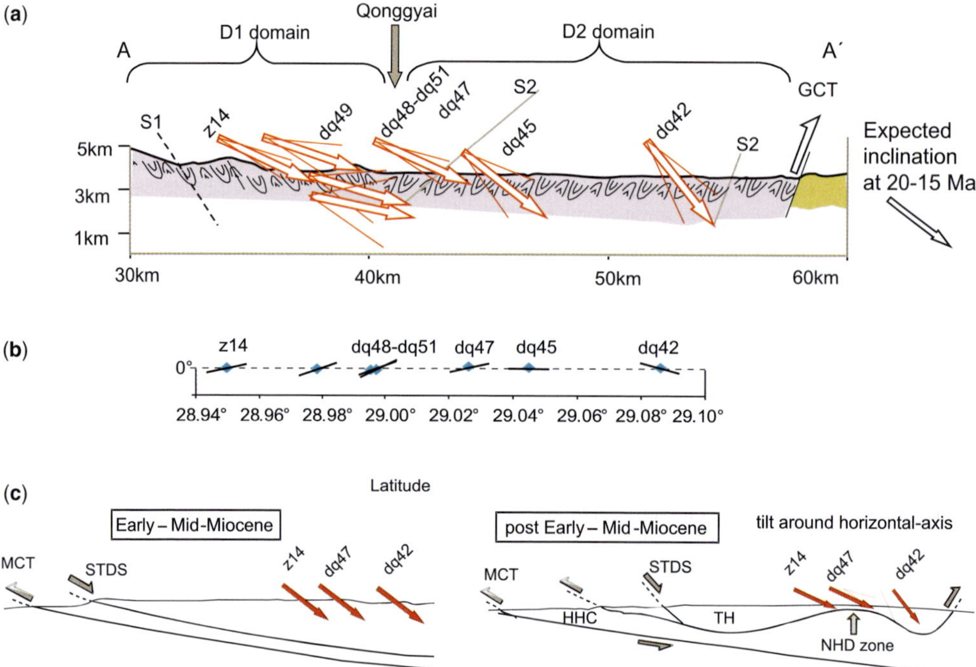

**Fig. 9.** Inclinations of *in situ* pyrrhotite mean directions in dioritic dykes along the Qonggai valley in SE Tibet (modified
after Antolin *et al.* 2010). (**a**) Observed inclinations (expected inclination shown as black arrow at the right side,
averaged from apparent polar wander paths of Acton (1999) and Besse & Courtillot (2002). (**b**) Expected inclination
subtracted from the observed values. (**c**) The systematic north–south trend can be explained by a tilting around a
horizontal axis due to the formation of North Himalayan Domes (NHD). MCT, Main Central Thrust; STDS, South
Tibetan Detachment System; TH, Tethyan Himalaya; HHC, Higher Himalayan Crystalline.

Unfortunately, exclusive use of the rock units of the TSS (which are known as good recorders of secondary pyrrhotite remanences) will only allow a limited distribution of palaeomagnetic working areas. Fortunately, stable secondary pyrrhotite remanences have also been found in the Higher Himalayan Crystalline. The HHC, with its widespread appearance along the Himalayan arc, may therefore represent a promising data repository for future work. It is clear, however, that the success rate from the HHC will be much lower and a large number of disappointing studies must be taken into account.

The reason why the HHC is less favourable for retrieving meaningful secondary pyrrhotite remanences is, on the one hand, related to large and therefore unsuitable multidomain pyrrhotite particles growing at higher temperatures. On the other hand, it may often also be caused by still ongoing ductile deformation at the time of remanence acquisition during the cooling process. The latter is very likely responsible for widely unsuccessful studies of the low-grade metamorphic flysch in the northern zone of the Tethyan Himalaya (unpublished). As mentioned above, the critical temperature for the transition from ductile to brittle deformation depends on the rock type and can vary considerably (from 200 to 400 °C; Carter & Tsenn 1987). It can be assumed that in the Triassic flysch of the TSS the conditions are unfavourable in this respect.

Pyrrhotite remagnetizations in the Himalaya could also be considered for more general research questions, for example, studying exhumation rates by investigation of systematic polarity sequences with altitude. For this purpose, sampling in Hidden Valley (Crouzet *et al.* 2001*b*) could be extended to higher altitudes from the presently top end at *c.* 5500 m to the summit of the Dampush Peak (*c.* 6100 m). Another interesting perspective is the resolution of Earth's magnetic field reversals in faster-cooled meta-carbonates near the contact to intrusions, as demonstrated by Wehland *et al.* (2005*b*) for the case of Elba Island. Pyrrhotite magnetizations can also be taken as a first-order indicator for metamorphic conditions in low-grade metasediments, where other classical methods are inefficient. In Hidden Valley for example, if antipodal components are recorded a TRM origin is clear: that is, $T_{max}$ certainly exceeded 325 °C ($T_c$ of pyrrhotite). If a primary magnetite component is preserved, such as in Dolpo area, $T_{max}$ was lower and likely *c.* 250 °C. Moreover, Schill *et al.* (2002*c*) successfully tested the possibility of using the pyrrhotite/magnetite (P/M) remanence intensity ratio as an indicator for peak metamorphic temperatures up to *c.* 300 °C, comparing the P/M ratio to the classical low-temperature calcite twin lamellae geothermometer.

For analysis of secondary pyrrhotite directions (and not only for these), small-circle analysis methods (Shipunov 1997; Waldhör *et al.* 2001; Henry *et al.* 2004; Waldhör & Appel 2006) often turned out to be helpful and even essential. The work in the Himalaya contributed to the recognition of this. On the other hand, the results from the Himalayan range can be helpful to further develop small-circle methods as a fundamental tool to better understand and analyse palaeomagnetic results from orogenic belts.

We thank the German Research Foundation (DFG) for their financial support of several projects, substantially contributing to the results of this review paper. We thank B. Henry and P. Rochette for helpful comments.

# References

ACTON, G. D. 1999. Apparent polar wander of India since the Cretaceous with implications for regional tectonics and true polar wander. *Memoirs of the Geological Society of India*, **44**, 129–175.

ANTOLIN, B., APPEL, E. *ET AL.* 2010. Paleomagnetic evidence for clockwise rotation and tilting in the eastern Tethyan Himalaya (SE Tibet): implications for the Miocene tectonic evolution of the NE Himalaya. *Tectonophysics*, **493**, 172–186, doi: 10.1016/j.tecto.2010.07.015

APPEL, E., MÜLLER, R. & WIDDER, R. W. 1991. Palaeomagnetic results from the Tibetan Sedimentary Series of the Manang area (north central Nepal). *Geophysical Journal International*, **104**, 255–266.

APPEL, E., PATZELT, A. & CHOUKER, C. 1995. Secondary palaeoremanence of Tethyan sediments from the Zanskar range (NW-Himalaya). *Geophysical Journal International*, **122**, 227–242.

AUBOURG, C. & POZZI, J.-P. 2010. Toward a new <250 °C pyrrhotite – magnetite geothermometer for claystones <250 °C pyrrhotite – magnetite geothermometer for claystones. *Earth and Planetary Science Letters*, **294**, 47–57.

BESSE, J. & COURTILLOT, V. 1991. Revised and synthetic apparent polar wander paths of the African, Eurasian, North American and Indian plates, and true polar wander since 200Ma. *Journal of Geophysical Research*, **96**, 4029–4050.

BESSE, J. & COURTILLOT, V. 2002. Apparent and true polar wander and the geometry of the geomagnetic field over the last 200 Myr. *Journal of Geophysical Research*, **B107**, 1–31.

BESSE, J., COURTILLOT, V., POZZI, J. P., WESTPHAL, M. & ZHOU, Y. 1984. Palaeomagnetic estimates of crustal shortening in the Himalayan thrusts and Zangpo suture. *Nature*, **311**, 621–626.

BEUGNIES, A. 1986. Le métamorphisme de l'aire anticlinale de l'Ardenne. *Hercynica*, **II**, 17–33.

BINA, M. & DALY, L. 1994. Mineralogical change and self-reversal magnetizations in pyrrhotite resulting from partial oxidation; geophysical implications. *Physics of the Earth and Planetary Interiors*, **85**, 83–99.

BINA, M., CORPEL, J., DALY, L. & DEBEGLIA, N. 1991. Transformation de la pyrrhotite sous l'effet de la température: une source d'anomalies magnétiques. *Comptes Rendu Académie des Sciences Paris*, **313**, 487–494.

BONHOMME, M. & GARZANTI, E. 1991. Age of metamorphism in the Zanskar Tethys Himalaya (India). *Géologie Alpine Mém. H.S.*, **16**, 15–16.

CARPENTER, R. H. 1974. Pyrrhotite isograde in the SE Tennessee and SW North Carolina. *Geological Society of America Bulletin*, **85**, 451–456.

CARTER, N. L. & TSENN, C. 1987. Flow properties of continental lithosphere. *Tectonophysics*, **136**, 27–63.

CROUZET, C., MENARD, G. & ROCHETTE, P. 1996. Post-Middle Miocene rotations recorded in the Bourg d'Oisans area (western Alps, France) by paleomagnetism. *Tectonophysics*, **263**, 137–148.

CROUZET, C., ROCHETTE, P. & MÉNARD, G. 2001a. Experimental evaluation of thermal recording of polarity reversals during metasediments uplift. *Geophysical Journal International*, **145**, 771–785.

CROUZET, C., STANG, H., APPEL, E., SCHILL, E. & GAUTAM, P. 2001b. Detailed analysis of successive pTRMs carried by pyrrhotite in Himalayan metacarbonates: an example from Hidden Valley, Central Nepal. *Geophysical Journal International*, **146**, 607–618.

CROUZET, C., GAUTAM, P., SCHILL, E. & APPEL, E. 2003. Multicomponent magnetization in western Dolpo (Tethyan Himalaya, Nepal): tectonic implications. *Tectonophysics*, **377**, 179–196.

CROUZET, C., DUNKL, I., PAUDEL, L., ÁRKAI, P., RAINER, T. M., BALOGH, K. & APPEL, E. 2007. Temperature and age constraints on the metamorphism of Tethyan Himalaya in Central Nepal: a multidisciplinary approach. *Journal of Asian Earth Sciences*, **30**, 113–130.

CROUZET, C., APPEL, E. *ET AL.* in press. Kinematics of the crust around the Ama Drime massif (southern Tibet) – constraints from paleomagnetic results. *Journal of Asian Earth Sciences*, doi: http://dx.doi.org/10.1016/j.jseaes.2012.06.010

D'AGRELLA-FILHO, M. S., BABINSKI, M., TRINDADE, R. I. F., VAN SCHMUS, W. R. & ERNESTO, M. 2000. Simultaneous remagnetization and U–Pb isotope resetting in Neoproterozoic carbonates of the São Francisco craton, Brazil. *Precambrian Research*, **99**, 179–196.

DEKKERS, M. J. 1989. Magnetic properties of natural pyrrhotite. II. High- and low-temperature behaviour of Jrs and TRM as function of grain size. *Physics of the Earth and Planetary Interiors*, **57**, 266–283.

DEKKERS, M. J. 1990. Magnetic monitoring of pyrrhotite alteration during thermal demagnetizations. *Geophysical Research Letters*, **17**, 779–782.

DUNKL, I., ANTOLIN, A. *ET AL.* 2011. Metamorphic evolution of the Tethyan Himalayan flysch in SE Tibet. *In*: GLOAGUEN, R. & RATSCHBACHER, L. (eds) *Growth and Collapse of the Tibetan Plateau*. Geological Society, London, Special Publications, **353**, 45–69, doi: 10.1144/SP353.4

EL BAY, R., APPEL, E., PAUDEL, L., NEUMANN, U. & SETZER, F. 2011. Palaeomagnetic remanences in high-grade metamorphic rocks of the Everest region: indication for Late Miocene crustal doming. *Geophysical Journal International*, **186**, 551–566, doi: 10.111/j.1365–246X.2011.05078.X

ELMORE, R. D., DULIN, S., ENGEL, M. H. & PARNELL, J. 2006. Remagnetization and fluid flow in the Old Red Sandstone along the Great Glen Fault, Scotland. *Journal of Geochemical Exploration*, **89**, 96–99.

FERRY, J. M. 1981. Petrology of graphitic sulfide-rich schists from south-central Maine: an example of desulfidation during prograde regional metamorphism. *American Mineralogist*, **66**, 908–930.

GARZANTI, E. 1999. Stratigraphy and sedimentary history of the Nepal Tethys Himalaya passive margin. *Journal of Asian Earth Sciences*, **17**, 805–827.

GARZANTI, E., GORZA, M., MARTELLINI, L. & NICORA, A. 1994. Transition from diagenesis to metamorphism in the Paleozoic to Mesozoic succession of the Dolpo-Manang synclinorium and Thakkola Graben (Nepal Tethys Himalaya). *Eclogae Geologicae Helvetiae*, **87**, 613–632.

GAUTAM, P. & FUJIWARA, Y. 2000. Magnetic polarity stratigraphy of Siwalik Group sediments of Karnali river section in western Nepal. *Geophysical Journal International*, **142**, 812–824.

GLEESON, T. P. & GODIN, L. 2006. The Chako antiform: a folded segment of the greater himalayan sequence, nar valley, central Nepal Himalaya. *Journal of the Asian Earth Sciences*, **27**, 717–734.

GODIN, L., GLEESON, T. P., SEARLE, M. P., ULLRICH, T. D. & PARRISH, R. R. 2006. Locking of southward extrusion in favour of rapid crustal-scale buckling of the Greater Himalayan sequence, Nar valley, central Nepal. *In*: LAW, R. D., SEARLE, M. P. & GODIN, L. (eds) *Channel Flow, Ductile Extrusion and Exhumation in Continental Collision Zones*. Geological Society, London, Special Publications, **268**, 269–292.

GRUJIC, D., HOLLISTER, L. S. & PARRISH, R. R. 2002. Himalayan metamorphic sequence as an orogenic channel: insight from Bhutan. *Earth and Planetary Science Letters*, **198**, 177–191.

GUILLOT, S., LE FORT, P., PÊCHER, A., ROY BARMAN, M. & APRAHAMIAN, J. 1995. Contact metamorphism and depth of emplacement of the Manaslu granite (central Nepal). Implications for Himalayan orogenesis. *Tectonophysics*, **241**, 99–119.

HALL, A. J. 1982. Gypsum as a precursor to Pyrrhotite in Metamorphic Rocks; Evidence from the Ballachulish slate, Scotland. *Mineral Deposita*, **17**, 401–409.

HALL, A. J. 1986. Pyrite-pyrrhotite redox reactions in nature. *Mineralogical Magazine*, **50**, 223–229.

HENRY, B., ROUVIER, H. & LE GOFF, M. 2004. Using syn-tectonic remagnetizations for fold geometry and vertical axis rotation: the example of the Cévennes border (France). *Geophysical Journal International*, **157**, 1061–1070.

HORNG, C. S. & ROBERTS, A. P. 2006. Authigenic or detrital origin of pyrrhotite in sediments? Resolving a paleomagnetic conundrum. *Earth and Planetary Science Letters*, **241**, 750–762.

JAMIESON, R. A., BEAUMONT, C., NGUYEN, M. H. & GRUJIC, D. 2006. Provenance of the Greater Himalayan Sequence and associated rocks: predictions of channel flow models. *In*: LAW, R. D., SEARLE, M. & GODIN, L. (eds) *Channel Flow, Ductile Extrusion and Exhumation of Lower-mid Crust in Continental Collision Zones*. Geological Society, London, Special Publications, **268**, 165–182.

KALI, E., LELOUP, P. H. *ET AL.* 2010. Exhumation history of the deepest central Himalayan rocks (Ama Drime range): key P–T–D–t constraints on orogenic models. *Tectonics*, **292**, 1–16.

KELLETT, D. A. & GODIN, L. 2009. Pre-Miocene deformation of the Himalayan superstructure, Hidden valley, central Nepal. *Journal of the Geological Society, London*, **166**, 261–275, doi: 10.1144/0016–76492008–097

KIM, W., DOH, S.-J., YU, Y., LEE, J. J. & SUK, D. 2009. Hydrothermal fluid controlled remagnetization of sedimentary rocks in Korea: Tectonic importance of pervasive Tertiary remagnetization. *Tectonophysics*, **474**, 684–695.

KLOOTWIJK, C. T. & BINGHAM, D. K. 1980. The extent of Greater India, III. Palaeomagnetic data from the Tibetan Sedimentary Series, Thakkola region, Nepal Himalaya. *Earth and Planetary Science Letters*, **51**, 381–405.

KLOOTWIJK, C. T. & RADHAKRISHNAMURTY, C. 1981. Phanerozoic palaeomagnetism of the Indian plate and the India-Asia collision. *In*: MCELHINNY, M. W. & VALENCIO, D. A. (eds) Paleoreconstruction of the Continents. *Geodynamic Series*, **2**. AGU, Washington, DC, 93–105, doi: 10.1029/GD002p0093

KLOOTWIJK, C. T., JAIN, A. K. & RAKESH, K. 1982. Palaeomagnetic constraints on allochthony and age of the Krol Belt Sequence, Garhwal Himalaya, India. *Journal of Geophysics*, **50**, 127–136.

KLOOTWIJK, C. T., SHAH, S. K., GERGAN, J., SHARMA, M. L., TIRKEY, B. & GUPTA, B. K. 1983. A palaeomagnetic reconnaissance of Kashmir, northwestern Himalaya, India. *Earth and Planetary Science Letters*, **63**, 305–324.

KLOOTWIJK, C. T., CONAGHAN, P. J. & POWELL, C. MCA. 1985. The Himalayan Arc: Large scale continental subduction, oroclinal bending and back-arc spreading. *Earth and Planetary Science Letters*, **75**, 167–183.

KLOOTWIJK, C. T., GEE, J. S., PEIRCE, J. W. & SMITH, G. M. 1991. Constraints on the India-Asia convergence: paleomagnetic results from Ninetyeast Ridge. *In*: WEISSEL, J., PEIRCE, J., TAYLOR, E. & ALT, J. (eds) *Proceedings ODP Scientific Results*, **121**. College Station, TX (Ocean Drilling Program), 777–881, doi: 10.2973/odp.proc.sr.121.121.1991

LAMBERT, I. B. 1973. Post-depositional avaibility of sulphur and metals and formation of secondary textures and structures in stratiform sedimentary sulphide deposits. *Journal of the Geological Society of Australia*, **20**, 205–215.

LELOUP, P. H., MAHEO, G. *ET AL.* 2010. The South Tibet detachment shear zone in the Dinggye area: time constraints on extrusion models of the Himalayas. *Earth and Planetary Science Letters*, **292**, 1–16, doi: 10.1016/j.epsl.2009.12.035

LIU, Y., SIEBEL, W., MASSONNE, H. J. & XIAO, X. C. 2007. Geochronological and petrological constraints for tectonic evolution of the central Greater Himalayan Sequence in the Kharta area, southern Tibet. *Journal of Geology*, **115**, 215–230, doi: 10.1086/510806

MCELHINNY, M. W. 1964. Statistical significance of fold test in paleomagnetism. *Geophysical Journal of the Royal Astronomical Society*, **8**, 338–340.

MCFADDEN, P. L. 1990. A new fold test for palaeomagnetic studies. *Geophysical Journal International*, **103**, 163–169.

MÉNARD, G. & ROCHETTE, P. 1992. Utilisation de réaimantation postmétamorphique pour une étude de l'évolution tectonique et thermique tardive dans les Alpes occidentales (France). *Bulletin de la Société Géologique de France*, **163**, 381–392.

OLIVER, G. J. H., JOHNSON, M. R. W. & FALLICK, A. E. 1995. Age of metamorphism in the Lesser Himalaya and the Main Central Thrust zone, Garhwal India: results of illite crystallinity, 40Ar-39Ar fusion and K–Ar studies. *Geological Magazine*, **132**, 139–149.

PATZELT, A., LI, H. M., WANG, J. D. & APPEL, A. 1996. Palaeomagnetism of Cretaceous to Tertiary sediments from southern Tibet: evidence for the extent of the northern margin of India prior to the collision with Eurasia. *Tectonophysics*, **259**, 259–284.

POULSON, S. R. & OHMOTO, H. 1989. Devolatilization equilibria in graphite-pyrite-pyrhotite bearing pelites with application to magma-pelite interaction. *Contributions to Mineralogy and Petrology*, **101**, 418–425.

ROBION, P., FRIZON DE LAMOTTE, D., KISSEL, C. & AUBOURG, C. 1995. Tectonic v. mineralogical fabrics of epimetamorphic slaty rocks: an example from the Ardennes Massif (France-Belgium). *Journal of Structural Geology*, **17/8**, 1111–1124.

ROBION, P., KISSEL, C., FRIZON DE LAMOTTE, D., LORAND, J. P. & GUÉZOU, J. C. 1997. Magnetic mineralogy and metamorphic zonation in the Ardennes Massif (France - Belgium). *Tectonophysics*, **271**, 231–248.

ROCHETTE, P. 1987. Metamorphic control of the magnetic mineralogy of black shales in the Swiss Alps: toward the use of magnetic isogrades. *Earth and Planetary Science Letters*, **84**, 446–456.

ROCHETTE, P. & LAMARCHE, G. 1986. Evolution des propriétés magnétiques lors des transformations minérales dans les roches: exemple du Jurassique Dauphinois (Alpes françaises). *Bulletin de Minéralogie*, **109**, 687–696.

ROCHETTE, R., FILLION, G., MATTÉI, J. L. & DEKKERS, M. J. 1990. Magnetic transition at 30–34 Kelvin in pyrrhotite: insight into a widespread occurrence of this mineral in rocks. *Earth and Planetary Science Letters*, **98**, 319–328.

ROCHETTE, P., MÉNARD, G. & DUNN, R. 1992. Thermochronometry and cooling rates deduced from single sample records of successive magnetic polarities during uplift of metamorphic rocks in the Alps (France). *Geophysical Journal International*, **108**, 491–501.

ROCHETTE, P., SCAILLET, B., GUILLOT, S., LEFORT, P. & PÊCHER, A. 1994. Magnetic properties of the High Himalayan leucogranites: structural implications. *Earth and Planetary Science Letters*, **126**, 214–234.

SCHILL, E., APPEL, E., GAUTAM, P. & SINGH, V. K. 1998. Preliminary palaeomagnetic results from medium grade metacarbonates of the Lesser Himalaya. *Journal of Nepal Geological Society*, **18**, 205–215.

SCHILL, E., APPEL, E., ZEH, O., SINGH, V. K. & GAUTAM, P. 2001. Coupling of late-orogenic tectonics and secondary pyrrhotite remanences: towards a separation of different rotation processes and quantification of rotational underthrusting in the western Himalaya (N-India). *Tectonophysics*, **337**, 1–21.

SCHILL, E., APPEL, E. & GAUTAM, P. 2002a. Thermotectonic history of the Tethyan Himalayas deduced from palaeomagnetic record of metacarbonates from

Central Nepal (Shiar Khola). *Journal of Asian Earth Sciences*, **20**, 203–210.

SCHILL, E., CROUZET, C., GAUTAM, P., SINGH, V. K. & APPEL, E. 2002*b*. Where did rotational shortening occur in the Himalaya? – Inferences from palaeomagnetic remagnetisations. *Earth and Planetary Science Letters*, **203**, 45–57.

SCHILL, E., APPEL, E. & GAUTAM, P. 2002*c*. Towards pyrrhotite/magnetite geothermometry in low-grade metamorphic carbonates of the Tethyan Himalayas (Shiar Khola, Central Nepal). *Journal of Asian Earth Sciences*, **20**, 195–201.

SCHILL, E., APPEL, E., GODIN, L., CROUZET, C., GAUTAM, P. & REGMI, K. R. 2003. Record of deformation by secondary magnetic remanences and magnetic anisotropy in the Nar/Phu valley (central Himalaya). *Tectonophysics*, **377**, 197–209.

SCHILL, E., APPEL, E., CROUZET, C., GAUTAM, P., WEHLAND, F. & STAIGER, M. 2004. Oroclinal bending and regional significant clockwise rotations of the Himalayan arc – constraints from secondary pyrrhotite remanences. *In*: SUSSMAN, A. J. & WEIL, A. B. (eds) *Orogenic Curvature: Integrating Paleomagnetic and Structural Analysis*. Geological Society of America, Boulder, Special Paper, **383**, 73–85.

SCHNEIDER, C. & MASCH, L. 1993. The metamorphism of the Tibetan Series from the Manang area Marsyangdi Valley, Central Nepal. *In*: TRELOAR, P. J. & SEARLE, M. P. (eds) *Himalayan Tectonics*. Geological Society, London, Special Publications, **74**, 357–374.

SHIPUNOV, S. V. 1997. Synfolding magnetization: detection, testing and geological applications. *Geophysical Journal International*, **130**, 405–410.

SOFFEL, H. C. 1977. Pseudo-single-domain effects and single-domain multidomain transition in natural pyrrhotite deduced from domain structure observations. *Journal of Geophysics*, **42**, 351–359.

SOFFEL, H. C. 1981. Domain structure of natural fine-grained pyrrhotite in a rock matrix (diabase). *Physics of the Earth and Planetary Interiors*, **26**, 98–106.

TAUXE, L. & OPDYKE, N. D. 1982. A time framework based on magnetostratigraphy for Siwalik sediments of the Khaur area, Northern Pakistan. *Paleogeography. Paleoclimatology, Paleoecology*, **37**, 43–61.

TORSVIK, T. H., PAULSEN, T. S., HUGHES, N. C., MYROW, P. M. & GANERODI, M. 2009. The Tethyan Himalaya: palaeogeographical and tectonic constraints from Ordovician palaeomagnetic data. *Journal of the Geological Society, London*, **166**, 679–687, doi: 10.1144/0016-76492008-123

TRACY, R. J. & ROBINSON, P. 1988. Silicate-sulfide-oxide-fluid reactions in granulite grade pelitic rocks. *Central Massachusetts. American Journal of Science*, **288**, 45–74.

WALDHÖR, M. & APPEL, E. 2006. Intersections of remanence small circles: new tools to improve data processing and interpretation in palaeomagnetism. *Geophysical Journal International*, **166**, 33–45.

WALDHÖR, M., APPEL, E., FRISCH, W. & PATZELT, A. 2001. Palaeomagnetic investigation in the Pamirs and its tectonic implications. *Journal of Asian Earth Sciences*, **19**, 429–451.

WANG, L., PAN, Y. X., LI, J. H. & QIN, H. F. 2008. Magnetic properties related to thermal treatment of pyrite. *Science in China Series D*, **51**, 1144–1153.

WATTS, D. R., HARRIS, N. B. W. & THE 2002 NASA GLENN SOARS WORKING GROUP 2005. Mapping granite and gneiss in domes along the North Himalayan antiform with ASTER SWIR band ratios. *The Geological Society of America Bulletin*, **117**, 879–886, doi: 10.1130/B25592.1

WEHLAND, F., STANCU, A., ROCHETTE, P., DEKKERS, M. & APPEL, E. 2005*a*. Experimental evaluation of magnetic interaction in pyrrhotite bearing samples. *Physics of the Earth and Planetary Interiors*, **153**, 181–190.

WEHLAND, F., ALT-EPPING, U., BRAUN, S. & APPEL, E. 2005*b*. Quality of pTRM acquisition in pyrrhotite bearing contact-metamorphic limestones: possibility of a continuous record of Earth magnetic field variations. *Physics of the Earth and Planetary Interiors*, **148**, 157–173.

WEHLAND, F., EIBL, O., KOTTHOFF, S., ALT-EPPING, U. & APPEL, E. 2005*c*. Pyrrhotite pTRM acquisition in metamorphic limestones in the light of microscopic observations. *Physics of the Earth and Planetary Interiors*, **151**, 107–114.

WIESMAYR, G. & GRASEMANN, B. 1999. Balanced cross-section and depth to detachment calculations for the Tethyan Himalaya (Spiti, N India): Where is the crystalline basement of the Higher Himalaya? *Terra Nostra*, **2**, 172–173.

WOODS, S. D., ELMORE, R. D. & ENGEL, M. H. 2000. The occurrence of pervasive chemical remanent magnetizations in sedimentary basins: implications for dating burial diagenetic events. *Journal of Geochemical Exploration*, **69–70**, 381–385.

YIN, A. 2006. Cenozoic evolution of the Himalayan orogen as constrained by along-strike variations of structural geometry, exhumation history, and foreland sedimentation. *Earth Science Reviews*, **76**, 1–134.

# Burial, claystones remagnetization and some consequences for magnetostratigraphy

CHARLES AUBOURG[1]*, JEAN-PIERRE POZZI[2] & MYRIAM KARS[1,2]

[1]*Laboratoire des Fluides Complexes et leurs Réservoirs,
Université de Pau, CNRS (UMR5150), 64000 Pau, France*

[2]*Laboratoire de Géologie, Ecole Normale Supérieure, CNRS (UMR 8538).
24 rue Lhomond, 75231 Paris cedex 05, France*

**Corresponding author (e-mail: charles.aubourg@univ-pau.fr)*


**Abstract:** We investigate the broad lines of magnetic mineral formation for non-metamorphic claystones. More particularly, we focus on the formation of magnetite from *c.* 20 day-experiments aiming to reproduce burial conditions. It is shown that the sole action of temperature from 50 °C to 250 °C leads to the formation of magnetite. The neoformed magnetites carry a chemical remanent magnetization which equates at least the natural remanent magnetization. We propose a schematic of burial with three magnetic windows where a chemical remanent magnetization superimposed the natural remanent magnetization. These are the greigite window (subsurface), the magnetite window (depth > 2 km) and the pyrrhotite window (depth > 6 km). The formation of magnetic minerals has profound consequences for the magnetostratigraphy record. We propose a conceptual model that shows that the continuous production of magnetite during burial may result in magnetozones that have no relation to the age of the sediment.

Widespread remagnetization is a common process in sedimentary rocks (Kligfield & Channell 1981; McCabe & Elmore 1989; Lu *et al.* 1990; Morris & Robertson 1993; Plattzman 1994). We focus here on the relationship between remagnetization and burial in claystones. The claystones are common targets for palaeomagnetic applications, and especially magnetostratigraphy. In addition, the claystones are potentially source rocks for oil and gas for burial conditions ranging *c.* 2–10 km in depth. The elucidation of remagnetization processes during burial in claystones may therefore have importance for academic and industrial applications.

During the early digenesis, many studies report on the dissolution of detrital iron oxides in the first metre of the sediments and the formation of iron sulphides such as greigite ($Fe_3S_4$) (Roberts & Turner 1993). The greigite forms within anoxic sedimentary environments that support active bacterial sulphate reduction. Formation of iron sulphides follows in a specific order: mackinawite ($FeS_{0.9}$), greigite or pyrrhotite ($Fe_7S_9$, $Fe_9S_{10}$) and finally pyrite ($FeS_2$) (e.g. Berner 1984). This process has some critical importance for dating sediments using magnetostratigraphy (Rowan & Roberts 2005). The prerequisite of magnetostratigraphy is that natural remanent magnetization (NRM) is properly isolated and is contemporaneous with deposit. The combined effect of iron oxide dissolution and greigite formation implies a magnetic

reset of the primary NRM. In addition, an early burial remagnetization carried by greigite delays the magnetostratigraphy record by tens–hundreds of thousands of years (Rowan *et al.* 2009). Larrasoana *et al.* (2007) demonstrated that the offset of magnetostratigraphy is possibly larger, from a few (<40) thousand years to a few (<3) million years. This early burial remagnetization is in essence a chemical remanent magnetization (CRM).

Jumping now to a depth >6 km, approaching 200 °C (with a geothermal gradient of 30 °C/km used throughout the text), several studies report on formation of pyrrhotite ($Fe_7S_8$) at the expense of magnetite ($Fe_3O_4$) and pyrite (Rochette & Lamarche 1986; Crouzet *et al.* 2001; Schill *et al.* 2002). More precisely, Rochette (1987) determined two isogrades based on the breakdown of magnetite (*c.* 250°C) and the breakdown of pyrite (>320 °C) into monoclinic pyrrhotite. For burial temperature >300 °C, the neoformed pyrrhotite completely replaces the magnetite, leading to a remagnetization which consists of an assemblage of CRM and a thermo remanent magnetization (TRM). The previous magnetic record, acquired at a temperature lower than 250 °C, is then completely removed. At this depth (*c.* 10 km), there is no longer primary magnetization contemporaneous with the deposit of the sediments.

Between these two end-members, limited by the formation of greigite within the first tens of metres and the formation of high-temperature pyrrhotite

*From*: ELMORE, R. D., MUXWORTHY, A. R., ALDANA, M. M. & MENA, M. (eds) 2012. *Remagnetization and Chemical Alteration of Sedimentary Rocks*. Geological Society, London, Special Publications, **371**, 181–188.
First published online June 26, 2012, http://dx.doi.org/10.1144/SP371.4

for depth >6 km, there should be another magnetic horizon that we focus on. We will use results from laboratory heating (50–250 °C) to better understand the nature of neoformed magnetic minerals.

## Experimental evidence for the formation of magnetite

For depth >2 km, widespread evidence points to the formation of magnetite. Illite-to-smectite transformation and alteration of pyrite are surely the most evoked processes to produce magnetite. Lu *et al.* (1990) first proposed that the Fe-smectite to illite transformation during burial leads to the formation of magnetite. Since this work, many authors envisaged a link between authigenic magnetite, smectite to illite transformation and remagnetization at depth >2 km (Banerjee *et al.* 1997; Katz *et al.* 2000; Tohver *et al.* 2008). In addition, Brothers *et al.* (1996) showed experimentally the development of a magnetite rim on pyrite at 90 °C in association with organic matter. It is common to observe an oxidation rim around pyrite grains (Suk *et al.* 1990).

Although the exact mechanism of formation of magnetite is not fully understood (Elmore *et al.* 2001), experimental heating studies in claystones aiming to reproduce burial confirmed that magnetite is produced on the sole action of heating. Cairanne *et al.* (2004), Moreau *et al.* (2005) and Aubourg *et al.* (2008) imparted a laboratory CRM at 250 °C, 150 °C and 95 °C, respectively. In these experiments, the production of magnetic minerals is monitored through the record of remanence acquired within an imposed magnetic field <5 mT. This remanence is by essence a CRM and a thermo viscous remanent magnetization (TVRM). The CRM is carried by neoformed grains with size larger than the blocking volume. The TVRM is essentially carried by former and neoformed magnetic grains below the blocking volume. The thermal demagnetization of CRM provides an unblocking temperature spectrum. The maximum unblocking temperature is then a good indication of the nature of neoformed magnetic minerals. For all experiments described here, the maximum unblocking temperature ranked between 400 and 580 °C (which is consistent with magnetite). After heating for 60 h at 250 °C in an argon atmosphere, Cairanne *et al.* (2004) reported the production of magnetite and some hematite. At 150 °C, Moreau *et al.* (2005) documented the production of magnetite after heating for at least one week in an argon atmosphere. In these two experiments, the claystones came from the same stratigraphic horizon from the Basin of Paris (Toarcian). These claystones are characterized by low NRM ($<0.5$ $\mu$A m$^2$ kg$^{-1}$). At the end of the heating experiment, Moreau *et al.* (2005) reported the production

of a CRM that is on average $9 \pm 14$ $\mu$A m$^2$ kg$^{-1}$ at 150 °C for 7 days. Cairanne *et al.* (2004) reported a CRM of *c.* 8 $\mu$A m$^2$ kg$^{-1}$ at 250 °C for 60 h. Assuming that CRM is proportional to the strength of the magnetic field (Stokking & Tauxe 1990) we calculate the expected $\mathrm{CRM_{EMF}}$ obtained at 50 $\mu$T, the value of the Earth's magnetic field (EMF). We obtain $\mathrm{CRM_{EMF}} = 0.4$ $\mu$A m$^2$ kg$^{-1}$ at 250 °C for 60 h and $\mathrm{CRM_{EMF}} = 4.5$ $\mu$A m$^2$ kg$^{-1}$ at 150 °C for 7 days. The $\mathrm{CRM_{EMF}}$ obtained at 150 °C and 250 °C are not directly comparable because the heating time is different. We notice however that $\mathrm{CRM_{EMF}} \geq$ NRM.

At 95 °C, Aubourg *et al.* (2008) and Aubourg & Pozzi (2010) reported the production of magnetite and some fine-grained pyrrhotite during heating in an open environment for 15 days or more. Here, we present a new set of results at 95 °C using the same claystones of these two studies.

## New results for laboratory heating at 95 °C

Two Bure and Opalinus claystones are studied. The Bure claystones (Callovo–Oxfordian) are from the borehole EST211 at 675 m depth from the Basin of Paris (ANDRA 2000). They experienced a peak burial temperature near 40 °C (Landais & Elie 1999). The Opalinus claystones (Aalenian) are from the Jura Fold Belt (Mont Terri laboratory, core BHE-D5). Opalinus claystones experienced a burial temperature near 85 °C (Mazurek *et al.* 2006). The NRM of Bure and Opalinus claystones are low ($<0.5$ $\mu$A m$^2$ kg$^{-1}$). We operated laboratory heating as described in Aubourg *et al.* (2008). Claystone fragments (few grams) are placed in a nylon holder where fresh air is slowly renewed. We used a homemade furnace, shielded against the Earth's magnetic field. The heating rate is about 2 °C min$^{-1}$, and the target temperature is reached within 20 min. During heating at 95 °C, the claystones are submitted to a vertical magnetic field of 2 mT. The remanence is monitored every few days for several weeks. To measure the remanence, we first cool down the sample to room temperature in a null magnetic field (residual field $<100$ nT). In doing this, we do not impart a thermo remanent magnetization in the sample. The sample is then measured using a Squid magnetometer 2G. Note that the sample is not removed from the holder during the measurement. The natural remanent magnetization of the samples is $<0.1$ $\mu$A m$^2$ kg$^{-1}$ after alternating field (AF) demagnetization at 100 mT. The isothermal remanent magnetization (IRM) acquired at 2 mT is $<0.5$ $\mu$A m$^2$ kg$^{-1}$. The residual NRM and $\mathrm{IRM_{2mT}}$ are negligible. The remanence is essentially the sum of a CRM and a TVRM. To remove the TVRM, the sample was left at temperature for at least 2 days within a null magnetic field.

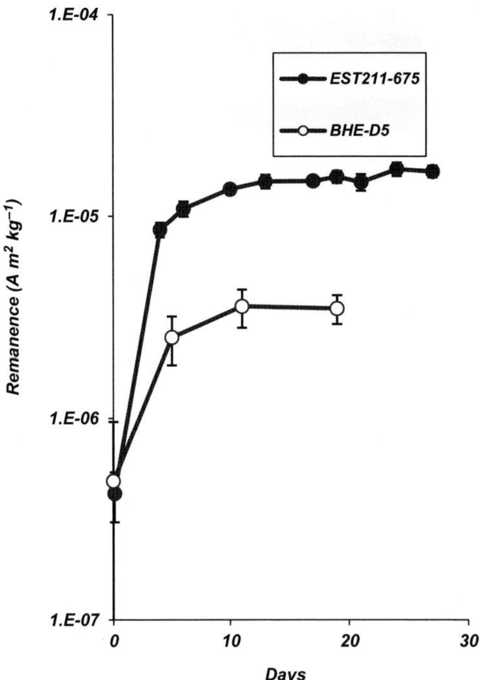


**Fig. 1.** Laboratory heating experiments at 95 °C. Bure claystones (EST-211–675 m): four samples. Opalinus claystones (BHE-D5): five samples. Error bars at 95% confidence.

Standard curves are shown in Figure 1, comparable to the curves obtained by Aubourg *et al.* (2008). Opalinus claystones (five samples) and Bure claystones (four samples) show the development of a plateau after *c.* 10 days of heating. The CRM is $5 \pm 2$ $\mu$A m$^2$ kg$^{-1}$ at 19 days for Opalinus claystones and $35 \pm 6$ $\mu$A m$^2$ kg$^{-1}$ at 27 days for Bure claystones. The TVRM represents about half of the remanence. We calculate CRM$_{\text{EMF}} = 0.1$ $\mu$A m$^2$ kg$^{-1}$ for Opalinus claystones and CRM$_{\text{EMF}} = 0.9$ $\mu$A m$^2$ kg$^{-1}$ for Bure claystones. We notice that CRM$_{\text{EMF}}$ equates NRM in both claystones.

## Toward a burial model

We define three magnetic windows, each characterized by a CRM related to the formation of greigite, magnetite and pyrrhotite (Fig. 2). This model is simplified because additional magnetic minerals can be formed under specific conditions, such as pyrrhotite (Gillett 2003; Wehland *et al.* 2005; Larrasoana *et al.* 2007; Aubourg & Pozzi 2010) or hematite and goethite (Evans & Elmore 2006). The participation of hematite and goethite can be neglected as magnetite, greigite and pyrrhotite have a stronger remanence by 1:100 (Maher & Thompson 1999).

Greigite, magnetite and pyrrhotite have distinct magnetic properties making the CRM distinguishable. Their Curie temperatures ($T_c$) are *c.* 340 °C, 585 °C and 320 °C, respectively (Hunt *et al.* 1995). The unblocking temperature ($T_{\text{ub}}$) spectrum of the CRM is distinct, as $T_{\text{ub}} \leq T_c$. A CRM carried by magnetite has unblocking temperature ranging between room temperature and 580 °C. A CRM carried by either greigite or pyrrhotite has an unblocking temperature between room temperature and *c.* 340 °C. The unblocking temperature spectrum of the CRM may therefore be a burial indicator. In our conceptual model represented in Figure 2, we plot the CRM unblocking spectra of neoformed magnetic minerals according to our proposed magnetic windows. When temperature increases, a thermo viscous remanent magnetization (TVRM) partially overlaps the CRM. The unblocking temperature of TVRM is larger than the burial temperature, from a few degrees Centigrade up to 200 °C depending on the time exposure to the temperature (Pullaiah *et al.* 1975; Dunlop & Özdemir 1997). We plot the TVRM to show its correspondence with the CRM. In our model, we follow the equation of Pullaiah *et al.* (1975) for magnetite at a given temperature to illustrate the unblocking spectra of the TVRM.

The greigite forms from the first centimetres to tens of metres, implying the delay of magnetostratigraphy (Rowan *et al.* 2009), the record of inconsistent polarities (Horng *et al.* 1998) or pervasive remagnetization several millions years after sediment deposition (Rowan & Roberts 2006). Within the greigite window, the CRM overlaps the primary magnetization contemporaneous with the deposit of the sediments. Depending on what proportion of the dissolution of iron oxides is driven by bacterial activity, there is the potential to erase the primary magnetization (Rowan & Roberts 2006). However, the original magnetostratigraphy can be preserved if some iron oxides (generally $>10$ $\mu$m) survive to dissolution (Roberts *et al.* 2010).

The boundary between the greigite and the magnetite windows is diffused, but it is well known that greigite is unstable for temperatures $>200$ °C (see Dekkers *et al.* 2000). It is generally assumed that the formation of magnetite starts for burial $>2$ km (Suk *et al.* 1993). Recent heating experiments (Kars *et al.* 2010, 2011) show that magnetite formed at temperatures as low as 50 °C. Taking this into consideration, we propose that the magnetite window starts for depth $>2$ km. We suggest that the CRM carried by greigite is disappearing progressively. For depth $>7$ km, the CRM carried by greigite may be no longer present (Fig. 2). The heating experiments on claystones suggest that the magnetite concentration is increasing with burial temperature. Considering comparable experiments at 95 °C (this study) and 150 °C

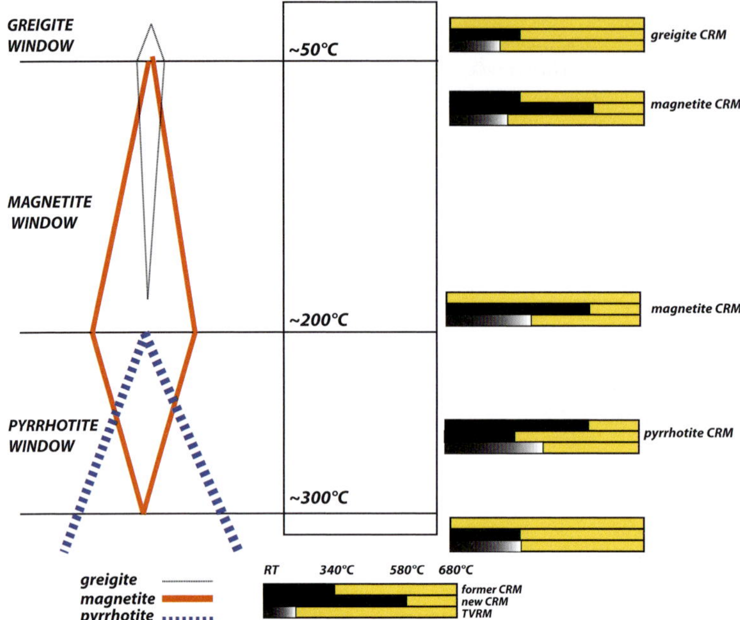


**Fig. 2.** Burial model for the formation of magnetic mineral in claystones. The central box represents a sedimentary column from the surface to 10 km depth. On the left, the concentration of neoformed magnetic mineral is indicated. On the right, the unblocking temperature spectra of chemical remanent magnetization (CRM) and thermo viscous remanent magnetization (TVRM) are indicated.

(Moreau *et al.* 2005), we observe that $CRM_{EMF}$ (150 °C) > $CRM_{EMF}$ (95 °C). Kars *et al.* (2011) observed the same trend from 50 °C to 130 °C using Bure claystones. This leads us to propose that the magnetite forms continuously from depth >2 km. Figure 2 illustrates the continuous production of magnetite, making a 'tie' shape (Fig. 2). The CRM within the magnetite window can be carried by greigite and magnetite. Because those minerals form at different depths, it is possible that CRM reveals distinct polarities in the course of thermal demagnetization.

The pyrrhotite window starts at *c.* 200 °C for depth >6 km. The pyrrhotite forms at the expense of magnetite and, for temperature >300 °C, it is generally accepted that only neoformed pyrrhotite carries the NRM (Rochette 1987). The progressive replacement of remanence carried by magnetite and pyrrhotite has been well observed by Schill *et al.* (2002). For burial temperature >320 °C, which is above the Curie temperature of pyrrhotite, the remanence is essentially a TRM that may record the successive change of magnetic polarities during cooling (Crouzet *et al.* 2001). Unlike magnetostratigraphy, where the oldest polarity is found in the oldest part of the sediment, the youngest polarity may be found at the lowest part of a metamorphic pile.

## The magnetostratigraphy in the magnetite window

Here we would like to focus on the consequence of magnetostratigraphy within the magnetite window as we suggest a continuous production of magnetite through burial >2 km. As a hypothesis, we envisage the case where the CRM overwhelms the NRM and the subsequent consequences for the magnetostratigraphy record. We make one additional assumption about magnetite formation: we assume that during grain growth through the critical single-domain blocking volume, the magnetic minerals record the magnetic polarity of the ambient field and that this record is retained even after subsequent mineral growth. The resulting polarity is then the sum of all polarities recorded by the neoformed grains (Crouzet *et al.* 2001). Cairanne (2003) and Cairanne *et al.* (2004) demonstrated this additive behaviour of magnetic polarities in a laboratory-induced CRM carried by magnetite.

We present two scenarios to illustrate the evolution of an initial magnetostratigraphy record through burial (Fig. 3). The initial sedimentary column has a three-part polarity zonation, with a normal polarity zone in the middle of the column.

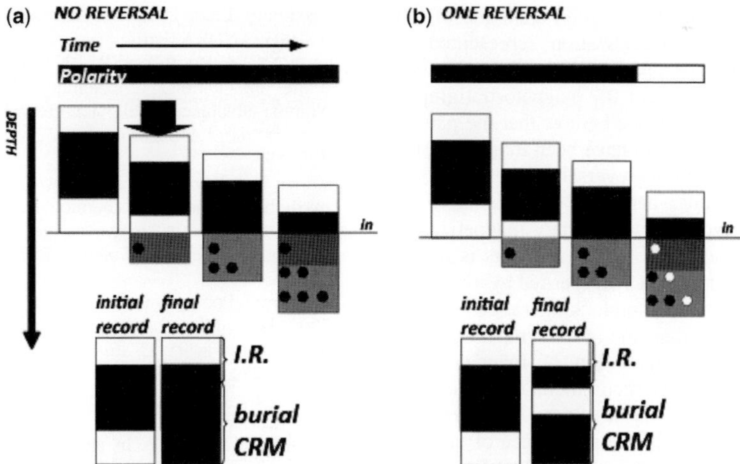


**Fig. 3.** Conceptual model of burial CRM. Horizontal scale is time. Vertical scale is depth. Line 'in' corresponds to the horizon where magnetite forms. Grey zone corresponds to the buried part of the sediment column below the line 'in'. Black (open) circles indicate a normal (reverse) polarity burial CRM. In the final record, we indicate the initial record of magnetostratigraphy (I.R.) and the burial CRM. (**a**) Scenario of mineral formation with one polarity. (**b**) Same scenario as A, but with a magnetic reversal occurring during the burial of the sediments. New magnetic grains record the reversed polarity magnetic field, while magnetic grains formed before the magnetic reversal retain a normal polarity magnetization.

In the first scenario, the magnetic field has the same polarity throughout the simulated burial period. During burial, the reversed-polarity magnetozone located in the lower part of the hypothesized sediment column is progressively remagnetized (Fig. 3a). At the end of the simulation, this magnetozone is completely remagnetized with normal polarity. The upper part of the stratigraphic column has preserved the initial polarity record. The final record is therefore a composite with an initially preserved polarity record and a later burial CRM. Misinterpretation of the polarity zonation will occur if the burial CRM carried by magnetite is not detected.

In a second scenario, we impose a magnetic-reversal sequence during the burial period (Fig. 2b). According to our second assumption above, the neoformed grains record the ambient magnetic polarity when they grow through their blocking volume. The resulting burial CRM at the end of the simulation is a sequence of normal and reversed polarities that has no correspondence with the initial record. Such a record of reversals during burial can be misinterpreted with respect to the primary record of magnetic reversals. Generally, the recognition of a sequence of magnetic reversals in a sedimentary record is considered as good evidence of a primary magnetization, particularly if a reversal test is satisfied. Our model contradicts this assumption because the continuous production of magnetic minerals during burial will disrupt the record of magnetic reversals.

The experimental data showed that CRM imparted during laboratory heating can be about the same order of NRM. It remains to be demonstrated whether burial conditions for depth >2 km trigger any substantial CRM in claystones. In this regard, it is interesting to look at the palaeomagnetic record of claystones that underwent burial >2 km. The Jurassic claystones from the western French Sub-Alpine chains represent the >4 km buried counterpart of Bure and Opalinus claystones. Along the eastern boarder of the Chartreuse massif, the Jurassic claystones experienced a maximum burial of *c.* 4 km (Deville & Sassi 2006). Aubourg & Rochette (1992) proposed that these claystones were fully remagnetized before the Alpine Miocene folding. This remagnetization overlaps the primary magnetization and is characterized by an absence of reverse polarity. Southwards of this area within the Vocontian trough, the burial of Jurassic claystones exceeded 5 km (Guilhaumou *et al.* 1996). There, Katz *et al.* (2000) documented a pervasive pre-Eocene folding remagnetization that they linked to peak burial during the long chrone of normal polarity 24 (83–124 Ma). Similarly to the remagnetization observed in the Chartreuse massif, the remagnetization within the Vocontian trough overlaps the primary magnetization and is characterized by an absence of reverse polarity. Both examples therefore suggest that CRM took place before Alpine folding, and are likely related to a >4 km burial that lasted essentially during the long chrone 24. These burial

CRMs probably constitute a good analogue of our burial model of remagnetization represented in Figure 3a. The lack of reverse polarity in the Jurassic claystones clearly rules out any magnetostratigraphy interpretation. However, we believe that the pattern of magnetic reversals may have been more complex if burial took place during reversal chrones. Within a sedimentary column, where various rocks alternate, we think the coexistence of primary magnetization in some rocks and burial CRMs in others is likely. The recognition of burial CRMs carried by magnetite is difficult because the unblocking range of temperature largely overlaps the burial temperature. To prevent incorrect interpretation of magnetostratigraphy in claystones, we recommend evaluating the burial temperature of claystones by using routine geothermometers such as the reflectance of vitrinite or the pyrolysis of organic matter (RockEval). For burial temperature $>100\,^{\circ}C$, some detailed rock magnetism studies should be carried out. The identification of stoichiometric magnetite, through the identification of Verwey transition at 120 K ($-153\,^{\circ}C$), may be a good indication that magnetite is neoformed (Aubourg & Pozzi 2010; Abdelmalak *et al.* 2012) allowing a CRM to be determined.

## Conclusion

We propose the existence of three magnetic windows where greigite, magnetite and pyrrhotite form successively. Greigite forms within the first tens of centimetres of the sedimentary column. Magnetite starts forming by at least 2 km and pyrrhotite forms at the expense of neoformed magnetite for depth $>6$ km. This leads to a continuous process of burial remagnetization that is by essence a chemical remanent magnetization. We investigate more specifically the consequences of the magnetostratigraphy record within the magnetite window, and propose that burial remagnetization may result in an alternation of normal and reverse polarities that has no correspondence with the age of the sediment.

We wish to thank D. Elmore for his contribution to this volume. Early drafts have benefited from attentive reading by A. Roberts and P. Rochette. This manuscript has benefited from the constructive reviews of A. Weil and E. Appel.

## References

ABDELMALAK, M. M., AUBOURG, C., GEOFFROY, L. & LAGGOUN-DEFFARGE, F. 2012. A new oil-window indicator? The magnetic assemblage of claystones from the Baffin Bay volcanic margin (Greenland). *AAPG Bulletin*, **96**, 205–215, doi:10.1306/07121111008.

ANDRA 2000. Thermicité du Callovo-Oxfordien du site de L'Est. Détermination d'un paléoenfouissement maximum. Etude complémentaire. Rep. D RP 0CRE 00–003, ANDRA, Paris.

AUBOURG, C. & ROCHETTE, P. 1992. Mise en évidence d'une aimantation pré-tectonique dans les Terres Noires subalpines (Callovien-Oxfordien). *Comptes Rendus de l'Académie des Sciences Paris*, **6**, 591–594.

AUBOURG, C. & POZZI, J. P. 2010. Toward a new <250 °C pyrrhotite-magnetite geothermometer for claystones <250 °C pyrrhotite–magnetite geothermometer for claystones. *Earth and Planetary Science Letters*, **294**, 47–57.

AUBOURG, C., POZZI, J.-P., JANOTS, D. & SARAHOUI, L. 2008. Imprinting chemical remanent magnetization in claystones at 95 °C. *Earth and Planetary Science Letters*, **272**, 172–180.

BANERJEE, S., ELMORE, R. D. & ENGEL, M. H. 1997. Chemical remagnetization and burial diagenesis: testing the hypothesis in the Pennsylvanian Belden Formation, Colorado. *Journal of Geophysical Research*, **102**, 24825–24842.

BERNER, R. A. 1984. Sedimentary pyrite formation. An update. *Geochimica et Cosmochimica Acta*, **48**, 605–615.

BROTHERS, L. A., ENGEL, M. H. & ELMORE, R. D. 1996. The late diagenetic conversion of pyrite to magnetite by organically complexed ferric iron. *Chemical Geology*, **130**, 1–14.

CAIRANNE, G. 2003. *Les réaimantations chimiques et le filtrage des polarités du champ magnétique terrestre. Expériences pressions-température et comparaisons avec des réaimantations naturelles.* University of Cergy Pontoise, Cergy Pontoise.

CAIRANNE, G., AUBOURG, C., POZZI, J.-P., MOREAU, M. G., DECAMPS, T. & MAROLLEAU, G. 2004. Experimental chemical remanent magnetization in a natural claystone: a record of two magnetic polarities. *Geophysical Journal International*, **159**, 907–916, doi: 10.1111/j.1365–246X.2004.02439x.

CROUZET, C., ROCHETTE, P. & MÉNARD, G. 2001. Experimental evaluation of thermal recording of polarity reversals during metasediments uplift. *Geophysical Journal International*, **145**, 771–785.

DEKKERS, M. J., PASSIER, H. F. & SCHOONEN, M. A. A. 2000. Magnetic properties of hydrothermally synthesized greigite (Fe3S4)-II High and low-temperature characteristics. *Geophysical Journal International*, **141**, 809–819.

DEVILLE, E. & SASSI, W. 2006. Contrasting thermal evolution of thrust systems: An analytical and modeling approach in the front of the western Alps. *AAPG Bulletin*, **90**, 887–907.

DUNLOP, D. J. & ÖZDEMIR, Ö. 1997. *Rock Magnetism: Fundamentals and Frontiers.* Cambridge University Press, Cambridge.

ELMORE, R. D., KELLEY, J., EVANS, M. & LEWCHUK, M. T. 2001. Remagnetization and orogenic fluids: testing the hypothesis in the central Appalachians. *Geophysical Journal International*, **144**, 568–576.

EVANS, M. & ELMORE, R. D. 2006. Fluid control of localized mineral domains in limestone pressure solution structures. *Journal of Structural Geology*, **26**, 284–301.

GILLETT, S. L. 2003. Paleomagnetism of the Notch Peak contact metamorphic aureole, revisited: pyrrhotite from magnetite +pyrite under submetamorphic conditions. *Journal of Geophysical Research*, **108**, 2446.

GUILHAUMOU, N., TOURAY, J. C., PERTHUISOT, V. & ROURE, F. 1996. Paleocirculation in the basin of southeastern France sub-alpine range: a synthesis from fluid inclusion studies. *Marine and Petroleum Geology*, **13**, 695–706.

HORNG, C. S., TORII, M., SHEA, K. S. & KAO, S. J. 1998. Inconsistent magnetic polarities between greigite- and pyrrhotite/magnetite-bearing marine sediments from the Tsailiaochi section, southwestern Taiwan. *Earth and Planetary Science Letters*, **164**, 467–481.

HUNT, C. P., BANERJEE, S. K., HAN, J., SOLHEID, P. A., OCHES, E., SUN, W. & LIU, T. 1995. Rock-magnetic proxies of climate change in the loess-palaeosol sequences of the western Loess Plateau of China. *Geophysical Journal Interior*, **123**, 232–244.

KARS, M., AUBOURG, C., POZZI, J.-P. & GIRARD, J.-P. 2010. *Neoformation of magnetic minerals in claystones during burial (<3 km)*. Paper presented at AGU Fall Meeting San-Francisco. GP11A-0753.

KARS, M., AUBOURG, C. & POZZI, J. P. 2011. Low temperature magnetic behaviour near 35 K in unmetamorphosed claystones. *Geophysical Journal International*, **186**, 1029–1035.

KATZ, B., ELMORE, R. D., COGOINI, M., ENGEL, M. H. & FERRY, S. 2000. Associations between burial diagenesis of smectite, chemical remagnetization, and magnetite authigenesis in the Vocontian trough, SE France. *Journal of Geophysical Research*, **105**, 851–868.

KLIGFIELD, R. & CHANNELL, J. E. T. 1981. Widespread remagnetization of Helvetic limestones. *Journal of Geophysical Research*, **86**, 1888–1900.

LANDAIS, P. & ELIE, M. 1999. *Utilisation de la géochimie organique pour la détermination du paléoenvironement et de la paléothermicité dans le Callovo-Oxfordien du site de l'Est de la France*. Acte des journées scientifiques CNRS/ANDRA.

LARRASOANA, J. C., ROBERTS, A. P., MUSGRAVE, R., GRACIA, E., PINERO, E., VEGA, M. & MARTNEZ-RUIZ, F. 2007. Diagenetic formation of greigite and pyrrhotite in gaz hydrate marine sedimentary systems. *Earth and Planetary Science Letters*, **261**, 350–366.

LU, G., MARSHAK, S. & KENT, D. V. 1990. Characteristics of magnetic carriers responsible for late Paleozoic remagnetization in carbonate strata of the Midcontinent, U.S.A. *Earth and Planetary Science Letters*, **99**, 351–361.

MAHER, B. A. & THOMPSON, R. 1999. *Quaternary Climates, Environments and Magnetism*. Cambridge University Press, Cambridge.

MAZUREK, M., HURFORDW, A. J. & LEUZ, W. 2006. Unravelling themulti-stage burial history of the Swiss Molasse Basin: integration of apatite fission track, vitrinite reflectance and biomarker isomerisation analysis. *Basin Research*, **18**, 27–50.

MCCABE, C. & ELMORE, R. D. 1989. The occurrence and origin of Late Paleozoic remagnetization in the sedimentary rocks of North America. *Reviews of Geophysics*, **27**, 471–494.

MOREAU, M. G., ADER, M. & ENKIN, R. J. 2005. The magnetization of clay-rich rocks in sedimentary basins: low-temperature experimental formation of magnetic carriers in natural samples. *Earth and Planetary Science Letters*, **230**, 193–210.

MORRIS, A. & ROBERTSON, A. H. F. 1993. Miocene remagnetization of carbonate platform and Antalya Complex units within the Isparta angle, SW Turkey. *Tectonophysics*, **220**, 243–266.

PLATTZMAN, 1994. Widespread Neogene remagnetization in Jurassic limestones of the South Iberian paleomargin (Western Betics, Gibraltar Arc). *Physics of the Earth and Planetary Interiors*, **85**, 3–15.

PULLAIAH, G., IRVING, E., BUCHAN, K. L. & DUNLOP, D. J. 1975. Magnetization changes caused by burial and uplift. *Earth and Planetary Science Letters*, **28**, 133–143.

ROBERTS, A. P. & TURNER, G. M. 1993. Diagenetic formation of ferrimagnetic iron sulphide minerals in rapidly deposited marine sediments, South Island, New Zealand. *Earth and Planetary Science Letters*, **115**, 257–273.

ROBERTS, A. P., FLORINDO, F., LARRASOANA, J. C., O'REGAN, M. A. & ZHAO, X. 2010. Complex polarity pattern at the former Plio–Pleistocene global stratotype section at Vrica (Italy): remagnetization by magnetic iron sulphides. *Earth and Planetary Science Letters*, **292**, 98–111.

ROCHETTE, P. 1987. Metamorphic control of the magnetic mineralogy of black shales in the Swiss Alps: toward the use of 'magnetic isogrades'. *Earth and Planetary Science Letters*, **84**, 446–456.

ROCHETTE, P. & LAMARCHE, G. 1986. Evolution des propriétés magnétiques lors des transformations minérales dans les roches: example du Jurassique Dauphinois (Alpes françaises). *Bulletin de Minéralogie*, **109**, 687–696.

ROWAN, C. J. & ROBERTS, A. P. 2005. Tectonic and geochronological implications of variably timed magnetizations carried by authigenic greigite in marine sediments from New Zealand. *Geology*, **33**, 553–556.

ROWAN, C. J. & ROBERTS, A. P. 2006. Magnetite dissolution, diachronous greigite formation, and secondary magnetizations from pyrite oxidation: unravelling complex magnetizations in Neogene marine sediments from New Zealand. *Earth and Planetary Science Letters*, **241**, 119–137.

ROWAN, C. J., ROBERTS, A. P. & BROADBENT, T. 2009. Reductive diagenesis, magnetite dissolution, greigite growth and paleomagnetic smoothing in marine sediments: a new view. *Earth and Planetary Science Letters*, **277**, 223–235.

SCHILL, E., APPEL, E. & GAUTAM, P. 2002. Towards pyrrhotite/magnetite geothermometry in low grade metamorphic carbonates of the Tethyan Himalayas (Shiar Khola, Central Nepal). *Journal of Asian Earth Sciences*, **20**, 195–201.

STOKKING, L. B. & TAUXE, L. 1990. Properties of chemical remanence in synthetic hematite: testing theoretical predictions. *Journal of Geophysical Research B: Solid Earth*, **95**, 12639–12652.

SUK, D., PEACOR, R. D. & VAN DER VOO, R. 1990. Replacement of pyrite framboid by magnetite in limestones and implication for paleomagnetism. *Nature*, **345**, 611–613.

SUK, D.-W., VAN DER VOO, R. & PEACOR, D. R. 1993. Origin of magnetite responsible for remagnetization of early Paleozoic limestones of New York State. *Journal of Geophysical Research B: Solid Earth*, **98**, 419–434.

TOHVER, E., WEILB, A. B., SOLUMC, J. G. & HALLD, C. M. 2008. Direct dating of carbonate remagnetization by 40Ar/39Ar analysis of the smectite–illite transformation. *Earth and Planetary Science Letters*, **274**, 524–530.

WEHLAND, F., EIBL, O., KOTTHOFF, S., ALT-EPPING, U. & APPEL, E. 2005. Pyrrhotite pTRM acquisition in metamorphic limestones in the light of microscopic observations. *Physics of the Earth and Planetary Interiors*, **151**, 107–114.

# Magnetic characterization of oil sands at Osmington Mills and Mupe Bay, Wessex Basin, UK

STACEY EMMERTON*, ADRIAN R. MUXWORTHY & MARK A SEPHTON

*Department of Earth Science and Engineering, Imperial College London,
South Kensington Campus, London, SW7 2AZ, UK*

*Corresponding author (e-mail: stacey.emmerton05@imperial.ac.uk)*

**Abstract:** This study combines magnetic experimentation and geochemical analysis on oil sands from Osmington Mills and Mupe Bay, Wessex Basin, UK to investigate the possibility of a relationship between hydrocarbons and magnetic mineralogy. Removal of hydrocarbons by chemical extraction was conducted to allow comparison of (1) oil sands and (2) cleaned sands. Detailed magnetic analysis including low-temperature and high-temperature experimentation revealed that all but one sample was dominated by siderite, identified by the Néel transition at 37–38 K as well as containing large grains of multidomain magnetite (Verwey transition 110 K) and hematite (Morin transition 250 K). Scanning electron microscopy and energy dispersive x-ray analysis confirmed the presence of iron oxides, in particular framboids 500 nm–45 μm in diameter, probably magnetite. Hysteresis parameters showed distinct grouping of oil sands compared to their clean counterparts and a negative linear regression in log space was observed ($R^2 = 0.7$) between the percentage of extractable organic matter and magnetic susceptibility. These results suggest a relationship exists between magnetic minerals and the alteration of oil due to biodegradation, which is not yet fully understood. Possible mechanisms are suggested to be due to anaerobic bacteria or the transportation of the oil as it migrates through the host rock.

Aeromagnetic surveys have established a general connection between magnetic anomalies and hydrocarbon accumulations on a regional scale but with a large degree of ambiguity (e.g. Donovan *et al.* 1979; Reynolds *et al.* 1991; Gay 1992; LeBlanc & Morris 1999). Several major remagnetizations worldwide (e.g. McCabe *et al.* 1984, 1987; Elmore & Leach 1990; Suk *et al.* 1993; Garner & Cioppa 2006) are associated with hydrocarbon provinces and anomalous aeromagnetic signals. This is thought to be due to chemical alteration of minerals as fluids such as hydrocarbons migrate through the host rock, but no definitive mechanism has been determined. Recent research by several groups has tried to identify the magnetic minerals themselves for hydrocarbon identification (e.g. Liu *et al.* 1998; Aldana *et al.* 1999; Diaz *et al.* 2000; Aldana & Costanzo-Alvarez 2003; Ivakhnenko & Potter 2004; Costanzo-Alvarez *et al.* 2006; Liu *et al.* 2006; Rijal *et al.* 2010; Aldana *et al.* 2011; Guzmán *et al.* 2011; Rijal *et al.* 2011). For example, Costanzo-Alvarez *et al.* (2000) and Liu *et al.* (2006) conducted rock magnetic measurements on producing and non-producing wells and identified anomalous values of magnetic susceptibility ($\chi$) at certain depth intervals within both oil-producing and non-producing wells characterized by a low-coercivity mineral, probably magnetite. Aldana *et al.* (1999) categorized these anomalies into two types: (1) Fe-rich spherical aggregates of authigenic origin as seen by scanning electron microscopy (SEM) analysis, only found in producing wells (Costanzo-Alvarez *et al.* 2000; Aldana & Costanzo-Alvarez 2003) and most likely magnetite; and (2) characterized by magnetite and hematite but no association with spherical aggregates, present in both non-producing and producing wells (Costanzo-Alvarez *et al.* 2000; Aldana & Costanzo-Alvarez 2003). Liu *et al.* (2006) also found variances between oil-producing and non-oil-producing layers using basic hysteresis parameters (saturation remanence magnetization $M_{rs}$, saturation magnetization $M_s$, coercivity $B_c$ and remanent coercivity $B_{cr}$). The results showed that within an oil-bearing layer its $\chi$, $M_s$ and $M_{rs}$ values were between 24 and 36 times higher than the corresponding layer obtained from a dry well with maghemite, siderite and pyrite contributing to the enhanced magnetic signature. Another study by Ivakhnenko & Potter (2004) investigated the magnetic susceptibility of crude oils, refined oil fractions and formation waters from sites worldwide. It was found that all the fluids were diamagnetic but had distinct differences in magnetic susceptibility, for example the crude oils were more negative than formation waters from the same location. The physical and chemical properties of the oils appeared to correlate with the magnetic susceptibility such as gravity, metal concentration density and API (American Petroleum Index).

*From*: ELMORE, R. D., MUXWORTHY, A. R., ALDANA, M. M. & MENA, M. (eds) 2012. *Remagnetization and Chemical Alteration of Sedimentary Rocks*. Geological Society, London, Special Publications, **371**, 189–198.
First published online June 26, 2012, http://dx.doi.org/10.1144/SP371.6

The aim of this paper is to complete full magnetic characterization of oil sands from the Wessex Basin, UK to try to determine if a relationship exists between the oil content of the samples and the magnetic signature. Understanding any association between hydrocarbons, their migration and the presence of magnetic minerals is of significant interest for two broad reasons: (1) it will help us to understand the conditions experienced by hydrocarbons during accumulation and migration as well as how the magnetic minerals are formed; and (2) it may lead to a proxy in which hydrocarbons can be identified by their magnetic signature. There have been several in-depth studies into magnetic minerals in proximity to hydrocarbon plumes (e.g. Machel &

Burton 1991; Machel 1995, 1996; Liu *et al.* 1998; Aldana *et al.* 1999; Diaz *et al.* 2000; Aldana & Costanzo-Alvarez 2003; Costanzo-Alvarez *et al.* 2006; Liu *et al.* 2006), but none of these have been able to relate these findings to the chemical composition and biodegradation of the hydrocarbons themselves.

## Sampling and laboratory methods

Six oil sands were collected from the Bencliff Grit of the Corallian Series (Upper Jurassic) at Osmington Mills (Fig. 1) and one sample from the sandstone formation of the Wealden Group (Lower Cretaceous) at Mupe Bay (Fig. 1, Table 1).

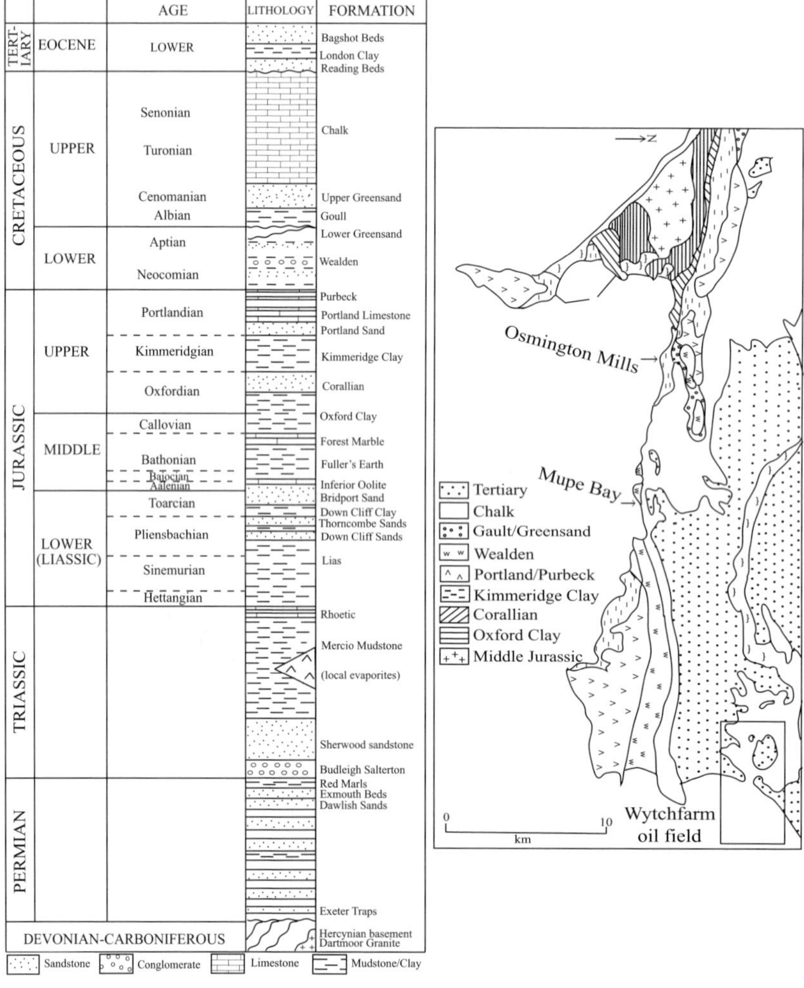


**Fig. 1.** Geological map (modified from Stoneley & Selley 1986) and stratigraphic column (modified from Underhill & Stoneley 1998) for the Wessex Basin. Six samples were collected from the Bencliff Grit (Upper Jurassic) at Osmington Mills (50°38'13.96"N, 2°22'29.22"W) and one sample from the Wealdon Group sandstone (Lower Cretaceous) at Mupe Bay (50°37'6.15"N, 2°13'24.86"W). Both localities are highlighted on the map.

**Table 1.** *Summary of samples collected at Osmington Mills (50°38'13.96"N, 2°22'29.22"W), all within a few metres of one another, and Mupe Bay (50°37'6.15"N, 2°13'24.86"W)*

| Sample | Age | Geological unit | Field observation | Description |
|---|---|---|---|---|
| OS1 | Jurassic | Bencliff Grit, Corallian Group | 1: Close to active oil seep | Unstained |
| OS2 | Jurassic | Bencliff Grit, Corallian Group | 1: Youngest (top) | Lightly stained |
| OS3 | Jurassic | Bencliff Grit, Corallian Group | 1: Oldest (bottom) | Heavily stained |
| OS4 | Jurassic | Bencliff Grit, Corallian Group | 1: Oil patch | Very heavily stained |
| OS5 | Jurassic | Bencliff Grit, Corallian Group | 2: Youngest (top) | Lightly stained |
| OS6 | Jurassic | Bencliff Grit, Corallian Group | 2: Oldest (bottom) | Heavily stained |
| MB | Cretaceous | Sandstone, Wealdon Group | Heavily biodegraded | Heavily stained |

Osmington Mills was selected because it is an exhumed (uplifted) reservoir which contains oil sands of variable staining within the two main cross-bedded sandstone units separated by a shale bed (Selley 1992; Bigge & Farrimond 1998; Watson *et al.* 2000). Mupe Bay is known to be more heavily biodegraded and have a lower maturity (extent of hydrocarbon generation) than Osmington Mills (Bigge & Farrimond 1998). The sample was taken from the oil-stained sandstone matrix. All exposures occur in northerly dipping beds, directly south of structural faults, and are located SE of the onshore oil-producing field of Wytchfarm.

To conduct a thorough analysis of the samples, three approaches were undertaken: (1) chemical analysis; (2) SEM imaging and energy dispersive X-ray (EDX) analysis; and (3) magnetic experimentation. For chemical experiments and SEM imaging it was necessary to chemically extract the oil from the sands. To remove the oil, powdered sands were weighed and placed into a solvent mixture of dichloromethane (DCM) and methanol (MeOH) in a 93:7 ratio and put into a ultrasonic bath at room temperature. The samples were then placed into a centrifuge at 1600 rpm for 5 min and the supernatant solvent (oil + solvent mixture) was extracted. The process was repeated until the supernatant solvent was clear, that is, at least four times. A rotary evaporator was used at 50 °C to reduce the supernatant fraction. The fraction was left to dry and the extracted organic matter (EOM) weighed. The extracted rock powder was also dried and weighed as 'clean' sands. Cleaned sands and their magnetic extracts were imaged on scanning electron microscopes (a Zeiss XB1540 in the Nanocentre at Cardiff University and on a JEOL6400 in the Department of Materials, Imperial College London). The SEM images were typically between 200 μm and 10 nm resolution. EDX analysis was conducted to determine the composition of particular grains.

Magnetic investigations were performed at the Institute of Rock Magnetism, University of Minneapolis and at Imperial College London. A Quantum Designs Magnetic Properties Measuring Systems (MPMS) cryogenic magnetometer was used to determine magnetic mineralogy and grain size information at sub-room temperatures. Two types of measurements were taken: (1) warming curves of saturation isothermal remanence magnetization (SIRM) induced in a field of 2.5 T at 10 K after field cooling (FC) (2.5 T) and zero-field cooling (ZFC) from 300 K in 10 K steps as well as cooling and warming curves of room-temperature (RT) induced SIRM to identify low-temperature crystallographic transitions; and (2) the measurement of AC magnetic susceptibility as a function of temperature (from 300 to 10 K in 10 K steps) and frequency (1.0, 32.2, 10.0, 31.6, 100.1, 316.7 and 997.3 Hz).

Hysteresis and backfield curves were measured at low temperature (10 K to 290 K in 20 K intervals) after ZFC and FC (1 T) on a Princeton Measurements Vibrating Sample Magnetometer (VSM). A high-temperature VSM was also used to measure cyclic warming and cooling curves evaluated over the range 200–350 °C at 50 °C intervals in a helium atmosphere, to help characterize the effect of oxidation at temperature. Due to the high organic content of the samples, they could not be heated beyond 350 °C on the VSM.

Samples were measured on an Agico KLY-2 KappaBridge AC Susceptibility Bridge to record susceptibility as a function of temperature using a CS-2 furnace and to estimate Curie temperatures. Samples were measured in an argon atmosphere to try to minimize chemical alteration and heated to 700 °C in one cycle. Magnetic susceptibility at room temperature was measured using a Bartington MS2B dual-frequency sensor.

## Results

### Organic matter

The percentage of EOM was determined by chemically removing the oil and measuring the yields at every step of separation (Table 2). Sample OS1 presented 0% EOM and is an oil-free sand sample. The variation of oil staining within the remaining five

**Table 2.** *Summary of chemical and magnetic experimental results for all the samples in this study (O: oil-stained; C: cleaned i.e. oil-free)*

| Sample | EOM% | $\chi\,(10^{-5}\,\mathrm{m^3\,kg^{-1}})$ | | $B_{\mathrm{cr}}$ (mT) | | $B_{\mathrm{c}}$ (mT) | | $M_{\mathrm{rs}}/M_{\mathrm{s}}$ | |
|---|---|---|---|---|---|---|---|---|---|
| | | O | C | O | C | O | C | O | C |
| OS1 | 0 | 7.70 | – | 29.6 | – | 12.9 | – | 0.11 | – |
| OS2 | 0.5 | 6.19 | 5.08 | 134 | 34.6 | 22.4 | 5.71 | 0.42 | 0.06 |
| OS3 | 4.8 | 1.49 | 1.30 | 42.8 | 14.9 | 14.7 | 7.03 | 0.16 | 0.07 |
| OS4 | 11.8 | 1.15 | 1.43 | 41.9 | 11.6 | 14.3 | 4.75 | 0.15 | 0.09 |
| OS5 | 2.9 | 1.60 | 1.76 | 47.4 | 15.9 | 14.6 | 2.98 | 0.15 | 0.05 |
| OS6 | 10.3 | 1.30 | 1.22 | 43.5 | 20.0 | 17.2 | 5.23 | 0.16 | 0.06 |
| MB | 10.4 | 0.33 | 0.19 | 34.6 | 41.4 | 0.91 | 7.34 | 0.01 | 0.08 |

samples from Osmington Mills was between 0.5% (OS2) and 11.8% (OS4) and the Mupe Bay sample had 10.4% EOM.

## SEM and EDX

Examples of SEM images and EDX spectra are shown in Figures 2 and 3. They show a variety of different minerals within the samples, all dominated by quartz, feldspars, pyroxenes and olivines, all of which are typical of sandstones. Magnetic extracts highlighted a number of different magnetic minerals and morphologies (Fig. 2). The magnetic minerals are mainly characterized by iron oxide grains from 200 nm to 60 μm (Fig. 2b) identified by EDX. Spherical clusters or framboids were also found in

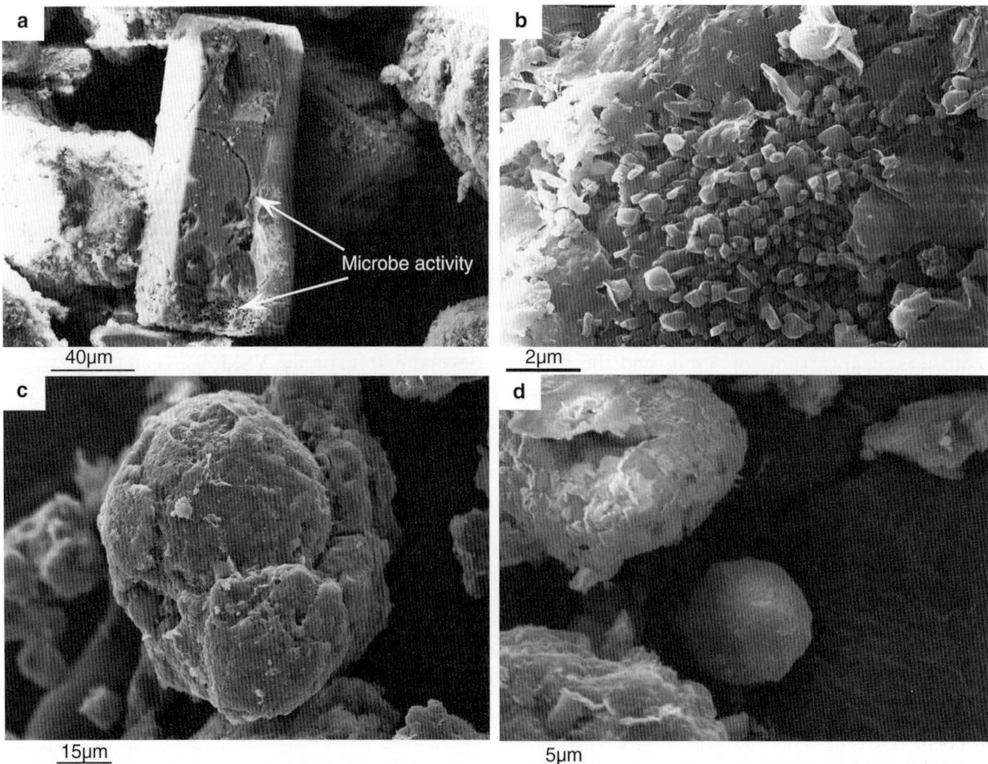


**Fig. 2.** SEM images showing (**a**) microbial activity within OS5c represented by the ectching pattern on the surface of the grain; (**b**) typical magnetic grains in OS2c with an EDX spectra of Fe and O indicating an iron oxide, magnetite or hematite ranging in size between 200 nm and 60 μm; and (**c, d**) magnetic framboids dominated by an iron oxide, magnetite or hematite ranging in size between 500 nm and 45 μm.

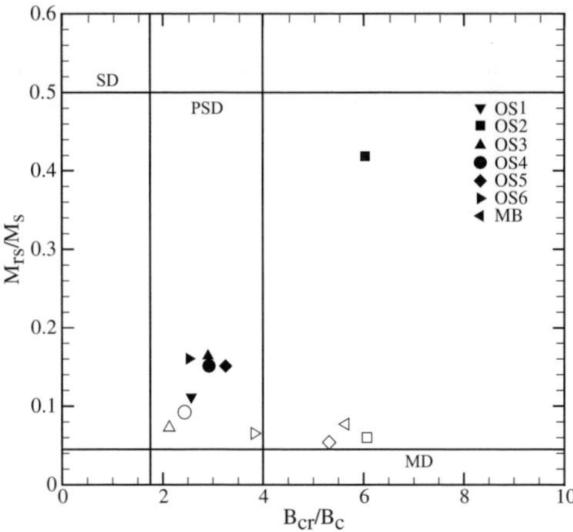



**Fig. 3.** A 'Day plot' (Day *et al.* 1977) of the ratios of the hysteresis parameters $M_{rs}/M_s$ v. $B_{cr}/B_r$ for the samples examined in this study. The regions commonly associated with SD, PSD and MD behaviour for magnetite are labelled for representative purposes. The oil-stained (closed symbols) and clean sands (open symbols) are plotted.

all the samples (e.g. Fig. 2c, d) ranging in size between 500 nm and 45 μm. Signs of microbe activity were identified within some of the grains by the etched trails on the grains, for example, OS5c (Fig. 2a).

EDX analysis revealed all the samples were dominated by silicon and oxygen as expected in sandstones, as well as several other elements in most spectra such as Fe, Mg, Al, K and S. The EDX analysis often gave complex mineralogical profiles due to the beam spot size. All framboids were dominated by iron and oxygen suggesting an iron oxide (magnetite, maghemite and/or hematite) as the main magnetic mineral, which is coherent with the findings of Aldana *et al.* (1999) and Costanzo-Alvarez *et al.* (2006).

## Magnetic investigation

Room-temperature hysteresis data are presented in Table 2. The oil sands were dominated by strong paramagnetic behaviour, limiting the quality of the data. In Figure 3, hysteresis parameters are represented on a 'Day plot' (Day *et al.* 1977) as ratios of saturation remanence magnetization to saturation magnetization ($M_{rs}/M_s$) and remanence coercivity to coercivity ($B_{cr}/B_c$). Day plots are used for distinguishing between domain states – single-domain (SD), pseudo-single-domain (PSD) and multi-domain (MD) – which can be utilized to infer grain sizes for a given mineralogy. Using domain boundaries suitable for magnetite, the Day plot

(Fig. 3) shows the predominance of PSD behaviour in nearly all the oil-saturated sands (Fig. 3) and highlights the differences between oil-stained sands and cleaned sands. After cleaning, the $M_{rs}/M_s$ ratios of the sands decrease significantly but have variable $B_{cr}/B_c$ behaviour.

Warming and cooling curves of SIRM induced in different fields are displayed in Figure 4. The warming curves of SIRM induced in a field of 2.5 T at 10 K after FC (field = 2.5 T) and ZFC, while cooling from 300 K, display steep decays in remanence at 35–40 K (Fig. 4); this is indicative of the Néel transition of siderite at 37–38 K (Jacobs 1963). At 10 K the FC SIRM intensity was 0.014 Am$^2$ kg$^{-1}$ for sample OS2 (Fig. 4a), which is significantly greater than the ZFC SIRM intensity at 10 K in the same sample (0.003 Am$^2$ kg$^{-1}$). Frederichs *et al.* (2003) attributed this difference to the acquisition of thermoremanence on cooling through the Néel temperature. OS6 shows a stronger remanence in the ZFC curve opposed to FC; FC remanences are usually larger than ZFC but the presence of multi-domain magnetite can counter this (Carter-Stiglitz *et al.* 2006; Fig. 4e). Cyclical cooling and warming curves of room-temperature (RT) SIRMs were also measured. Typically, the two curves diverged at 98–100 K on heating, indicating the presence of multi-domain magnetite (Fig. 4e) due to the Verwey transition (100–120 K depending on stoichiometry; Walz 2002). A transition step at 250 K can also be identified, likely due to hematite's Morin transition (Morin 1950; Fig. 4b).

High-temperature mass susceptibility measurements are shown in Figure 5. Sample behaviour was classified into three types: (A) reversible Curie curves; (B) the formation of a stronger (purer) ferromagnetic (*sensu latu*) mineral on heating typically at 400–550 °C and represented by sharply increasing susceptibilities (Fig. 5); and (C) similar behaviour to type B but also with the formation of a new magnetic mineral on cooling. Sample OS4 (Fig. 5c) is the only type A susceptibility curve which showed noisy but reversible heating and cooling curves with a Curie temperature 560 °C (magnetite); when cleaned however, it showed behaviour type B. Samples OS3, OS5 and MB are type B with a more ferromagnetic mineral being formed at 400–550 °C as well as an additional mineral formation at 250–350 °C in OS3 (Fig. 5b). On cooling from 700 °C, the change in slope between 530 and 590 °C indicates a Curie temperature for magnetite in all type B samples. OS2 and OS6 are type C, with OS6 (Fig. 5e) showing multiple formations of magnetic minerals on heating as well as the formation of a new mineral at 260–300 °C during cooling. After heating, all the samples showed enhanced susceptibility values (e.g. the magnetic

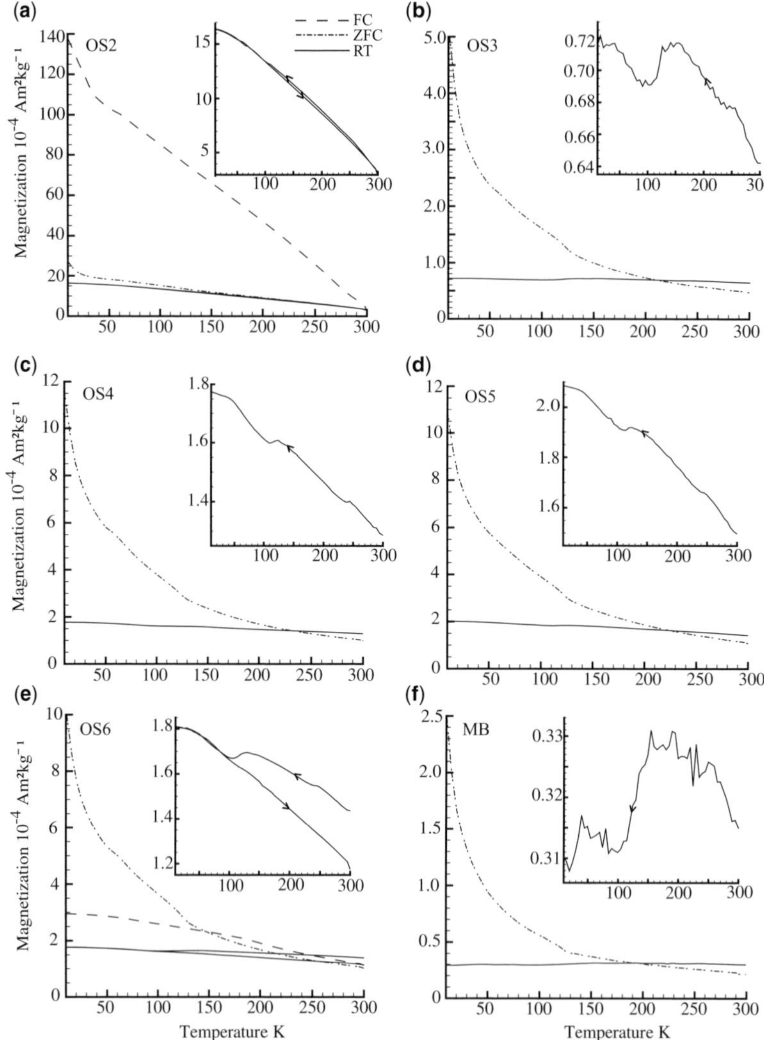

**Fig. 4.** MPMS measurements for zero field cooling (ZFC), field cooling (FC) curves and room temperature remanence curves (RT) for cooling ( ← ) and warming ( → ) from 10 K to 300 K, giving information on mineralogy and grain sizes: (**a**) OS2; (**b**) OS3; (**c**) OS4; (**d**) OS5; (**e**) OS6; and (**f**) MB.

susceptibility of OS5 increased by 50% on heating) with most samples displaying no noticeable difference in trend between the oil and clean samples (with the exception of OS4). OS4 displayed type A behaviour when oil stained and type B behaviour when cleaned.

The RT susceptibility measurements were recorded both before and after cleaning. The original oil sands had magnetic susceptibilities between $3.3 \times 10^{-6}$ $m^3$ $kg^{-1}$ and $7.7 \times 10^{-5}$ $m^3$ $kg^{-1}$ and the clean sands between $1.9 \times 10^{-6}$ $m^3$ $kg^{-1}$ and $5.1 \times 10^{-5}$ $m^3$ $kg^{-1}$ (Table 2). Samples OS2, OS3, OS6 and MB displayed decreased susceptibilities after cleaning; samples OS4 and OS5 demonstrated the opposite trend, however.

## Discussion

Magnetic characterization reveals that nearly all the oil-stained sands from the Wessex Basin (except OS4) demonstrate the presence of large amounts of siderite. Siderite ($FeCO_3$) is an iron-bearing carbonate of authigenic origin and is common in sedimentary settings, specifically in marine and lacustrine environments, and is typical of anoxic

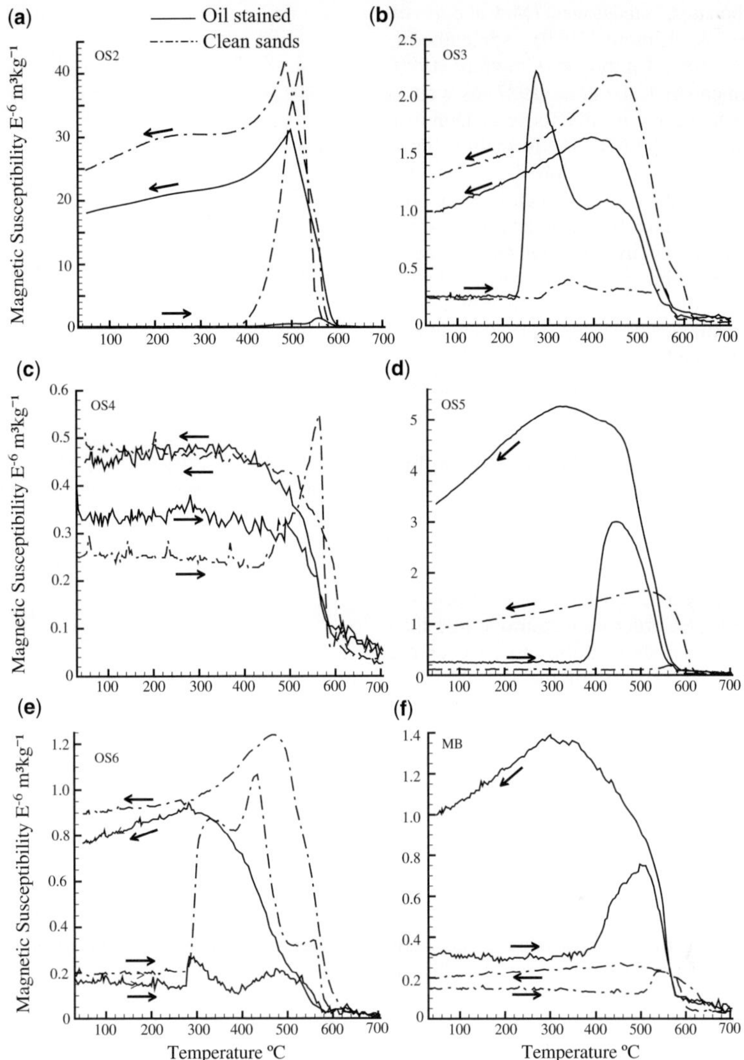



**Fig. 5.** Magnetic susceptibility measurements as a function of temperature (up to 700 °C) for oil sands and clean sands on warming ( → ) and cooling ( ← ), giving information on magnetic mineralogy through Curie temperatures: (**a**) OS2; (**b**) OS3; (**c**) OS4; (**d**) OS5; (**e**) OS6; and (**f**) MB.

environments. It is paramagnetic at room temperature but can be identified with low-temperature experiments (<40 K) due to its Néel transition at 37–38 K (Jacobs 1963; Pan *et al.* 2000; Frederichs *et al.* 2003). Siderite is well known to oxidize to magnetite, maghemite and hematite at high temperatures (from 490 °C; Pan *et al.* 2000) as seen by the formation of a new ferromagnetic mineral at 400–550 °C in the high-temperature mass-susceptibility measurements (Fig. 5).

SEM images of magnetic extracts from the Wessex Basin oil sands revealed the presence of magnetite and hematite grains from 200 nm to 60 μm. The observed magnetic grains relate well to the magnetic measurements (Figs 2, 4 & 5) which indicate multi-domain magnetite and hematite as contributors to the magnetic signal in both OS and MB samples (Fig. 5). Siderite was not identified in the SEM images because it is paramagnetic at room temperature. The occurrence of framboidal material, most likely magnetite (Fig. 2c, d) ranging between 500 nm and 45 μm in diameter, was also observed. Magnetic spherical aggregates have been found in several different geological units for example, carbonates, sandstones (McCabe *et al.* 1987; Kilgore & Elmore 1989), speleothems (Elmore *et al.* 1987; Elmore & Crawford 1990) and solid bitumen (McCabe *et al.* 1987) as well as within oil-producing environments (e.g. Donovan *et al.* 1979; Aldana *et al.* 1999). The formation mechanism for these magnetite framboids is not yet certain and a relationship to hydrocarbons is only alluded to; however, it has been argued that the magnetite spherules found in hydrocarbon environments are the oxidation product of epigenetic sulphide framboids formed during the migration of hydrocarbons through sediments and therefore a secondary mineralogy (Reynolds *et al.* 1991). On the other hand, Suk *et al.* (1990) claim it is the reverse of pyritization (Canfield & Berner 1987) which can occur at later stages of diagenesis in limestones. The samples in this study were taken free of drilling and other man-made contaminations.

The sample from Mupe Bay displayed similar magnetic mineralogy to the six Osmington Mills samples, but revealed some different magnetic characteristics (Table 2). MB room-temperature susceptibility was an order of magnitude less than any OS samples, and showed the largest decrease in susceptibility in 72% of all the samples after cleaning. It is possible these differences are due to the original sandstone mineralogy, but it could also be due to the heavily biodegraded nature of MB compared to OS (Bigge & Farrimond 1998).

## Relationship between oil content and magnetic signature

Differences between the oil-stained and clean sands are observed in Figures 3–6. The Day plot (Fig. 3) shows the predominance of PSD behaviour (for magnetite) in all oil sands with very little difference between samples; when compared to their clean sand counterparts a marked contrast is apparent, however. In all cleaned samples the $M_{rs}/M_s$ ratios decrease significantly but with variable $B_{cr}/B_c$ behaviour, generally becoming more PSD/MD. This indicated that the oil contains more single-domain grains and smaller grain sizes (based on magnetite as the major magnetic mineralogy).

Figures 5 and 6 show some marked differences between oil-stained and clean sand behaviour. Samples OS2, OS3, OS6 and MB display a decrease in magnetization in the clean sands, implying some magnetic minerals have been incorporated into the oil. Other samples (OS4 and OS5) have higher magnetization in their clean counterparts (Table 2) however, implying a possible change in magnetic mineralogy or grain size distribution during oil extraction. In contrast to the other samples, OS4 only showed the formation of a ferromagnetic

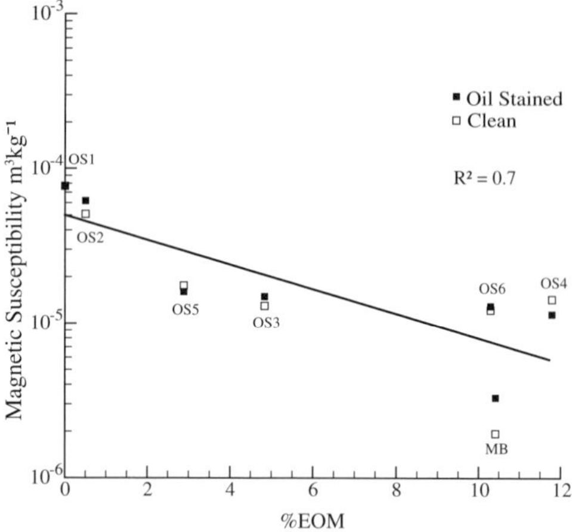



**Fig. 6.** Plot of magnetic susceptibility and the percentage of extractable organic matter (EOM) for the oil sands from Osmington Mills and Mupe Bay. Samples are represented by both oil-stained and clean oil-free sands.

mineral on heating in the clean sample (Fig. 5) and had little siderite present before cleaning (Figs 4 & 5). It also had the largest increase (24%) in room-temperature susceptibility when cleaned (Table 2), inferring a significant change during oil extraction. OS4 was selected because it was an extremely oil-stained patch within the oil-stained sandstone unit, and reflects an area of severe oil impregnation and possibly severe biodegradation.

By combining magnetic susceptibility at room temperature and the percentage of EOM (Table 1, Fig. 6) from the oil-stained and clean sands, a relationship can be observed. Figure 6 displays a negative linear regression in log space ($R^2 = 0.7$), that is, there is an inverse relationship between magnetic material content and EOM content. There are two possible mechanisms for this relationship. (1) There are several well-known bacterium (Heider *et al.* 1998) that readily use Fe ions in anaerobic processes within oil biodegradation which could be responsible for the relationship observed, implying a bacterial response (Bigge & Farrimond 1998). For example, McCabe *et al.* (1987) indicated that the composition of solid bitumen samples was a result of microbial attack at shallow depths on what was once liquid crude oil. (2) It could be due to the nature of the oil itself. For example, as the oil is transported from its source, the magnetic minerals are dispersed into the reservoir implying the relationship could be due to porosity and/or transportation. This has a significant impact on our understanding of hydrocarbons and magnetic minerals, although further chemical analysis is needed to determine if oil-sand magnetization is due to bio-degradation and/or transport of the oil.

## Conclusions

Oil sands from the Wessex Basin, UK have been magnetically characterized to determine if a relationship exists between oil chemistry and magnetic mineralogy. The oil sands were found to be magnetically dominated by siderite (expect sample OS4), which was identified by low-temperature magnetometry (Fig. 4) and its distinctive behaviour during heating to 700 °C (Fig. 5). Further low-temperature results indicated the presence of multi-domain magnetite and hematite marked by the presence of Verwey transitions (110–120 K) and Morin transitions (250 K), respectively, as well as iron-oxide identification within SEM images through EDX analysis (Fig. 2). In comparison to their oil-stained counterparts, chemically cleaned sands displayed variable behaviour in magnetic susceptibility after oil removal (Fig. 6) but did have lower $M_{rs}/M_s$ ratios (Fig. 4) indicating differences in oil chemistry as they belong to the same geological formation (except MB). Osmington Mills and Mupe Bay samples revealed similar magnetic mineralogy, but MB displayed magnetic susceptibility an order of magnitude lower than OS and showed the largest difference after oil removal; this may be related to the high levels of biodegradation of MB. OS4 displayed significantly different magnetic properties to the other samples, which may be correlated to its high oil impregnation and therefore its differing chemical and biodegradation levels.

By combining the magnetic data and chemical analysis, a negative linear regression in log space ($R^2 = 0.7$) between the magnetic susceptibility and the percentage of EOM in the oil-stained and clean samples (Fig. 6) was revealed. It is suggested that this relationship may be due to a bacterial response. These results indicate a complex relationship between existing magnetic minerals within the sandstones, and the alteration of these magnetic minerals due to the multiplex biological activity and biodegradation of the oil.

This work was funded by a NERC studentship to SE and supported by the Engineering and Physical Sciences Research Council (EP/F056745/1). Thank you to the Institute of Rock Magnetism, University of Minnesota for a visiting fellowship to conduct the magnetic experimentation and G. Lalev at the Cardiff Nanocentre for the use of SEM equipment and SEM imaging.

## References

ALDANA, M. & COSTANZO-ALVAREZ, V. 2003. Magnetic and mineralogical studies to characterize oil reservoirs in Venezuela. *The Leading Edge*, **22**, 526–529, doi: 10.1190/1.1587674.

ALDANA, M., COSTANZO-ALVAREZ, V., VITIELLO, D., COLMENARES, L. & GOMEZ, L. 1999. Framboidal magnetic minerals and their possible association to hydrocarbons: La Victoria oil field, south-western Venezuela. *Geofisica International*, **38**, 137–152.

ALDANA, M., COSTANZO-ALVAREZ, V., GOMEZ, L., GONZÁLEZ, C., DÍAZ, M., SILVA, P. & RADA, M. 2011. Identification of magnetic minerals related to hydrocarbon authigenesis in Venezuelan oil fields using an alternative decomposition of isothermal remanence curves. *Studia Geophysica et Geodaetica*, **55**, 343–358.

BIGGE, M. A. & FARRIMOND, P. 1998. Biodegradation of seep oils in the Wessex basin–a complication for correlation. *In*: UNDERHILL, J. R. (ed.) *Development, Evolution and Petroleum Geology of the Wessex Basin*. Geological Society, London, Special Publications, **133**, 373–386, doi: 10.1144/GSL.SP.1998.133.01.19.

CANFIELD, D. E. & BERNER, R. A. 1987. Dissolution and pyritization of magnetite in anoxic marine sediments. *Geochima et Cosmochimica Acta*, **51**, 645–659, doi: 10.1016/0016–7037(87)90076-7.

CARTER-STIGLITZ, B., MOSOWTIZ, B., SOLHEID, P., BERQUÓ, T. S., JACKSON, M. & KOSTEROV, A. 2006. Low-temperature magnetic behaviour of multi-domain titanomagnetites: TM0, TM16, and TM35. *Journal of Geophysical Research*, **111**, B12S05, doi: 10.1029/2006JB004561.

COSTANZO-ALVAREZ, V., ALDANA, M., ARISTEQUIETA, O., MARCANO, M. C. & ACONCHA, E. 2000. Study of magnetic contrasts in the Guafita oil field (South-Western Venezuela). *Physics and Chemistry of the Earth, Part A: Solid Earth and Geodesy*, **25**, 437–445, doi: 10.1016/S1464-1895(00)0068-5.

COSTANZO-ALVAREZ, V., ALDANA, M., DIAZ, M., BAYONA, G. & AYALA, C. 2006. Hydrocarbon-induced magnetic contrasts in some Venezuelan and Columbian oil wells. *Earth, Planets and Space*, **58**, 1401–1410.

DAY, R., FULLER, M. & SCHMIDT, V. A. 1977. Hysteresis properties of titanomagnetites: grain size and composition dependence. *Physics of the Earth and Planetary Interiors*, **13**, 260–267, doi: 10.1016/0031-9201(77)90108-X.

DIAZ, M., ALDANA, M., COSTANZO-ALVAREZ, V., SILVA, P. & PEREZ, A. 2000. EPR and Magnetic susceptibility studies in well samples from some Venezuelan oil fields. *Physics and Chemistry of the Earth (A)*, **25**, 447–453, doi: 10.1016/S1464-1895(00)00069-7.

DONOVAN, T. J., FORGEY, R. L. & ROBERTS, A. A. 1979. Aeromagnetic detection of diagenetic magnetite over oil fields. *AAPG Bulletin*, **63**, 245–248.

ELMORE, R. D. & CRAWFORD, L. 1990. Remanence in authigenic magnetite: testing the hydrocarbon-magnetite hypothesis. *Journal of Geophysical Research*, **95**, 4539–4549, doi: 10.1029/JB095iB04 p04539.

ELMORE, R. D. & LEACH, M. C. 1990. Remagnetization of the rush springs formation, cement, Oklahoma: implications for dating hydrocarbon migration and aeromagnetic exploration. *Geology*, **18**, 124–127.

ELMORE, R. D., ENGEL, M. H., CRAWFORD, L., MICK, K., IMBUS, S. & SOFER, S. 1987. Evidence for a relationship between hydrocarbons and authigenic magnetite. *Nature*, **325**, 428–430, doi: 10.1038/325428a0.

FREDERICHS, T., DOBENECK, T., BLEIL, U. & DEKKERS, M. J. 2003. Towards the identification of siderite, rhodochrosite, and vivianite in sediments by their low-temperature magnetic properties. *Physics and Chemistry of the Earth*, **28**, 669–679, doi: 10.1016/S1474-7065(03)00121-9.

GARNER, N. & CIOPPA, M. T. 2006. Late Paleozoic remagnetization of the Trenton Formation in Ordovician petroleum reservoirs of southwestern Ontario. *Journal of Geochemical Exploration*, **89**, 119–123.

GAY, S. P. Jr 1992. Epigenetic v. syngenetic magnetite as a cause of magnetic anomalies. *Geophysics*, **57**, 60–68, doi: 10.1190/1.1443189.

GUZMÁN, O., COSTANZO-ALVAREZ, V., ALDANA, M. & DIAZ, M. 2011. Study of magnetic contrasts applied to hydrocarbon exploration in the Maturin sub-basin (Eastern Venezuela). *Studia Geophysica et Geodaetica*, **55**, 359–376.

HEIDER, J., SPORMANN, A. M., BELLER, H. R. & WIDDEL, F. 1998. Anaerobic bacterial metabolism of hydrocarbons. *FEMS Microbiology Reviews*, **22**, 459–473, doi: 10.1016/S0168-6445(98)00025-4.

IVAKHNENKO, O. P. & POTTER, D. K. 2004. Magnetic susceptibility of petroleum reservoir fluids. *Physics and Chemistry of the Earth*, **29**, 899–907, doi: 10.1016/j.pce.2004.06.001.

JACOBS, I. S. 1963. Metamagnetism of siderite (FeCO3). *Journal of Applied Physics*, **34**, 1106–1107.

KILGORE, B. & ELMORE, R. D. 1989. A study of the relationship between hydrocarbon migration and the precipitation of authigenic minerals in the Triassic Chugwater Formation, southern Montana. *Geological Society of America Bulletin*, **101**, 1280–1288.

LEBLANC, G. E. & MORRIS, W. A. 1999. Aeromagnetics of Southern Alberta within areas of hydrocarbon accumulation. *Canadian Petroleum Geology Bulletin*, **47**, 439–454.

LIU, Q. S., LIU, S. G., QU, Z. & HOU, W. G. 1998. Magnetic and mineralogical characteristics of hydrocarbon microseepage above oil/gas reservoirs of the Tuoko region, northern Tarim Basin, China. *Science in China Series (D)*, **41**, 121–129, doi: 10.1007/BF02 932430.

LIU, Q., LUI, Q., CHAN, L., YANG, T., XIA, X. & CHENG, T. 2006. Magnetic enhancement caused by hydrocarbon migration in the Mawangmiao Oil Field, Jianghan Basin, China. *Journal of Petroleum Science and Engineering*, **53**, 25–33.

MACHEL, H. G. 1995. Magnetic mineral assemblages and magnetic contrasts in diagenetic environments—with implications for studies of paleomagnetism, hydrocarbon migration and exploration. *In*: TURNER, P. & TURNER, A. (eds) *Palaeomagnetic Applications in Hydrocarbon Exploration and Production*. Geological Society, London, Special Publications, **98**, 9–29.

MACHEL, H. G. 1996. Magnetic contrasts as a result of hydrocarbon seepage and migration. *In*: SCHUMACHER, D. & ABRAMS, M. A. (eds) *Hydrocarbon Migration and its Near-Surface Expression*. American Association for Petroleum Geologists, Tulsa, Memoir, **66**, 99–109.

MACHEL, H. G. & BURTON, E. A. 1991. Causes and spatial distribution of anomalous magnetisation in hydrocarbon seepage environments. *AAPG Bulletin*, **75**, 1864–1876.

MCCABE, C. R., VAN DER VOO, R. & BALLARD, M. M. 1984. Late paleozoic remagnetisation of the Trenton limestone. *Geophysical Research Letters*, **11**, 979–982.

MCCABE, C. R., SASSEN, R. & SAFFER, B. 1987. Occurrence of secondary magnetite within biodegraded oil. *Geology*, **15**, 7–10.

MORIN, J. 1950. Magnetic susceptibility of $\alpha$-Fe2O3 and Fe2O3 added titanium. *Physics Review*, **78**, 819–820.

PAN, Y., ZHU, R., BANERJEE, S. K., GILL, J. & WILLIAMS, Q. 2000. Rock magnetic properties related to thermal treatment of siderite: behaviour and interpretation. *Journal of Geophysical Research*, **105**, 783–794.

REYNOLDS, R. L., FISHMAN, N. S. & HUDSON, M. R. 1991. Sources of aeromagnetic anomalies over cement oil field (Oklahoma), Simpson oil field (Alaska) and the Wyoming-Idaho-Utah thrust belt. *Geophysics*, **65**, 606–617.

RIJAL, M. L., APPEL, E., PETROVSKÝ, E. & BLAHA, U. 2010. Change of magnetic properties due to fluctuations of hydrocarbon contamination groundwater in unconsolidated sediments. *Environmental Pollution*, **158**, 1756–1762.

RIJAL, M. L., PORSCH, K., APPEL, E. & KAPPLER, A. 2011. Magnetic signatures of hydrocarbon contaminated soils and sediments at the former oil-field Hänigsen, Germany. *Studia Geophysica et Geodaetica*, **56**.

SELLEY, R. C. 1992. Petroleum seepages and impregnations in Great Britain. *Marine and Petroleum Geology*, **9**, 226–244.

STONELEY, R. & SELLEY, R. C. 1986. *A field guide to the petroleum geology of the Wessex basin*. Imperial College, London.

SUK, D., PEACOR, D. R. & VAN DER VOO, R. 1990. Replacement of pyrite framboids by magnetite in limestone and implications for paleomagnetism. *Nature*, **345**, 611–613, doi: 10.1038/345611a0.

SUK, D., VAN DER VOO, R., PEACOR, D. R. & LOHMAN, K. C. 1993. Late Paleozoic remagnetisation and its carrier in the Trenton and black river carbonates from the Michigan basin. *Geology*, **101**, 795–808.

UNDERHILL, J. R. & STONELEY, R. 1998. Introduction to the development, evolution and petroleum geology of the Wessex Basin. *In*: UNDERHILL, J. R. (ed.) *Development, Evolution and Petroleum Geology of the Wessex Basin*. Geological Society, London, Special Publications, **133**, 1–18.

WALZ, F. 2002. The verwey transition – a topical review. *Journal of Physics-Condensed Matter*, **14**, R285–R340.

WATSON, D. F., HINDE, A. D. & FARRIMOND, D. 2000. Organic geochemistry of petroleum seepages within the Jurassic Bencliff grit, Osmington Mills, Dorset, UK. *Petroleum Geoscience*, **6**, 289–297, doi: 10.1144/petgeo.6.4.289.

# Rock magnetic characterization of early and late diagenesis in a stratigraphic well from the Llanos foreland basin (Eastern Colombia)

VINCENZO COSTANZO-ÁLVAREZ[1]*, MILAGROSA ALDANA[1], GERMÁN BAYONA[2], DIEGO LÓPEZ-RODRÍGUEZ[3,4] & JOAN MARIE BLANCO[1]

[1]*Departamento de Ciencias de la Tierra, Universidad Simón Bolívar, Caracas, Venezuela*

[2]*Corporación Geológica Ares, Bogotá, Colombia*

[3]*Laboratorio de Física Teórica de Sólidos, Escuela de Física, Universidad Central de Venezuela, Caracas, Venezuela*

[4]*Present address: Facultad de Ciencias Físicas, Universidad Complutense de Madrid, Ciudad Universitaria, 28040 Madrid, Spain*

**Corresponding author (e-mail: vcosta@usb.ve)*

**Abstract:** We have carried out rock magnetic characterizations of different lithofacies along the stratigraphic well Saltarín 1A, in order to learn about the various diagenetic events that could have affected the Miocene sequence of the Llanos foreland basin (Colombia). Thermomagnetic and low-temperature susceptibility measurements performed on some selected samples were complemented with analyses of Scanning Electron Microscopy (SEM), Energy Dispersive X-ray (EDX) and isothermal remanent magnetization (IRM) acquisition curves. The identification of the magnetic mineral assemblages at each depth level analysed, as well as their relative concentrations, were determined from a direct signal analysis (DSA) of the IRM curves. Samples from the top of the Guayabo formation reveal the presence of hydrocarbons-related microscopic framboids of pyrite with partial replacement of magnetite. The bottom of the Guayabo formation shows hematite and goethite and appears to record a thoroughly documented Middle Miocene global regression. In samples from the León formation, pyrrhotite could have resulted from an early diagenesis that took place in a lacustrine environment via sulphate reduction. Traces of crude oil in samples from the Carbonera formation, and the additional occurrence of hematite and magnetite, suggest that a hydrocarbons-mediated late diagenesis could also have affected the lowermost levels of Saltarín 1A.

Oil exploration at the Llanos foreland basin (eastern Colombia) has thus far been constrained to 2D seismic surveys, the acquisition of petrophysical logging data in reservoir units and the study of drill cuttings and short cores taken from the producing levels of Cretaceous and Palaeogene sandstones. More recently, the drilling of the Saltarín 1A, an approximately 700-m-long stratigraphic well located at the distal part of the Llanos foreland (Fig. 1), shed new light on the way fluctuations in subsidence and climate conditions can affect the sedimentary filling of this continental basin. The samples obtained cover the whole sequence of the Guayabo and León formations, plus the upper part of Carbonera (Fig. 2). All of these units are allegedly of Miocene age.

Bayona *et al.* (2008*a*) integrated sedimentological, stratigraphic, biostratigraphic and provenance results for this well with the purpose of characterizing major sedimentary environments and correlation surfaces, useful in the interpretation of seismic lincs.

In addition, Da Silva *et al.* (2010) showed how variations in magnetic properties along a section of Saltarín 1A relate to proxies of global climate changes during the Miocene. In order to either monitor or preclude late digenesis, they introduced the discussion about the extent of tardy alteration effects upon primary magnetic minerals.

In the present work, we seek to characterize different magnetic minerals at various stratigraphic levels of this well in an attempt to unravel the geochemical conditions that prevailed during the diagenetic events that must have affected these rocks. Such conditions were related to either palaeoenvironmental changes during deposition of the different sedimentary layers or to the ensuing migration and accumulation of oil and gas in the region. As well as the significance of this study in understanding the expected magnetic overprints contained in these rocks, we believe that it also improves the description of major lithofacies along the sedimentary column and has important implications for further

*From*: ELMORE, R. D., MUXWORTHY, A. R., ALDANA, M. M. & MENA, M. (eds) 2012. *Remagnetization and Chemical Alteration of Sedimentary Rocks*. Geological Society, London, Special Publications, **371**, 199–216.
First published online September 20, 2012, http://dx.doi.org/10.1144/SP371.13

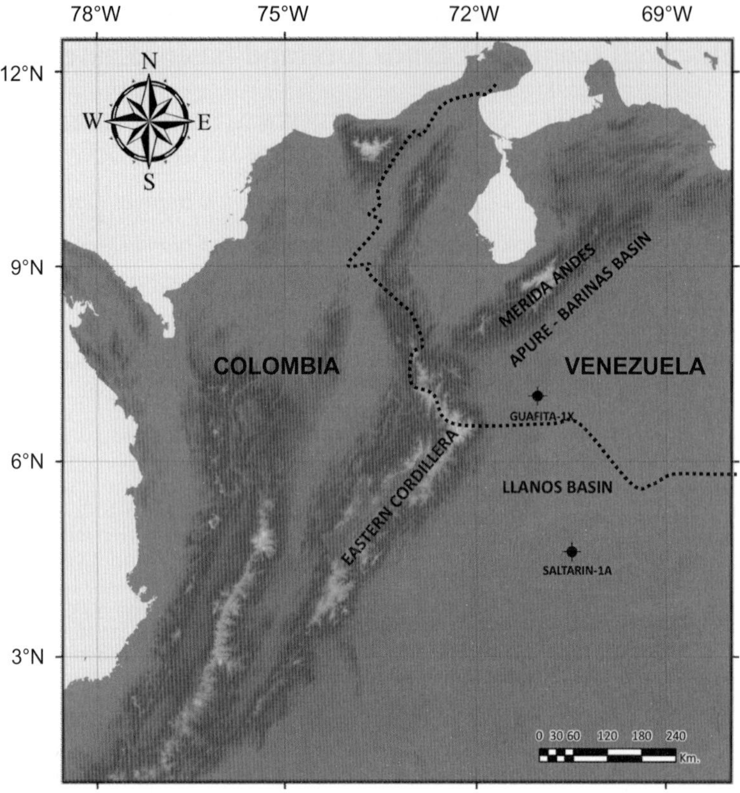



**Fig. 1.** Geographical setting of the stratigraphic well Saltarín 1A in the Llanos Foreland Basin (eastern Colombia). This map is modified from figure 1 in Da Silva *et al.* (2010).

hydrocarbon exploration ventures in the Llanos foreland basin.

The discussion of the magnetic properties of some representative samples from this stratigraphic well has been enhanced by our experimental results, including scanning electron microscopy (SEM), energy dispersive X-ray spectroscopy (EDX) analyses, isothermal remanent magnetization (IRM) acquisition curves and thermomagnetic and low-temperature susceptibility measurements. To decompose the IRM curves and to identify the number and type of magnetic minerals that contribute into their signals, we used a numerical approach based on a novel application of the Direct Signal Analysis (DSA) method developed by Aldana *et al.* (1994, 2011).

Our first aim is to explore the scope of late diagenetic events in Saltarín 1A, resulting from the chemical alteration of early Fe-oxides in redox conditions related to the presence of crude oils (e.g. Foote 1984, 1987, 1992, 1996; Saunders & Terry 1985; Benthiem & Elmore 1987; Elmore *et al.* 1987, 1993; McCabe *et al.* 1987; Saunders *et al.*

1991, 1999; Hall & Evans 1995; Porsch *et al.* 2010; Rijal *et al.* 2010, among others). Hydrocarbons-related chemical alteration has also been studied in oil fields from eastern and western Venezuela by Aldana *et al.* (1999, 2003, 2011), Guzmán *et al.* (2011) and Costanzo-Álvarez *et al.* (2006), among others. Some results from the present work are compared with those previously obtained at the neighbouring Guafita oil field (southwestern Venezuela) by Aldana *et al.* (1999, 2003) and Costanzo-Álvarez *et al.* (2000, 2006). The Guafita field seems to share with Saltarín 1A similar lithological features and the 'oil magnetic signature' of a subsurface hydrocarbon reservoir.

We also anticipate that some rock magnetic properties of this well could contribute to a better understanding of the lithological contrasts produced by diverse geochemistry and palaeoenvironmental changes along the whole stratigraphic column (Thompson *et al.* 1975; King *et al.* 1982; Robinson 1986; Doh *et al.* 1988; Tarduno 1994; Stoner *et al.* 1996; Dekkers 1997; Evans *et al.* 1997; Bloemendal *et al.* 1998).

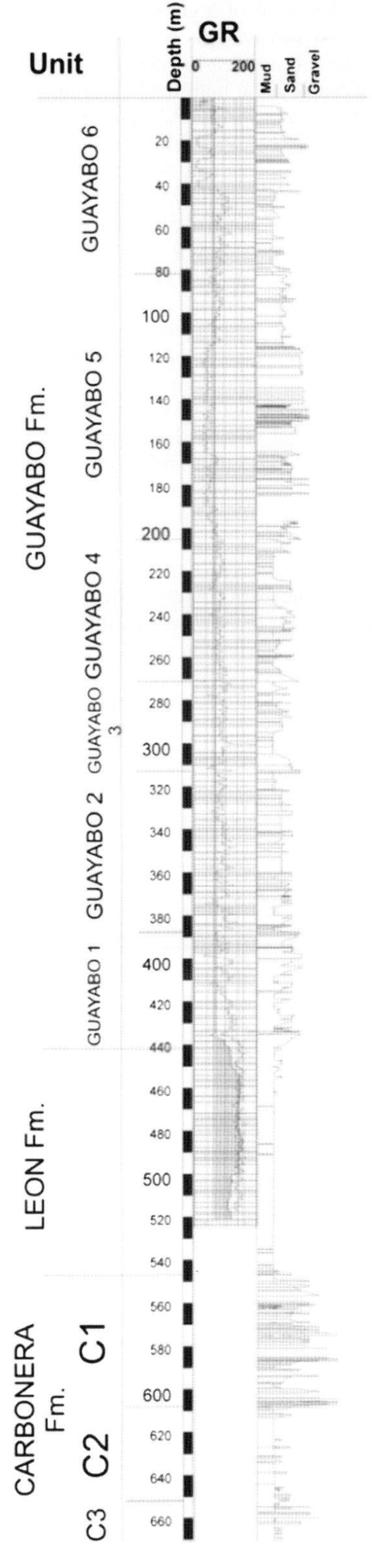



**Fig. 2.** Lithological column of the stratigraphic well Saltarín 1A with the limits of the main formations, textures and sedimentary structures.

## Geological setting

The Upper Cretaceous–Pliocene succession in the Llanos basin is an eastwards-thinning clastic wedge that records a shift in depositional environments from dominantly marine during the latest Cretaceous, to marginal and continental during the Palaeogene, and to continental in the Neogene (Bayona *et al.* 2007). The Upper Eocene–Pliocene sedimentary sequence overlies unconformably Paleocene, Cretaceous and Palaeozoic rocks in the proximal, middle and distal segments of the Llanos basin, respectively. The onset of deposition in the proximal foreland began in the Late Eocene in the southern and central Llanos, in the Middle Oligocene in the northern Llanos, and in the Early Miocene in the distal Llanos (Bayona *et al.* 2011). The Saltarín 1A is located in the distal segment of the central Llanos basin (Fig. 1). This vertical succession documents the multi-phase growth of the adjacent eastern Cordillera (Bayona *et al.* 2008*b*) to the west, intraplate uplifts to the south that separate this basin from the Putumayo–Amazonas basin (Bayona *et al.* 2011) and mantle-driving tectonic subsidence processes since the Oligocene that controlled the tabular geometry of the clastic wedge in the distal Llanos basin (Bayona *et al.* 2011; Farris *et al.* 2011).

The deepest rocks drilled in Saltarín 1A are from the Upper Carbonera formation that includes a lower sandstone unit (654.6–671 m) accumulated in a fluvial system, a middle mudstone unit accumulated in a lacustrine system (608.2–654.6 m) and an upper sandstone unit that records sedimentation in a fluvial-deltaic system (546.9–608.2 m). Overlying the Carbonera formation is the León formation (441.8–546.9 m), a muddy sequence of sediments that accumulated in a freshwater lacustrine system (Fig. 2). Finally, the youngest unit is the Guayabo formation that was divided into six lithological units (Bayona *et al.* 2008*a*).

G1 (388–441.8 m) and G2 (312.9–388 m), the two lower units, consist of green-coloured laminated mudstones grading to sandstones interbedded with light-coloured massive mudstones with ferruginous nodules. These lithologies were interpreted as the sedimentation from a fluvial-deltaic system changing to more continental sediment accumulation in fluvial floodplains.

G3 (271.5–312.9 m) and G4 (205.5–271.5 m), the overlying units, are dominantly mudstones and siltstones that accumulated in fluvial floodplains. Unit G3 has more evidence of sub-aerial exposure (light-coloured mudstones, formation of ferruginous nodules), whereas preservation of coal beds and laminated mudstones in unit G4 indicates less sub-aerial exposure of the floodplains.

G5 (81.6–205.5 m) consists of feldspar-rich sandstones that record the filling of fluvial channels.

G6 (0–81.6 m), the uppermost unit of the Guayabo formation, records a change to the floodplain with evidence of sub-aerial exposure.

## Methods

Room-temperature magnetic susceptibility $\kappa$ and S-ratios were measured in 90 cored samples provided by Hocol S.A. (Bogotá, Colombia). They were taken at approximately every 7 m, the usual spacing between samples cored for petrophysical and conventional geological purposes (i.e. from 5 up to 15 m).

The measurements of $\kappa$ were performed in a Bartington susceptometer with dual frequency (i.e. 4.65 and 0.465 Hz) of the applied field. Frequency-dependent susceptibility is less than 1% in most cases, precluding the presence of a significant fraction of superparamagnetic magnetite in these samples. In this study, $\kappa$ readings for each of the 90 samples analysed were repeated five times and a

standard deviation of less than 10% was obtained for every sample average.

For the S-ratio measurements of these 90 samples, we employed a high-field pulse magnet and a 2 G Enterprises cryogenic magnetometer with a sensitivity of $10^{-11}$ A m$^{-1}$. The S-ratio, a rock magnetic index that accounts for the relative contributions of low- and high-coercivity material to the total saturation isothermal remanent magnetization, was determined according to the definition by Bloemendal *et al.* (1992). Figure 3 presents the profiles of $\kappa$ and S-ratios showing the limits of the geological formations as well as the eight selected stratigraphic levels where we carried out more detailed experiments.

SEM and EDX analyses were performed in magnetic separates of samples 127.4 and 147.25 m from Guayabo G5, 541.75 m from the contact León/Carbonera and 616.48 from Carbonera C1/C2. These separates were obtained by applying a hand magnet to a suspension in acetone of a finely ground rock. They were then mixed with ethanol and exposed to an ultrasonic bath for *c.* 5 min. A drop of this preparation was placed on a thin layer of polyethylene set on a carbon holder, and covered by a carbon film (*c.* 10 nm). The scanning electron microscope used for SEM studies was a JEOL JSM–6390 and the X-ray probe and energy disperser was an

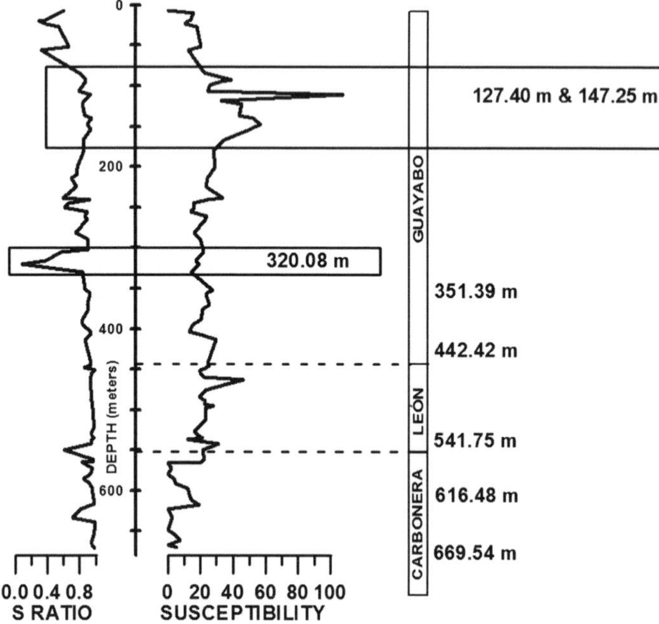


**Fig. 3.** Profiles of S-ratio and magnetic susceptibility ($\kappa$) values along the 670 m of the stratigraphic well Saltarín 1A. Boundaries between formations (i.e. Guayabo, León and Carbonera) and depth levels for some selected samples chosen for further detailed analyses are also shown. Part of these data, for the depth interval 304.75 and 610.20 m, was previously reported by Da Silva *et al.* (2010).

Oxford Instruments model 7582 (spot size 10 mm$^2$, window ATW2, resolution at 5.9 keV = 137 V).

In addition, we obtained thermomagnetic susceptibility curves for heating (up to 700 °C) and cooling cycles. These experiments were carried out in air atmosphere for samples 127.4 and 147.25 m (Guayabo G5), 320.08 m (Guayabo G2/G3), 351. 39 m (Guayabo G2), 442.42 m (contact Guayabo/ León), 541.75 m (contact León/Carbonera), 616. 48 m and 669.54 m (Carbonera). Low-temperature susceptibility experiments were performed on samples 127.4 m, 147.25 m, 541.75 m and 616.48 m within a temperature range between 275 and 75 K attained with liquid nitrogen. In all these experiments we used a Geofyzika KLY-2 KappaBridge AC Susceptibility meter.

Rock magnetic results were complemented by the analysis of IRM acquisition curves obtained for 127.4 m, 147.25 m, 320.08 m, 351.39 m, 541.75 m and 669.54 m. These samples were stepwise magnetized in an ASC IM-10–30 pulse magnet with exchangeable coils, up to 2.5 T (*c.* 10 mT increments for the first 100 mT, 25 mT increments between 100 and 200 mT, 50 mT increments between 200 and 500 mT and 500 mT increments for the last steps up to 2.5 T).

In order to analyse the IRM curves, we used a method based on DSA (Aldana *et al.* 1994) previously proposed and described by Aldana *et al.* (2011) and Rada *et al.* (2011). In these works, the method has been applied to analyse IRM curves from drill cuttings and archaeological pottery samples. The experimental IRM is described as the sum of N elementary curves modelled using the formula of Robertson & France (1994):

$$\text{IRM}(B)\!:\!\sum_{i=1}^{N}\frac{\text{Mrm}_i}{DP_i(2\pi)}\int\limits_{-\infty}^{\infty}\exp\left\{-\frac{[\log_{10}(B)-\log_{10}(B_{1/2i})]^2}{2(DP_i)^2}\right\}d[\log_{10}(B)]$$

The log $B_{1/2}$ varies in a window that includes all the possible magnetic phases present in the sample. The algorithm adjusts the contribution Mrm$_i$ (heights) of all elementary curves to the experimental IRM curve. The result is a spectral histogram of log $B_{1/2}$ from which we can obtain the number of main contributions, their widths and the mean coercivities associated with the number and type of magnetic minerals (Aldana *et al.* 2011).

The DSA is an alternative approach to that proposed by Heslop *et al.* (2002). One of the advantages of the method used here is that in the Heslop *et al.* (2002) approach we have to provide beforehand the expected number of magnetic phases in the sample, an initial estimation of the parameters that characterize each component and some information about saturation of the IRM. For the DSA method however, we only adjust the contribution of each

elementary curve. There is no need to provide initial parameters or information regarding the IRM saturation as the number of magnetic phases and their parameters are obtained from the adjusted spectral histogram of weights.

## Experimental results

The magnetic susceptibility profile of Figure 3 shows a strip of anomalous high values that spans *c.* 75–175 m. The lithologies of these stratigraphic levels are characterized by the fine- to medium-grained sandstones of the G5 informal unit at the Upper Guayabo formation (Fig. 2). IRM acquisition curves obtained for samples 127.40 m and 147.25 m within this strip are shown in Figure 4, together with their corresponding fittings and after applying the direct signal analysis (DSA) method proposed by Aldana *et al.* (1994, 2011). The spectral histograms and the parameters of the main peaks are also presented in this figure and in Table 1.

The component peak at log $B_{1/2}$ = 1.17 for 127.40 m (Table 1) could be due to the presence of either magnetite or pyrrhotite, since both minerals have low coercivities and therefore similar log $B_{1/2}$ ranges. Hence, further experiments (i.e. SEM, EDX, thermomagnetic susceptibility curves in air atmosphere and low-temperature susceptibility measurements) are required to better identify such magnetic phases. Peaks at log $B_{1/2}$ = 1.92, 2.65 and 3.58 in this sample (Table 1) could be indicative of pyrrhotite, hematite and goethite, respectively (e.g. Kruiver *et al.* 2001; Heslop *et al.* 2002). Similarly to 127. 40 m, either magnetite or pyrrhotite would produce the peak centred at log $B_{1/2}$ = 1.29 in 147.25 m; peaks at log $B_{1/2}$ = 2.67 and 3.57 suggest the occurrence of both hematite and goethite. It is important to bear in mind that the areas under each Gaussian envelope in the spectral histograms of Figure 4 are proportional to the relative amounts of their corresponding magnetic phases (Aldana *et al.* 2011). Thus, for samples at 127.4 m and 147.25 m (Fig. 4a, b and Table 1), the envelopes of the lowest log $B_{1/2}$ peaks (i.e. 1.17 and 1.29) could be related to their most abundant magnetic minerals (68% and 74%, respectively, of the magnetic fractions of these two samples; Table 1).

SEM and EDX analyses in magnetic separates from 127.4 m and 147.25 m (Fig. 5a, b) show microscopic (over 6 μm diameter) Fe and sulphur-rich framboids (spherical aggregates of submicronic crystals). The Verwey transition between 110 and 120 K (low temperature) and the sharp reversible drop at 580 °C, preceded by a progressive decrease of the susceptibility signal and a Hopkinson peak around 500 °C (heating curve), confirm the presence of mostly stoichiometric magnetite in these

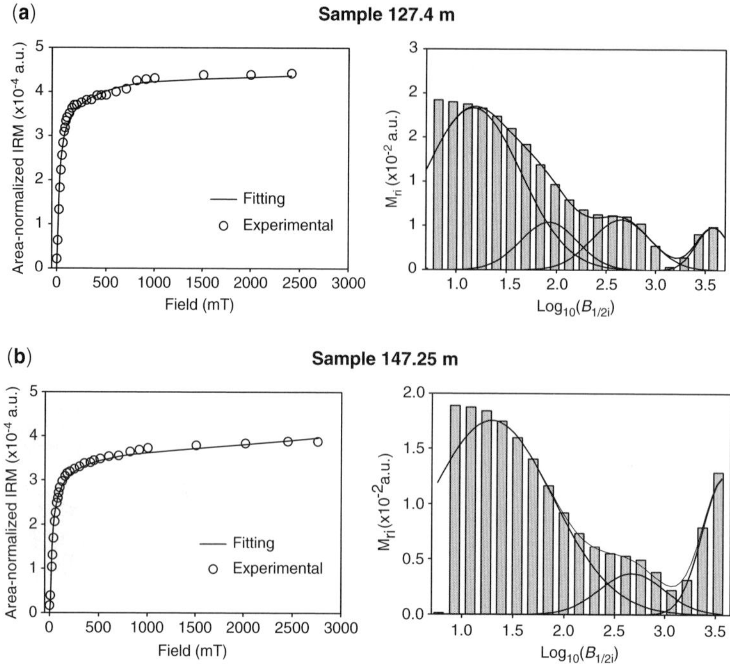

**Fig. 4.** DSA decomposition of the IRM curves for samples (**a**) 127.40 m and (**b**) 147.20 m at the Upper Guayabo formation (G5). In each case, the experimental data (circles) and the results of the DSA fitting (smooth line) are presented. The resulting spectral histogram and the adjusted Gaussian envelopes corresponding to the magnetic phases identified in these samples are also shown.

**Table 1.** *Parameters of the main peaks identified in the spectral histograms obtained after applying a DSA to the IRM curves of six selected samples.*

| Depth level (m) | Peak log $B_{1/2}$ | Magnetic phase | Area (%) |
|---|---|---|---|
| 127.4 | 1.17 | Magnetite | 68 |
|  | 1.92 | Pyrrhotite | 13 |
|  | 2.65 | Hematite | 13 |
|  | 3.58 | Goethite | 6 |
| 147.25 | 1.29 | Magnetite | 74 |
|  | 2.67 | Hematite | 8 |
|  | 3.57 | Goethite | 18 |
| 320.08 | 2.89 | Hematite | 43.5 |
|  | 3.68 | Goethite | 56.5 |
| 351.39 | 1.32 | Magnetite | 31 |
|  | 1.80 | Pyrrhotite | 33 |
|  | 2.58 | Hematite | 18 |
|  | 3.65 | Goethite | 18 |
| 541.75 | 1.38 | Magnetite | 30 |
|  | 1.86 | Pyrrhotite | 27 |
|  | 2.59 | Hematite | 8 |
|  | 3.74 | Goethite | 35 |
| 669.54 | 1.25 | Magnetite | 40 |
|  | 1.83 | Pyrrhotite | 23.5 |
|  | 2.62 | Hematite | 10 |
|  | 3.66 | Goethite | 26.5 |

The log $B_{1/2}$ centres of the peaks allow the identification of the different magnetic minerals. The area of each main contribution allows the calculation of the relative proportion of these minerals in each sample.

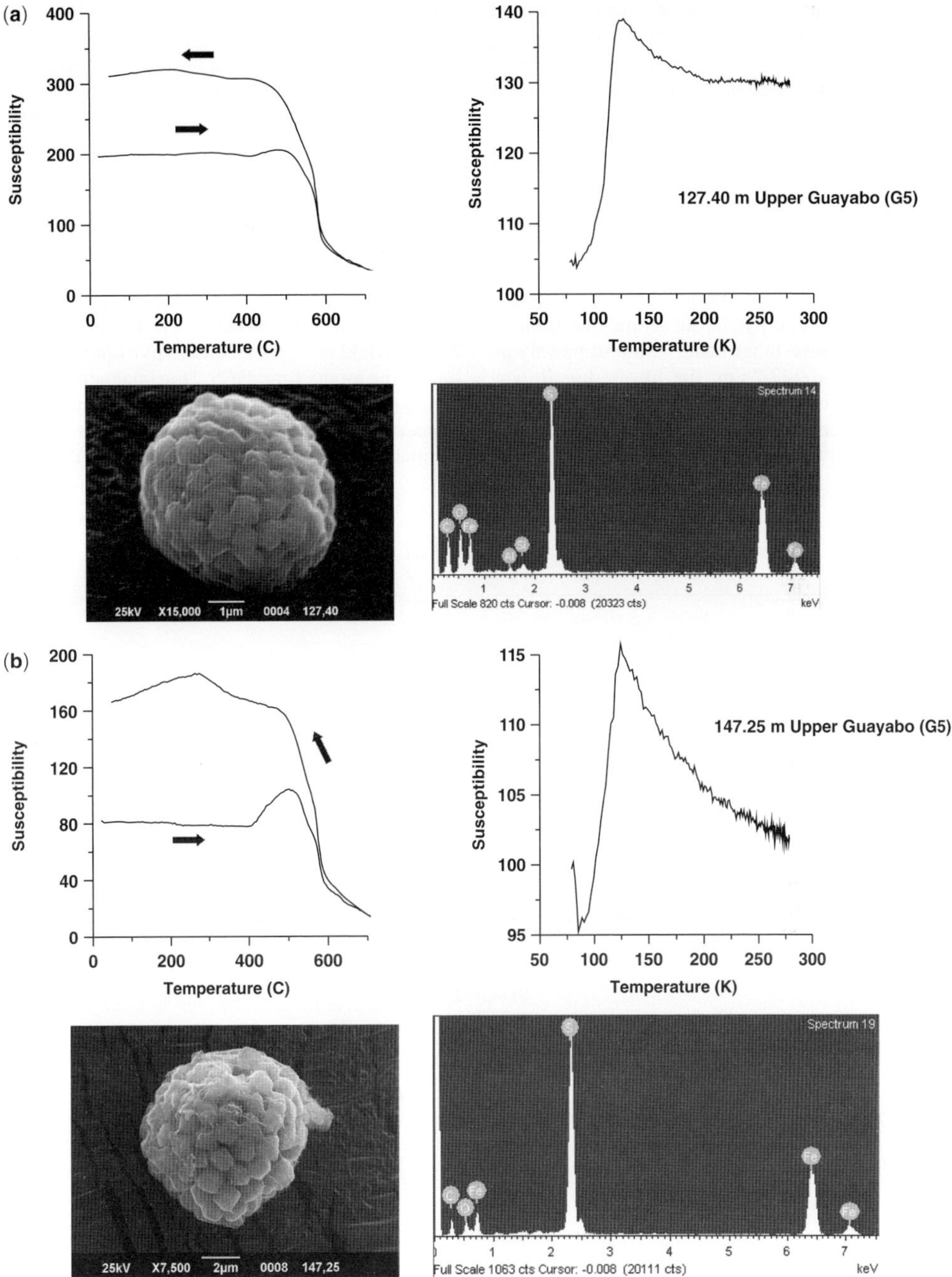



**Fig. 5.** Thermomagnetic susceptibility curves (heating and cooling) and low-temperature magnetic susceptibility measurements carried out in air atmosphere for samples (**a**) 127.40 and (**b**) 147.20 m, located within Guayabo's G5 anomalous magnetic susceptibility strip (*c.* 100–150 m). SEM photomicrographs of magnetic separates and EDX results for these same levels are also shown, revealing the presence of Fe and S-rich framboids of authigenic origin.

samples. Hematite is also present as revealed by the significant magnetic susceptibility signal above 580 °C (Fig. 5). On the other hand, these experiments do not show evidence for pyrrhotite and goethite. This is probably due to the small concentrations of these two minerals contrasting with the overwhelming prominence of magnetite. Based on these results, we argue that the lowest log $B_{1/2}$ peak centres identified in 127.4 m and 147.25 m, with the highest percentage areas, may be most likely due to magnetite rather than to pyrrhotite (Table 1).

IRM curves were also obtained at the bottom of the Guayabo formation for samples 320.08 m (G2/G3) and 351.39 m (G2) (Fig. 6a, b, respectively). The spectral histogram of 320.08 m has a peak centred at log $B_{1/2} = 2.89$ produced by hematite. A second peak at log $B_{1/2} = 3.68$ could be due to goethite. Neither magnetite or pyrrhotite seems to appear in this sample. On the other hand, in the same way as for 127.4 m and 147.25 m, the lowest log $B_{1/2} = 1.32$ peak centre in sample 351.39 m is associated with the presence of magnetite (Table 1). In this same sample, peak centres at log $B_{1/2} = 1.80, 2.58$ and 3.65 are most likely related to pyrrhotite, hematite and goethite, respectively.

Thermomagnetic susceptibility curves from levels 320.08 m (G2/G3) and 351.39 m (G2) are presented in Figure 7a, b (after Da Silva *et al.* 2010). When these curves are measured in ambient atmosphere, a sharp increase of the signal above 400 °C usually indicates the formation of new magnetic minerals during heating. In some samples, the susceptibility might also show a small bump around 220 °C that could be due to the shift ($\lambda$ transition) of antiferromagnetic hexagonal pyrrhotite to a ferromagnetic phase that takes place between 200 and 265 °C (Dekkers 1989). Natural pyrrhotite is a mixture of the monoclinic $Fe_7S_8$ (ferrimagnetic) and the hexagonal $Fe_9S_{10}$. A $\lambda$ transition within the low-temperature range of the heating curve is therefore distinctive and diagnostic of pyrrhotite (Dunlop &

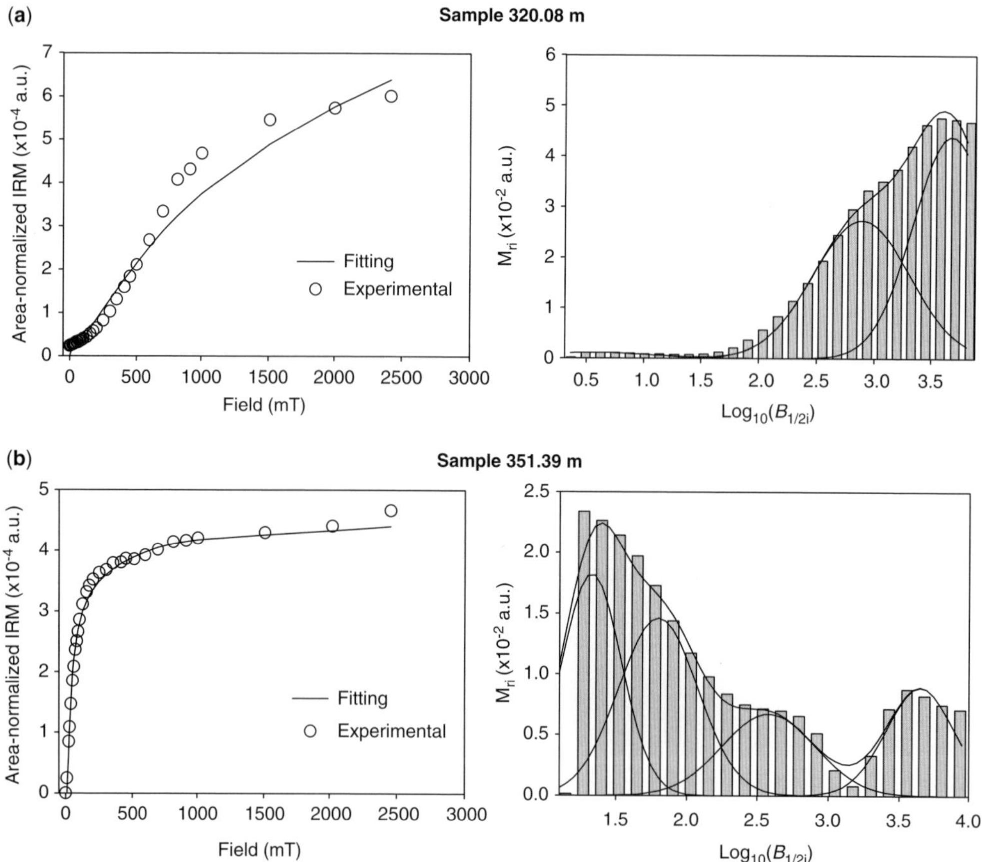

**Fig. 6.** DSA decomposition of the IRM curves for samples (**a**) 320.08 m (boundary G2/G3) and (**b**) 351.39 m (unit G2). In each case, the experimental data (circles) and the results of the DSA fitting (smooth line) are presented. The resulting spectral histogram and the adjusted Gaussian envelopes corresponding to the magnetic phases identified in these samples are also shown.

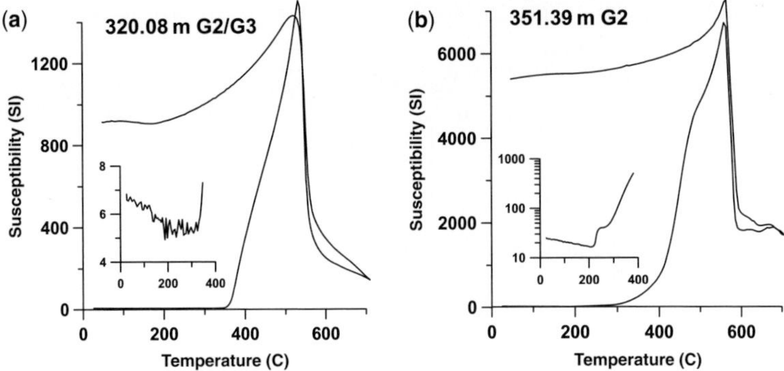

**Fig. 7.** Thermomagnetic susceptibility curves (heating and cooling) for samples (**a**) 320.08 m (boundary G2/G3) and (**b**) 351.39 m (unit G2) within the Guayabo formation. The magnetic susceptibility curve for 320.08 m suggests the presence of hematite. Whereas the thermomagnetic susceptibility curve for 351.39 m suggests the presence of hexagonal pyrrhotite with a λ phase above 200 °C, the same feature is not observed in 320.08 (see insets). This figure is equivalent to figure 4 in Da Silva *et al.* (2010).

Özdemir 1997). Hence, the small increase of the magnetic susceptibility (over 200 °C) that is observed in sample 351.39 m, when the scale is blown up for a temperature interval between *c.* 100 and 300 °C, is probably due to the presence of pyrrhotite. Such a minute rise is augmented in Figure 7b by a logarithmic scale. Pyrrhotite could also account for the peak centred at $\log B_{1/2} = 1.84$. In contrast, there is not an equivalent increase in sample 320.08 m that is also magnetically weaker, suggesting the presence of only hematite and goethite for this sample as previously inferred from the analysis of its corresponding IRM curve.

Both 320.08 m and 351.39 m samples show a rise of the magnetic susceptibility at temperatures of *c.* 400 °C, followed by a drop above 500 °C. In the case of 320.08 m, this behaviour is the likely result of hematite reduction to magnetite upon heating. For the 351.39 m sample, such a feature is probably due to the presence of pyrrhotite that irreversibly transforms to magnetite (Dunlop & Özdemir 1997). Moreover, the susceptibility signal in this sample seems to be enhanced by a Hopkinson peak at *c.* 500 °C. Conversely, the presence of hematite and goethite in 351.39 m is not evident from this curve (Fig. 7b).

Underlying the Guayabo sequence is the León formation, which appears to have accumulated in a freshwater lacustrine system (Bayona *et al.* 2008*a*). Figure 8a (after Da Silva *et al.* 2010) shows a thermomagnetic curve for sample 442.42 m located at the Guayabo/León contact. This figure also presents a thermomagnetic and a low-temperature susceptibility curve for 541.75 m at the León/Carbonera contact, together with results from its corresponding SEM and EDX analyses (Fig. 8b). In both thermomagnetic susceptibility curves there is

a small bump of the susceptibility values within the 200–265 °C interval, followed by a second progressive increase of the signal that starts at *c.* 400 °C. Finally, both curves drop above the 500 °C temperature (insets Fig. 8a, b). As in sample 351.39 m, the first bump is probably due to the presence of hexagonal pyrrhotite whereas the second bump might indicate neoformation of magnetite during laboratory heating. Contrasting with 127.40 m and 147.25 m (Fig. 5), the low-temperature susceptibility curve of 541.75 m in Figure 8b does not display a Verwey transition between 110 and 120 K, probably suppressed by the presence of non-stoichiometric or partially oxidized magnetite in this sample (Özdemir *et al.* 1993).

The bottom of the Saltarín 1A stratigraphic column encompasses part of the Carbonera formation, a sequence of sandstones accumulated in a fluvial-deltaic system. EDX analyses of magnetic separates obtained from sample 616.48 m reveal a dominant presence of sulphur and minute amounts of Fe (Fig. 9a). Additionally, the thermomagnetic susceptibility curves for samples 616.98 m and 669.54 m, at the lowermost level of this formation (Fig. 9), behave in the same way as those for 351.39 m (Fig. 7b), 442.42 m and 541.75 m (Fig. 8a, b); namely, there is a small increase of the magnetic susceptibility at *c.* 220 °C (Fig. 9a and inset in Fig. 9b) due to the presence of pyrrhotite and the possible formation of magnetite upon heating. As for 541.75 m, the behaviour of the low-temperature susceptibility curve for 616.48 m in Figure 9a precludes the presence of stoichiometric magnetite.

To further characterize the magnetic mineral assemblages that are present in samples from León and Carbonera, we also analysed the IRM

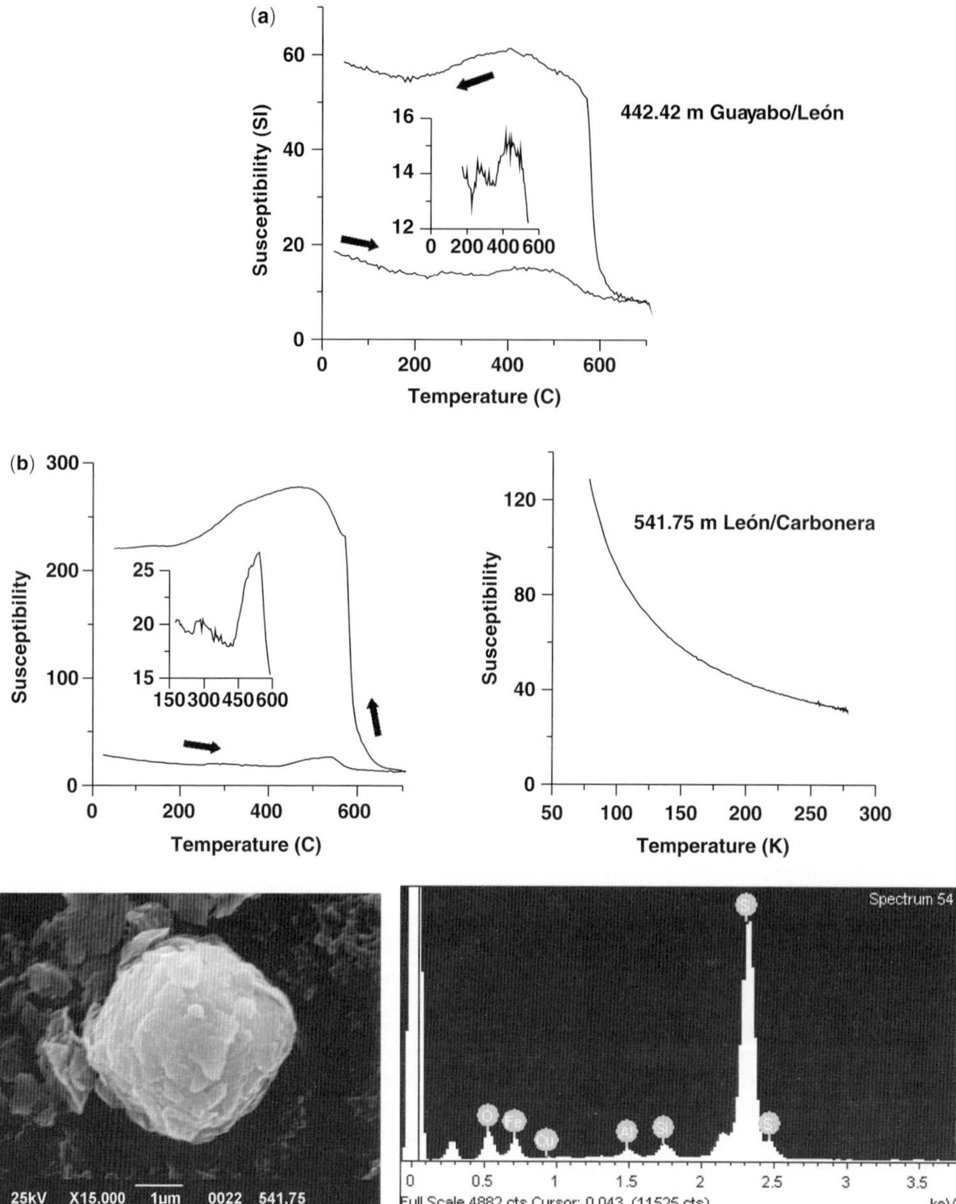



**Fig. 8.** (**a**) Thermomagnetic susceptibility curve (heating and cooling) for sample 442.42 m at the boundary between Guayabo and León formations (after fig. 5a in Da Silva *et al.* 2010). (**b**) Thermomagnetic susceptibility curves (after fig. 6a in Da Silva *et al.* 2010) and low-temperature magnetic susceptibility measurements for sample 541.75 m, located on the boundary between León and Carbonera. All the experiments were performed in air atmosphere. SEM photomicrograph and EDX spectra from magnetic separates of 541.75 m are also shown. Magnetic susceptibility heating curves for (a) and (b) suggest the presence of hexagonal pyrrhotite with a λ phase transition above 200 °C (see insets). SEM photomicrograph and EDX analyses reveal the presence of sulphur and minute amounts of Fe in 541.75 m.

acquisition curves of 541.75 m at the León/Carbonera contact (Fig. 10a) and 669.54 m (Fig. 10b) within Carbonera itself. The histogram for the 541.75 m sample reveals magnetite and hematite (i.e. $\log B_{1/2}$ peaks centred at 1.38 and 2.59, respectively) as well as pyrrhotite associated with a log

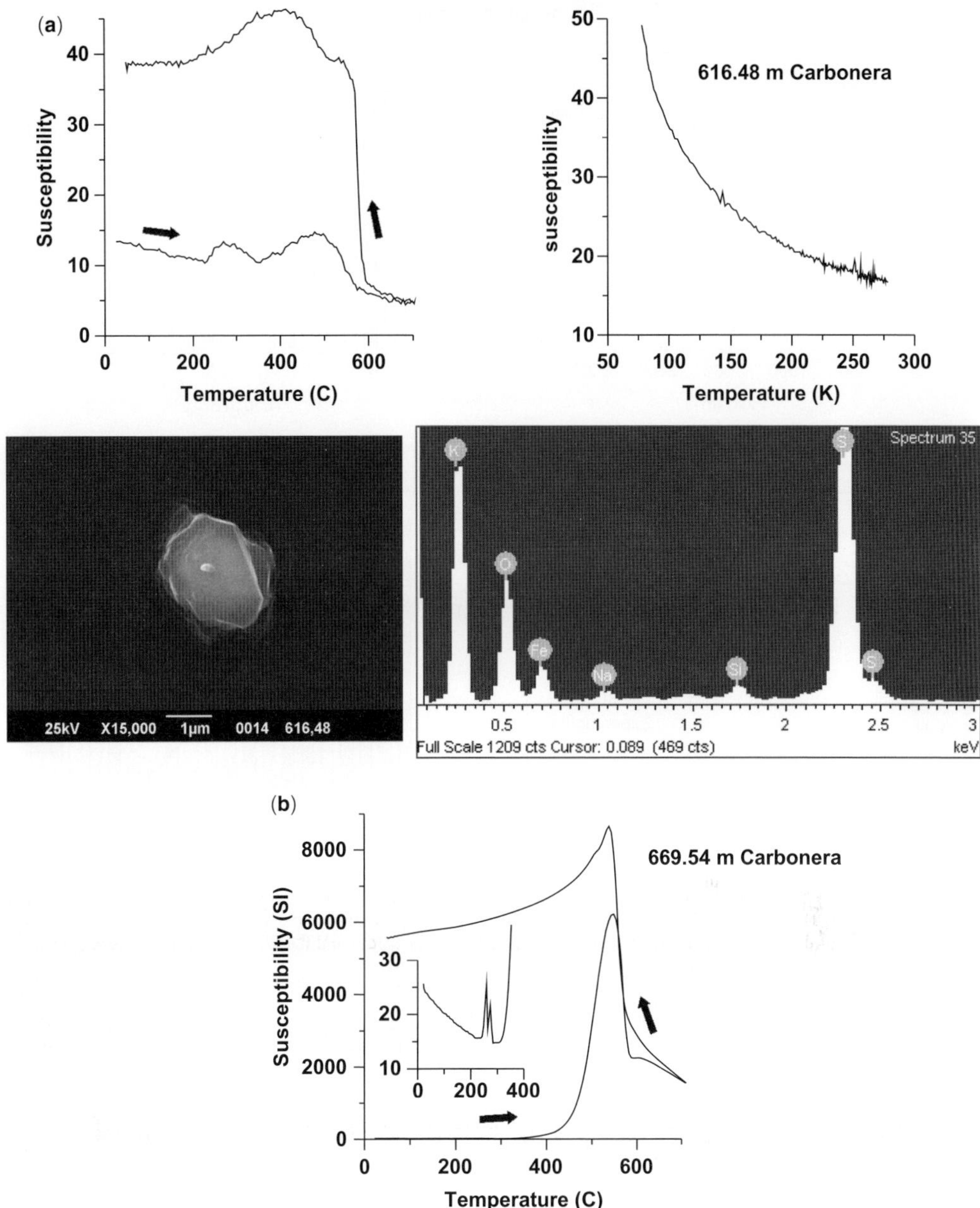

**Fig. 9.** (**a**) Thermomagnetic susceptibility curve (after fig. 6b in Da Silva *et al.* 2010) and low-temperature magnetic susceptibility measurements for sample 616.48 m at the boundary C1/C2 (Upper Carbonera formation). These results are accompanied by the corresponding SEM photomicrograph and EDX spectra from magnetic separates of 616.48 m. (**b**) Thermomagnetic susceptibility curve (after fig. 6c in Da Silva *et al.* 2010) for sample 669.54 m at C3. Magnetic susceptibility heating curves for (a) and (b) suggest the presence of hexagonal pyrrhotite with a $\lambda$ phase transition above 200 °C (see insets). SEM photomicrograph and EDX analyses reveal the presence of sulphur and minute amounts of Fe in 616.48 m. All the thermomagnetic susceptibility experiments were performed in air atmosphere.

$B_{1/2} = 1.86$ peak centre. Goethite also seems to be present in this sample with a peak at $\log B_{1/2} = 3.74$. Similarly to 541.75 m, sample 669.54 m clearly shows the coexistence of magnetite ($\log B_{1/2} = 1.25$), pyrrhotite ($\log B_{1/2} = 1.83$), hematite ($\log B_{1/2} = 2.62$) and goethite ($\log B_{1/2} = 3.66$).

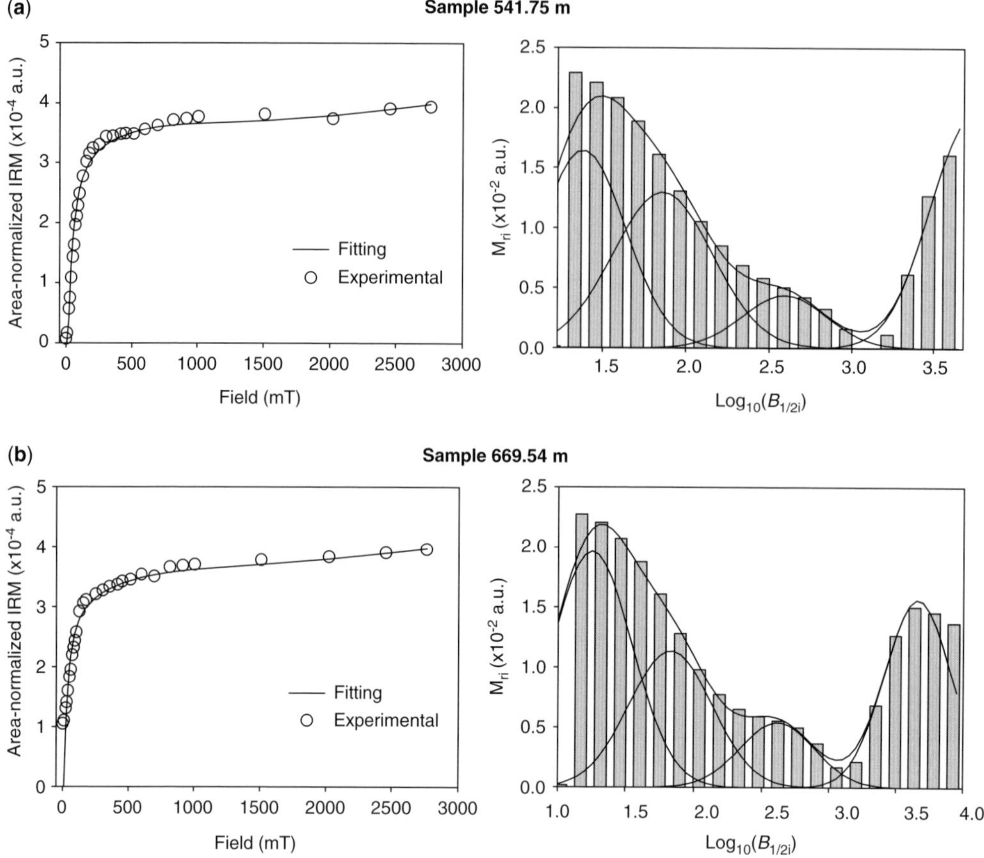

**Fig. 10.** DSA decomposition of the IRM curves for samples (**a**) 541.75 m, located on the boundary between León and Carbonera, and (**b**) 669.54 m at the Carbonera formation. In each case, the experimental data (circles) and the results of the DSA fitting (smooth line) are presented. The resulting spectral histogram and the adjusted Gaussian envelopes corresponding to the magnetic phases identified in these samples are also shown.

## Discussion

With magnetic Fe and sulphur-rich microscopic framboids, samples 127.40 m and 147.25 m come from a *c.* 100-m-thick strip of anomalous high susceptibility values, far above from the oil producer horizons of the Carbonera formation. The depth interval where this strip is located (i.e. 75–175 m) shares some lithological similarities with those depth levels where a reducing zone was previously recognized in oil wells from the neighbouring Guafita field (e.g. Costanzo-Álvarez *et al.* 2000; Díaz *et al.* 2000, 2006). In both cases in fact, these sediments are mostly immature sandstones of fluvial provenance that belong to the Guayabo formation in Colombia and to the Guayabo group of molasses (Rio Yuca/Parángula formation) in Venezuela. Since the reducing zone recognized in the Guafita oil field seems to be related to the

underlying petroleum reservoir, and is also characterized by botryoidal magnetic minerals, it is reasonable to believe that framboids in Saltarín 1A are also linked to the presence of hydrocarbons. Authigenic magnetic spherules and pseudoframboids of sulphide/iron oxides might however be formed in the absence of crude oils (e.g. Reynolds *et al.* 1990; Suk *et al.* 1990, 1992, 1993). However, it has been widely documented that either biodegradation or inorganic chemical processes can also form these types of minerals in the proximity (and/or above) of a leaking reservoir. Recently, Porsch *et al.* (2010) have conducted magnetic susceptibility measurements in soil samples following the signature of microbial Fe-mineral transformation that could be associated with hydrocarbon contamination.

Benthiem & Elmore (1987), McCabe *et al.* (1987) and Elmore *et al.* (1987, 1993) proposed a causal link between crude oils and spherical aggregates of submicronic magnetite in samples of solid bitumen and speleothems. In addition, Jiang *et al.* (2001) found authigenic framboids of Fe-sulphides related to sulphur-rich hydrocarbons in mudstones from the Lower Gutingkeng formation in southwestern Taiwan. Reynolds *et al.* (1993) also recognized post-depositional Fe-sulphides (i.e. greigite and pyrrhotite) linked to oil microseepage over Cement (Oklahoma) and Simpson (Alaska) fields, as well as the Edwards deep gas trend in southern Texas coastal plains. In the particular case of a number of Venezuelan and Colombian oil wells, Aldana *et al.* (1999, 2003, 2011), Guzmán *et al.* (2011) and Costanzo-Álvarez *et al.* (2006) (among others) have put forward a likely relationship between crude oils and framboids of magnetite and/or Fe-sulphides.

Conversely, Suk *et al.* (1990, 1992, 1993) have reported the presence of magnetic framboids in some environments that show no association with hydrocarbons. They actually argued that, in remagnetized rocks from the Onondaga, Helderberg and Trenton limestones in New York, framboidal magnetite seems to replace pyrite by fluid-mediated oxidation; however, in their unremagnetized counterparts (i.e. Wabash formation in Indiana and the Mississippian Pride Mountain formation in Alabama), both hematite and magnetite can be present. Although we should not exclude beforehand a hydrocarbon-independent process to explain the origin of the spherical aggregates in Saltarín 1A, such a possibility therefore seems to be remote considering the geological setting of the oil-rich Llanos foreland basin.

For the specific cases of the Fe and sulphur framboids of Saltarín 1A and the Guafita counterparts, chemical differences seem to suggest distinct hydrocarbons-induced magnetic diagenesis pathways. In fact, SEM and EDX analyses of magnetic separates from Guafita's oil well samples at their susceptibility anomalous levels (between 200 and 300 m) reveal the restrictive presence of magnetite spherules with no sulphur or Ti content at all (Costanzo-Álvarez *et al.* 2000, 2006; Díaz *et al.* 2006). Díaz *et al.* (2006) and Costanzo-Álvarez *et al.* (2006) proposed a possible mechanism to explain the formation of these sulphur-free framboids in Guafita. They argue that, at certain shallow depth levels, a hydrocarbons-mediated transfer of electrons from the altered organic matter to $Fe^{3+}$ yields extractable organic matter (EOM) and $Fe^{2+}$ magnetic minerals (e.g. magnetite), defining a zone of anomalous magnetic susceptibility, EOM and organic-matter-free radical concentration (OMFRC) values. Such a zone is a rather wide strip where the appropriate reducing conditions for diagenesis of the earliest Fe-oxides have been achieved, namely

type A magnetic anomalies as referred to by Costanzo-Álvarez *et al.* (2000) and Díaz *et al.* (2000).

On the other hand, magnetic framboids with sulphur content were previously reported by Guzmán *et al.* (2011) and Aldana *et al.* (2011) in samples from high magnetic susceptibility anomalies at shallow depth levels of some eastern Venezuela oil wells. They claim that, in these cases, a complex process might have taken place involving a series of chemical reactions between the $H_2S$, the organic matter and the earliest Fe-oxides. The by-products of these reactions would be both Fe-oxides and sulphides in a near-surface reducing zone. Although in Guafita's oil wells a hydrocarbons-mediated transfer of electrons from the organic matter must have occurred, producing magnetite (Díaz *et al.* 2006), we argue that a higher concentration of $H_2S$ in the crude oils of the Colombian Llanos foreland basin might account for the formation of secondary Fe-sulphides in Saltarín 1A.

A comparison between Saltarín 1A and Guafita oil wells, regarding the different hydrocarbons-related process that resulted in magnetite and/or Fe-sulphide diagenesis, must take into account both the depth of the reservoir and the quantity of sulphur in the crude oils. In the case of Guafita, the producing levels are located at depths lower than 1800 m (Guafita formation); in the distal part of the Llanos foreland basin, the petroleum reservoir at the Carbonera formation has maximum depths of *c.* 1000 m. Moreover, in a geochemical study of light crude oils (gravity values from 30 to 36° API) from the Guafita field, Labrador *et al.* (1995) found that their sulphur content varies between 0.54 and 0.64%. On the other hand, geochemical studies of hydrocarbons from the Llanos (Palmer & Russell 1988) and other Colombian sedimentary basins (Rangel *et al.* 2003) show a progressive increase of the concentration of sulphur with the rise of the specific gravity. In the Llanos basin, for instance, Palmer & Russell (1988) recognized five families of crude oils (gravity values from 40 to 15° API) with amounts of sulphur as high as 1.8% for 15° API (family 4). In the particular case of Saltarín 1A, hydrocarbons seem to be heavier than in Guafita (Diego García, pers. comm., 2011). Providing that such a variation in specific gravity is accompanied by a corresponding increase in the concentration of sulphur, and that Saltarín 1A strata were exposed to a digenetic setting enriched in strong reductants (i.e. a proximal reservoir), it would be possible to explain the presence of sulphur in these framboids. Burton *et al.* (1993) argued that under certain thermochemical conditions that include the closeness of leaking hydrocarbon accumulations and high total sulphur content, pyrite could be formed at shallow depth levels as a result of hydrocarbons-induced hematite reduction. The hematite identified in samples 127.40 m and 147.25 m from the analyses of their corresponding IRM curves could therefore be an earlier magnetic phase, whereas partial replacement of magnetite, in pyrite framboids with an origin related to the presence of hydrocarbons (McCabe *et al.* 1987; Machel 1995), would be the result of a late oxidation event comparable to that described by Suk *et al.* (1990). The modelling of mineral assemblages proposed by Burton *et al.* (1993) could also apply to Guafita under different thermochemical conditions (i.e. distal leaking hydrocarbon accumulations with low sulphur content). In this case, magnetite would be formed at shallow depth levels via hydrocarbons-induced reduction of hematite.

Further down along Saltarín 1A, sample 320.08 m lies within a depth interval (300–350 m) of low S-ratios (Fig. 3) that nearly coincides with a palynological age reported for the top of the Middle Miocene (Carlos Jaramillo, pers. comm., 2009). This interval also encompasses Guayabo's G3 and G2 informal units (Bayona *et al.* 2008*a*). According to Da Silva *et al.* (2010), low S-ratios at these depth levels could be associated with an important transition between G3 and G2 (Figs 2 & 3), from oxidized palaeosols to alluvial plains accumulated in reducing conditions, and a thoroughly documented global regression that occurred at the Serravallian stage (Vail *et al.* 1977). The coexistence of both hematite and goethite, inferred from the analyses of the IRM acquisition curves (Fig. 6a) and the thermomagnetic susceptibility curves (Fig. 7a), could serve as additional evidence for the G3/G2 pedoclimatic change or for a late alteration of hematite to goethite, related to the variation of the weathering conditions (i.e. from warmer and drier to cooler and moister environments).

For the rest of the samples that we analysed in detail (i.e. 351.39 m in Lower Guayabo, 442.42 m in Guayabo/León, 541.75 m in León/Carbonera and 616.48 m and 669.54 m in Carbonera), rock magnetic experiments show that pyrrhotite is an important magnetic phase. Pyrrhotite might have formed in these samples during the early stages of sedimentation of the sedimentary sequence. However, traces of crude oil observed in a few stratigraphic levels of the Carbonera formation cast some doubts about the early origin of this mineral (Alejandro Mora, pers. comm., 2011).

Because of the type of palaeoenvironments associated with the León formation, it is likely that (at least in samples 442.42 m and 541.74 m) pyrrhotite would be a by-product of an early diagenesis not related to the presence of hydrocarbons. Indeed, authigenic formation of magnetic Fe-sulphides in lacustrine environments, such as those that seem to have prevailed during the accumulation of the León strata, usually occurs at the first stages of sedimentation via sulphate reduction, taking advantage of native sulphate and organic carbon (Tric *et al.* 1991; Shouyun *et al.* 2002; Sagnotti 2007). Euxinic lacustrine environments with restricted water circulation, high production of organic matter and high sedimentation rates also favour the formation of Fe sulphides. Sediments in oxygen-depleted bottom waters become anoxic and rich in organic matter due to the rapid burial that inhibits their oxidation. Providing there is a limited amount of sulphide, reactive iron and/or organic matter available, greigite and pyrrhotite would be preserved (e.g. Weaver *et al.* 2002; Otamendi *et al.* 2006; Babinszki *et al.* 2007). Pyrrhotite is usually more stable than greigite in most reducing environments with a higher concentration of $H_2S$ (i.e. higher consumption of organic carbon). Moreover, according to Burton *et al.* (1993), under certain appropriated conditions of pressure and temperature that apply to deep geological settings, originally formed pyrrhotite could remain stable even at the proximity of a leaking hydrocarbons accumulation with crude oils of high total sulphur content.

On the other hand, the occurrence of magnetite and hematite in Lower Guayabo, León and Carbonera samples, revealed by the analysis of their corresponding IRM curves (Figs 6, 10 & 11), draws our attention to the fact that some hydrocarbons-mediated late diagenesis cannot be completely ruled out either. Indeed, Burton *et al.* (1993) argued that, under the same thermochemical conditions that favour the stability of originally formed pyrrhotite at deep geological settings, hematite could also reduce to secondary pyrite, magnetite and/or pyrrhotite. However, the rather good correlation obtained by Da Silva *et al.* (2010) between S-ratios, magnetic susceptibility and global climate proxies ($\delta^{18}O$) at the stratigraphic levels that include the Lower Guayabo, León and Carbonera formations seems to preclude the possibility of an overwhelming late obliteration of the earliest magnetic minerals due to the presence of hydrocarbons. It therefore remains unclear in what extent pyrrhotite is either a late or a stable early diagenesis magnetic phase.

The Guayabo, León and Carbonera formations, encompassed by the different stratigraphic levels of Saltarín 1A, show not only lithological differences (Bayona *et al.* 2008*a*) but also significant magnetic variability. Horng *et al.* (1992) and Kao *et al.* (2004) have previously argued that there is a relationship between lithology and magnetic mineral assemblages. This fact is clearly illustrated by Figure 11 which depicts the relative amounts of magnetite, pyrrhotite, hematite and goethite for each of the six stratigraphic levels where IRM experiments were carried out (percentage areas listed in Table 1). Magnetite seems to be the dominant mineral in the fine- to medium-grained continental sediments

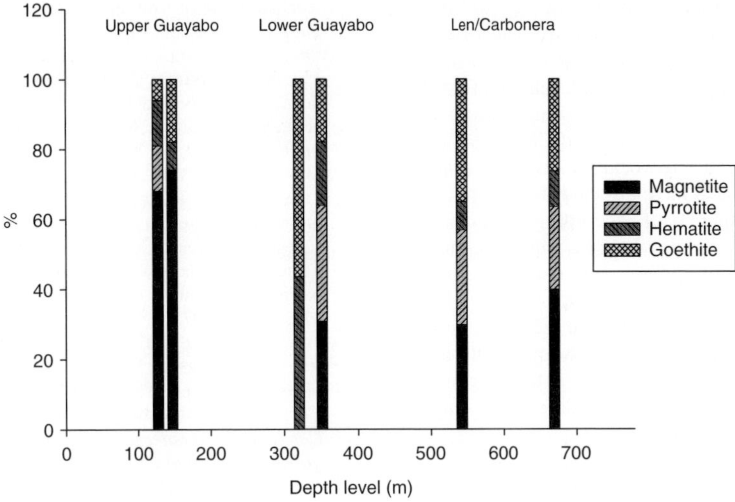



**Fig. 11.** Bar diagram showing the relative amounts of magnetite, pyrrhotite, hematite and goethite over the total fraction of magnetic minerals for each depth level where IRM experiments were carried out (percentage areas are listed in Table 1).

of the Upper Guayabo sandstones (G5), characterized by an anomaly of relatively high magnetic susceptibility values. On the other hand, sample 320.08 m coincides with a palaeoenvironmental change from oxidized palaeosols to alluvial plains accumulated in reducing conditions, and shows only hematite and goethite. Finally, pyrrhotite appears as an important magnetic mineral in the rest of the vertical succession analysed for the Lower Guayabo (G2), León and Carbonera formations (medium- to fine-grained lacustrine mudstones and fluvial-deltaic units). In almost all these samples, goethite must be the latest by-product resulting from the weathering of the earliest magnetic minerals such as pyrrhotite.

## Conclusions

Rock magnetic characterizations of some depth levels in the stratigraphic well Saltarín 1A provide important information about the early and late diagenesis events that affected the Upper Cretaceous–Pliocene sequence of the distal Llanos foreland basin. The record of these events seems to be encrypted in the way the magnetic mineral assemblages change down the sedimentary sequence.

At shallow depth levels (75–175 m) within a strip of anomalous high susceptibility values, partial replacement of magnetite (about 70% of the total magnetic fraction) in most probably hydrocarbons-produced pyrite framboids seems to be the result of a late oxidation event that affected the fine- to

medium-grained continental sandstones of the Upper Guayabo formation. Although the oil wells at the nearby Guafita field show lithological and rock magnetic features that resemble those of Saltarín 1A, their magnetite spherules have no sulphur at all; different hydrocarbons-related magnetic diagenesis pathways must therefore have taken place. As a matter of fact, opposite to the Guafita oil field the distal part of the Llanos foreland basin seems to be characterized by a relatively shallow reservoir and crude oils of elevated sulphur content. Although in Saltarín 1A a higher concentration of $H_2S$ allowed the formation of secondary pyrite, in the Guafita oil wells a net transfer of electrons from the organic matter, induced by the hydrocarbons, must have occurred producing $Fe^{2+}$ magnetic minerals (i.e. magnetite spherules).

Further down on the Guayabo formation, a low S-ratio anomaly (300–350 m) could be associated with a thoroughly documented global regression that took place at the end of Middle Miocene times. This anomaly also agrees with an important pedoclimatic variation at the bottom of Guayabo, from oxidized palaeosols to alluvial plains accumulated in reducing conditions that could also have account for the sole presence of both hematite and goethite.

The rest of the depth levels analysed, including the lowermost units of Guayabo plus the León and Carbonera formations, are mainly characterized by a significant fraction of pyrrhotite together with magnetite and hematite. In samples from the León formation (medium- to fine-grained freshwater lacustrine mudstones) pyrrhotite could be the by-product of an early diagenesis that took place via sulphate reduction in euxinic conditions. However, traces of crude oil in some samples from Carbonera, and the presence of hematite and magnetite at the lower Guayabo, León and Carbonera formations, suggest that a hydrocarbons-mediated late diagenesis could also have affected these depth levels of the stratigraphic well. According to Burton *et al.* (1993), under certain thermochemical conditions that include the proximal accumulation of crude oils with high total sulphur content, originally formed pyrrhotite could remain stable whereas hematite would reduce to both secondary magnetite and/or pyrrhotite. It is unclear in what extent pyrrhotite is either a late or a stable early diagenesis magnetic phase in these samples.

Samples were generously provided by A. Mora (Hocol S.A., Bogotá, Colombia). We are also grateful to D. García (ECOPETROL, Colombia), C. Jaramillo (Smithsonian Tropical Research Institute, Panamá), W. Williams and J. Tait (University of Edinburgh, Scotland), M. Rada and G. Rodríguez (Universidad Simón Bolivar, Venezuela) and Corporación Geológica ARES (Colombia). This research was partially funded by the Decanato de Investigación y Desarrollo, the Dirección de Desarrollo Profesoral (Universidad Simón Bolívar, Venezuela) and LOCTI Research Grants (Venezuelan Ministry of Science and Technology) to MA and VC-A.

# References

ALDANA, M., LAREDO, E., BELLO, A. & SUAREZ, N. 1994. Direct signal analysis applied to the determination of the relaxation parameters from TSDC spectra of polymers. *Journal of Polymer Science Part B, Polymer Physics*, **32**, 2197–2206.

ALDANA, M., COSTANZO-ÁLVAREZ, V., VITIELLO, D., COLMENARES, L. & GÓMEZ, G. 1999. Framboidal magnetic minerals and their possible association to hydrocarbons: La Victoria oil field (South-western Venezuela). *Geofísica Internacional*, **38**, 137–152.

ALDANA, M., DÍAZ, M., COSTANZO-ÁLVAREZ, V., GONZÁLEZ, F. & ROMERO, I. 2003. EPR studies in soil samples from a prospective area at the Andean Range, Venezuela. *Revista Mexicana de Física*, **49**(S3), 4–6.

ALDANA, M., COSTANZO-ÁLVAREZ, V., GÓMEZ, L., GONZÁLEZ, C., DÍAZ, M., SILVA, P. & RADA TORRES, M. A. 2011. Identification of magnetic mineralogies associated to hydrocarbon microseepage applying the DSA method to IRM curves from Venezuelan oil fields. *Studia Geophysica et Geodaetica*, **55**, 343–358.

BABINSZKI, E., MÁRTON, E., MÁRTON, P. & KISS, L. F. 2007. Widespread occurrence of greigite in the sediments of Lake Pannon: implications for environment and magnetostrtigraphy. *Palaeogeography, Palaeoclimatology, Palaeoecology*, **252**, 626–636.

BAYONA, G., JARAMILLO, C., RUEDA, M., REYES-HARKER, A. & TORRES, V. 2007. Paleocene–middle Miocene flexural-margin migration of the no marine Llanos foreland basin of Colombia. *CT&F Ciencia, Tecnología y Futuro*, **3**, 141–160.

BAYONA, G., VALENCIA, A., MORA, A., RUEDA, M., ORTIZ, J. & MONTENEGRO, O. 2008a. Estratigrafía y procedencia de las rocas del Mioceno en la parte distal de la cuenca antepais de los Llanos de Colombia. *Geología Colombiana*, **33**, 23–46.

BAYONA, G., CORTÉS, M., JARAMILLO, C., OJEDA, G., ARISTIZABAL, J. & REYES-HARKER, A. 2008b. An integrated analysis of an orogen–sedimentary basin pair: Latest Cretaceous–Cenozoic evolution of the linked Eastern Cordillera orogen and the Llanos foreland basin of Colombia. *Geological Society of America*, **120**, 1171–1197.

BAYONA, G., BLANCO, Y., GARCÍA, D., MORA, A., JARAMILO, C., DE ARMAS, M. & MORA, A. 2011. Evolution of the Late Eocene-Pliocene Llanos foreland basin of Colombia: Local to lithospheric tectonic controls. *14 Congreso Latinoamericano de Geología y 13 Congreso Colombiano de Geología, August 29th to September 2nd*, Medellín Colombia.

BENTHIEM, R. & ELMORE, R. 1987. Origin of magnetization in the Phosphoria Formation at Sheep Mountain, Wyoming: a possible relationship with hydrocarbons. *Geophysical Research Letters*, **14**, 323–326.

BLOEMENDAL, J., KING, J. W., HALL, F. R. & DOH, S. J. 1992. Rock magnetism of Late Neogene and Pleistocene deep-sea sediments: relationship to sediment source, diagenetic processes and sediment lithology. *Journal of Geophysical Research*, **97**, 4361–4375.

BLOEMENDAL, J., LAMB, B. & KING, J. W. 1998. Paleoenvironmental implications of rock magnetic properties of Late Quaternary sediment cores from the eastern equatorial Atlantic. *Paleoceanography*, **3**, 61–87.

BURTON, E. A., MACHEL, H. G. & QI, J. 1993. Thermodynamic constraints on anomalous magnetization in shallow and deep hydrocarbon seepage environments. *In*: AÏSSAUOUR, D. M., MCNEILL, D. F. & HURLEY, N. F. (eds) *Applications of Paleomagnetism to Sedimentary Geology*. Society for Sedimentary Geology, Tulsa, Special Publication, **49**, 193–207.

COSTANZO-ÁLVAREZ, V., ALDANA, M., ARISTEGUIETA, O., MARCANO, M. C. & ACONCHA, E. 2000. Studies of magnetic contrasts in the Guafita oil field (southwestern Venezuela). *Physics and Chemistry of the Earth (A)*, **25**, 437–445.

COSTANZO-ÁLVAREZ, V., ALDANA, M., DÍAZ, M., BAYONA, G. & AYALA, C. 2006. Hydrocarbon-induced magnetic contrasts in some Venezuelan and Colombian oil wells. *Earth, Planets and Space*, **58**, 1401–1410.

DA SILVA, A., COSTANZO-ÁLVAREZ, V., HURTADO, N., ALDANA, M., BAYONA, G., GUZMÁN, O. & LÓPEZ-RODRÍGUEZ, D. 2010. Possible correlation between Miocene global climatic changes and magnetic proxies, using neuro fuzzy logic analysis in a stratigraphic well at the Llanos foreland basin, Colombia. *Studia Geophysica et Geodaetica*, **54**, 607–631.

DEKKERS, M. J. 1989. Magnetic properties of natural pyrrhotite II: high and low temperature behaviour of JRS and TRM as function of grain size. *Physics of the Earth and Planetary Interiors*, **57**, 266–283.

DEKKERS, M. J. 1997. Environmental magnetism: an introduction. *Geologie en Mijnbouw*, **76**, 163–182.

Díaz, M., Aldana, M., Costanzo-Álvarez, V., Silva, P. & Pérez, A. 2000. EPR and Magnetic Susceptibility studies in well samples from some Venezuelan oil fields. *Physics and Chemistry of the Earth (A)*, **25**, 447–453.

Díaz, M., Aldana, M., Jiménez, S. M., Sequera, P. & Costanzo-Álvarez, V. 2006. EPR and EOM studies in well samples from some Venezuelan oil fields: correlation with magnetic authigenesis. *Revista Mexicana de Física*, **52**(S3), 63–65.

Doh, S. J., King, J. W. & Leinen, M. 1988. A rock magnetic study of Giant Piston Core LL44-GPC from the central north Pacific and its Paleoceano graphic significance. *Paleoceanography*, **3**, 89–111.

Dunlop, D. J. & Özdemir, Ö. 1997. *Rock Magnetism Fundamentals and Frontiers*. Cambridge University Press, Cambridge, UK.

Elmore, R. D., Engel, M. H., Crawford, L., Nick, K., Imbus, S. & Soler, Z. 1987. Evidence for a relationship between hydrocarbons and authigenic magnetite. *Nature*, **325**, 428–430.

Elmore, R. D., Imbus, D., Engel, S. W. & Fruit, M. H. 1993. Hydrocarbons and magnetizations in magnetite. *In*: Aïssaoui, D. M., McNeill, D. & Hurley, N. (eds) *Applications of Paleomagnetism to Sedimentary Geology*. Society for Sedimentary Geology, Tulsa, Special Publication, **49**, 181–191.

Evans, M. E., Heller, F., Bloemendal, J. & Thouveny, N. 1997. Natural magnetic archives of past global change. *Surveys in Geophysics*, **18**, 183–196.

Farris, D. W., Jaramillo, C. *et al.* 2011. Fracturing of the Panamanian Isthmus during initial collision with South America. *Geology*, **39**, 1007–1010, http://dx.doi.org/10.1130/G32237.1

Foote, R. S. 1984. Significance of near-surface magnetic anomalies. *In*: Davidson, M. J. & Gottlied, B. M. (eds) *Unconventional Methods in Exploration for Petroleum and Natural Gas*. Southern Methodist University, Institute for Study of Earth and Man, Dallas, 12–24.

Foote, R. S. 1987. Correlations of borehole rock magnetic properties with oil and gas producing areas. *Association of Petroleum Geochemical Explorationists Bulletin*, **3**, 114–134.

Foote, R. S. 1992. Use of magnetic field aids oil search. *Oil & Gas Journal*, **May**, 137–141.

Foote, R. S. 1996. Relationship of near-surface magnetic anomalies to oil- and gas-producing areas. *In*: Schumacher, D. & Abrams, M. A. (eds) *Hydrocarbon Migration and its Near-surface Expression*. AAPG, Tulsa, Memoir, **66**, 111–126.

Guzmán, O., Costanzo-Álvarez, V., Aldana, M. & Díaz, M. 2011. Study of magnetic constrasts applied to hydrocarbon exploration in the Maturin sub-basin (Eastern Venezuela). *Studia Geophysica et Geodaetica*, **55**, 359–376.

Hall, S. A. & Evans, I. 1995. Palaeomagnetic and rock magnetic properties of hydrocarbon reservoir rocks from the Permian Basin, southeastern New Mexico. *In*: Turner, P. & Turner, A. (eds) *Paleomagnetic Applications in Hydrocarbon Exploration and Production*. Geological Society, London, Special Publication, **98**, 79–95.

Heslop, D., Dekkers, M. J., Kruiver, M. & Van Oorschot, H. 2002. Analysis of isothermal remanent magnetization acquisition curves using the expectation-maximization algorithm. *Geophysical Journal International*, **148**, 58–64.

Horng, C.-S., Chen, J.-C. & Lee, T.-Q. 1992. Variations in magnetic minerals from two Plio-Pleistocene marine-deposited sections. Southwestern Taiwan. *Journal of the Geological Society of China*, **35**, 323–335.

Jiang, W., Horng, C., Roberts, A. P. & Peacor, D. R. 2001. Contradictory magnetic polarities in sediments and variable timing of neoformation of authigenic greigite. *Earth and Planetary Science Letters*, **193**, 1–12.

Kao, S.-J., Horng, C.-S., Roberts, A. P. & Liu, K.-K. 2004. Carbon–sulphur–iron relationships in sedimentary rocks from southwestern Taiwan: influence of geochemical environment on greigite and pyrrhotite formation. *Chemical Geology*, **203**, 153–158.

King, J. W., Banerjee, S. K., Marvin, J. & Özdemir, Ö. 1982. A comparison of different magnetic methods for determining the relative grain size of magnetite in natural minerals, some results for lake sediments. *Earth and Planetary Science Letters*, **59**, 404–419.

Kruiver, M., Dekkers, P. & Heslop, D. 2001. Quantification of magnetic coercivity components by the analysis of acquisition curves of isothermal remanent magnetization. *Earth and Planetary Science Letters*, **189**, 269–276.

Labrador, H., López, L. & Galarraga, F. 1995. Estudio geoquímico de crudos del campo Guafita, estado Apure, Venezuela. *Interciencia*, **20**, 30–36.

Machel, H. G. 1995. Magnetic mineral assemblages and magnetic contrasts in diagenetic environments with implications for studies of paleomagnetism, hydrocarbon migration and exploration. *In*: Turner, P. & Turner, A. (eds) *Paleomagnetic Applications in Hydrocarbon Exploration and Production*. Geological Society, London, Special Publication, **98**, 9–29.

McCabe, C., Sassen, R. & Saffer, B. 1987. Occurrence of secondary magnetite within biodegraded oil. *Geology*, **15**, 7–10.

Otamendi, A. M., Díaz, M., Costanzo-Álvarez, V., Aldana, M. & Pilloud, A. 2006. EPR Stratigraphy applied to the study of two marine sedimentary sequences in Southwestern Venezuela. *Physics of the Earth and Planetary Interiors*, **154**, 243–254.

Özdemir, Ö., Dunlop, D. & Moskowitz, B. M. 1993. The effect of oxidation on the Verwey transition in magnetite. *Geophysical Research Letters*, **20**, 1671–1674.

Palmer, S. E. & Russell, J. A. 1988. The Five Oil Families of the Llanos Basin. *III Simposio Bolivariano de Exploración Petrolera en las Cuencas Subandinas*. Caracas, Venezuela, 724–754.

Porsch, K., Dippon, U., Rijal, M. L., Appel, E. & Kappler, A. 2010. *In-situ* magnetic susceptibility measurements as a tool to follow geomicrobiological transformation of Fe minerals. *Environmental Science & Technology*, **44**, 3846–3852, http://dx.doi.org/10.1021/es903954u

Rada Torres, M. A., Costanzo-Álvarez, V., Aldana, M., Suárez, N., Campos,, Mackowiak – Antczak, M. M. & Brandt, M. C. 2011. Petrographic, rock magnetic and dielectric characterization of prehistoric

Amerindian potsherds from Venezuela. *Studia Geophysica et Geodaetica*, **55**, 717–736, doi: 10.1007/s11200-010-9021-1.

RANGEL, A., ESCALANTE, C. & MORA, C. 2003. Evaluación Geoquímica Integrada de los Gases y Crudos Colombianos: un nuevo enfoque para la Exploración de Hidrocarburos. *VIII Simposio Bolivariano de Exploración Petrolera en las Cuencas Subandinas*, Cartagena de Indias, Colombia, 285–295.

REYNOLDS, R. L., FISHMAN, N. S. & HUDSON, M. R. 1990. Sources of aeromagnetic anomalies over Cement oil field (Oklahoma), Simpson oil field (Alaska), and Wyoming-Idaho-Utah thrust belt. *Geophysics*, **56**, 606–617.

REYNOLDS, R. L., GOLDHABER, M. B. & TUTTLE, M. L. 1993. Sulfidization and magnetization above hydrocarbon reservoirs. *In*: AÏSSAOUI, D. M., MCNEILL, D. & HURLEY, N. (eds) *Applications of Paleomagnetism to Sedimentary Geology*. Society for Sedimentary Geology, Tulsa, Special Publication, **49**, 167–179.

RIJAL, M. L., APPEL, E., PETROVSKÝ, E. & BLAHA, U. 2010. Change of magnetic properties due to fluctuations of hydrocarbon contaminated groundwater in unconsolidated sediments. *Environmental Pollution*, **158**, 1756–1762.

ROBERTSON, D. J. & FRANCE, D. E. 1994. Discrimination of remanence-carrying minerals in mixture, using isothermal remanent magnetisation acquisition curves. *Physics of the Earth and Planetary Interiors*, **84**, 223–234.

ROBINSON, S. G. 1986. The Late Pleistocene paleoclimatic record of North Atlantic deep sea sediments revealed by mineral magnetic measurements. *Physics of the Earth and Planetary Interiors*, **42**, 22–46.

SAGNOTTI, L. 2007. Iron sulphides. *In*: GUBBINS, D. & HERRERA-BERVERA, E. (eds) *Encyclopedia of Geomagnetism and Paleomagnetism*. Springer, Heidelberg, 454–459.

SAUNDERS, D. F. & TERRY, S. A. 1985. Onshore exploration using the new geochemistry and geomorphology. *Oil & Gas Journal*, **September**, 126–130.

SAUNDERS, D. F., BURSON, K. R. & THOMPSON, C. K. 1991. Observed relation of soil magnetic susceptibility and soil gas hydrocarbon analyses to subsurface hydrocarbon accumulations. *The American Association of Petroleum Geologists Bulletin*, **75**, 389–408.

SAUNDERS, D. F., BURSON, K. R. & THOMPSON, C. K. 1999. Model for hydrocarbon microseepage and related near-surface alterations. *The American Association of Petroleum Geologists Bulletin*, **83**, 170–185.

SHOUYUN, H. U., APPEL, E., HOFFMANN, V. & SCHMAHL, W. 2002. Identification of greigite in lake sediments and its magnetic significance. *Science in China Series D: Earth Sciences*, **45**, 81–87.

STONER, J. S., CHANNELL, J. E. T. & HILLAIRE-MARCEL, C. 1996. The magnetic signature of rapidly deposited detrital layers from the deep Labrador Sea. Relationship to North Atlantic Heinrich layers. *Paleoceanography*, **11**, 309–325.

SUK, D., PEACOR, D. R. & VAN DER VOO, R. 1990. Replacement of pyrite framboids by magnetite in limestone and implications for paleomagnetism. *Nature*, **345**, 611–613.

SUK, D., VAN DER VOO, R. & PEACOR, D. R. 1992. SEM/STEM observation of magnetic minerals in presumably unremagnetized Paleozoic carbonates from Indiana and Alabama. *Tectonophysics*, **215**, 255–272.

SUK, D., VAN DER VOO, R. & PEACOR, D. R. 1993. Origin of magnetite responsible for remagnetization of Early Paleozoic limestones of New York State. *Journal of Geophysical Research*, **98**, 419–434.

TARDUNO, J. A. 1994. Temporal trends of magnetic dissolution in the pelagic realm: gauging paleoproductivity? *Earth and Planetary Science Letters.*, **123**, 39–48.

THOMPSON, R., BATTARBEE, R. W., O'SULLIVAN, P. E. & OLDFIELD, F. 1975. Magnetic susceptibility of lake sediments. *Limnology & Oceanography*, **20**, 687–698.

TRIC, E., LAJ, C., JEHANNO, C., VALET, J. P., KISSEL, C., MAZAUD, A. & IACCARINO, S. 1991. High resolution record of the upper Olduvai polarity transition from Po Valley (Italy) sediments:support fordipolar transition geometry? *Physics of the Earth and Planetary Interiors*, **65**, 319–336.

VAIL, P. R., MITCHUN, R. M. JR., TODD, R. G., WIDMIER, J. M., THOMPSON, S. & SANGREE, J. R. 1977. Seismic stratigraphy and global changes of sea level. *In*: PAYTON, C. E. (ed.) *Seismic Stratigraphy – Application to Hydrocarbon Exploration*. AAPG, Tulsa, Memoir, **26**, 49–212.

WEAVER, R., ROBERTS, A. P. & BARKER, A. J. 2002. A late diagenetic (synfolding) magnetization carried by pyrrhotite: implications for paleomagnetic studies from magnetic iron sulphide bearing sediments. *Earth and Planetary Science Letters*, **200**, 371–386.

# Rock magnetic properties of drill cutting from a hydrocarbon exploratory well and their relationship to hydrocarbon presence and petrophysical properties


M. MENA* & A. M. WALTHER

*Dpto. Ciencias Geológicas, Facultad de Ciencias Exactas y Naturales,
Universidad de Buenos Aires, Argentina*

**Corresponding author (e-mail: mena@gl.fcen.uba.ar)*



**Abstract:** The Golfo San Jorge Basin is one of the most important hydrocarbon-producing basins in Argentina. A study of magnetic properties performed on drill cutting from an oil well drilled in this basin was carried out. The cutting samples, taken from an interval of about 400 m thickness, correspond to the upper units of the Pozo D-129 Formation, the main source rock of the basin. Magnetic susceptibility measurements were made and rock-magnetism studies were conducted. Concentration indices of the different magnetic species determined were calculated based on isothermal remanent magnetization acquisition curves. A correlation analysis among magnetic properties, hydrocarbon content and well logs (sonic, neutron, density, induction, resistivity and photoelectric factor) was performed. Several kinds of significant correlations were found: a positive correlation between susceptibility and relative hydrocarbon content; a positive correlation between magnetic properties and porosities (especially good with the neutron log porosity); and a negative correlation between the concentration indices of some magnetic species (magnetite and pyrrhotite) and resistivity. Pyrrhotite could be directly related to the presence or migration of hydrocarbons through the porous units.

The qualitative correlations between magnetic data and key petrophysical parameters such as porosity, along with the association of magnetic mineralogy to hydrocarbon presence or migration, suggest the potential usefulness of these techniques for subsurface exploration.


The magnetic properties of rocks are dominated by the presence and the concentration of the minerals that contain iron, mainly iron oxides, oxyhydroxides and sulphides. Some magnetic properties are also influenced by the grain size of these minerals.

Although in most of the sedimentary rocks Fe is not a major component in quantitative terms, it should be considered as an important element because the different forms in which it occurs are closely related to the depositional or diagenetic environment in which these rocks were formed as well as to the variations of physical–chemical and biological conditions during their later geological history. As a result, a change in the provenance area can be reflected in the change of the magnetic mineralogy that may be seen in the contrasts in magnetic parameters.

Considering that magnetic mineralogy is very sensitive to changes in temperature, pressure, pH and Eh, the magnetic parameters may help differentiate environmental changes due to the presence or migration of hydrocarbons.

Magnetic susceptibility anomalies caused by enhanced concentrations of authigenic magnetite and/or Fe-sulphides have been found above subsurface hydrocarbon reservoirs as a result of changes in the redox conditions due to hydrocarbon migration or to seepage of $H_2S$ from the underlying layers (Donovan *et al.* 1984; Foote 1984, 1992; Saunders & Terry 1985; Elmore *et al.* 1987, 1989, 1993; Saunders *et al.* 1993; Mello *et al.* 1996). Several authors have identified, by scanning electronic microscopy (SEM), spherical aggregates of submicronic crystals of magnetite associated with solid bitumen and speleothems with hydrocarbon inclusions. These aggregates appear to be linked to crude oil biodegradation (Elmore *et al.* 1987; McCabe *et al.* 1987) or to inorganic chemical processes due to the presence of hydrocarbons (Elmore *et al.* 1993). Magnetic mineralogy has been used as non-conventional methodology for hydrocarbon exploration in Venezuelan and Colombian oil fields (Aldana *et al.* 1999, 2003, 2011; Costanzo *et al.* 2000, 2006; Guzmán *et al.* 2011).

Rock-magnetism methods, based on the investigation of non-directional magnetic properties of the materials, are especially appropriate for the study of unconsolidated material. They can provide economic methods for investigating fluid migration and correlation between different layers, supplementing other geological and geophysical studies.

In this study, rock-magnetism methods were used on drill cutting for the determination of the magnetic mineralogy of subsurface sedimentary


*From*: Elmore, R. D., Muxworthy, A. R., Aldana, M. M. & Mena, M. (eds) 2012. *Remagnetization and Chemical Alteration of Sedimentary Rocks*. Geological Society, London, Special Publications, **371**, 217–228.
First published online October 01, 2012, http://dx.doi.org/10.1144/SP371.14

rocks and to investigate the possible correlation between magnetic and petrophysical well data.

## Sampling and methodology

The rock-magnetism studies were performed on drill cutting samples, coming from an exploratory oil well drilled on the Golfo San Jorge Basin, Chubut province, Argentina. This basin is one of the most important hydrocarbon-producing basins in the country. It is characterized by good source rock, favourable structural complexity and multiple thin tuffaceous sandstones as reservoirs.

The sampled interval, with a thickness of about 400 m, corresponds to the upper levels of the Pozo D-129 Formation. This lacustrine-fluvial formation of Early Cretaceous age is located at more than 1500 m of depth in most of the basin. Lithologically, it is composed mainly by tuffs and siltstones including bituminous black shales in the lower section and dark-grey tuffaceous sandstone intercalated in the upper section. Thin levels of oolitic and pisolitic limestones are characteristic for this formation (Uliana & Legarreta 1999). The lacustrine black shales, with about 4% of total organic carbon (TOC), are the main hydrocarbon source rocks of the basin (Baldi & Nevistic 1996) while in some areas the sandstones in the upper section act as reservoir rocks.

Measurements of magnetic susceptibility and acquisition of isothermal remanent magnetization (IRM) were used to estimate non-directional magnetic parameters for the sampled lithological levels. A total of 126 cutting samples, each representative of *c.* 3 m of thickness, were analysed. The petrophysical description of the cuttings indicates that the sampled rocks are tuffs, tuffaceous sandstones and siltstones containing pyrite and hydrocarbons, with some thin intercalated layers of oolitic limestones and claystones.

Previously homogenized and quartered, and with the objective of ensuring representative measurements, three specimens were taken from each sample. In a few cases the scarcity of material only permitted the extraction of one or two specimens. A total of 356 specimens were taken. Each specimen is a cylindrical plastic tube of 2.54 cm of diameter by 2.2 cm height, containing *c.* 15 g of cutting.

The specimens were weighed to within 1 mg of precision and their bulk magnetic susceptibility (MS) was measured using a Bartington dual MS2 susceptibilimeter working at 470 Hz and 4700 Hz. In order to analyse the measurement dispersion, the MS of each specimen was measured six times at 470 Hz and six times at 4700 Hz. Before each measurement, the equipment was zeroed to avoid measurement errors due to drift related to changes in laboratory environmental conditions. The MS values were expressed as mass magnetic susceptibility ($\chi$).

A peculiar behaviour of MS values was found in the upper 36 m of the sampled section. This effect (as described below) was avoided by consolidating all the specimens with vinyl adhesive.

For each consolidated specimen, six measurements at each frequency were performed. For each specimen the mean MS at low frequency ($\chi_{lf}$), the mean MS at high frequency ($\chi_{hf}$) and the respective standard deviations were calculated. From these values, the related average frequency-dependent susceptibility factor FDF, defined

$$\mathrm{FDF} = 100\left(\frac{\chi_{lf} - \chi_{hf}}{\chi_{lf}}\right),$$

the mean mass susceptibility at low frequency, the standard deviation (S) and the coefficients of variation (CV = S/mean) for each level were calculated. The FDF characterizes the presence of small particles, with sizes just below the limit between the superparamagnetic (SP) and the stable single-domain (SD) particles. When the operation frequency is comparable with the grain relaxation frequency, the magnetization of fine grains can be blocked and these particles with SP behaviour at low frequency behave like SD at high frequency. Susceptibility values at high frequency will therefore be proportionally lower than those measured at low frequency. High FDF values, especially those larger than 10%, suggest a significant presence of magnetite SP (Maher & Thompson 1999). The $\chi$ and FDF variations in the profile were analysed.

Detailed measurements of isothermal remanent magnetization (IRM) acquisition were performed, in order to determine the coercivity spectrum. Using a pulser magnetometer, one specimen by level was subjected to uniaxial fields applied in 15 steps from 20 to 2000 mT. After each pulse the intensity of the acquired IRM was measured with a 2G-Enterprise cryogenic magnetometer. The IRM acquisition curve can be 'unmixed' into the components that contribute to the bulk IRM curve (Robertson & France 1994). From the analysis of IRM acquisition curves, the magnetic coercivity components were quantified using three parameters: the mean coercivity ($B_{1/2}$) estimated by the applied field at which the mineral phase acquires half of its saturation (SIRM), the dispersion parameter (DP) that provides the half-width of the coercivity distribution and the relative contribution (RC) of the component to the bulk IRM curve. To estimate these parameters, the procedures of Kruiver *et al.* (2001) and Heslop *et al.* (2002) were used.

Hysteresis measurements at room temperature in fields up to 1 T were performed on 18 specimens from lithological representative levels using a

Molspin vibrating sample magnetometer (VSM) at the Magnetic Anisotropy and Rock Magnetism Laboratory of the Geoscience Institute, Sao Paolo University. The saturation magnetization ($M_s$), the coercitive force ($H_c$) and saturation remanent magnetization ($M_{rs}$) were determined after the paramagnetic correction. The coercivity of remanence ($H_{cr}$) was determined by applying progressively increasing backfield after saturation.

Finally, a correlation analysis among the obtained magnetic parameters, the cutting hydrocarbon content and the petrophysical properties determined from the cutting or well logs was made.

## Results

### $\chi$ from unconsolidated material

A peculiar behaviour of the measured magnetic susceptibility (MS) was found in the samples from the upper 36 m of the section. For these lesser depth levels, the dispersions within each specimen are larger with coefficient of variation (CV) between 10 and 25%; these dispersions are lower on the remaining parts, with CV generally <7% and <5% in most of the specimens.

The levels with high dispersion are characterized by an increased contribution of claystones in relation to the rest of the section. The MS measured on the specimens from these levels gradually increased until it reaches a stable value as the sequence of six measurements progressed. This did not happen in the remaining levels. To prove the hypothesis that this increment is caused by the particles accommodation, the following procedure was used. Starting from a new homogenized initial state, three new specimens were taken from cutting samples belonging to the upper 46 m of the section. These comprise the upper part with high dispersion and the following two lower levels.

Trying to move the material as little as possible, the MS of each specimen was measured three times at low frequency ($\chi_{lf}$) and three times at high frequency ($\chi_{hf}$). The measurements were repeated inverting the specimen position. To favour the particle arrangement, each specimen was subjected to 10 seconds of vertical mechanical shaking, again measuring $\chi_{lf}$ and $\chi_{hf}$ in normal and inverted positions. The whole process was repeated twice more. The values obtained after shaking were larger than the initial values. This effect is not related to the selection of materials based on their weight during the shaking, since normal and inverted positions give similar values. The first six measurements of the specimens of each level and for each frequency were averaged obtaining the means before shaking ($\chi_{lfb}$ and $\chi_{hfb}$). The same procedure was performed with the six measurements made after the last shaking step ($\chi_{lfa}$ and $\chi_{hfa}$).

For the upper part of section, the differences between the mean values before and after shaking are significant except for two samples at 10 m and 28 m depth (lithological levels 5 and 11, respectively); the same is true for the two samples below the sector with claystones (lithological levels 15 and 16). These four samples do not show appreciable differences at low or high frequency in the means after and before shaking. When shaking increases the values, the differences at high and at low frequency are similar (Fig. 1). In general there are larger differences with larger susceptibilities, which suggest the existence of a proportional effect. These differences have a certain relationship with the lithology (Fig. 2). Moreover, samples with lower differences also have the lowest susceptibilities.

Tuffs dominate in this section; sandy tuffs and tuffaceous sandstones have abundant pyrite, low porosity and rakes of dark brown hydrocarbons. Oolitic limestones and compact claystones with pyrite also appear in this sector. The largest susceptibility variation seems to be related to the presence of claystones. A certain positive correlation between the susceptibility differences and the FDF is also seen (Fig. 2).

In specimens from rocks of fine grain such as the claystones, small-size magnetic particles with SP behaviours are expected. Most of the claystones are largely composed by clay minerals. In some of these, the phyllosilicate sheets are not electrically neutral due to the substitutions of some cations for others, such as $Al^{3+}$, $Fe^{3+}$ and $Fe^{2+}$. Due to the presence of $Fe^{3+}$ and $Fe^{2+}$, clays are paramagnetic minerals with variable susceptibilities according to the iron content and in general range from 1.3 to 1.5 × $10^{-7}$ m$^3$ kg$^{-1}$ (Dunlop & Özdemir 1997). Due to their planar habit, they present anisotropy of magnetic susceptibility as crystalline as shape anisotropy. Although their paramagnetic properties can be masked by ferromagnetic (*s.l.*) mineral presence, even being inside the rock in very low proportions, their magnetic anisotropy is the strongest.

Additionally, during diagenesis many iron silicates can form ferromagnetic minerals keeping the shape and crystalline orientations of the minerals they replace (Tarling & Hrouda 1993). The results here suggest that the phenomenon of post-shaking susceptibility increase registered in the specimens is probably due to the paramagnetic and ferromagnetic minerals realignment that would allow or increase the magnetic interaction among adjacent particles.

### $\chi$ from consolidated material

The $\chi$ measured on the consolidated specimens from the upper sector were lower than those measured on

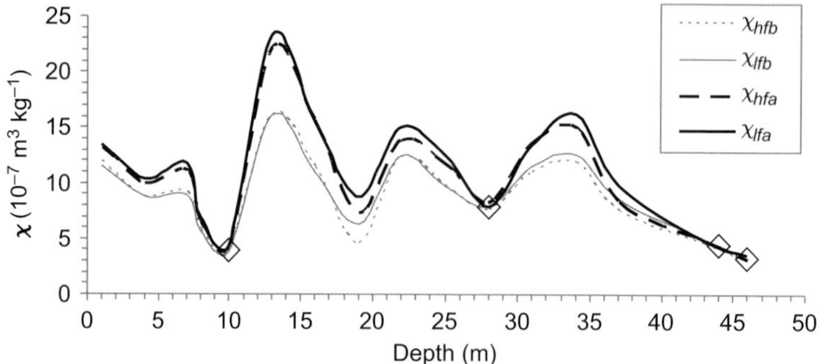

**Fig. 1.** Mass magnetic susceptibility at low and high frequency measured before ($\chi_{lfb}$ and $\chi_{hfb}$) and after ($\chi_{lfa}$ and $\chi_{hfa}$) shaking the specimens. Diamonds show levels without differences.

unconsolidated material, with an average decrease of 22%. A specimen plastic tube filled only with vinyl adhesive has $\chi$ of $-2 \times 10^{-9}$ m$^3$ kg$^{-1}$. The maximum $\chi$ decrease explained by the diamagnetic susceptibility of the adhesive is therefore only 0.11–2.37% of each susceptibility. Therefore, the additional effect can be due to the adhesive decreasing or annulling a magnetic couple effect among neighbouring particles. The similarity between the curves of $\chi$ increase due to shaking of the specimens and those of $\chi$ decrease when consolidating them support the existence of this magnetic couple effect (Fig. 3).

Although the $\chi$ values measured (1.39 to 22.67 $\times$ 10$^{-7}$ m$^3$ kg$^{-1}$, Fig. 4) are within the theoretical range for sedimentary rocks, they do cover the region of high values that have smaller probability in these rocks. These relatively high values can be related to the combined effect of the high content of pyroclastic material in these rocks and with authigenic magnetic minerals generated under environmental reducing conditions derived by hydrocarbon presence.

An analysis of variance (ANOVA) at 95% confidence performed on the measurements suggests that, along the section, the $\chi$ configure a signal significantly differentiated from the noise introduced by sample internal variations or experimental errors. Based on this, level mean values of $\chi$ and FDF were calculated.

Curves of $\chi$ and FDF were drawn and analysed together with the lithological percent composition of each level determined by the petrographic description of the cutting samples (Fig. 5). In many cases the increase in the tuff content is related to a local decrease of $\chi$, but there is no conspicuous relationship between $\chi$ and lithology. There is no clear relationship between the $\chi$ and the limestone presence although, in many sectors with greater limestone content, a local decrease of the $\chi$ appears. Since calcium and magnesium carbonates are diamagnetic minerals, where limestones are present lower $\chi$ are expected. The regions with greatest FDF are located in levels with tuff preponderance, suggesting a greater presence of SP particles in tuffs. These particles could be small crystals of ferrimagnetic minerals present in the volcanic ashes, originating from the rapid cooling of the magma.

To analyse the possible relationship between the hydrocarbon (HC) presence and $\chi$, the relative

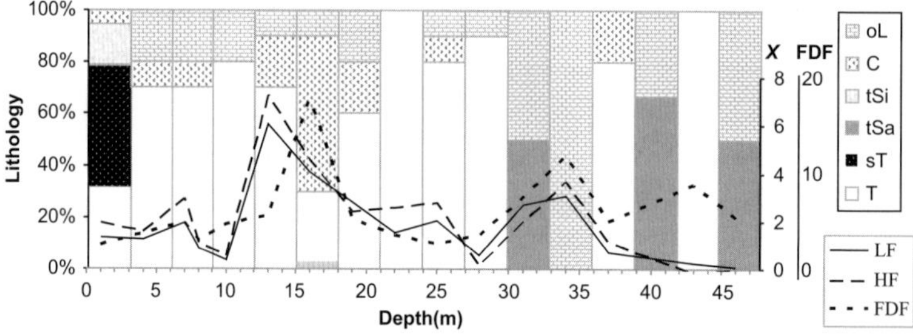

**Fig. 2.** Percent lithology and curves of mean mass magnetic susceptibility at low (LF) and high (HF) frequency and frequency dependence factor (FDF) for the upper section. oL: oolitic limestone; C: claystone; tSi: tuffaceous siltstone; tSa: tuffaceous sandstone; sT: sandy tuff; T: tuff.

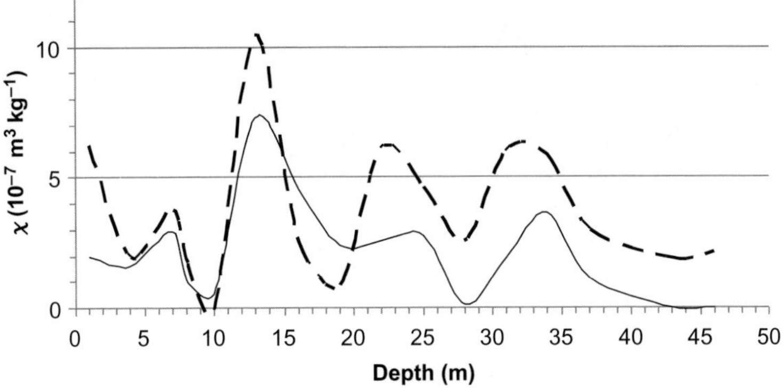

**Fig. 3.** Absolute values of magnetic susceptibility increase after mechanical shaking (solid line) and absolute values of susceptibility decrease after cohesion (dotted line).

contents of HC in the cutting were determined and quantified by a qualitative score scale. The assigned scores were: $8 =$ total impregnation; $7 =$ partial impregnation; $6 =$ scarce partial impregnation; $5 =$ filled fissures; $4 =$ abundant rakes; $3 =$ rakes; $2 =$ isolated rakes; and $1 =$ very isolated rakes.

The relative content of each sample was calculated by multiplying the HC score by the percentage of each lithology containing HC and dividing by 100. Considering only the levels where hydrocarbons were registered in cutting, a trend of increasing susceptibility with greater relative content in HC was found (Fig. 6). This relation has a significant lineal correlation (Pearson correlation coefficient $r = 0.6645$, $p_{value} < 0.001$).

## IRM acquisition and hysteresis loops

The IRM acquisition curves obtained are typical for magnetite-rich rocks, with abrupt slopes from the beginning (Fig. 7a). In most cases, the saturation is reached in fields of 1 T or smaller, indicating the scarce or null presence of hematite. Many of the analysed specimens show complex curves with more than a plateau (Fig. 7b), indicating the existence of minerals with different coercivity spectra. SIRM/$\chi$ values are lower, ranging from 1.2 to 20 kA m$^{-1}$.

IRM acquisition curve modelling was used as a first approach of magnetic mineralogy. Between one and three components with different coercivity spectra were defined through this analysis. The distribution of the $B_{1/2}$ values defined is multimodal, with prevalence of the lesser coercivities, with a principal mode centred at 25 mT and secondary modes at 63 mT and 400 mT. These components are characterized by reduced dispersion (DP < 0.3). The modelling of the IRM acquisition curves is facilitated by these well-separated coercivity distributions.

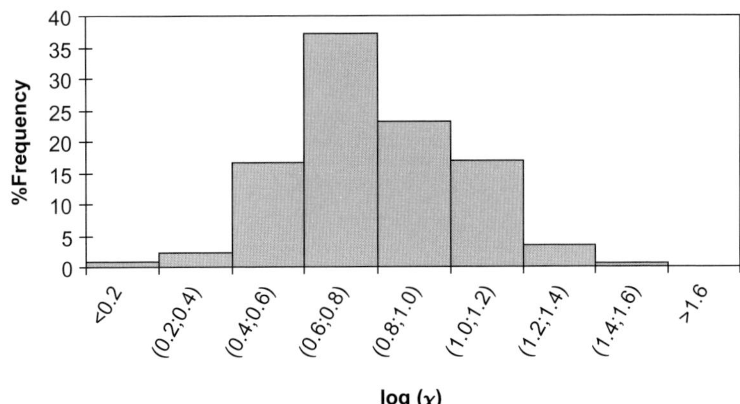

**Fig. 4.** Relative frequency histogram for the decimal logs of the mass susceptibility ($0 \times 10^{-7}$ m$^3$ kg$^{-1}$).

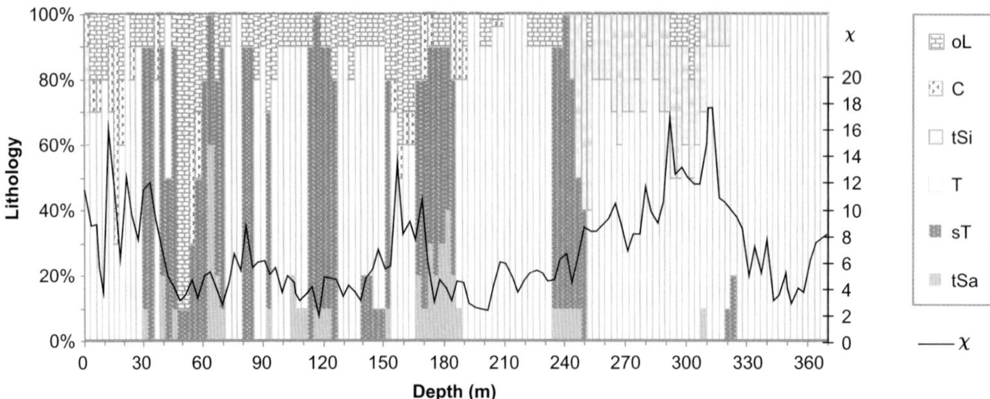

**Fig. 5.** Mean level susceptibility ($\chi$) curves and diagram of percent lithology from cutting samples. oL: oolitic limestones, C: claystones, tSi: tuffaceous siltstones, T: tuff, sT: sandy tuff, tSa: tufaceous sandstones.

These $B_{1/2}$ modal values indicate the presence of at least three magnetic mineralogies: one of low, one of intermediate and another of high coercivity ($H_{cr}$). As a first approximation they could be attributed to titanomagnetites, magnetite and pyrrhotite presence.

The hysteresis loops (Fig. 8) can be divided into three different general patterns: (1) typical hysteresis loop from a multi-domain assemblage, saturating under 300–400 mT; (2) wider loops with higher $H_c$, saturating at $c$. 500 mT; and (3) loops with sharp slopes, which are common when the sample has little ferromagnetic material and is rich in iron-bearing phases such as clay minerals.

In general, the type (1) loops are typical of levels whose specimens have IRM acquisition curves with saturation under 500 mT fields and with components with low and intermediate $H_{cr}$. Lithologically, this type occurs in samples of tuffs and siltstones. The type (2) loops appear in levels with three $H_{cr}$ components corresponding to tuffaceous sandstones, sandy tuffs and limestones. Those of type (3), much scarcer, are characteristic of levels with a prevalence of tuffs and a strong presence of low $H_{cr}$ components.

Most of the specimens (78%) show two coercivity components, three components are defined from 18%, only two specimens have four coercivity components and there is only one single-component specimen. All specimens show the low $H_{cr}$ component saturating below 100 mT, except that single-component with intermediate $H_{cr}$ that saturates at 300 mT. The 65% of samples with two components have a second component of high $H_{cr}$ that saturates at $c$. 500–800 mT, while the remainder have a second component of intermediate $H_{cr}$ saturating at <300 mT in most cases. The two specimens with four components have two of low $H_{cr}$ with near $B_{1/2}$ values, one of intermediate and another of high $H_{cr}$.

These data suggest that practically all the specimens contain detectable quantities of titanomagnetite. Half of them have high coercivity particles

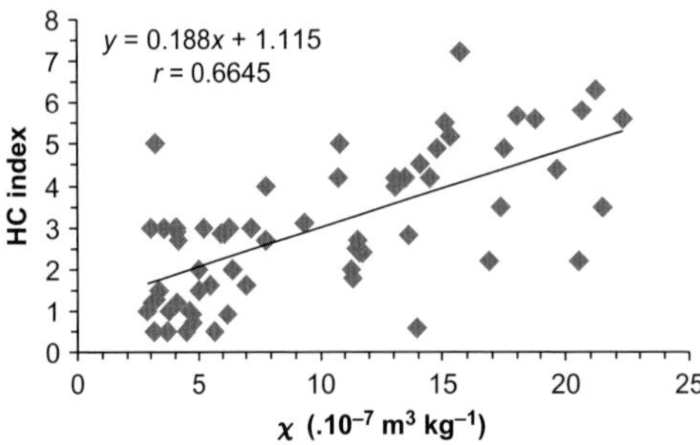

**Fig. 6.** Scatter plot for hydrocarbon (HC) concentration index v. mass susceptibility.

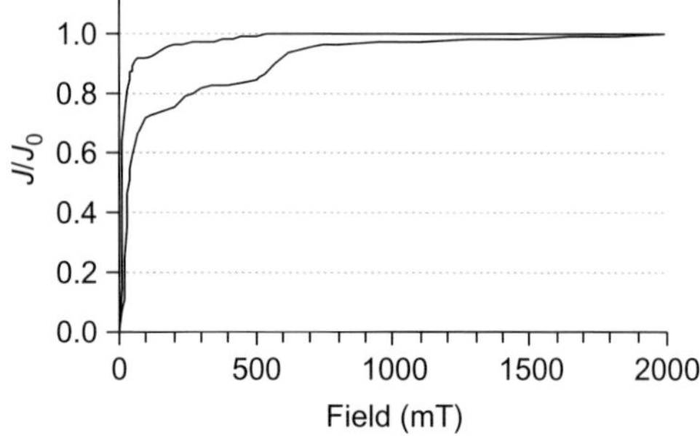

**Fig. 7.** Typical isothermal remanent magnetization (IRM) acquisition curves for the two types found. $J/J_0$, normalized magnetization.

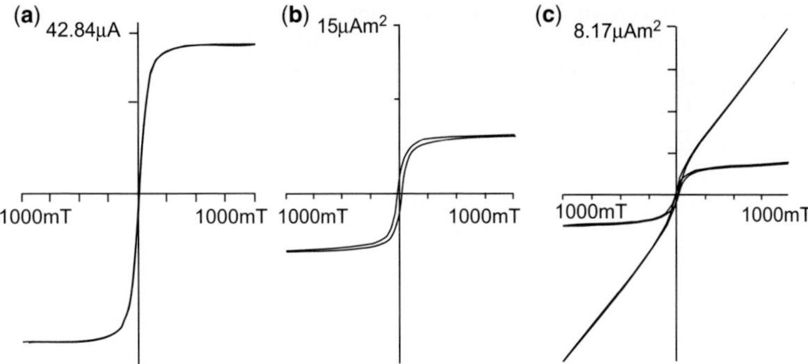



**Fig. 8.** Typical hysteresis loop from (**a**) assemblage of low $H_c$ minerals, (**b**) greater $H_c$ minerals and (**c**) high paramagnetic proportion.

that could correspond to pyrrhotite. In 28% of them, the titanomagnetite accompanies magnetite or poor Ti-titanomagnetite and 20% have all three magnetic mineralogies.

A ternary diagram (Fig. 9) of the component percent contributions to the IRM acquisition curve shows the general prevalence of the lower coercivities. The different lithologies do not form clusters in this graph, but fall mixed at all positions. In the group with a component of low and another of high $H_{cr}$, the biggest contribution is made by the component of low $H_{cr}$. If this is interpreted in terms of remanence intensity acquired by each mineral species, it implies a greater intensity for minerals of low $H_{cr}$ than for high $H_{cr}$. This result is coherent with considering titanomagnetites as the minerals of low $H_{cr}$ and pyrrhotite as those of high $H_{cr}$, since the relationship of pyrrhotite SIRM to magnetite SIRM is of the order 2:3 (c. $80 \times 10^3$ A m$^{-1}$ for pyrrhotite and c. $125 \times 10^3$ A m$^{-1}$ for titanomagnetite 60; McElhinny & McFadden 2000).

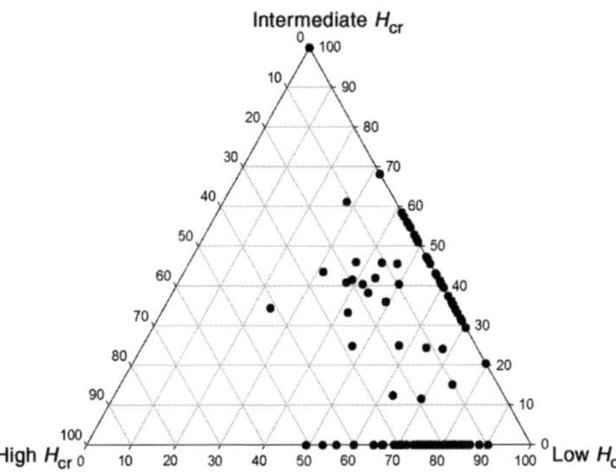



**Fig. 9.** Ternary diagram for the percent contributions of the coercivity components defined from analysis of IRM acquisition curves.

The contributions are more balanced for the group with a low and another intermediate $H_{cr}$ component. This is also coherent with considering them representative of titanomagnetites with different titanium contents, which is the reason why they would have more similar SIRM values (in which case the RC would be more influenced by the relative concentration of these minerals). Finally, the specimens with three or four components are located in the intermediate region with contributions under 30% for the high $H_{cr}$ and relatively balanced for the low and intermediate $H_{cr}$.

The specimens containing each component of coercivity (low, intermediate and high) were pooled in three groups. From each group, the 25% of the specimens with higher relative content of its component were formed into a subgroup. Lithological content and HC relative content averages were calculated for each subgroup (Table 1).

There seems to be some lithological control on the predominance of coercivity components, because the average content of tuff and the average content of claystones and siltstone are greater when the low $H_{cr}$ component is abundant. Moreover, the sandstone average content increases when the high $H_{cr}$ component is abundant. The contents of the different lithologies are intermediate in the subgroup of intermediate $H_{cr}$ component. The difference of the HC mean relative content is notable. These HC contents are higher for the high $H_{cr}$ subgroup and similarly low in the other two.

In addition, the samples were separated into three groups: the first consisted of the specimens that had the three components of coercivity; the second was composed of specimens that had only low and intermediate $H_{cr}$; and the third included specimens with low and high $H_{cr}$ components. The lithological contents were analysed for these groups (Fig. 10). It was found that the largest average content in the tuff appears in samples with the three components; this

**Table 1.** *Lithological and HC content for specimens carrying different coercivity components. Each subgroup is formed with the specimens with content within the 25% of higher values of their group.*

| Subgroup | | High $H_{cr}$ (%) | Intermediate $H_{cr}$ (%) | Low $H_{cr}$ (%) |
|---|---|---|---|---|
| Mean content | Tuff | 63 | 63 | 65 |
| | Tuffaceous sandstone + sandy tuff | 20 | 18 | 15 |
| | Claystones + tuffaceous siltstone | 10 | 12 | 16 |
| | Limestone | 7 | 7 | 4 |
| | HC mean relative content | 4.7 | 1.62 | 1.80 |

is also true for the clays. Conversely, the highest average content from sandy tuffs and tuffaceous sandstones appears in the samples containing the components of low and high $H_{cr}$.

The distribution in these three groups of samples with and without HC was also analysed. As stated above, except for one sample lacking HC all others samples contain the low $H_{cr}$ component. The 85% of the samples containing HC carry the high $H_{cr}$ component (65% accompanied by the lower $H_{cr}$ component and 20% with the three components). This sample percentage decreased for the samples without HC, as the high $H_{cr}$ component appears in 60% of these samples (40% also have the low $H_{cr}$ component and 20% with the three components). The relationship is reversed with the intermediate component that appears in 35% of the samples with HC and 60% of the samples without HC. This can be interpreted as evidence of a direct correlation between the presence of the high coercivity phase and the presence of HC.

Considering that each component contributes to the specimen SIRM (SIRMe) with an intensity of magnetization given by the product of the relative contribution (RC) and the SIRMe, then dividing this quantity by the theoretical magnetization of saturation ($M_s$) of each component results in a concentration index (CI) for this component in the sample (Mena 2007). Concentration indices (CI$_l$, CI$_m$, CI$_h$) were calculated for the three coercivity components (low, intermediate and high) determined in the profile. Theoretical values of $M_s$ tabulated for titanomagnetite 60, magnetite and pyrrhotite (24, 92 and 15 Am$^2$ kg$^{-1}$, respectively; Dunlop & Özdemir 1997) were used as approximations.

Comparing mineral concentration indices and lithology shows that a silty-tuffaceous claystone location corresponds to the highest concentrations of magnetic minerals in general.

In accordance with the indices distribution in the profile, the predominant magnetic minerals are titanomagnetites with a smaller yet important presence

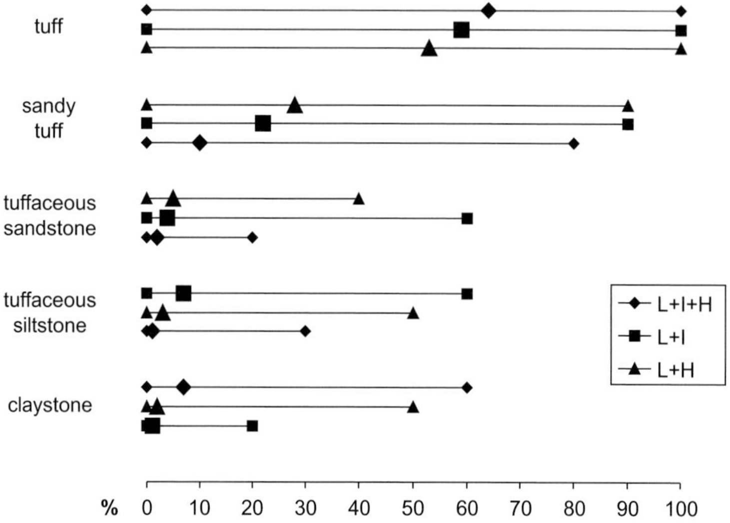


**Fig. 10.** Range of lithological contents in the groups of specimens which carried the three coercivity components (L + I + H), only the low and intermediate $H_{cr}$ components (L + I) and only the low and high $H_{cr}$ components (L + H).

**Table 2.** *Significant correlations between magnetic and geophysical parameters ($\rho$: Spearman correlation coefficient; p: p = value). Significant correlations are formatted in bold.*

| | | DFL | HDRS | RHOB | PE | DPHI | NPHI | SPHI |
|---|---|---|---|---|---|---|---|---|
| $\chi$ | $\rho$ | **− 0.295** | **− 0.281** | 0.081 | **0.337** | − 0.081 | **0.377** | **0.236** |
| | $p$ | **$3.60 \times 10^{-4}$** | **$6.70 \times 10^{-4}$** | 0.182 | **$5.10 \times 10^{-5}$** | 0.181 | **$5.70 \times 10^{-6}$** | **0.004** |
| FDF | $\rho$ | **0.158** | 0.11 | − 0.081 | **− 0.131** | 0.082 | **− 0.128** | − 0.036 |
| | $p$ | **0.039** | 0.11 | 0.184 | **$7.20 \times 10^{-2}$** | 0.183 | **$7.70 \times 10^{-2}$** | 0.344 |
| SIRM | $\rho$ | **− 0.305** | **− 0.335** | **0.226** | **0.217** | − 0.226 | **0.377** | − 0.054 |
| | $p$ | **$7.80 \times 10^{-4}$** | **$2.40 \times 10^{-4}$** | **0.010** | **0.013** | **0.010** | **$3.70 \times 10^{-5}$** | 0.291 |
| $CI_l$ | $\rho$ | **− 0.237** | **− 0.282** | **0.252** | 0.223 | **0.251** | **0.312** | − 0.043 |
| (titanomagnetite) | $p$ | **0.007** | **0.002** | **0.005** | 2.322 | **0.005** | **0.001** | 0.333 |
| $CI_m$ **(magnetite)** | $\rho$ | − 0.151 | − 0.244 | **0.337** | **0.396** | **− 0.338** | 0.082 | − 0.027 |
| | $p$ | 0.149 | 0.046 | **0.009** | **0.002** | **0.009** | 0.289 | 0.428 |
| $CI_h$ **(pyrrhotite)** | $\rho$ | **− 0.267** | **− 0.3** | 0.111 | 0.195 | − 0.112 | **0.415** | 0.086 |
| | $p$ | **0.031** | **0.017** | 0.222 | 0.089 | 0.221 | **0.001** | 0.277 |

of magnetite (especially in the inferior third of the profile) and with pyrrhotite in smaller quantities and in a more sporadic way. The relationship between the concentration indices and the lithology is consistent with the relationships between components of coercivity, lithologies and presence of hydrocarbons, and together these suggest a diagenetic origin for the pyrrhotite and a volcanic origin for the titanomagnetites.

## Correlation analysis

A correlation analysis between magnetic parameters and a group of lithological and geophysical variables obtained from well logs was carried out. The presence of magnetic minerals was characterized with $\chi$, FDF, SIRM, the relationship SIRM/$\chi$ and the indices $CI_l$, $CI_m$, $CI_h$. The variables from the well include: cutting description, sonic (DT), induction (DFL and HDRS), coefficient of photoelectric absorption (PE), density (RHOB) and porosity calculated from the density (DPHI), neutron (NPHI) and sonic (SPHI) logs.

Averages of data for each well log included in 3-m-thick windows (equivalent to the thickness represented by each cutting sample) were calculated in order to adapt the density of cutting sampling and well logs. In this way, 126 multivariate data were related.

Bivariate correlations among the different variables were analysed and the Spearman non-parametric correlation coefficients with their respective significance levels were calculated. The correlations that exceeded 95% confidence were considered significant. The significant correlations between magnetic and geophysical parameters are summarized in Table 2. From this analysis, the following conclusions could be drawn.

- The mass magnetic susceptibility $\chi$ has a positive significant correlation with SIRM, the concentration indexes and the porosities NPHI and SPHI, and a negative significant correlation with FDF and resistivity. The relationship between the SIRM and $\chi$ is expected since, in all levels, the dominant magnetic mineral is titanomagnetite.
- The correlation of $\chi$ with SPHI suggests that more porous rocks will have higher susceptibility. The correlation $\chi$ with NPHI suggests that rocks with a larger content of fluids, either water or oil, will have higher $\chi$. If the water is housed in pores, NPHI does measure porosity. NPHI can also represent the content of water retained in clays, however, in which case the real porosity will be lower than that indicated by the NPHI. These differences may cause the higher correlation between $\chi$ and NPHI than $\chi$ and SPHI. The apparent relationship 'the

greater the porosity, the higher the susceptibility' can be attributed to a purely lithological effect. It can also be caused by the presence of fluids with high content of organic matter generators of a reductive environment that favours the formation of Fe minerals such as magnetite or pyrrhotite.

- There is a negative correlation between $\chi$ and FDF. This suggests that the presence of abundant magnetic particles of very small sizes is associated with a smaller concentration of magnetic minerals.
- The positive correlation between $\chi$ and PE can be related to the relatively larger molecular mass of Fe minerals.
- There are significant negative correlations between SIRM and resistivity, both HDRS and DFL. This suggests that the magnetic mineral content will be larger as the rock resistivity decreases, which can be related to greater porosity associated with salinity in formation waters. On the other hand, there is a positive correlation between DFL and FDF which suggests that, in more resistive rocks, the proportion of SP particles is larger.
- There are positive correlations among $CI_l$ and the porosities NPHI and DPHI and with RHOB, and negative with both resistivities. $CI_m$ has positive correlations with PE and RHOB, and negative with DPHI. These correlations suggest that titanomagnetite has the main presence in densest rocks or with high clay content. Magnetite also appears in greater proportion in the denser rocks. This analysis supports a volcanic origin for these minerals.
- $CI_h$ has a positive correlation with NPHI and a negative with DFL and HDRS. These suggest a greater presence of pyrrhotite in rocks with a high fluid content, especially with saline formation waters, suggesting a diagenetic origin by alteration of pre-existing Fe minerals associated with the circulation of fluids.

## Conclusions

The analyses described here aim to contribute to the feasible application of magnetic mineralogy techniques to supplement other geological and geophysical studies in order to characterize and to correlate different levels or geological units. On the other hand, the joint interpretation of magnetic parameters curves with the section depth and the well logs reinforce the previous results and lead to the following conclusions.

The magnetic property curves permit the observation of progressive changes in the tuffs near areas with more hydrocarbons, changes that are due to the decrease of the titanomagnetite content and the increase of pyrrhotite content, indicating variations in the diagenetic mineral participation. These changes are not seen in the petrological analysis of the cutting.

Strong coincidences exist between the depth intervals with higher porosity and higher magnetic mineral concentration (especially pyrrhotite). The hydrocarbon presence in cutting is frequently related to locally higher values of the FDF that could indicate deposits of authigenic magnetite with SP sizes and therefore not detectable by concentration indices. In areas where the relationship between DPHI and NPHI values suggests gas presence, the FDF decreases; c. 6 m above these depths the pyrrhotite CI increases. All of this suggests that the pyrrhotite appearance could be directly related to the presence or to the migration of hydrocarbons through the porous units.

The qualitative correlations between magnetic data and key petrophysical parameters such as porosity, along with the association of magnetic mineralogy to hydrocarbon presence or migration, suggest the potential usefulness of these techniques for subsurface exploration.

The authors are grateful to E. Petrovsky and A. Sinito for constructive reviews that significantly improved the quality of the manuscript. The measurements were performed at Laboratorio de Paleomagnetismo 'Daniel A. Valencio', Dpto Ciencias Geológicas, FCEN, UBA-CONICET. This study was supported by the Buenos Aires University, UBACYT X466 (2008–2010) and UBACYT 20020090200640 (2010–2012) projects.

# References

ALDANA, M., COSTANZO-ALVAREZ, V., VITIELLO, D., COLMENARES, L. & GOMEZ, G. 1999. Framboidal magnetic minerals and their possible association to hydrocarbons: La Victoria oil field (Southwestern Venezuela). *Geofisica Internacional*, **38**, 137–152.

ALDANA, M., DÍAZ, M., COSTANZO-ALVAREZ, V., GONZÁLEZ, F. & ROMERO, I. 2003. EPR studies in soil samples from a prospective area at the Andean Range, Venezuela. *Revista Mexicana de Física*, **49**(Suplemento 3), 4–6.

ALDANA, M., COSTANZO-ÁLVAREZ, V., GÓMEZ, L., GONZÁLEZ, C., DÍAZ, M., SILVA, P. & RADA, M. 2011. Identification of magnetic minerals related to hydrocarbon authigenesis in venezuelan oil fields using an alternative decomposition of isothermal remanence curves. *Studia Geophysica et Geodaetica*, **55**, 343–358.

BALDI, J. E. & NEVISTIC, V. A. 1996. Cuenca costa afuera del Golfo San Jorge. Cap. X. *In*: RAMOS, V. & TURIC, M. A. (eds) *Relatorio XIII Congreso Geológico Argentino y II Congreso de exploración de* Hidrocarburos, BsAs. *Geología y Recursos Naturales de la Plataforma Continental Argentina*. Asociación Geológica Argentina Press, Buenos Aires.

COSTANZO-ALVAREZ, V., ALDANA, M., ARISTEGUIETA, O., MARCANO, M. C. & ACONCHA, E. 2000. Studies of magnetic constrast in the Guafita oil field (Southwestern Venezuela). *Physics and Chemistry of the Earth A*, **25**, 437–445.

COSTANZO-ALVAREZ, V., ALDANA, M., DIAZ, M., BAYONA, G. & AYALA, C. 2006. Hydrocarbon-induced magnetic contrast in some Venezuelan and Colombian oil wells. *Earth Planets Space*, **58**, 1401–1410.

DONOVAN, T. J., HENDRICKS, J. D., ROBERTS, A. A. & ELIASON, P. T. 1984. Low-altitude aeromagnetic reconnaissance for petroleum in the Artic National Wildlife Refuge, Alaska. *Geophysics*, **49**, 1338–1353.

DUNLOP, D. & ÖZDEMIR, O. 1997. *Rock Magnetism: Fundamentals and Frontiers*. Cambridge University Press, UK.

ELMORE, R. D., ENGEL, M. H., CRAWFORD, L., NICK, K., IMBUS, S. & SOFER, Z. 1987. Evidence for a relationship between hydrocarbons and authigenic magnetite. *Nature*, **325**, 428–430.

ELMORE, R. D., McCOLLUM, R. & ENGEL, M. H. 1989. Evidence for a relationship between hydrocarbon migration and diagenetic magnetic minerals: implications for petroleum exploration. *Bulletin Association of Petroleum Geochemical Explorationists*, **5**, 1–17.

ELMORE, R. D., IMBUS, S. W., ENGEL, M. H. & FRUIT, D. 1993. *Hydrocarbons and Magnetizations in Magnetite, Applications of Paleomagnetism to Sedimentary Geology*. SEPM, Tulsa, Special Publication, **49**, 181–191.

FOOTE, R. S. 1984. Significance of near-surface magnetic anomalies. *In*: DAVIDSON, M. J. & GOTTLIEB, B. M. (eds) *Unconventional Methods in Exploration for Petroleum and Natural Gas*. Institute for the study of Earth and Man, Southern Methodist University, Dallas, 12–24.

FOOTE, R. S. 1992. Use of magnetic field aids oil search. *Oil & Gas Journal*, **May**, 137–141.

GUZMÁN, O., COSTANZO-ALVAREZ, V., ALDANA, M. & DIAZ, M. 2011. Study of magnetic contrasts applied to hydrocarbon exploration in the Maturín sub-basin (Eastern Venezuela). *Studia Geophysica et Geodaetica*, **55**, 359–376.

HESLOP, D., DEKKERS, M. J., KRUIVER, P. P. & VAN OORSCHOT, I. H. M. 2002. Analysis of isothermal remanent magnetization acquisition curves using the expectation–maximization algorithm. *Geophysical Journal International*, **148**, 58–64.

KRUIVER, P. P., DEKKERS, M. J. & HESLOP, D. 2001. Quantification of magnetic coercivity components by the analysis of acquisition curves of isothermal remanent magnetization. *Earth and Planetary Science Letters*, **189**, 269–276.

MAHER, A. B. & THOMPSON, R. (eds) 1999. *Quaternary Climates, Environments and Magnetism*. Cambridge University Press, UK, 390.

McCABE, C., ROGER, S. & SAFFER, B. 1987. Occurrence of secondary magnetite within biodegraded oil. *Geology*, **15**, 7–10.

McELHINNY, M. W. & McFADDEN, P. L. 2000. *Paleomagnetims. Continents and Oceans*. International Geophysics Series, Academic Press, San Diego, USA, 73.

MELLO, M. R., GONCALVEZ, F. T., BABINSKU, N. A. & MIRANDA, F. P. 1996. Hydrocarbon prospecting in the amazon rain forest: application of surface geochemical, microbiological and remote sensing methods. *In*: SCHUMACHER, D. & ABRAMS, M. A. (eds) *Hydrocarbon Migration and its Near Surface Expression*. AAPG, Tulsa, Memoir, **66**, 401–411.

MENA, M. 2007. *Propiedades magnéticas de rocas: su correlación con registros de pozo*. PhD thesis, Universidad de Buenos Aires, Buenos Aires.

ROBERTSON, D. J. & FRANCE, D. E. 1994. Discrimination of remanence-carryng mineral in mixtures, using isothermal remanent magnetisation curves. *Physics of the Earth and Planetary Interiors*, **84**, 223–234.

SAUNDERS, D. F. & TERRY, S. A. 1985. Onshore exploration using the new geochemistry and geomorphology. *Oil and Gas Journal*, **September**, 126–130.

SAUNDERS, D. F., BURSON, K. R. & THOMPSON, C. K. 1993. Observed relation of soil magnetic susceptibility and soil gas hydrocarbon analyses to subsurface hydrocarbon accumulations. *The American Association of Petroleum Geologists Bulletin*, **75**, 389–408.

TARLING, D. H. & HROUDA, F. 1993. *The Magnetic Anisotropy of rocks*. Chapman & Hall, London.

ULIANA, M. A. & LEGARRETA, L. 1999. Jurásico y Cretácico de la Cuenca del Golfo San Jorge. *In*: CAMINOS, R. (ed.) *Geología Argentina*. SEGEMAR, Argentina.

# Rock magnetism of remagnetized carbonate rocks: another look

MIKE JACKSON* & NICHOLAS L. SWANSON-HYSELL

*Institute for Rock Magnetism, Winchell School of Earth Sciences,
University of Minnesota, Minnesota, US*

*Corresponding author (e-mail: jacks057@umn.edu)*

**Abstract:** Authigenic formation of fine-grained magnetite is responsible for widespread chemical remagnetization of many carbonate rocks. Authigenic magnetite grains, dominantly in the superparamagnetic and stable single-domain size range, also give rise to distinctive rock-magnetic properties, now commonly used as a 'fingerprint' of remagnetization. We re-examine the basis of this association in terms of magnetic mineralogy and particle-size distribution in remagnetized carbonates having these characteristic rock-magnetic properties, including 'wasp-waisted' hysteresis loops, high ratios of anhysteretic remanence to saturation remanence and frequency-dependent susceptibility. New measurements on samples from the Helderberg Group allow us to quantify the proportions of superparamagnetic, stable single-domain and larger grains, and to evaluate the mineralogical composition of the remanence carriers. The dominant magnetic phase is magnetite-like, with sufficient impurity to completely suppress the Verwey transition. Particle sizes are extremely fine: approximately 75% of the total magnetite content is superparamagnetic at room temperature and almost all of the rest is stable single-domain. Although it has been proposed that the single-domain magnetite in these remagnetized carbonates lacks shape anisotropy (and is therefore controlled by cubic magnetocrystalline anisotropy), we have found strong experimental evidence that cubic anisotropy is not an important underlying factor in the rock-magnetic signature of chemical remagnetization.

The usefulness of a palaeomagnetic remanence is directly related to the accuracy with which its age is known (e.g. Van der Voo 1990). Primary remanence, acquired during or very soon after rock formation, provides direct information on palaeofield orientation and strength at that time. The possibility of partial or complete remagnetization at a later time complicates palaeomagnetic interpretation, as has long been recognized (e.g. Graham 1949). Every robust palaeomagnetic study must therefore include some effort to constrain the ages of identified components of the natural remanent magnetization (NRM). Relative dating with respect to sedimentary processes, structural tilting or cross-cutting relationships is the basis of the classical geometric palaeomagnetic tests (fold test and conglomerate test, Graham 1949; baked contact test, Everitt & Clegg 1962; unconformity test, Kirschvink 1978). The special circumstances required for application of these tests are often not available however, or positive test results may provide relatively loose constraints allowing magnetization ages that significantly post-date the age of the rock formation. As a result, other evidence (although generally more indirect) becomes essential in evaluating the age, origin and significance of NRM components. Such evidence commonly includes petrographic observations, isotopic and geochemical data and rock magnetic characterization.

By their very existence, magnetic overprints provide evidence of some process or event of potential geological significance (e.g. McCabe & Elmore 1989; Elmore & McCabe 1991). The 'usual suspects' in remagnetization are a set of mechanisms involving elevated temperature, stress and chemical activity, acting alone or in concert, and producing changes in the magnetic minerals themselves and/ or in the remanence that they carry. When the NRM and the population of magnetic carriers are both modified by some process or event, careful rock magnetic analysis may shed light on the culprit.

Our focus here is chemically remagnetized carbonate rocks where any primary remanence is completely obscured by ancient yet secondary magnetizations. Many such units have been documented in recent decades (e.g. McCabe & Elmore 1989). With the advent of superconducting magnetometers in the 1970s and 1980s (Goree & Fuller 1976), carbonate rocks represented a new frontier for palaeomagnetic research. These and other weakly magnetic sedimentary rock strata had been unmeasurable during progressive demagnetization with the limited-sensitivity spinner magnetometers then available. Superconducting magnetometers enabled a rapid expansion into these untapped palaeomagnetic archives, but by the mid 1980s it was already evident that many Palaeozoic carbonate rocks had been completely remagnetized in late Palaeozoic time and it became important to understand the processes responsible (e.g. McCabe & Elmore 1989).

*From*: ELMORE, R. D., MUXWORTHY, A. R., ALDANA, M. M. & MENA, M. (eds) 2012. *Remagnetization and Chemical Alteration of Sedimentary Rocks*. Geological Society, London, Special Publications, **371**, 229–251.
First published online June 26, 2012, http://dx.doi.org/10.1144/SP371.3

## Review of previous work

### Origins of the secondary remanence and its carriers

Widespread remagnetization of Palaeozoic carbonate strata in the Appalachian Basin was initially recognized and documented through combined palaeomagnetic, structural and petrographic observations. These observations included the recognition that the characteristic remanent magnetization (ChRM) directions sometimes clustered significantly more tightly prior to correction for structural tilting, and sometimes clustered most tightly at intermediate unfolding. This led to interpretations that the magnetizations were acquired after and/or during folding (McCabe & Elmore 1989). Petrographic work led to the observation of spheroidal, botryoidal and framboidal magnetite grains of pure end-member $Fe_3O_4$ composition that were interpreted to be of low-temperature diagenetic origin (McCabe et al. 1983). Early rock-magnetic characterization focused on acquisition and demagnetization of isothermal remanent magnetization (IRM) and anhysteretic remanent magnetization (ARM), leading in many cases to the conclusion that the remanence carriers were dominantly pseudo-single-domain (PSD) or multi-domain (MD) magnetite (e.g. McElhinny & Opdyke 1973; Kent 1979). Hysteresis measurements eventually led to the recognition that the dominant magnetic carriers in many remagnetized carbonates are in fact very fine grained, mainly in the superparamagnetic (SP) and stable single-domain (SSD) size range (Jackson 1990; McCabe & Channell 1994; Suk & Halgedahl 1996; Dunlop 2002b). Moreover, the high proportions of SP and SSD particles in the ferrimagnetic inventory of these rocks result in unusual combinations of the summary parameters conventionally used to represent hysteresis behaviour, namely coercivity $B_c$, remanent coercivity $B_{cr}$, saturation magnetization $M_s$, saturation remanence $M_{rs}$ and the ratios $B_{cr}/B_c$ and $M_{rs}/M_s$. On the 'Day plot' ($M_{rs}/M_s$ v. $B_{cr}/B_c$; Day et al. 1977; Dunlop 2002a, b), remagnetized carbonates are largely isolated in a region that is distinct from the regions occupied by most other rocks, sediments and synthetic materials (Fig. 1), leading to the suggestion that hysteresis properties could be used as a fingerprint for recognition of remagnetized strata (Jackson 1990; Channell & McCabe 1994; McCabe & Channell 1994). We will explore the basis and the limitations of this idea in some depth in the present paper.

The origin of the ancient yet secondary characteristic remanence was attributed by McCabe et al. (1983) to chemical processes (growth of new minerals and/or replacement of pre-existing minerals). The spatial/temporal association of widespread remagnetization with Appalachian–Alleghenian–Variscan–Hercynian orogenesis (for summaries see e.g. Weil & Van der Voo 2002; Tohver et al. 2008) soon led to various hypotheses involving continental-scale flow of orogenic fluids (Oliver 1986; Bethke & Marshak 1990; Garven 1995) to account for not only remagnetization, but also ore mineralization and hydrocarbon distribution in a number of different basins. Isotopic and geochemical evidence in some cases ruled out a significant role for exotic brines (e.g. Elmore et al. 1993; Ripperdan et al. 1998), and pointed instead to more closed-system mechanisms including clay diagenesis (Katz et al. 2000; Woods et al. 2002), organic maturation (Blumstein et al. 2004) and pressure solution (Zegers et al. 2003; Elmore et al. 2006; Oliva-Urcia et al. 2008). Kent (1985) questioned whether the remagnetization necessarily involved chemical mechanisms at all, suggesting that viscous acquisition during the Kiaman superchron was a viable alternative. Noting that the range of observed unblocking temperatures could be accounted for by the Viscous Remanent Magnetization (VRM) model of Walton (1980), Kent (1985) demonstrated moreover that the Brunhes-aged viscous overprint in these rocks had unblocking temperatures similar to those predicted by the same model. However, this model was later shown to be inappropriate for predicting unblocking temperatures (Dunlop et al. 1997a, b), undermining the theoretical basis for a viscous origin but not the empirical observations. Similarly, Kligfield & Channell (1981) invoked thermoviscous mechanisms to explain widespread Brunhes-age remagnetization of limestones from the Swiss Alps, in which the ChRM was carried chiefly by pyrrhotite with fine grain sizes spanning the SP–SSD range. They noted that the pyrrhotite may have formed by alteration of early-diagenetic pyrite, but did not address the possibility of a chemical origin for the remanence.

Subsequent work on the remagnetized Appalachian Basin carbonates and remagnetized carbonates in other basins has largely supported the idea that the remanence carriers are predominantly of late diagenetic origin, and that one or more chemical mechanisms were involved in producing the secondary characteristic components of natural remanence (Reynolds 1990; Suk et al. 1990a, b, 1991, 1993; Weil & Van der Voo 2002; Zwing et al. 2005). Our focus here is on the important role that rock magnetic studies have played in the understanding of carbonate remagnetization, but of course we recognize that a complete picture also includes careful petrographic, geochemical, isotopic, microstructural and analytical microscopic work (Elmore & McCabe 1991; Elmore 2001).

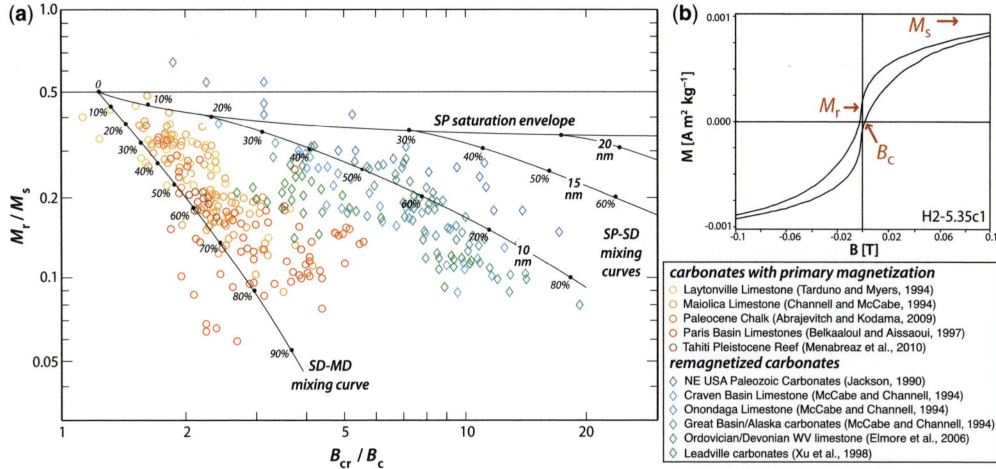



**Fig. 1.** (**a**) Summary Day plot of published hysteresis parameters for carbonate rocks, in which the ratio of remanent coercivity ($B_{cr}$) to coercivity ($B_c$) is plotted against the ratio of saturation remanence ($M_r$) to saturation magnetization ($M_s$). Data from carbonates interpreted to carry primary remanence are marked with circles, and data from carbonates interpreted to be remagnetized are marked with diamonds. (**b**) An example hysteresis loop from sample H2-5.35 of the Manlius Formation of the Helderberg Group is shown with $B_c$, $M_r$ and $M_s$ marked. Note that the loop had not reached saturation by 100 mT (the maximum of the *x*-axis), such that $M_s$ is not reached until higher applied fields (off the plot).

In the interest of clarity, we will take a specific and limited set of previously studied remagnetized carbonate units from the Appalachian Basin as prototypes of the distinctive set of rock-magnetic properties and NRM characteristics that we wish to understand. These include the Devonian Onondaga Limestone (Kent 1979; Jackson *et al.* 1988; Suk *et al.* 1990*a*; Lu *et al.* 1991) and Helderberg Group (Scotese *et al.* 1982) and the Ordovician Trenton Limestone (McElhinny & Opdyke 1973; McCabe *et al.* 1984) and Knox Dolomite (Bachtadse *et al.* 1987; Suk *et al.* 1990*b*). All of these carry a two-component NRM with a well-documented reverse-polarity Permian-aged ChRM and a normal-polarity recent viscous overprint, and all exhibit rock-magnetic properties dominated by SP and SSD ferrimagnets (Jackson 1990; Jackson *et al.* 1993; Jackson & Worm 2001; Dunlop 2002*b*). By understanding as well as possible the relationship between the ChRM and the magnetic phases and their origins in these prototypical units, our goal is to improve our ability to interpret similarities and differences in properties observed in other carbonate strata for which the magnetization history is less well known.

## Unresolved questions

*Nature of grain-scale anisotropy.* The hysteresis ratios of several of the remagnetized Palaeozoic carbonate units from the Appalachian Basin follow a power-law trend, visible as a linear array of points on a bilogarithmic Day plot as in Figure 1. This trend is best explained as resulting from a mixture of SP and SSD ferrimagnets (Dunlop 2002*a*, *b*). Further, it has been suggested (Jackson 1990) that the SSD fraction (at least) in remagnetized Appalachian carbonates has properties dominated by cubic magnetocrystalline (multiaxial) anisotropy rather than by shape anisotropy. In a population of randomly oriented single-domain ferrimagnetic particles dominated by uniaxial shape anisotropy, the ratio $M_r / M_s$ should approach 0.5 as the coercivitiy ratio approaches 1 (Stoner & Wohlfarth 1948). In contrast, the $M_r / M_s$ intercept for the remagnetized Appalachian carbonates is *c.* 0.89, as expected for randomly oriented single-domain (SD) grains with cubic anisotropy (Wohlfarth & Tonge 1957). It should be noted that the error bars on this intercept value are relatively large, and moreover that linear extrapolation is probably not valid (Dunlop 2002*a*, *b*). Nevertheless, the suggestion of dominantly cubic anisotropy is interesting and has important implications. Populations of ferrimagnetic grains that are dominated by multiaxial anisotropy are unusual in nature, because even a slight elongation (e.g. length to width ratio of *c.* 1.1 and greater) is sufficient to make magnetostatic (shape) anisotropy the dominant factor controlling the hysteresis and remanence behaviour of magnetite or maghemite. The strong implication of a nearly complete lack of particle elongation is that the remanence-carrying particles have authigenic/diagenetic morphologies similar to those observed for the much larger particles retrieved by magnetic extraction of insoluble residues (e.g. McCabe *et al.* 1983) and that the unobserved, much finer, SSD particles originated in the same way.

On the basis of hysteresis parameters, a similar argument for multiaxial anisotropy was later made by Gee & Kent (1995) for mid-ocean-ridge basalts (MORB), whose remanence resides chiefly in titanomagnetites of composition $Fe_{3-x}Ti_xO_4$ with $x$ c. 0.6 (TM60). This argument has been criticized by Fabian (2006) on two grounds. First, theoretical considerations, combined with known intrinsic material properties, suggest that cubic magnetocrystalline (multiaxial) anisotropy should almost always be less important than magnetostrictive and shape (uniaxial) anisotropies for titanomagnetites. Second, careful measurements showed that magnetization of MORB samples is often not fully saturated in the field range used for hysteresis loop measurements (typically not exceeding 1 Tesla (T), and incomplete saturation in combination with conventional linear high-field slope calculations yield erroneously low values for $M_s$ (and thus erroneously high $M_r/M_s$ ratios). Fabian (2006) showed for several MORB samples that even although loops measured on a vibrating-sample magnetometer (with peak fields of 1 T) yielded apparent $M_r/M_s$ ratios well above 0.5, measurement of the same samples in fields up to 7 T in a superconducting magnet resulted in significantly higher $M_s$ estimates and $M_r/M_s$ ratios below the uniaxial limit. Fabian (2006) concluded that uniaxial stress anisotropy exerts the dominant control on the hysteresis behaviour of MORBs. Although he did not discuss remagnetized carbonates, his findings raised questions about the identification of multiaxial anisotropy in any material through standard hysteresis experimental protocols.

Fabian's (2006) arguments and experimental results appeared conclusive, but the potential importance of magnetocrystalline anisotropy in MORBs has not been put to rest. A strong counterargument in favour of such multiaxial anisotropy in MORB titanomagnetites was recently made by Lanci (2010), by means of a clever technique based on measurements of field-impressed anisotropy of susceptibility (illustrated in Fig. 2a). The essence of Lanci's approach involves the 'inverse' anisotropy of magnetic susceptibility (AMS) of SD particles: an individual SD grain has minimum (essentially zero) susceptibility parallel to its remanent moment (because it is already magnetized to saturation in that direction) and maximum susceptibility perpendicular to that easy axis. For a population of uniaxial SD grains this results in the well-known 'inverse-fabric' effect (Potter & Stephenson

1988); a preferential alignment of long axes gives minimum susceptibility parallel to lineation. When uniaxial particles are randomly oriented, the acquisition of a strong-field remanence has only very weak effects on AMS, because particle moments are spread out all over the directional hemisphere around the net remanence (i.e. around the applied field orientation). Lanci's (2010) idea is that multiaxial particle anisotropy makes remanence effects much more extreme: even if easy axes are randomly oriented, acquisition of a strong-field remanence causes the particle moments to be confined to a much smaller range of orientations close to the applied field direction because there are more easy axes per particle. This therefore becomes the minimum susceptibility direction (Fig. 2a). Very large anisotropies of 200–300% can be produced by acquisition of a saturation isothermal remanent magnetization (SIRM) in populations of randomly oriented SD grains with multiaxial anisotropy. Lanci's experimental results showed that imprinting a 1 T IRM on several MORB samples with very weak initial AMS resulted in a strong anisotropy of susceptibility (10–25%) with $k_{min}$ parallel to the IRM, and showed that the effect could be erased by tumbling alternating-field (AF) demagnetization. He calculated that 20–30% of the susceptibility in those MORB samples can be attributed to titanomagnetite particles with multiaxial anisotropy.

Further evidence for multiaxial anisotropy in MORB samples has been presented by Mitra *et al.* (2011) using another novel experimental approach, involving asymmetry of forward and reverse IRM acquisition (illustrated in Fig. 2b). An individual uniaxial SD grain has the same switching threshold in fields of either polarity. A population of uniaxial SD grains, initially in a weakly magnetized state (e.g. NRM state, or after tumbling AF), will acquire an IRM of intensity $M_r^+$ in a non-saturating applied field $B_{app}$. Application of an equal-strength field in the opposite direction will produce an IRM of equal intensity ($M_r^- = M_r^+$) antiparallel to the positive-field IRM, as each uniaxial particle that reversed its moment into alignment with the positive field reverses again into alignment with the negative field. This equivalence between the $M_r^-$ and $M_r^+$ does not hold for multiaxial SD grains (Fig. 2b). In contrast to uniaxial grains where the moment is confined to the long axis of the grain and only two possible remanent orientations exist, multiaxial grain moments have a number of possible remanent orientations only one of which is antiparallel to the original moment orientation. Application of a non-saturating field may cause the moment of a particular grain to jump from one easy axis to another, closer to the applied field direction and thus contributing to $M_r^+$. Subsequent application of a reverse

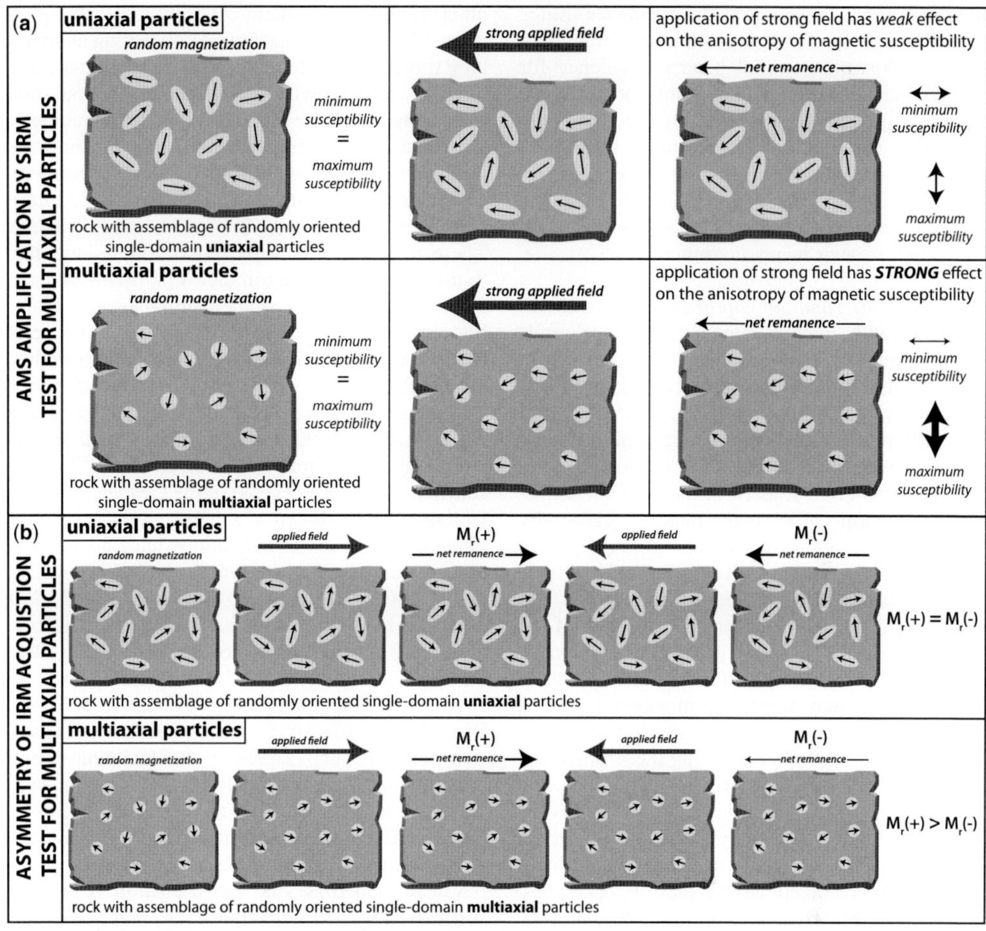


**Fig. 2.** Schematic illustration of tests for multiaxial particle anisotropy in single-domain populations. The test depicted in (**a**) involves the effects of strong-field remanence acquisition on anisotropy of low-field susceptibility (Lanci 2010); with multiaxial particle anisotropy, a strong alignment of particle moments results in a strong anisotropy of magnetic susceptibility (AMS). The test depicted in (**b**) involves asymmetric stepwise acquisition of isothermal remanent magnetization (IRM) in positive and negative fields of equal strength (Mitra *et al.* 2011). See text for detailed explanation of both tests.

field of equal strength to the population of multiaxial grains will not necessarily cause the moment of the same grain to reorient, because the angle between the moment and the field is now more acute than it was in the initial state. $M_r^-$ is consequently of lower intensity than $M_r^+$ for equal and opposite applied non-saturating fields (Fig. 2b). Mitra *et al.* (2011) demonstrated such positive/negative asymmetry in IRM acquisition experiments for MORB samples and for hematite samples (with multi-axial basal-plane grain-scale anisotropy) with inten-sity differences reaching 100% for intermediate applied fields, decreasing to zero for saturating fields. Similar measurements on other control samples showed essentially identical forward and

reverse IRM acquisition, as expected for samples with uniaxial remanence carriers.

Although these arguments over multiaxial aniso-tropy have focused almost entirely on MORB samples, the same questions apply to remagnetized carbonates. Suk & Halgedahl (1996) pointed out that observed coercivities of remagnetized Appla-chian Basin carbonates were two to three times higher than can be theoretically produced by magne-tocrystalline anisotropy in magnetite, but they suggested that a mixture of uniaxial and multiaxial particles could account for all observed properties. Evidence for multiaxial grain-scale anisotropy comes almost entirely from the intercept value of the trend line on the bilogarithmic Day plot; few individual loops have yielded remanence ratios above 0.5 and, for those, non-saturation may be a problem. Moreover, no independent evidence for multiaxial anisotropy has been found for carbonate rocks. One goal of the present study is to apply the new techniques of Lanci (2010) and Mitra *et al.* (2011) to some of the well-studied Palaeozoic carbonates of the Appalachian Basin, to test the hypothesis that multiaxial anisotropy is a characteristic feature of their magnetic mineralogy.

*Mineral mixtures and the importance of pyrrhotite.* Wasp-waisted hysteresis loops and Day plot trends similar to those of the remagnetized carbonates in Figure 1 require the presence of two or more distinct coercivity fractions, which may correspond to either different remanence-carrying minerals with strongly contrasting coercivities (Jackson *et al.* 1990; Muttoni 1995; Roberts *et al.* 1995) or different size fractions of a single mineral (especially if one is superparamagnetic; Tauxe *et al.* 1996; Dunlop 2002*a, b*; Lanci & Kent 2003). These two scenarios have different implications for the origins of the carriers and, correspondingly, of the natural remanence. A single process or remagnetization event could account for the co-occurrence of SP and SSD magnetites in chemically overprinted rocks, with a unimodal size distribution of monogenetic carriers naturally producing the strong coercivity contrast (Lanci & Kent 2003). Polymineralic carrier populations imply a more complicated history where either a primary phase was preserved with a secondary authigenic phase, or where there are multiple preserved generations of secondary authigenic phases.

The magnetic mineralogy of many previously studied remagnetized carbonates is thought to comprise primarily magnetite, largely on the basis of unblocking temperatures, but remains incompletely known. In some cases pyrrhotite (Xu *et al.* 1998; Zegers *et al.* 2003) or hematite (see list in McCabe & Elmore 1989) have been documented as overprint carriers. Multiple studies have argued for the preservation of primary magnetite remanence in carbonates that have a secondary remanence held by pyrrhotite (Muttoni & Kent 1994; Symons *et al.* 2010). Strong-field thermomagnetic data are generally rare, due to the difficulty of obtaining measurable signals from small quantities of this weakly magnetic material. While a thermomagnetic curve on a sample chip of the Manlius Formation of the Helderberg Group revealed a smooth reversible curve with a Curie temperature at *c.* 565 °C (Scotese *et al.* 1982), this result has not been confirmed for other remagnetized carbonates or for other Helderberg Group samples. Laboratory heating also typically results in some degree of alteration of the magnetic mineralogy, especially for pyrrhotite and other sulphides, adding uncertainty to the interpretation. More readily measurable unblocking temperature spectra of NRM and of artificial remanences have generally shown low-to-moderate thermal stability, with typical median demagnetization temperatures (i.e. temperatures required to erase half the initial remanence) of 200–300 °C and maximum unblocking temperatures *c.* 450–500 °C (e.g. Jackson *et al.* 1993; Zegers *et al.* 2003; Elmore *et al.* 2006). These data allow multiple possibilities, including (but not requiring) significant contributions from pyrrhotite or detrital titanomagnetites as well as fine-grained pure magnetite (with $T_{ub} < T_c$) and/or magnetite with dominantly magnetocrystalline anisotropy.

Low-temperature remanence measurements (Jackson *et al.* 1993) on samples of the Devonian Onondaga Limestone and Ordovician Trenton Limeston of New York and of the Ordovician Knox Dolomite have shown a general absence of clear transitions, but subtle indications have been found in some samples of the magnetite Verwey transition ($T_v$ *c.* 120 K) (Verwey & Haayman 1941; Walz 2002) and/or of the pyrrhotite Besnus transition ($T_{Bs}$ *c.* 32 K) (Besnus 1966; Dekkers *et al.* 1989; Rochette *et al.* 1990). As has long been known (Verwey & Haayman 1941; Özdemir *et al.* 1993; Walz 2002; Özdemir & Dunlop 2010), the Verwey transition is very sensitive to deviations from perfect stoichiometry. Small degrees of cation substitution or moderate cation deficiency (due to low-temperature oxidation) depress the transition temperature and diminish the changes in measurable physical properties, and the transition is entirely suppressed by just a few percent content of substituted metal cations or vacancies due to high-temperature oxidation (Verwey & Haayman 1941). Low-temperature oxidation produces a maghemite shell on a stoichiometric magnetite core, and some expression of the Verwey transition persists to high overall degrees of oxidation (Özdemir & Dunlop 2010). The variables controlling the Besnus transition are much less well known, and even its fundamental mechanism remained unclear until quite recently (Wolfers *et al.* 2011). Nevertheless, there is a clear grain-size dependence of the magnetic behaviour across the transition. The drop in magnetization (or susceptibility) on cooling through $T_{Bs}$ increases strongly with increasing particle size, and the fractional recovery of remanence on rewarming through the transition is greater for the finer sizes (Dekkers *et al.* 1989). The smallest fraction studied by Dekkers *et al.* (1989) was 0–5 μm and for this fraction the drop in remanence on cooling across $T_{Bs}$ was about 15% of the initial room-temperature intensity, with nearly 100% recovery on rewarming. Finer-grained pyrrhotites (but of uncontrolled and imprecisely known submicron sizes) in claystones exhibit what has been termed the P-transition (Aubourg & Pozzi 2010), a ramp-like and perfectly reversible change in magnetization over the temperature range 50–10 K. This is in contrast to the step-like change in larger pyrrhotite particles over a very narrow temperature range around $T_{Bs}$. The P-transition is quite sensitive to even very weak ambient fields, and may be manifested as an increase or a decrease in magnetization while cooling in the imperfectly controlled zero field ($\pm 500$ nT after careful adjustment) of the Quantum Designs Magnetic Property Measurement System (MPMS) instrument commonly used for low-temperature measurements. Depending on the relative abundances of paramagnetic iron-bearing minerals, submicron pyrrhotite and other remanence-bearing phases, the P-transition may be difficult or impossible to discern.

Given the possibility of poorly expressed or entirely suppressed low-temperature transitions in magnetite and pyrrhotite, any definitive quantification of their respective contributions to natural and artificial remanences remains challenging. The large spheroidal particles extracted from a number of remagnetized carbonates (McCabe *et al.* 1983; Suk *et al.* 1990*a*, *b*, 1993; Suk & Halgedahl 1996) have been definitively shown – by x-ray diffraction and scanning electron microscopy (SEM) energy-dispersive analysis – to consist in most cases of nearly pure magnetite. No evidence of magnetic sulphides was found in any of these studies on Appalachian Palaeozoic carbonates, although some grain morphologies such as framboids almost certainly indicate formation of magnetite by replacement of early-diagenetic pyrite (e.g. Reynolds 1990; Suk *et al.* 1990*a*) or perhaps greigite (e.g. Roberts *et al.* 2011). Pyrrhotite has been observed petrographically in carbonates from the lower Carboniferous Leadville Formation of central Colorado that also contain authigenic magnetite (Xu *et al.* 1998). By measuring hysteresis loops of individual extracted spheroidal particles using an extremely sensitive alternating gradient magnetometer, Suk & Halgedahl (1996) made two key additional observations. First, the saturation magnetic moment of many of the spherules was much less than it would be for the same mass of pure magnetite; these must contain large portions of void space, non-magnetic minerals (e.g. silicates or pyrite) and/or weakly magnetic phases such as hematite or goethite. Second, although some relatively large (up to 100 μm) spherules yielded $M_r/M_s$ ratios of 0.5 or more (generally those with quite low $M_s$), the trend for the spherule samples on the Day plot differed strongly from that of the bulk rocks; these can be interpreted as SSD–MD mixtures and SP–SSD mixtures, respectively (Suk & Halgedahl 1996; Dunlop 2002*b*). The bulk rocks contain a large population of ultrafine SP and SSD particles that evade or are destroyed by the extraction process (e.g. Sun & Jackson 1994; Weil & Van der Voo 2002) and whose mineralogy is not constrained by any direct evidence.

# New results

We will now attempt to address some of the outstanding questions described above through the results of experiments conducted on a new sample set collected from Devonian carbonates of the Helderberg Group in New York State (Fig. 3a). These new samples were collected in the Hudson Valley fold-thrust belt just to the north of Kingston, NY at sites along roadcuts on the north side of Route 199 that correspond to two of the localities sampled by Scotese *et al.* (1982) (Fig. 3b, c). Oriented cores were collected from: limestone mudstone with fine-scale laminations of the Manlius Formation (our stratigraphic section H2; site 24 of Scotese *et al.* 1982); limestone packstone comprised predominantly of brachiopod shells with a fine-grained mud matrix of the Becraft Limestone (our stratigraphic section H1 metre levels 0 to 1.4; site 23 of Scotese *et al.* 1982); and from limestone mudstone and wackestone of the Alsen Formation (our stratigraphic section H1 metre levels above 1.4; also site 23 of Scotese *et al.* 1982). The thermal demagnetization of Scotese *et al.* (1982) isolated a characteristic remanence direction from these localities interpreted to be held by magnetite whose direction corresponds with that of other remagnetized Appalachian basin carbonates.

In addition to analyses on these new samples, we also reanalyse some of the earlier datasets (Jackson 1990; Jackson *et al.* 1993) in light of more recent developments in hysteresis loop processing (Fabian 2006; Jackson & Sølheid 2010), with particular attention to the question of undersaturation.

## *Reanalysis of previous hysteresis loops*

The measurements of Jackson (1990) on samples of the Trenton, Onondaga and Knox carbonates used maximum fields of 1 T or less; in some cases 500 mT or even 300 mT was the maximum applied field. With fields of this magnitude it is legitimate to question whether the magnetization was saturated over the region used for high-field slope fitting. Although a few of those samples were archived, most were not, so we cannot remeasure them in higher fields. However, the original loop measurement data are still available and can be reprocessed using statistical tests for non-saturation and non-linear approach-to-saturation fitting (Fabian 2006; Jackson & Sølheid 2010) in the high-field interval.

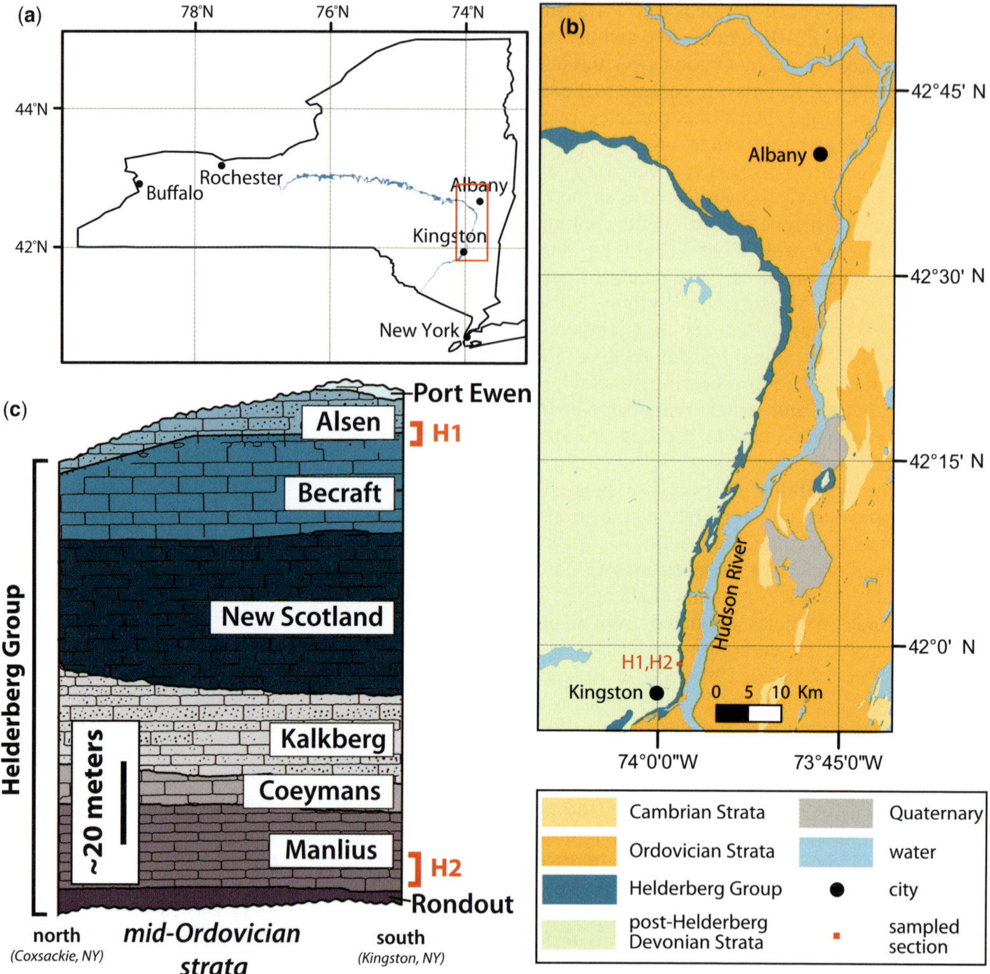



**Fig. 3.** (**a**) Map of New York State with the Helderberg Group mapped out in blue (bedrock geology from Fisher *et al.* 1970). (**b**) Simplified geological map in the vicinity of Kingston and Albany, NY (modified from Fisher *et al.* 1970). The sites sampled in this study to the north of Kingston, New York comprised a stratigraphic section across the contact between the Becraft and Alsen Formations (H1; 41.97604°N, 073.97581°W) and a stratigraphic section within the Manlius Formation (H2; 41.97604°N, 073.97581°W). (**c**) Stratigraphic relationships between the formations that comprise the Helderberg Group between Coxsackie and Kingston, NY above its basal unconformity with the underlying Ordovician flysch of the Austin Glen Formation (modified from Fisher 1987).

The highest $M_r/M_s$ ratios in the original study (in some cases exceeding 0.5) were obtained for the Knox Dolomite and, in a number of cases, these samples were measured in maximum fields of 300 mT under the assumption that this would be sufficient to saturate magnetite. Reprocessing shows that this was far from a safe assumption: F tests (Jackson & Sølheid 2010) comparing total-misfit variance with pure-error variance (quantified by departures from inversion symmetry) indicate that we must reject the hypothesis of linearity over the interval from 70% to 100% of the maximum applied field,

with a high degree of certainty (F ranges from 300 to over 1000; compared to a critical value of about 1.5). Recalculation using a non-linear fitting method yields significantly higher estimates of $M_s$ and accordingly diminished estimates of $M_r/M_s$ (Fig. 4). The calculated coercivity is also affected by the slope correction, and both coordinates on the Day plot therefore differ from those originally published by Jackson (1990). This revised analysis undermines the original argument for multiaxial anisotropy that was based on the apparent occurrence of $M_r/M_s$ ratios above 0.5, the theoretical

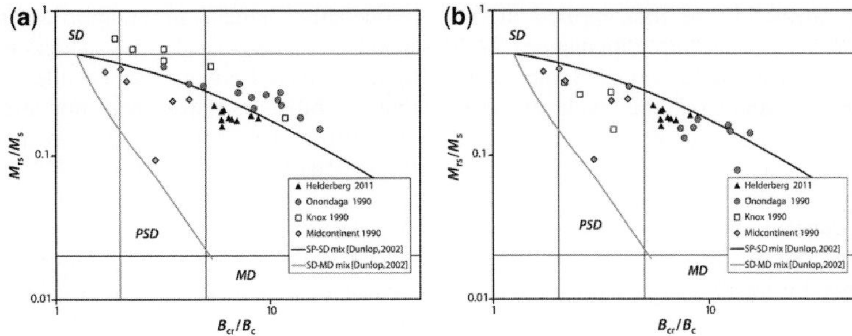

**Fig. 4.** (a) Day plot for Onondaga and Knox samples of Jackson (1990), together with new data for mid-continent samples (Jackson *et al.* 1992) and for new Helderberg Group samples. (b) Recalculated parameters for the Knox Dolomite and Onondaga Limestone loops, using an approach-to-saturation fit to estimate $M_s$ (see text).

upper limit for uniaxial anisotropy. However, these lowered recalculated $M_r/M_s$ ratios alone do not rule out the possibility of a significant contribution from such particles.

## Tests for multiaxial anisotropy

The experimental approach of Lanci's (2010) test for multiaxial anisotropy (Fig. 2) is straightforward, but for instrumental reasons we have modified it slightly. Lanci's technique began with an initial AMS determination followed by pulse magnetizing in 1 T to produce SIRM, and then involved repeated AMS measurements following tumbling AF demagnetization steps to document the gradual removal of the strong field-impressed AMS in multiaxial SD grains. Lacking a tumbling AF instrument, we have slightly modified the experiment for the Helderberg Group samples and have worked in the opposite direction. We begin in the NRM state where moments are very weakly aligned and, after measuring initial

AMS, we imprinted an IRM in one or more field steps, remeasuring AMS after each step.

As shown in Table 1, the effect of exposure to strong fields on the AMS of the Helderberg Group samples was negligible. Diagonal tensor elements change by much less than 1%. Because most of the susceptibility comes from paramagnetic and superparamagnetic sources, there is a relatively large correction that must be applied. Consequently there are rather large uncertainties in the corrected degree of anisotropy, but it is nevertheless clear that the AMS of the potential remanence carriers (i.e. the non-SP ferrimagnetic population) is strongly controlled by uniaxial grain-scale anisotropy.

The dominance of uniaxial anisotropy in the analysed Helderberg Group samples is also confirmed in experiments using the approach of Mitra *et al.* (2011) (Fig. 2). These experiments began in the NRM state with no prior exposure to DC or AC fields, and continued with stepwise application of fields of increasing strength ultimately to 1 T. For

**Table 1.** *Field-impressed changes in AMS on Helderberg Group limestones*

| Specimen_ID | $k_{xx}$ | $k_{yy}$ | $k_{zz}$ | Treatment |
|---|---|---|---|---|
| H1-2.3C (Alsen Formation) | 54.25 | 56.09 | 59.73 | None (NRM state) |
| H1-2.3C | 54.37 | 56.28 | 59.81 | 1 T field along $z$ |
| Percent change | 0.22% | 0.34% | 0.13% | |
| H1-2.6B (Alsen Formation) | 147.63 | 154.00 | 156.18 | None (NRM state) |
| H1-2.6B | 147.42 | 154.77 | 156.37 | 100 mT along $x$ |
| Percent change | 0.14% | 0.50% | 0.12% | |
| H1-2.6B | 147.68 | 154.93 | 156.63 | 200 mT along $x$ |
| Percent change | 0.04% | 0.60% | 0.29% | |
| H2-5.35B (Manlius Formation) | 216.98 | 226.28 | 222.78 | None (NRM state) |
| H2-5.35B | 216.59 | 225.94 | 221.99 | 1 T field along $z$ |
| Percent change | 0.18% | 0.15% | 0.35% | |

$k_{xx}$, $k_{yy}$, $k_{zz}$: diagonal elements of susceptibility tensor ($\mu$SI) in the specimen coordinate system

each field step, the field was first applied in the positive direction; the resulting remanence ($M_r^+$) was measured. The same field was then applied in the negative direction and again the resulting remanence ($M_r^-$) was measured. For several samples the experiment was carried out using a pulse magnetizer and a 2G superconducting magnetometer. For a few samples, an equivalent experiment was conducted, using a vibrating-sample magnetometer to measure a series of minor hysteresis loops with increasing peak fields (of equal strength in the positive and negative directions). In all cases, the difference between $M_r^+$ and $M_r^-$ was negligible, thereby demonstrating that the anisotropy of the particles in the analysed Helderberg Group carbonates is overwhelmingly uniaxial (Fig. 5).

## Superparamagnetic fraction

Although the SP population plays no role in carrying the NRM, it is of considerable interest in understanding the origins and significance of the (presumably) cogenetic SSD particles that do carry the palaeomagnetic information in these rocks. Further, any quantitative analysis and interpretation of in-field (hysteresis and susceptibility) measurements must properly account for the SP fraction in order to isolate the contribution of the stable remanence carriers.

Magnetic properties change dramatically with increasing size within the SP range and across the SP–SSD threshold (or blocking volume), and are strongly temperature dependent (Néel 1949; Dunlop & Özdemir 1997). SP particles that are much smaller than the blocking volume at a given temperature remain in equilibrium with ambient fields, even those that alternate at high frequency; they therefore have frequency-independent susceptibility that is fully in-phase, and not especially high (e.g. Worm 1998). During hysteresis experiments, these SP particles approach saturation slowly in relatively high fields (Tauxe et al. 1996). Larger particles near the SP–SSD threshold show more strongly time- or frequency-dependent magnetization on laboratory measurement timescales; susceptibilities are very high with in-phase ($k'$) and quadrature ($k''$) components that vary with frequency. These 'viscous superparamagnetic' (VSP) particles saturate in relatively low fields, thereby 'shearing' the loop of accompanying stably magnetic phases during hysteresis experiments to generate 'wasp-waisted' loops (Fig. 1b; Tauxe et al. 1996; Fabian 2003).

At room temperature, a relatively strong (10–20% per decade) frequency dependence of susceptibility and high ratios of $\chi_f/M_s$ (c. 50 μm A$^{-1}$) in previously studied remagnetized carbonates (Jackson et al. 1993) and in our new Helderberg Group samples clearly indicates the presence of a significant population of grains with $T_b$ c. 295 K (for magnetite, this $T_b$ implies grains with diameters of c. 15–25 nm; Worm 1998). During cooling, as the size window for VSP behaviour shifts to finer sizes, a strong frequency dependence of $k'$ and a significant $k''$ persist down to 20 K for the Helderberg Group samples (Fig. 6), indicating a broad distribution of nanometric particle sizes.

Hysteresis loops measured on a Quantum Designs MPMS, using maximum fields of 2.5 T and temperatures from 300 K down to 10 K, show strong, progressive increases in $M_r$ and $B_c$ on cooling (Fig. 7) as the SP fraction becomes thermally stable. The loop constriction that is so prominent at room temperature gradually gives way to more 'rectangular' or 'pot-bellied' shapes (Tauxe et al. 1996; Fabian 2003). Quantitative loop analysis requires initial separation of the ferrimagnetic signal from the dia/paramagnetic background signal. This becomes somewhat problematic at the lowest temperatures however, where paramagnetic magnetization becomes a distinctly non-linear function of applied field (especially in fields over 1 T). Further, antiferromagnetic nanoparticles of materials such as ferrihydrite may contribute to low-temperature high-field non-linearity. Here we use a consistent fitting method, in which the high-field interval is tested for statistically significant deviations from linearity (Jackson & Sølheid 2010). An exponential approach-to-saturation law (Fabian 2006) is applied if the linearity test fails. We recognize that this method overestimates $M_s$ for the lowest temperatures, because it incorporates some of the non-linear paramagnetic and/or antiferromagnetic magnetization into the modelled ferrimagnetic loop; however

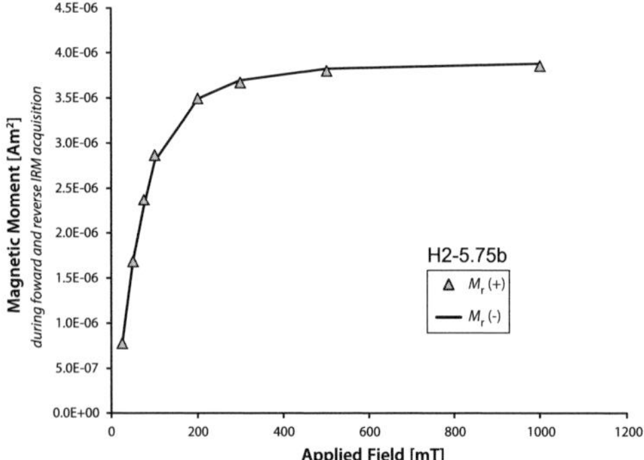


**Fig. 5.** Results of a test for multiaxial anisotropy based on forward and reverse IRM acquisition (Fig. 2; Mitra et al. 2011). At each field step a positive field is applied, the resulting remanence ($M_r^+$) is measured, a negative field of equal strength is applied and the remanence ($M_r^-$) is measured. Intensity differences for this sample of the Manlius Formation of the Helderberg Group are negligible, showing that the particle anisotropy is overwhelmingly uniaxial.

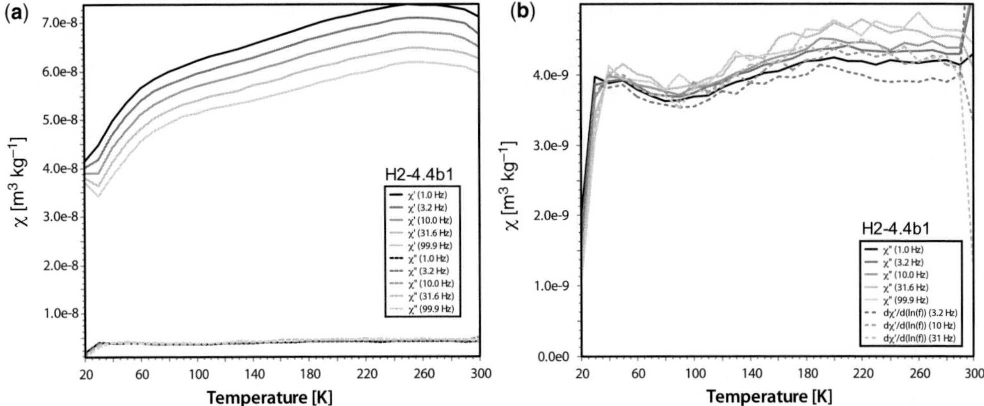

**Fig. 6.** (**a**) In-phase and quadrature susceptibility at five frequencies for sample H2-4.4 (Manlius Formation, Helderberg Group), measured from 20 to 300 K. Strong frequency dependence over the whole temperature range indicates a broad size distribution of nanoparticles (e.g. Worm 1998). (**b**) Expanded view of quadrature susceptibilities, compared to the derivative of in-phase susceptibility with respect to $\ln(f)$.

this will not be critical to our analysis. For temperatures down to about 40 K, $M_s$ changes by only a very slight amount; $M_r$ however increases to 3 times its room-temperature value and $B_c$ increases by a factor of 6 (Figs 8a, b), both as a result of SP–SSD transitions in progressively finer particles on cooling. The shape parameter $\sigma_{hys}$ defined by Fabian (2003) quantifies the dramatic change from strongly wasp-waisted loop shapes to more rectangular or pot-bellied shapes at lower temperatures (Fig. 8c). The sharp increase in calculated $M_s$ values below 40 K is no doubt partly due to the aforementioned non-linear paramagnetic signal affecting the approach-to-saturation calculations, but may also be partly due to ordering of some iron-bearing phase. Jackson & Worm (2001) saw evidence of

this ordering in low-temperature susceptibility data for specimens of the Trenton Limestone.

Thermal demagnetization of a low-temperature strong-field isothermal remanence can be used, together with some assumptions, to quantify the SP content in a sample (Banerjee *et al.* 1993) and to calculate the size distribution of particles that unblock while warming to room temperature (e.g. Worm & Jackson 1999). The key assumption is that the loss of remanence is related to the SSD-SP transition, rather than to multi-domain phenomena such as wall unpinning or domain reorganization due to the changes in temperature-dependent anisotropy constants (e.g. Moskowitz *et al.* 1998). This assumption is well justified for the remagnetized carbonates that we are discussing, as detailed

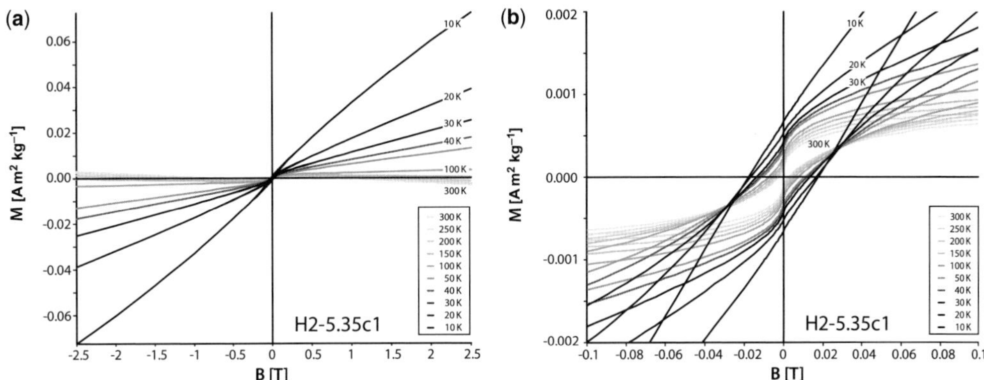

**Fig. 7.** (**a**) Low-temperature high-field hysteresis loops for sample H2-5.35 (Manlius Formation, Helderberg Group), uncorrected for dia- and paramagnetic background. (**b**) Expanded view of low-field portion of the loops, showing the characteristic wasp-waisted shape at room temperature (light grey). This wasp-waisted shape dissipates significantly on cooling as the superparamagnetic particles become progressively blocked in.

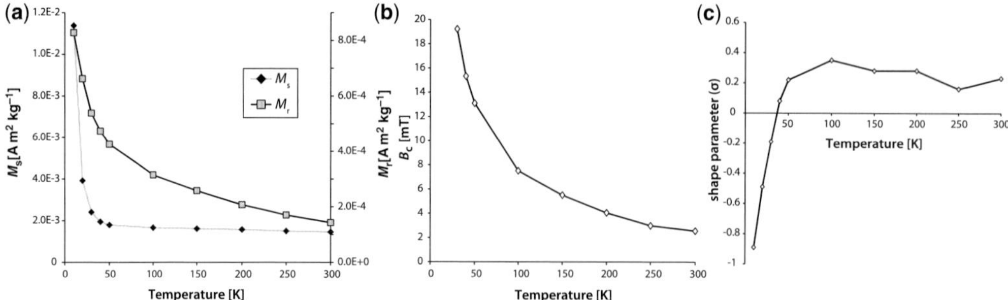

**Fig. 8.** (**a**) Saturation remanence, (**b**) coercivity and (**c**) shape parameter as a function of temperature as calculated for the low-temperature hysteresis loops of sample H2-5.35 (Fig. 7). (a) Saturation remanence increases significantly on cooling, as SP particles become thermally stable. Saturation magnetization changes little, at least down to 40 K, which appears to be the ordering temperature of an unidentified iron-bearing phase. Loop shapes are constricted (positive value of $\sigma$) down to 40 K, below which most of the ferrimagnetic signal comes from thermally stable carriers.

below. For the analysed Helderberg Group samples, roughly three-quarters of the low-temperature remanence imposed isothermally by a 2.5 T field at 20 K is erased by zero-field warming to room temperature. This result implies that approximately 75% of the ferrimagnetic material present is in the SP state at 300 K and does not contribute to the ancient NRM (Fig. 9a). This result agrees well with the estimate of Dunlop (2002*b*) for the Onondaga and Trenton Limestones that was based on modelling of room-temperature hysteresis data. The uninflected demagnetization curve suggests a broad, unimodal size distribution of the SP particles, and simple calculations using the method of Worm & Jackson (1999) make this explicit (Fig. 9b). The distribution peaks at or below $10^{-24}$ m$^3$ (approximately 10 nm diameter), in very good accord with the modelled SSD-SP mixing curves of Dunlop (2002*b*).

*Stable SD fraction*

Whereas in-field measurements such as hysteresis and susceptibility are strongly influenced by SP particles, we can isolate the properties of the SSD and larger fraction by measuring remanent magnetizations. IRM acquisition and AF demagnetization curves for the new Helderberg Group samples are essentially identical to those obtained previously for the Onondaga and Trenton limestones (Jackson 1990), intersecting at fields of about 50 mT, with crossover parameters (Cisowski 1981) of $R$ c. 0.5 (Fig. 10). Roughly 10% of the 1 T SIRM survives AF demagnetization at 200 mT. The acquisition curves strongly resemble those found by Elmore *et al.* (2006) in Helderberg Group samples from West Virginia. They also exhibit some similarity to those defined by unmixing analysis (Gong *et al.* 2009) as the dominant component in remagnetized Cretaceous carbonates from the Organyá Basin of northern Spain, showing significant lack of saturation in 300 mT. Gong *et al.* (2009) interpreted this component as due to very fine magnetite, near the SP–SSD transition. Low-temperature cycling of a room-temperature SIRM (Fig. 11) suggests the presence of minor amounts of remanence-carrying hematite and pyrrhotite, with subtle inflections

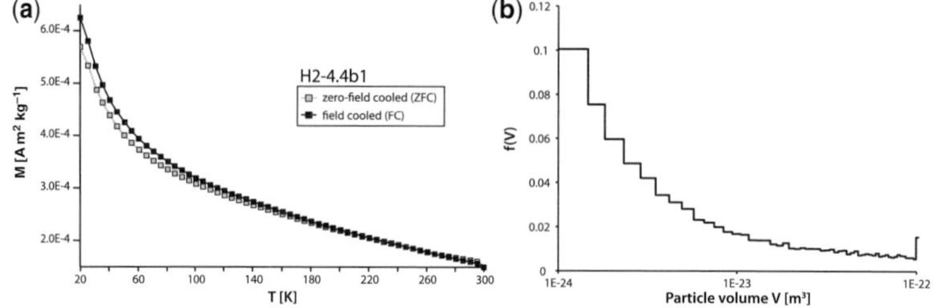

**Fig. 9.** (**a**) Thermal demagnetization of low-temperature remanence for sample H2-4.4 shows progressive unblocking of the nanophase ferrimagnets, with no indication of the Verwey transition. FC remanence was imprinted by cooling in a 2.5 T field from 300 K to 20 K; ZFC remanence was imparted isothermally at 20 K by application and removal of a 2.5 T field (after zero-field cooling from 300 K). Both remanences were measured while warming in zero field. (**b**) Particle-size distribution calculated for the ZFC curve in (a).

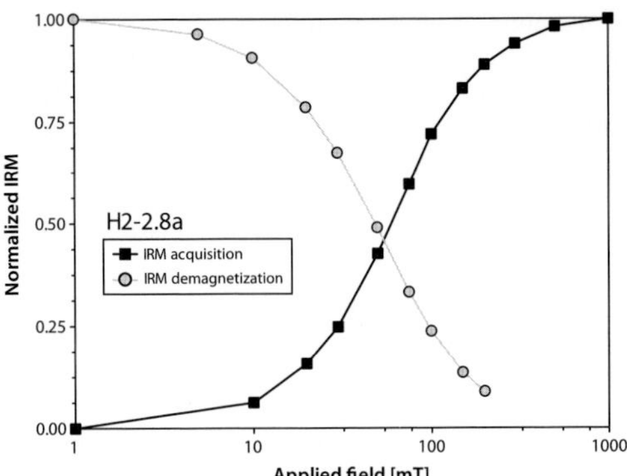



**Fig. 10.** Acquisition and AF demagnetization of IRM for sample H2-2.8 (Manlius Formation, Helderberg Group) shows near-ideal non-interacting SD behaviour (Cisowski 1981). The crossover point between the acquisition and demagnetization curves is 54 mT, $R = 0.46$.

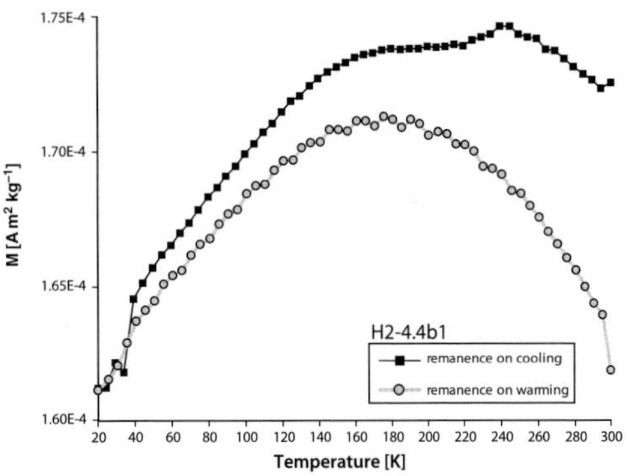



**Fig. 11.** Low-temperature demagnetization (LTD) cycle for H2-4.4. The sample was magnetized isothermally at 300 K in a 2.5 T field, then measured in zero field while cooling to 20 K and rewarming to room temperature. The broad maximum around 150–200 K resembles that observed by Özdemir & Dunlop (2010) in oxidized submicron magnetite; however, here we see no distinct indication of the Verwey transition. The Morin transition (250 K) and the Besnus transition (32 K) indicate small remanence contributions from hematite and pyrrhotite, respectively.

near the Morin transition (*c.* 250 K; e.g. Özdemir & Dunlop 2006) and near the Besnus transition (32 K; Besnus 1966; Dekkers *et al.* 1989; Rochette *et al.* 1990). However, the magnetization changes very little overall (less than 8%) during cooling and rewarming, with some small irreversible loss of remanence. The major feature of the rewarming curve is a broad maximum between 150 and 200 K, resembling the hallmark features noted for oxidized magnetite by Özdemir & Dunlop (2010) and for pyrrhotite by Dekkers (1989).

Anhysteretic remanent magnetization (ARM) acquisition characteristics provide additional evidence that the remanence carriers in many remagnetized carbonate units overwhelmingly comprise non-interacting SSD particles. ARM intensity increases rapidly as a function of DC bias field ($B_{dc}$), with ratios of ARM/SIRM reaching nearly 30% for $B_{dc} = 0.2$ mT (Fig. 12; see also Jackson *et al.* 1992, 1993). The initial slope (calculated for $0 \leq B_{dc} \leq 0.01$ mT) is the normalized anhysteretic susceptibility $\chi_a$/SIRM with a value of 2.1 mm A$^{-1}$, comparable to theoretical values for non-interacting SD magnetite (Egli & Lowrie 2002; Egli 2006), to values observed for cultured magnetotactic bacteria (Moskowitz *et al.* 1993) and for igneous materials such as the Tiva Canyon Tuff (Till *et al.* 2010) and the Lambertville plagioclase (Dunlop & Özdemir 1997, fig. 11.6) that are known to contain dominantly non-interacting SSD magnetic particles.

To summarize, ARM and IRM characteristics together indicate that the thermally stable (non-SP) room-temperature remanence carriers in the new sample set from the Helderberg Group, and from similar previously analysed remagnetized carbonates in the Appalachian Basin and the

mid-continental USA, are almost entirely in the SD state with very minimal contributions from PSD and MD grains. These SSD particles are dominated by uniaxial anisotropy and constitute roughly a quarter of the ferrimagnetic material

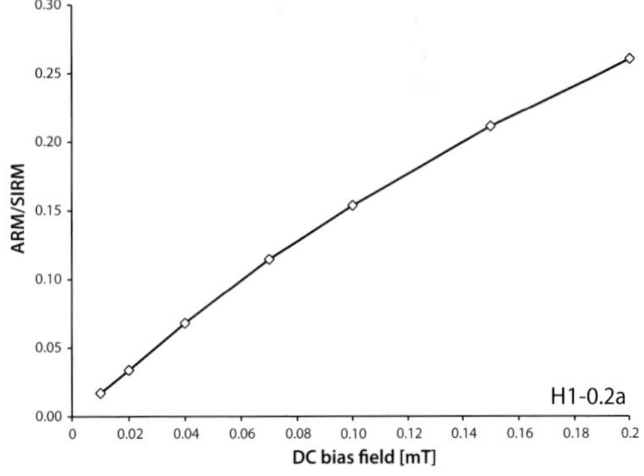



**Fig. 12.** Non-linear anhysteretic remanent magnetization (ARM) acquisition for sample H1-0.2 (Becraft Formation, Helderberg Group) shows non-interacting SD behaviour (Dunlop & Özdemir 1997, fig. 11.8; Egli & Lowrie 2002; Egli 2006). Initial slope gives $\chi_a$/SIRM = 2.1 × 10$^{-3}$ m A$^{-1}$, comparable to theoretical values for non-interacting SD magnetite (Egli & Lowrie 2002; Egli 2006), to values observed for cultured magnetotactic bacteria (Moskowitz *et al.* 1993) and for igneous materials such as the Tiva Canyon Tuff (Till *et al.* 2010) and the Lambertville plagioclase (Dunlop & Özdemir 1997, fig. 11.6)

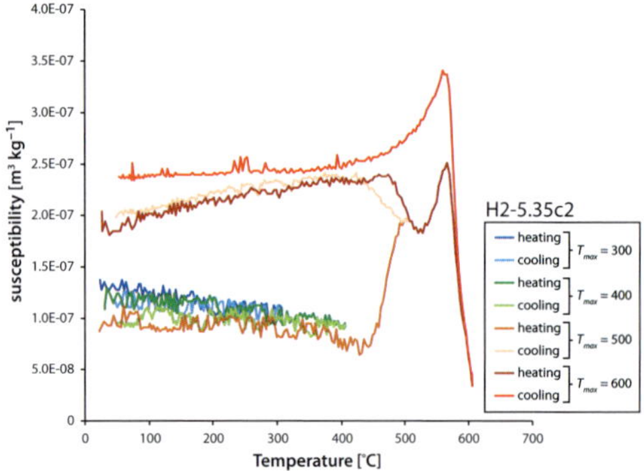

**Fig. 13.** Multicycle high-temperature susceptibility measurements for sample H2-5.35 (Manlius Formation, Helderberg Group) show nearly reversible behaviour below 400 °C (with some decrease due to the 400 degree step), followed by a major increase on heating to 500 °C.

in these samples. Almost all of the rest of the magnetic mineralogy is superparamagnetic at room temperature, as shown by low-temperature hysteresis, low-temperature remanence, frequency-dependent susceptibility and mixing models of room-temperature hysteresis. From these experiments the distribution of domain states is well defined, but specifics of the mineralogy remain elusive.

## Ferrimagnetic mineral composition

Given the lack of pronounced diagnostic low-temperature transitions, the best remaining hope for definitive identification of the magnetic mineralogy is through measurements above room temperature. Low-field high-temperature AC susceptibility of the new Helderberg Group samples, measured on a KLY-2 Kappabridge in an argon atmosphere in a series of heating–cooling cycles with successively higher peak temperatures, shows a weak and noisy signal but essentially reversible behaviour up to 400 °C followed by sharp increases in susceptibility for higher-temperature cycles (Fig. 13). This behaviour is virtually identical to that observed by Zegers *et al.* (2003), which they attributed to formation of new magnetite by oxidation of pyrite in the temperature range 420–500 °C. It differs significantly from the results obtained for the Trenton Limestone by Jackson & Worm (2001), in which significant mineral alteration also began for heating runs to maximum temperatures above 400 °C, but the new mineral was most likely pyrrhotite with a Curie temperature near 320 °C.

High-temperature hysteresis measurements (Fig. 14) also had rather marginal signal/noise ratios, due to the combination of weak intensities and small allowable sample sizes. Loops and back-field remanence curves were measured between room temperature and 400 °C, in order to focus on the naturally occurring ferrimagnetic population and avoid the formation of new magnetic material at higher temperatures. Saturation magnetization decreases by about 50% between 20 and 400 °C, significantly more than the drop of $c.$ 38% expected for pure magnetite over this temperature interval. This may be due to a small contribution from ferrimagnetic pyrrhotite as a slight dip in the $M_s(T)$ curve near 300 °C suggests; however, the data are not of sufficient quality to allow decisive determination. Nevertheless, the trend suggests that monoclinic pyrrhotite is not a major phase in these

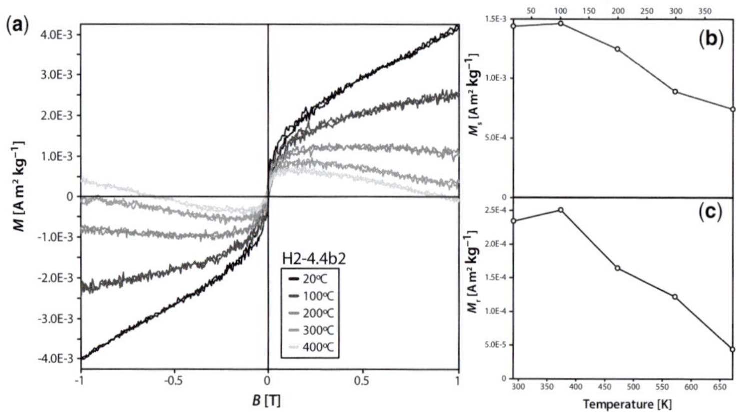

**Fig. 14.** (**a**) High-temperature hysteresis loops for sample H2-4.4, measured on a Princeton Measurements MicroMag VSM (temperatures from 20 to 400 °C). (**b**) Calculated saturation magnetization and (**c**) saturation remanence as functions of temperature.

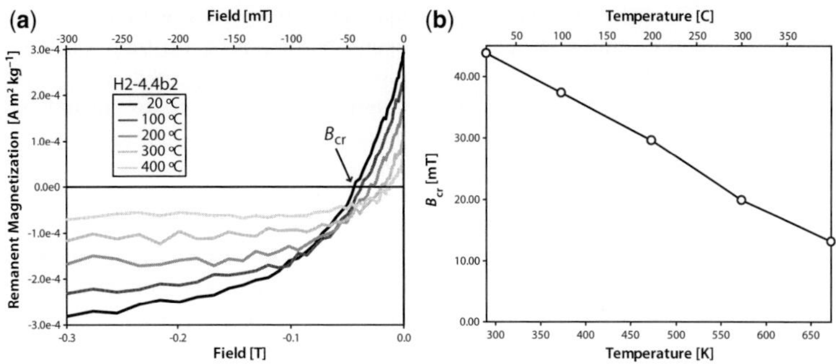

**Fig. 15.** (**a**) High-temperature back-field remanence curves for sample H2-4.4, measured on a Princeton Measurements MicroMag VSM (temperatures from 20 to 400 °C). (**b**) Remanent coercivity determined from the back-field remanence curves as a function of temperature.

samples. Saturation remanence drops by nearly 80% over this temperature range, and the ferrimagnetic population is almost entirely in the SP state by 400 °C. The backfield remanence curves (Fig. 15) show a corresponding progressive shift of the coercivity spectrum to lower fields, with a decrease in $B_{cr}$ by roughly two-thirds over this temperature range.

The method of Lowrie (1990) for identifying remanence carriers according to both coercivity and unblocking-temperature ranges yields similar results (Fig. 16). Orthogonal IRMs were imprinted with successively decreasing pulsed fields: 1 T along the specimen $z$ axis, 300 mT along $x$ and 100 mT along $y$. The two lower-field treatments each reoriented most of the remanence acquired in previous steps (Fig. 16a), confirming that hard antiferromagnets such as hematite could only be making a minor contribution to the IRM. Stepwise thermal demagnetization shows that the three coercivity

fractions all unblock at approximately the same rate, with more than 80% of the remanence erased by 400 °C and nearly 100% demagnetized by 500° (Fig. 16b). A slight change in slope in the intermediate-coercivity $x$ component from 300 to 350 °C may be due to pyrrhotite, but overall the remanence appears to reside in a magnetite-like phase with relatively low unblocking temperatures.

The low-temperature data clearly show that stoichiometric magnetite is nearly absent from these samples. Cation-deficient magnetite is a good candidate to account for the low-T properties, but is less successful in accounting for $M_s(T)$ above room temperature since oxidation increases the Curie temperature (Özdemir & Banerjee 1984). The observed $M_s(T)$ trend suggests a Curie temperature below that of pure magnetite. Cation substitution could account for this, but would be quite surprising as it would seem to be inconsistent with a low-temperature chemical origin for the

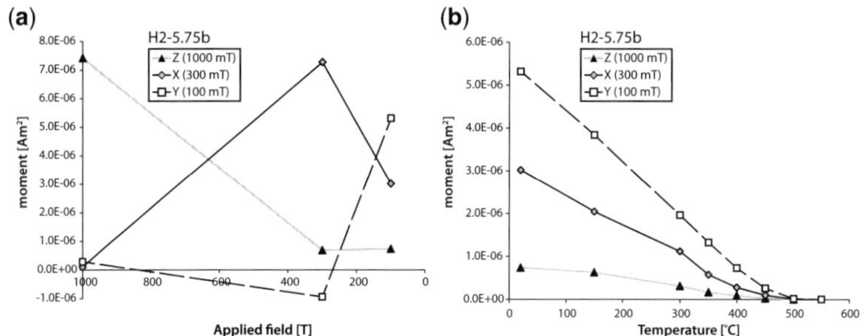

**Fig. 16.** (**a**) Acquisition and (**b**) thermal demagnetization of a three-component IRM for sample H2-5.75 (Manlius Formation, Helderberg Group). Acquisition involved application of a 1 T pulsed field along the specimen's $z$ axis, followed by a 300 mT field along the $x$ axis and a 100 mT field along the $y$ axis. The two latter treatments each reorient most of the remanence, thus shown to be carried predominantly by relatively soft (ferrimagnetic) phases. The three components are removed at roughly the same rates by thermal demagnetization, and are almost completely removed by 500 °C. A dip in the intermediate-coercivity $x$ component at 300–350 °C may be due to pyrrhotite.

dominant magnetic phase. On the other hand, Xu *et al.* (1998) found magnetite grains of clearly authigenic origin that contained minor amounts of substituted Ti and Mn ($x < 0.15$ in $Fe_{3-x}Ti_xO_4$) in the Leadville carbonates of central Colorado. Such compositions would suppress the Verwey transition and also depress the Curie temperature and unblocking temperatures. However, authigenic cation-substituted magnetite has not (to our knowledge) been found in other studies of remagnetized carbonates. Probably the best way to explain both the low- and high-temperature data is through a mixture of maghematized magnetite and smaller amounts of monoclinic pyrrhotite nanoparticles, but cation-substituted authigenic magnetite may also be important.

## Discussion

Do hysteresis properties provide reliable indications of the presence or absence of substantial chemical overprints in carbonate rocks? Certainly *in situ* growth of a significant population of SP–SSD particles should be expected to exert a strong influence on both the NRM and on the rock-magnetic properties, and a number of studies have confirmed this relationship. But how often might we encounter 'false positives' (hysteresis ratios along the trend defined by the remagnetized Appalachian basin carbonates, in rocks where the characteristic remanence is primary) or 'false negatives' (hysteresis ratios along the trend for SD–MD magnetites, in rocks that are substantially or completely remagnetized)? How might we recognize these situations? Given the variety of possible remagnetization mechanisms and triggers (e.g. Elmore 2001), as well as the variety of possible primary iron phases in carbonates, there are a number of important points that have to be considered.

### *Primary magnetic particles in carbonates: sources, characteristics, preservation*

Major proportions of SP material are relatively rare in geological materials, with notable exceptions mostly including those in which the SP material formed *in situ*: soils and palaeosols (e.g. Mullins 1977; Maher & Taylor 1988; Dearing *et al.* 1997; Guyodo *et al.* 2006); rapidly cooled volcanics (e.g. Schlinger *et al.* 1991; Gee & Kent 1999; Till *et al.* 2010); and marine or lacustrine sediments where SP magnetite is produced in the sediment column by dissimilatory iron-reducing bacteria (e.g. Moskowitz *et al.* 1989; Maloof *et al.* 2007). The high surface area to volume ratios of SP magnetite likely contribute to their geochemical instability in detrital environments (e.g. Li *et al.* 2009). Carbonates with primary

magnetization from which hysteresis data have been obtained reveal a trend on the compiled Day plot of hysteresis parameters (Fig. 1) that indicates magnetite populations dominated by SD and MD particles without significant SP contributions. These units are predominantly deepwater pelagic limestones (Laytonville Limestone, Maiolica Limestone, ODP 728) and, in such lithologies, an SD–MD grain size distribution is consistent with detrital input of magnetite and either a lack of SP input or alteration of any SP particles. However, in shallow-water carbonate depositional environments that are isolated from aqueous detrital input and where there is early cementation, is it possible that there could be significant contributions of both ultrafine SP and SSD ferrimagnets? If so, they may yield false positive indications of remagnetization. In the absence of aqueous detrital input there are two main potential sources of magnetic minerals to carbonate environments: biogenic precipitation and aeolian delivery (including extraterrestrial fluxes).

Freeman (1986) documented the common occurrence of 'cosmic spherules' in magnetic extracts from pelagic limestones by detailed SEM work, but was unable to quantify the proportions of ultrafine material in these extraterrestrial magnetic particles. It is not uncommon for airborne dust collections to show significant populations of SP–SSD ferrimagnets, especially when no major local dust sources are present (e.g. Oldfield *et al.* 1985). Magnetic material found in dust layers in the Greenland and Antarctic ice cores contained significant proportions of SP–SSD nanoparticles whose origins were ascribed to ablation of meteorites in Earth's upper atmosphere (Lanci & Kent 2006; Lanci *et al.* 2007, 2008). Detrital inputs of ferrimagnetic nanoparticles could therefore potentially mimic the hysteresis properties of the SP–SSD grains that are produced through authigenic processes associated with remagnetization.

Extracellular formation of magnetite during microbial iron reduction can lead to the *in situ* formation of prodigious amounts of SP magnetite during deposition (Frankel 1987; Maloof *et al.* 2007). Combined with SD grains formed by magnetotactic bacteria during or immediately following deposition (e.g. Kopp & Kirschvink 2008; Moskowitz *et al.* 2008), these populations could also give false indications of remagnetization based on hysteresis properties. With the current published datasets predominantly coming from deepwater carbonates, the possibility that there can be preservation of a significant primary SP population in early cemented shallow-water carbonates is difficult to evaluate. This reality points to the need for more detailed hysteresis data from shallow-water carbonates to see if the preservation of primary SP grains can have a significant influence on hysteresis parameters.

Primary detrital magnetic oxide grains in the PSD and MD size range can moreover be diminished to SP–SSD sizes by reductive dissolution (e.g. Karlin & Levi 1983; Tarduno 1995; Smirnov & Tarduno 2001). New iron sulphide minerals (predominantly greigite) commonly also form in the sediment column during early diagenetic sulphate reduction (Maloof *et al.* 2007; Roberts *et al.* 2011). The combination of such sulphides with residual magnetite could conceivably mimic the hysteresis signature of the remagnetized carbonates, whether or not the new sulphides carry a significant portion of the remanence.

It has long been known that constricted ('wasp-waisted') hysteresis loops and high coercivity ratios can arise from mixtures of hard (e.g. imperfect antiferromagnets such as hematite and goethite) and soft (e.g. ferrimagnets such as magnetite and maghemite) phases (e.g. Nagata & Carleton 1987). There are many situations imaginable (e.g. detrital magnetite and late weathering products such as goethite) where Day plot locations and constricted loops might incorrectly suggest an ancient remagnetization. In the case of a mix of soft ferrimagnetic and hard antiferromagnetic carriers, values of $B_c$ and $B_{cr}$ would be higher than in the case of a dominant SP–SSD population. Additionally, if the loop constriction is due to multiple mineralogies there may be discernible inflections in the gradient of magnetization acquisition and backfield demagnetization of an SIRM. Given the common occurrence of goethite as a surface weathering product and the precipitation of secondary iron sulphides through fluid flow or early diagenesis, interpretations of hysteresis loop parameters should be made cautiously and be paired with other rock magnetic experiments that can demonstrate a largely mono-mineralic population of ferrimagnetic grains.

## *Sources of iron for formation of new magnetic phases and mechanisms and triggers for authigenic iron oxide growth*

Oxidation of pyrite or other sulphides to magnetite in remagnetized carbonates has been documented through detailed analytical microscopy (e.g. Fruit *et al.* 1995; Weil & Van der Voo 2002). Early diagenetic pyrite framboids, aggregates up to 100 μm in diameter comprised of large numbers of small (<1 μm) crystallites, have frequently been found in North American Palaeozoic carbonates to be completely altered by later diagenesis to magnetite (Suk *et al.* 1990*a, b*, 1991, 1993). Suk & Halgedahl (1996) found a variety of other iron-oxide spheroid morphologies that may have formed in this way or by different processes (Freeman 1986), many of which exhibited the MD magnetic behaviour

expected for particles of that size but some of which had PSD or SSD hysteresis properties; a few even had wasp-waisted loops. They argued that despite the large particle dimensions, such spheroids may be the major remanence carriers in the remagnetized carbonates of the Appalachian Basin and elsewhere. This argument is difficult to refute, especially if we recognize that the spheroid population recovered by magnetic extraction from the insoluble residue is likely to be strongly biased in favour of the larger and more highly magnetic particles, and that therefore the spherules found with relatively low $M_s$ (and high $M_r/M_s$) may be the most representative of those in the bulk rock. However, it is not clear that the preponderance of SP magnetite in bulk-rock remagnetized carbonate samples has any direct association with pre-existing sulphides or with the observed spheroids. Replacement of iron sulphides by magnetite is certainly involved in many cases and may be a chief mechanism of remagnetization when large-scale fluid flow is involved. It remains difficult to predict with any certainty that a distinctive hysteresis signature must necessarily accompany this process, so the possibility of false negative tests cannot be ruled out. Suk & Halgedahl (1996) found 'normal' unconstricted hysteresis loops in remagnetized Onondaga Limestone samples from western New York, suggesting that SP magnetite was not formed by the remagnetization process or not preserved.

Many studies have convincingly demonstrated a connection between remagnetization and clay diagenesis, particularly the conversion of smectite to illite (Jackson *et al.* 1988; Hirt *et al.* 1993; Katz *et al.* 1998; Elliott *et al.* 2006*a, b*; Tohver *et al.* 2008). Illitization of smectite occurs at relatively low temperatures (between *c.* 60 °C and 120 °C) and liberates both water and cations such as iron, creating conditions suitable for the *de novo* formation of magnetite or other remanence-carrying minerals. Nucleation and growth processes can be expected to produce nanoparticle populations with size distributions that are ultimately governed by factors including element availability, diffusion rates, reactivity, temperature and time. The growth of magnetite through this process is likely to produce populations that span the SP–SSD size range. In such a scenario, the formation of both abundant SP grains and the grains that reached the critical size for thermal stability (and thus acquisition of a secondary remanence) can be attributed to a single chemical overprinting event. This process seems to us to be the situation in which unique hysteresis properties may most reliably indicate strong chemical remagnetizations analogous to those seen in North American Palaeozoic carbonates.

## The rock-magnetic signature of remagnetization in carbonate rocks

Given the wide variety of possible remagnetization mechanisms, parent phases and final carriers, and the range of scenarios discussed above for potential false positives and false negatives, it is somewhat surprising that a rock-magnetic signature of remagnetization would have any generality. This is even more true now that we have eliminated one of the proposed rationales for a distinctive hysteresis signature, based on multiaxial individual-particle anisotropy in equidimensional authigenic magnetites. However, many examples are now available for which demonstrably remagnetized units have some or all of the putative rock-magnetic fingerprints (e.g. Channell & McCabe 1994; McCabe & Channell 1994; Molina Garza & Zijderveld 1996; Suk & Halgedahl 1996; Banerjee *et al.* 1997; Butler *et al.* 1997; Xu *et al.* 1998; d'Agrella Filho *et al.* 2000; Dinares-Turell & Garcia-Senz 2000; Enkin *et al.* 2000; Zegers *et al.* 2003; Trindade *et al.* 2004; Zwing *et al.* 2005; Elmore *et al.* 2006; Soto *et al.* 2008). We are not aware of any clear-cut false positives (hysteresis characteristics indicating complete remagnetization in carbonate rocks bearing a primary NRM) and very few false negatives (hysteresis behaviour indicating primary NRM carriers in fully chemically overprinted carbonate rocks). However, there is a number of rock units with 'intermediate' or ambiguous hysteresis characteristics, some of which clearly have multi-mineralic remanence-carrying populations (e.g. Weil & Van der Voo 2002; Zegers *et al.* 2003; Zwing *et al.* 2005; Oliva-Urcia & Pueyo 2007).

Zwing *et al.* (2005) noted a strong association of lithology and hysteresis properties of remagnetized rocks: reef carbonates, with low primary detrital content, exhibited hysteresis properties similar to those of the archetypal remagnetized carbonates; platform carbonate samples had intermediate hysteresis ratios; and remagnetized siliciclastics had hysteresis properties suggestive of primary detrital mineralogy. They reached the reasonable conclusion that a remagnetization test based on hysteresis behaviour only works when authigenic phases represent an overwhelming majority of the magnetic mineralogy, which in most cases means that detrital phases cannot be present in any great abundance. Indeed, our compilation in Figure 1 also appears to show a systematic dependence on depositional environment, with samples that plot along the 'primary' SD–MD mixing line coming largely from pelagic limestones (Channell & McCabe 1994; Tarduno & Myers 1994; Abrajevitch & Kodama 2009) while the Pleistocene Tahiti reef (Ménabréaz *et al.* 2010) had significant input of volcaniclastic sands. In each of these cases, it

appears magnetite from a detrital or bacterial source is holding a primary remanence.

In our view, the most diagnostic indicator of magnetite authigenesis in carbonate rocks is the occurrence of a unimodal particle-size distribution that peaks well below the room-temperature SP–SSD threshold. Such distributions are relatively rare in geological materials and, when they do occur, are almost always associated with *in situ* processes of origin (nucleation and growth in volcanic glass, metabolic electron-transfer reactions mediated extracellularly by bacteria, pedogenesis or various candidate processes in carbonate rocks). Particles that grow through the SP–SSD threshold acquire a stable CRM, and the population characteristics account for the distinctive hysteresis behaviour, high ARM/SIRM ratios and elevated frequency-dependence of susceptibility. Further work is needed, however, to evaluate whether a similar particle-size distribution could arise in the primary magnetic mineralogy of shallow-water carbonates that are isolated from detrital input.

## Summary and conclusions

In the prototypical remagnetized carbonates that have been the principal focus of this study, both the ChRM and the rock-magnetic properties are intimately related to the SP–SSD nanoparticle population that formed *in situ* during the Permian, many tens of millions of years after the strata were deposited. Clay diagenesis and maturation of organic matter, driven by moderately elevated burial temperatures as well as changes in geochemical conditions possibly due to introduction of tectonically driven brines, resulted in neoformation of these nanoparticles and alteration of pre-existing iron sulphides. Large proportions of the authigenic magnetite never grew to sizes larger than 20 nm; these SP particles are responsible for the frequency-dependent susceptibility and, to a large extent, for the wasp-waisted hysteresis loops and strongly elevated $H_{cr}/H_c$ ratios that constitute the proposed rock-magnetic fingerprint of remagnetization. Some fraction of the authigenic magnetite grew through the SP–SSD threshold, acquiring the ChRM and accounting for the elevated ARM/SIRM ratios that help to identify the carriers as authigenic. In the Helderberg Group samples of this study, some 75% of the magnetite is superparamagnetic at room temperature, and almost all of the rest is SSD. The exact mineralogical composition remains unclear, but it is certainly not pure magnetite; some small degree of cation substitution or a larger degree of cation deficiency is required for the almost complete lack of a Verwey transition.

One of the original explanations offered for a correlation between hysteresis properties and strong chemical remagnetization involved authigenic SD particles having an almost complete lack of shape anisotropy (i.e. spheroidal morphologies resembling those observed in large extracted magnetite particles), and therefore having magnetic properties governed primarily by the cubic magnetocrystalline (multiaxial) anisotropy of magnetite (Jackson 1990). We have now shown unequivocally that this is not correct for the newly analysed samples of the Helderberg Group. Two new approaches (Lanci 2010; Mitra *et al.* 2011) applied to these samples to test for the presence of multiaxial anisotropy have revealed that the magnetite particles are instead controlled by uniaxial anisotropy. The fundamental basis of the distinctive hysteresis behaviour of these remagnetized carbonates is the dominant contribution from the SP and SSD size fractions.

It should be borne in mind that although there is a strong link between rock-magnetic behaviour and size distributions in these rocks, the link between particle sizes and a late authigenic origin is more indirect. It is conceivable that dominant SP–SSD size fractions could also originate biogenically or by detrital input, and that rocks with a primary or very early ChRM may therefore sometimes exhibit the rock-magnetic characteristics associated with remagnetization. At this time, we are unaware of any such 'false positives' in the published literature where carbonates with primary magnetization yield hysteresis parameters in the range typically restricted to remagnetized carbonates. Nevertheless, there is a paucity of hysteresis data from shallow-water carbonates with primary ChRMs, and data are needed from such units to further evaluate whether the primary coexistence of SP and SSD grains ever leads to false positives for remagnetization. 'False negative' results (in which remagnetized units have the rock-magnetic characteristic of unremagnetized carbonates) appear to be more common, when detrital materials occur in sufficient abundance to shift the peak of the size distribution into the PSD–MD range.

We thank A. Hirt and an anonymous reviewer for their helpful suggestions and S. Swanson-Hysell for assistance with fieldwork. This is contribution 1106 of the Institute for Rock Magnetism, which is supported by grants from the Instruments and Facilities Program, Division of Earth Science, National Science Foundation.

# References

ABRAJEVITCH, A. & KODAMA, K. 2009. Biochemical v. detrital mechanism of remanence acquisition in marine carbonates: A lesson from the K-T boundary interval. *Earth and Planetary Science Letters*, **286**, 269–277.

AUBOURG, C. & POZZI, J. P. 2010. Toward a new <250 °C pyrrhotite-magnetite geothermometer for claystones <250 °C pyrrhotite-magnetite geothermometer for claystones. *Earth and Planetary Science Letters*, **294**, 47–57.

BACHTADSE, V., VAN DER VOO, R., HAYES, F. M. & KESLER, S. E. 1987. Late Paleozoic remagnetization of mineralized and unmineralized Ordovician carbonates from east Tennessee: evidence for a post-ore chemical event. *Journal of Geophysical Research B: Solid Earth*, **92**, 14 165–14 176.

BANERJEE, S. K., HUNT, C. P. & LIU, X.-M. 1993. Separation of local signals from the regional paleomonsoon record of the Chinese loess plateau: a rock-magnetic approach. *Geophysical Research Letters*, **20**, 843–846.

BANERJEE, S., ELMORE, R. D. & ENGEL, M. H. 1997. Chemical remagnetization and burial diagenesis: testing the hypothesis in the Pennsylvanian Belden Formation, Colorado. *Journal of Geophysical Research*, **102**, 24 825–24 842.

BELKAALOUL, N. K. & AISSAOUI, D. M. 1997. Nature and origin of magnetic minerals within the Middle Jurassic shallow-water carbonate rocks of the Paris Basin, France: implications for magnetostratigraphic dating. *Geophysical Journal International*, **130**, 411–421.

BESNUS, M. J. 1966. *Propriétés magnétiques de la pyrrhotite naturelle*. PhD thesis, University of Strasbourg.

BETHKE, C. M. & MARSHAK, S. 1990. Brine migrations across North America: the plate tectonics of groundwater. *Annual Reviews of Earth and Planetary Sciences*, **18**, 287–315.

BLUMSTEIN, A. M., ELMORE, R. D., ENGEL, M. H., ELLIOT, C. & BASU, A. 2004. Paleomagnetic dating of burial diagenesis in Mississippian carbonates, Utah. *Journal of Geophysical Research*, **109**, 1–16.

BUTLER, R. F., GEHRELS, G. E. & BAZARD, D. R. 1997. Paleomagnetism of Paleozoic strata of the Alexander terrane, southeastern Alaska. *GSA Bulletin*, **109**, 1372–1388.

CHANNELL, J. E. T. & MCCABE, C. 1994. Comparison of magnetic hysteresis parameters of unremagnetized and remagnetized limestones. *Journal of Geophysical Research B: Solid Earth*, **99**, 4613–4623.

CISOWSKI, S. 1981. Interacting v. non-interacting single-domain behavior in natural and synthetic samples. *Physics of the Earth and Planetary Interiors*, **26**, 77–83.

D'AGRELLA FILHO, M. S., BABINSKI, M., TRINDADE, R. I. F., VAN SCHMUS, W. R. & ERNESTO, M. 2000. Simultaneous remagnetization and U–Pb isotope resetting in Neoproterozoic carbonates of the Sao Francisco Craton, Brazil. *Precambrian Research*, **99**, 179–196.

DAY, R., FULLER, M. & SCHMIDT, V. A. 1977. Hysteresis properties of titanomagnetites: grain-size and compositional dependence. *Physics of the Earth and Planetary Interiors*, **13**, 260–266.

DEARING, J. A., BIRD, P. M., DANN, R. J. L. & BENJAMIN, S. F. 1997. Secondary ferrimagnetic minerals in Welsh soils: a comparison of mineral magnetic detection methods and implications for mineral formation. *Geophysical Journal International*, **130**, 727–736.

DEKKERS, M. J. 1989. Magnetic properties of natural pyrrhotite. II. High- and low-temperature behavior of Jrs and TRM as a function of grain size. *Physics of the Earth and Planetary Interiors*, **57**, 266–283.

DEKKERS, M. J., MATTÉI, J.-L., FILLION, G. & ROCHETTE, P. 1989. Grain-size dependence of the magnetic behavior of pyrrhotite during its low-temperature transition at 34 K. *Geophysical Research Letters*, **16**, 855–858.

DINARES-TURELL, J. & GARCIA-SENZ, J. 2000. Remagnetization of Lower Cretaceous limestones from the southern Pyrenees and relation to the Iberian plate geodynamic evolution. *Journal of Geophysical Research*, **105**, 19 405–19 418.

DUNLOP, D. J. 2002a. Theory and application of the Day plot (Mrs/Ms v. Hcr/Hc). 1. Theoretical curves and tests using titanomagnetite data. *Journal of Geophysical Research*, **107**, doi:10.1029/2001JB000487.

DUNLOP, D. J. 2002b. Theory and application of the Day plot (Mrs/Ms v. Hcr/Hc). 2. Application to data for rocks, sediments, and soils. *Journal of Geophysical Research*, **107**, 1. doi:10.1029/2001JB000486.

DUNLOP, D. J. & ÖZDEMIR, Ö. 1997. *Rock Magnetism. Fundamentals and Frontiers*. Cambridge University Press, Cambridge.

DUNLOP, D. J., ÖZDEMIR, Ö. & SCHMIDT, P. W. 1997a. Paleomagnetism and paleothermometry of the Sydney Basin. 2. Origin of anomalously high unblocking temperatures. *Journal of Geophysical Research*, **102**, 27285–27295.

DUNLOP, D. J., SCHMIDT, P. W., OZDEMIR, O. & CLARK, D. A. 1997b. Paleomagnetism and paleothermometry of the Sydney Basin. 1. Thermoviscous and chemical overprinting of the Milton Monzonite. *Journal of Geophysical Research*, **102**, 27271–27283.

EGLI, R. 2006. Theoretical considerations on the anhysteretic remanent magnetization of interacting particles with uniaxial anisotropy. *Journal of Geophysical Research B. Solid Earth*, **111**, doi:10.1029/2006 JB004577.

EGLI, R. & LOWRIE, W. 2002. Anhysteretic remanent magnetization of fine magnetic particles. *Journal of Geophysical Research-Solid Earth*, **107**, 2209, doi:10.1029/2001JB000671.

ELLIOTT, W. C., BASU, A., WAMPLER, J. M., ELMORE, R. D. & GRATHOFF, G. H. 2006a. Comparison of K-Ar ages of diagenetic illite-smectite to the age of a chemical remanent magnetization (CRM): an example from the Isle of Skye, Scotland. *Clays and Clay Minerals*, **54**, 314–323.

ELLIOTT, W. C., OSBORN, S. G., O'BRIEN, V. J., ELMORE, R. D., ENGEL, M. H. & WAMPLER, J. M. 2006b. On the timing and causes of illite formation and remagnetization in the Cretaceous Marias River Shale, Disturbed Belt, Montana. *Journal of Geochemical Exploration*, **89**, 92–95.

ELMORE, R. D. 2001. A review of palaeomagnetic data on the timing and origin of multiple fluid-flow events in the Arbuckle Mountains, southern Oklahoma. *Petroleum Geoscience*, **7**, 223–229.

ELMORE, R. D. & MCCABE, C. 1991. The occurrence and origin of remagnetization in the sedimentary rocks of North America. *Reviews of Geophysics*, **29** (Suppl.), 377–383.

ELMORE, R. D., LONDON, D., BAGLEY, D., FRUIT, D. & GAO, G. 1993. Remagnetization by basinal fluids: testing the hypothesis in the Viola Limestone, southern Oklahoma. *Journal of Geophysical Research B, Solid Earth*, **98**, 6237–6254.

ELMORE, R. D., FOUCHER, J. L. E., EVANS, M., LEWCHUK, M. & COX, E. 2006. Remagnetization of the Tonoloway Formation and the Helderberg Group in the Central Appalachians: testing the origin of syntilting magnetizations. *Geophysical Journal International*, **166**, 1062–1076.

ENKIN, R. J., OSADETZ, K. G., BAKER, J. & KISILEVSKY, D. 2000. Orogenic remagnetizations in the Front Ranges and Inner Foothills of the southern Canadian Cordillera: Chemical harbinger and thermal handmaiden of Cordilleran deformation. *Geological Society of America Bulletin*, **112**, 929–942.

EVERITT, C. W. F. & CLEGG, J. A. 1962. A field test of palaeomagnetic stability. *Geophysical Journal of the Royal Astronomical Society*, **6**, 312–319.

FABIAN, K. 2003. Some additional parameters to estimate domain state from isothermal magnetization measurements. *Earth and Planetary Science Letters*, **213**, 337–345.

FABIAN, K. 2006. Approach to saturation analysis of hysteresis measurements in rock magnetism and evidence for stress dominated magnetic anisotropy in young mid-ocean ridge basalt. *Physics of the Earth and Planetary Interiors*, **154**, 299–307.

FISHER, D. W. 1987. Lower Devonian limestone, Helderberg Escarpment. *In*: ROY, D. C. (ed.) *Geological Society of America Centennial Field Guide – Northeastern Section*. Geological Society of America, Boulder, 119–122.

FISHER, D. W., ISACHSEN, Y. W. & RICKARD, L. V. 1970. Geologic Map of New York State (1:250000) New York State Museum Map and Chart Series No. 15.

FRANKEL, R. B. 1987. Anaerobes pumping iron. *Nature*, **330**, 208.

FREEMAN, R. 1986. Magnetic mineralogy of pelagic limestones. *Geophysical Journal of the Royal Astronomical Society*, **85**, 433–452.

FRUIT, D., ELMORE, R. D. & HALGEDAHL, S. 1995. Remagnetization of the folded Belden formation, northwest Colorado. *Journal of Geophysical Research B: Solid Earth*, **100**, 15009–15024.

GARVEN, G. 1995. Continental-scale groundwater flow and geological processes. *Annual Review of Earth and Planetary Sciences*, **24**, 89–117.

GEE, J. & KENT, D. V. 1995. Magnetic hysteresis in young mid-ocean ridge basalts: dominant cubic anisotropy? *Geophysical Research Letters*, **22**, 551–554.

GEE, J. & KENT, D. V. 1999. Calibration of magnetic granulometric trends in oceanic basalts. *Earth and Planetary Science Letters*, **170**, 377–390.

GONG, Z., DEKKERS, M. J., HESLOP, D. & MULLENDER, T. A. T. 2009. End-member modelling of isothermal remanent magnetization (IRM) acquisition curves: a novel approach to diagnose remagnetization. *Geophysical Journal International*, **178**, 693–701.

GOREE, W. S. & FULLER, M. D. 1976. Magnetometers using r-f driven SQUIDs and their application in rock magnetism and paleomagnetism. *Reviews of Geophysics and Space Physics*, **14**, 591–608.

GRAHAM, J. W. 1949. The stability and significance of magnetism in sedimentary rocks. *Journal of Geophysical Research*, **54**, 131–167.

GUYODO, Y., LaPARA, T. M., ANSCHUTZ, A. J., PENN, R. L., BANERJEE, S. K., GEISS, C. E. & ZANNER, W. 2006. Rock magnetic, chemical and bacterial community analysis of a modern soil from Nebraska. *Earth and Planetary Science Letters*, **251**, 168–178.

HIRT, A. M., BANIN, A. & GEHRING, A. U. 1993. Thermal generation of ferromagnetic minerals from iron-enriched smectites. *Geophysical Journal International*, **115**, 1161–1168.

JACKSON, M. J. 1990. Diagenetic sources of stable remanence in remagnetized Paleozoic cratonic carbonates: a rock magnetic study. *Journal of Geophysical Research B: Solid Earth*, **95**, 2753–2761.

JACKSON, M. J. & WORM, H.-U. 2001. Anomalous unblocking temperatures, viscosity and frequency-dependent susceptibility in the chemically-remagnetized Trenton Limestone. *Physics of the Earth and Planetary Interiors*, **126**, 27–42.

JACKSON, M. J. & SØLHEID, P. 2010. On the quantitative analysis and evaluation of magnetic hysteresis data, Geochemistry, Geophysics. *Geosystems*, **11**, doi:10.1029/2009GC002932.

JACKSON, M. J., McCABE, C., BALLARD, M. M. & VAN DER VOO, R. 1988. Magnetite authigenesis and diagenetic paleotemperatures across the northern Appalachian Basin. *Geology*, **16**, 592–595.

JACKSON, M. J., WORM, H.-U. & BANERJEE, S. K. 1990. Fourier analysis of digital hysteresis data: rock magnetic applications. *Physics of the Earth and Planetary Interiors*, **65**, 78–87.

JACKSON, M. J., SUN, W.-W. & CRADDOCK, J. P. 1992. The rock magnetic fingerprint of chemical remagnetization in midcontinental Paleozoic carbonates. *Geophysical Research Letters*, **19**, 781–784.

JACKSON, M. J., ROCHETTE, P., FILLION, G., BANERJEE, S. K. & MARVIN, J. A. 1993. Rock magnetism of remagnetized Paleozoic carbonates: low-temperature behavior and susceptibility characteristics. *Journal of Geophysical Research B: Solid Earth*, **98**, 6217–6225.

KARLIN, R. & LEVI, S. 1983. Diagenesis of magnetic minerals in recent haemipelagic sediments. *Nature*, **303**, 327–330.

KATZ, B., ELMORE, R. D. & ENGEL, M. H. 1998. Authigenesis of magnetite in organic-rich sediment next to a dike: implications for thermoviscous and chemical remagnetizations. *Earth and Planetary Science Letters*, **163**, 221–234.

KATZ, B., ELMORE, R., COGOINI, M., ENGEL, M. & FERRY, S. 2000. Associations between burial diagenesis of smectite, chemical remagnetization, and magnetite authigenesis in the Vocontian trough, SE France. *Journal of Geophysical Research*, **105**, 1.

KENT, D. V. 1979. Paleomagnetism of the Devonian Onondaga limestone revisited. *Journal of Geophysical Research B: Solid Earth*, **84**, 3576–3588.

KENT, D. V. 1985. Thermoviscous remagnetization in some Appalachian limestones. *Geophysical Research Letters*, **12**, 805–808.

KIRSCHVINK, J. L. 1978. The Precambrian-Cambrian boundary problem: paleomagnetic directions from the Amadeus Basin, Central Australia. *Earth and Planetary Science Letters*, **40**, 91–100.

KLIGFIELD, R. & CHANNELL, J. E. T. 1981. Widespread remagnetization of Helvetic limestones. *Journal of Geophysical Research*, **86**, 1888–1900.

KOPP, R. E. & KIRSCHVINK, J. L. 2008. The identification and biogeochemical interpretation of fossil magnetotactic bacteria. *Earth-Science Reviews*, **86**, 42–61.

LANCI, L. 2010. Detection of multi-axial magnetite by remanence effect on anisotropy of magnetic susceptibility. *Geophysical Journal International*, **181**, 1362–1366.

LANCI, L. & KENT, D. V. 2003. Introduction of thermal activation in forward modeling of hysteresis loops for single-domain magnetic particles and implications for the interpretation of the Day diagram. *Journal of Geophysical Research*, **108**, doi:10.1029/2001JB000944.

LANCI, L. & KENT, D. V. 2006. Meteoric smoke fallout revealed by superparamagnetism in Greenland ice. *Geophysical Research Letters*, **33**, doi:10.1029/2006GL026480.

LANCI, L., KENT, D. V. & BISCAYE, P. E. 2007. Meteoric smoke concentration in the Vostok ice core estimated from superparamagnetic relaxation and some consequences for estimates of Earth accretion rate. *Geophysical Research Letters*, **34**, doi: 10.1029/2007gl029811.

LANCI, L., DELMONTE, B., MAGGI, V., PETIT, J. R. & KENT, D. V. 2008. Ice magnetization in the EPICA-Dome C ice core: Implication for dust sources during glacial and interglacial periods. *Journal of Geophysical Research*, **113**, D14207, doi:14210.11029/12007JD009678.

LI, Y.-L., PFIFFNER, S. M. *ET AL.* 2009. Degeneration of biogenic superparamagnetic magnetite. *Geobiology*, **7**, 25–34.

LOWRIE, W. 1990. Identification of ferromagnetic minerals in a rock by coercivity and unblocking temperature properties. *Geophysical Research Letters*, **17**, 159–162.

LU, G., McCABE, C., HANOR, J. S. & FERRELL, R. E. 1991. A genetic link between remagnetization and potassic metasomatism in the Devonian Onondaga formation, northern Appalachian basin. *Geophysical Research Letters*, **18**, 2047–2050.

MAHER, B. A. & TAYLOR, R. M. 1988. Formation of ultrafine-grained magnetite in soils. *Nature*, **336**, 368–371.

MALOOF, A. C., KOPP, R. E. *ET AL.* 2007. Sedimentary iron cycling and the origin and preservation of magnetization in platform carbonate muds, Andros Island, Bahamas. *Earth and Planetary Science Letters*, **259**, 581–598.

McCABE, C. & ELMORE, R. D. 1989. The occurrence and origin of Late Paleozoic remagnetization in the sedimentary rocks of North America. *Reviews of Geophysics*, **27**, 471–494.

McCABE, C. & CHANNELL, J. E. T. 1994. Late Paleozoic remagnetization in limestones of the Craven basin (northern England) and the rock magnetic fingerprint of remagnetized sedimentary carbonates. *Journal of Geophysical Research B: Solid Earth*, **99**, 4603–4612.

McCabe, C., Van der Voo, R., Peacor, D. R., Scotese, C. R. & Freeman, R. 1983. Diagenetic magnetite carries ancient yet secondary remanence in some Paleozoic sedimentary carbonates. *Geology*, **11**, 221–223.

McCabe, C., Van der Voo, R. & Ballard, M. M. 1984. Late Paleozoic remagnetization of the Trenton limestone. *Geophysical Research Letters*, **11**, 979–982.

McElhinny, M. W. & Opdyke, N. D. 1973. Remagnetization hypothesis discounted: a paleomagnetic study of the Trenton limestone, New York State. *Geological Society of America Bulletin*, **84**, 3697–3708.

Ménabréaz, L., Thouveny, N., Camoin, G. & Lund, S. P. 2010. Paleomagnetic record of the late Pleistocene reef sequence of Tahiti (French Polynesia): a contribution to the chronology of the deposits. *Earth and Planetary Science Letters*, **294**, 58–68.

Mitra, R., Tauxe, L. & Gee, J. 2011. Detecting uniaxial single domain grains with a modified IRM technique. *Geophysical Journal International*, **187**, 1250–1258, doi: 10.1111/j.1365-246X.2011.05224.x.

Molina Garza, R. S. & Zijderveld, J. D. A. 1996. Paleomagnetism of Paleozoic strata, Brabant and Ardennes Massifs, Belgium: implications of prefolding and postfolding Late Carboniferous secondary magnetizations for European apparent polar wander. *Journal of Geophysical Research, B, Solid Earth and Planets*, **101**, 15799–15818.

Moskowitz, B. M., Frankel, R. B., Bazylinski, D. A., Jannasch, H. W. & Lovley, D. R. 1989. A comparison of magnetite particles produced anaerobically by magnetotactic and dissimilatory iron-reducing bacteria. *Geophysical Research Letters*, **16**, 665–668.

Moskowitz, B. M., Frankel, R. & Bazylinski, D. 1993. Rock magnetic criteria for the detection of biogenic magnetite. *Earth and Planetary Science Letters*, **120**, 283–300.

Moskowitz, B. M., Jackson, M. & Kissel, C. 1998. Low-temperature magnetic behavior of titanomagnetites. *Earth and Planetary Science Letters*, **157**, 141–149.

Moskowitz, B. M., Bazylinski, D. A., Egli, R., Frankel, R. B. & Edwards, K. J. 2008. Magnetic properties of marine magnetotactic bacteria in a seasonally stratified coastal pond (Salt Pond, MA, USA). *Geophysical Journal International*, **174**, 75–92.

Mullins, C. E. 1977. Magnetic susceptibility of the soil and its significance in soil science – a review. *Journal of Soil Science*, **28**, 294–306.

Muttoni, G. 1995. 'Wasp-waisted' hysteresis loops from a pyrrhotite and magnetite-bearing remagnetized Triassic limestone. *Geophysical Research Letters*, **22**, 3167–3170.

Muttoni, G. & Kent, D. V. 1994. Paleomagnetism of Latest Anisian (Middle Triassic) Sections of the Prezzo Limestone and the Buchenstein Formation, Southern Alps, Italy. *Earth and Planetary Science Letters*, **122**, 1–18.

Nagata, T. & Carleton, B. J. 1987. Magnetic remanence coercivity of rocks. *Journal of Geomagnetism and Geoelectricity*, **39**, 447–461.

Néel, L. 1949. Théorie du traînage magnétique des ferromagnétiques en grains fins avec applications aux terres cuites. *Annales de Géophysique*, **5**, 99–136.

Oldfield, F., Hunt, A., Jones, M. D. H., Chester, R., Dearing, J. A., Olsson, L. & Prospero, J. M. 1985. Magnetic differentiation of atmospheric dusts. *Nature*, **317**, 516–518.

Oliva-Urcia, B. & Pueyo, E. L. 2007. Rotational basement kinematics deduced from remagnetized cover rocks (Internal Sierras, southwestern Pyrenees). *Tectonics*, **26**, 1.

Oliva-Urcia, B., Pueyo, E. L. & Larrasoana, J. C. 2008. Magnetic reorientation induced by pressure solution: a potential mechanism for orogenic-scale remagnetizations. *Earth and Planetary Science Letters*, **265**, 525–534.

Oliver, J. 1986. Fluids expelled tectonically from orogenic belts: their role in hydrocarbon migration and other geological phenomena. *Geology*, **14**, 99–102.

Özdemir, Ö. & Banerjee, S. K. 1984. High temperature stability of maghemite ($\gamma$-Fe$_2$O$_3$). *Geophysical Research Letters*, **11**, 161–164.

Özdemir, Ö. & Dunlop, D. J. 2006. Magnetic memory and coupling between spin-canted and defect magnetism in hematite. *Journal of Geophysical Research-Solid Earth*, **111**, 1.

Özdemir, Ö. & Dunlop, D. J. 2010. Hallmarks of maghemitization in low-temperature remanence cycling of partially oxidized magnetite nanoparticles. *Journal of Geophysical Research B: Solid Earth*, **115**, 1, doi:10.1029/2009JB006756.

Özdemir, Ö., Dunlop, D. J. & Moskowitz, B. M. 1993. The effect of oxidation of the Verwey transition in magnetite. *Geophysical Research Letters*, **20**, 1671–1674.

Potter, D. K. & Stephenson, A. 1988. Single-domain particles in rocks and magnetic fabric analysis. *Geophysical Research Letters*, **15**, 1097–1100.

Reynolds, R. L. 1990. Paleomagnetism – a polished view of remagnetization. *Nature*, **345**, 579–580.

Ripperdan, R. L., Riciputi, L. R., Cole, D. R., Elmore, R. D., Banerjee, S. & Engel, M. H. 1998. Oxygen isotope ratios in authigenic magnetites from the Belden Formation, Colorado. *Journal of Geophysical Research*, **103**, 1.

Roberts, A. P., Cui, Y.-L. & Verosub, K. L. 1995. Wasp-waisted hysteresis loops: mineral magnetic characteristics and discrimination of components in mixed magnetic systems. *Journal of Geophysical Research B: Solid Earth*, **100**, 17909–17924.

Roberts, A. P., Chang, L., Rowan, C. J., Horng, C.-S. & Florindo, F. 2011. Magnetic properties of sedimentary greigite (Fe3S4). *An update: Reviews of Geophysics*, **49**, RG1002.

Rochette, P., Fillion, G., Mattéi, J.-L. & Dekkers, M. J. 1990. Magnetic transition at 30–34 Kelvin in pyrrhotite: insight into a widespread occurrence of this mineral in rocks. *Earth and Planetary Science Letters*, **98**, 319–328.

Schlinger, C. M., Veblen, D. R. & Rosenbaum, J. G. 1991. Magnetism and magnetic mineralogy of ash flow tuffs from Yucca Mountain, Nevada. *Journal of Geophysical Research B: Solid Earth*, **96**, 6035–6052.

Scotese, C. R., Van der Voo, R. & McCabe, C. 1982. Paleomagnetism of the Upper Silurian and Lower Devonian carbonates of New York State: evidence for secondary magnetizations residing in magnetite. *Physics of the Earth and Planetary Interiors*, **30**, 385–395.

SMIRNOV, A. V. & TARDUNO, J. A. 2001. Estimating superparamagnetism in marine sediments with the time dependency of coercivity of remanence. *Journal of Geophysical Research*, **106**, 16135–16143.

SOTO, R., VILLALAIN, J. J. & CASAS-SAINZ, A. M. 2008. Remagnetizations as a tool to analyze the tectonic history of inverted sedimentary basins: a case study from the Basque-Cantabrian basin (north Spain). *Tectonics*, **27**, doi: 10.1029/2007TC002208.

STONER, E. C. & WOHLFARTH, E. P. 1948. A mechanism of magnetic hysteresis in heterogeneous alloys. *Philosophical Transactions of the Royal Society of London, Series A*, **240**, 599–602.

SUK, D.-W. & HALGEDAHL, S. L. 1996. Hysteresis properties of magnetic spherules and whole-rock specimens from some Paleozoic platform carbonate rocks. *Journal of Geophysical Research B: Solid Earth*, **101**, 25053–25076.

SUK, D.-W., PEACOR, D. R. & VAN DER VOO, R. 1990a. Replacement of pyrite framboids by magnetite in limestone and implications for paleomagnetism. *Nature*, **345**, 611–613.

SUK, D.-W., VAN DER VOO, R. & PEACOR, D. R. 1990b. Scanning and transmission electron microscope observations of magnetite and other iron phases in Ordovician carbonates from east Tennessee. *Journal of Geophysical Research B. Solid Earth*, **95**, 12327–12336.

SUK, D., VOO, R. V. D. & PEACOR, D. R. 1991. SEM/STEM observations of magnetite in carbonates of eastern North America: evidence for chemical remagnetization during the Alleghenian Orogeny. *Geophysical Research Letters*, **18**, 939–942.

SUK, D.-W., VAN DER VOO, R. & PEACOR, D. R. 1993. Origin of magnetite responsible for remagnetization of early Paleozoic limestones of New York State. *Journal of Geophysical Research B: Solid Earth*, **98**, 419–434.

SUN, W.-W. & JACKSON, M. J. 1994. Scanning electron microscopy and rock magnetic studies of magnetic carriers in remagnetized Early Paleozoic carbonates from Missouri. *Journal of Geophysical Research B: Solid Earth*, **99**, 2935–2942.

SYMONS, D. T. A., KAWASAKI, K. & PANNALAL, S. J. 2010. Paleomagnetic mapping of the regional fluid flow event that mineralized the Upper Mississippi Valley Zn-Pb ore district, Wisconsin, USA. *Journal of Geochemical Exploration*, **106**, 188–196.

TARDUNO, J. A. 1995. Superparamagnetism and reduction diagenesis in pelagic sediments: enhancement or depletion? *Geophysical Research Letters*, **22**, 1337–1340.

TARDUNO, J. A. & MYERS, M. 1994. A primary magnetization fingerprint from the Cretaceous Laytonville Limestone: further evidence for rapid oceanic plate velocities. *Journal of Geophysical Research B, Solid Earth*, **99**, 21691–21703.

TAUXE, L., MULLENDER, T. A. T. & PICK, T. 1996. otbellies, wasp-waists, and superparamagnetism in magnetic hysteresis. *Journal of Geophysical Research B, Solid Earth*, **101**, 571–583.

TILL, J. L., JACKSON, M. J., ROSENBAUM, J. G. & SOLHEID, P. 2010. Magnetic properties in an ashflow tuff with continuous grain size variation: A natural reference for magnetic particle granulometry. *Geochemistry, Geophysics, Geosystems*, **12** (Q07Z26), doi: 10.1029/2011GC003648.

TOHVER, E., WEIL, A. B., SOLUM, J. G. & HALL, C. M. 2008. Direct dating of carbonate remagnetization by 40Ar/39Ar analysis of the smectite–illite transformation. *Earth and Planetary Science Letters*, **274**, 524–530.

TRINDADE, R. I. F., D'AGRELLA, M. S., BABINSKI, M., FONT, E. & NEVES, B. B. B. 2004. Paleomagnetism and geochronology of the Bebedouro cap carbonate: evidence for continental-scale Cambrian remagnetization in the Sao Francisco craton, Brazil. *Precambrian Research*, **128**, 83–103.

VAN DER VOO, R. 1990. The reliability of paleomagnetic data. *Tectonophysics*, **184**, 1–9.

VERWEY, E. J. & HAAYMAN, P. W. 1941. Electronic conductivity and transition point in magnetite. *Physica*, **8**, 979–982.

WALTON, D. 1980. Time-temperature relations in the magnetization of assmblies of single-domain grains. *Nature*, **286**, 245–247.

WALZ, F. 2002. The Verwey transition – a topical review. *Journal of Physics: Condensed Matter*, **14**, R285–340.

WEIL, A. B. & VAN DER VOO, R. 2002. Insights into the mechanism for orogen-related carbonate remagnetization from growth of authigenic Fe-oxide: A scanning electron microscopy and rock magnetic study of Devonian carbonates from northern Spain. *Journal of Geophysical Research*, **107**, 1, doi:10.1029/2001JB000200.

WOHLFARTH, E. P. & TONGE, D. G. 1957. The remanent magnetization of single-domain ferromagnetic particles. *Philosophical Magazine*, **2**, 1333–1344.

WOLFERS, P., FILLION, G., OULADDIAF, B., BALLOU, R. & ROCHETTE, P. 2011. The Pyrrhotite 32 K magnetic transition. *Solid State Phenomena*, **170**, 174–179.

WOODS, S. D., ELMORE, R. D. & ENGEL, M. H. 2002. Paleomagnetic dating of the smectite-to-illite conversion: testing the hypothesis in Jurassic sedimentary rocks, Skye, Scotland. *Journal of Geophysical Research*, **107**, doi: 10.1029/2000JB000053.

WORM, H.-U. 1998. On the superparamagnetic-stable single domain transition for magnetite, and frequency dependence of susceptibility. *Geophysical Journal International*, **133**, 201–206.

WORM, H.-U. & JACKSON, M. 1999. The superparamagnetism of Yucca Mountain Tuff. *Journal of Geophysical Research B: Solid Earth*, **104**, 25415–25425.

XU, W., VAN DER VOO, R. & PEACOR, D. R. 1998. Electron microscopic and rock magnetic study of remagnetized Leadville carbonates, central Colorado. *Tectonophysics*, **296**, 333–362.

ZEGERS, T. E., DEKKERS, M. J. & BAILLY, S. 2003. Late Carboniferous to Permian remagnetization of Devonian limestones in the Ardennes: role of temperature, fluids, and deformation. *Journal of Geophysical Research*, **108**, doi: 10.1029/2002JB002213.

ZWING, A., MATZKA, J., BACHTADSE, V. & SOFFEL, H. C. 2005. Rock magnetic properties of remagnetized Palaeozoic clastic and carbonate rocks from the NE Rhenish massif, Germany. *Geophysical Journal International*, **160**, 477–486.

# End-member modelling as an aid to diagnose remagnetization: a brief review

MARK J. DEKKERS

*Department of Earth Sciences, Paleomagnetic Laboratory 'Fort Hoofddijk',
Faculty of Geoscience, Utrecht University, Budapestlaan 17, 3584 CD Utrecht,
The Netherlands (e-mail: m.j.dekkers@uu.nl)*

**Abstract:** Remagnetization of a palaeomagnetic signal is difficult to recognize independently of directional information. The situation becomes more complex when remagnetized rocks pass palaeomagnetic field tests, for example when the remagnetization of a rock sequence has occurred before folding. It is evident that palaeogeographic reconstructions are seriously flawed when actually remagnetized rocks are not identified as such. Here we discuss the merits and pitfalls of so-called end-member modelling of acquisition curves of the isothermal remanent magnetization (IRM) to recognize remagnetized strata. The technique requires no *a priori* information about the IRM acquisition curves. The algorithm unmixes a set of IRM acquisition curves into a number of invariant curves termed end members and calculates the mixing proportions of the end members for each sample. Since primary natural remanent magnetization (NRM) and remagnetized NRM are acquired by different processes, their signatures can be recognized from subtle differences in the magnetic properties. We illustrate the potential of the approach by three case studies, one from Spain and two from Turkey, in which the magnetic properties of remagnetized and non-remagnetized rocks are evaluated.

The natural remanent magnetization (NRM) of a rock is described as remagnetized when it represents a palaeomagnetically younger age than the sediment age of the rock formation under study. Remagnetization interferes with palaeogeographic reconstructions, as is increasingly being appreciated. The 'Late Palaeozoic remagnetization event' was firmly recognized in the Appalachians in the USA in the early 1980s (e.g. McCabe *et al.* 1983; see the review of Van der Voo & Torsvik 2012) while the implications of late Palaeozoic remagnetization in European counterparts, for example the Rhenohercynian belt and the Cantabrian mountain chain in Spain, were fully realized about a decade later (e.g. Molina Garza & Zijderveld 1996; Van der Voo *et al.* 1997; Weil *et al.* 2000; Weil & Van der Voo 2002; Zwing *et al.* 2002, 2009; Zegers *et al.* 2003). Yet remagnetization does not appear to be restricted to certain time periods in certain regions. Remagnetized rocks are being documented in rocks of all geological ages – including very young rocks – in orogenic belts and forelands across the entire globe. It can occur any time during the geological history, during either pre-folding (e.g. Perroud & Van der Voo 1984), syn-folding (e.g. Kent & Opdyke 1985) or post-folding (e.g. Stearns & Van der Voo 1987). It should therefore be tested whether or not a certain rock unit is remagnetized despite passing palaeomagnetic field tests. Remagnetized rocks obviously cannot be used for classic palaeogeographic reconstruction, a considerable disadvantage that has led

to serious disappointment. However, the remagnetization process can offer geological information that is difficult to acquire otherwise. This includes spatial and temporal delineation of (approximate) palaeotemperatures that can be tied to maturation histories of potential hydrocarbon source rocks (Font *et al.* 2006) and the definition of 'remagnetized aureoles' that may indicate telethermal ore bodies, for example, Pb–Zn Mississippi Valley Type ores. Further, temporally slightly different remagnetization events could be demonstrated for several Iberian Mesozoic sedimentary basins, linked to the Aptian rotation of Iberia during a series of basin remagnetizations (Gong *et al.* 2009*a*).

While thermoviscous resetting (Kent 1985) is a plausible scenario to explain remagnetization processes in a number of occasions, in many other situations the rocks have never been heated for a sufficient amount of time at a sufficiently elevated temperature to make this mechanism viable. Grain growth at low temperature, (far) below the magnetic ordering temperature of the magnetic particles involved, assisted by fluids which are either internally buffered or externally derived, is the currently prevailing mechanism to explain remagnetization (e.g. McCabe & Elmore 1989; Elmore *et al.* 2006). The role of tectonic stress which is sometimes advocated (e.g. Evans *et al.* 2003; Evans & Elmore 2006; Oliva-Urcia *et al.* 2008; Zwing *et al.* 2009) is rather elusive. Substantial effort has gone into the visualization of magnetic particles

*From*: ELMORE, R. D., MUXWORTHY, A. R., ALDANA, M. M. & MENA, M. (eds) 2012. *Remagnetization and Chemical Alteration of Sedimentary Rocks*. Geological Society, London, Special Publications, **371**, 253–269.
First published online August 22, 2012, http://dx.doi.org/10.1144/SP371.12

by scanning and transmission electron microscopy that are argued to carry the remagnetization (e.g. Suk *et al.* 1993; Sun & Jackson 1994). Rocks that are considered to be remagnetized by a chemical remanent magnetization (CRM) contain abundant very small particles, straddling the nominally super-paramagnetic (SP) to the single-domain (SD) size range. The focus of this contribution is on remagnetizations residing in magnetite, but remagnetizations that resides in greigite are also possible (e.g. Rowan & Roberts 2006, 2008).

Ideally, we would have criteria at hand to recognize remagnetization without the need to use palaeomagnetic directional information. In this case, a researcher could identify remagnetized rocks on the basis of their magnetic properties. The identification of the 'remagnetized' and 'non-remagnetized' trends based on hysteresis parameters plotted on a Day plot (Day *et al.* 1977) is an example of this approach (Channell & McCabe 1994). Modelling of SP particles on the Day plot (Dunlop 2002*a, b*; Lanci & Kent 2003) indeed shows positions on the Day plot perfectly compatible with the remagnetized and non-remagnetized trends of Channell & McCabe (1994). However, in a large number of studies (e.g. Zegers *et al.* 2003), actually measured hysteresis parameters appear to plot in between the two trend lines complicating the interpretation in terms of remagnetization. This could be due to mixed magnetic mineralogy not accounted for in Day plots or to actual contributions from a detrital component and a remagnetized component, leading to positions on the Day plot between the two trend lines. Further, for weakly magnetic rocks, it is not always straightforward to determine meaningful hysteresis parameters because of low signals despite the very sensitive instrumentation available. The correction of the high-field slope of the hysteresis loop becomes critical in cases of a small ferromagnetic (*sensu lato*) contribution and a large paramagnetic contribution.

Acquisition curves of isothermal remanent magnetization (IRM) can always be measured faithfully so that limit-of-detection problems inherent to hysteresis loop measurement are avoided. In remagnetization studies, we are primarily interested in the remanence carriers and so a remanence-based criterion to recognize remagnetization, for example an IRM-based criterion, would be preferable. In addition, chemical remagnetization adds magnetic particles to an existing particle suite in a rock often without actually replacing it. This follows from the notion that it is very difficult to envisage that fine-grained diagenetic magnetite would precipitate while coarser-grained detrital magnetite would dissolve at the same time. However, in cases where the sequence of rocks has undergone an 'intermediate' diagenetic greigite formation phase (e.g. Rowan *et al.*

2009), the originally present detrital magnetite could have been (largely) reductively dissolved before the later precipitation of diagenetic magnetite. In the cases discussed here greigite has not been determined, but it cannot be excluded at this stage that the diagenetic magnetite could be a product of greigite reactions. The resulting magnetic properties of a remagnetized rock sequence can be considered a composite of two processes: (1) detrital deposition/ early biogenic formation and (2) later diagenetic formation, with the former potentially partially leached from the rock. These are often difficult to untangle by hysteresis parameter analysis.

It may be clear that in chemically remagnetized rocks the original 'detrital' NRM and the remagnetized NRM are acquired by very different processes that have led to different magnetic particle distributions. Such differences can be measured by means of detailed IRM acquisition curves. Since pervasive remagnetization is anticipated to have acted to a variable extent on a collection of samples, it is mathematically possible to unmix or decompose a collection of IRM acquisition curves into two or more invariant IRM acquisition curves termed end members (e.g. Weltje 1997; Heslop & Dillon 2007). All measured IRM acquisition curves are then expressed as the result of mixing of those end members. Here we briefly summarize the principles of the end-member unmixing algorithm of Weltje (1997) and demonstrate the potential of the end-member modelling approach to diagnose remagnetized rocks. It should be realized that a purely thermoviscous resetting of the NRM cannot be assessed with this technique because the grain-size distribution is not changed. However, in hydrous pro-grade high-diagenetic temperature regimes or very low-grade anchimetamorphic conditions in sediments it is hard to conceive that there would be no changes in the grain-size distribution of the magnetic particles as a result of diagenesis and metamorphism.

## End-member modelling

End-member modelling is an inverse mathematical technique, that is, the model is derived from (measured) data (in the present contribution a collection of IRM acquisition curves). The measured data are considered to be a linear mixture of a number of invariant constituent components termed end members. The mixing space or polytope is subject to a constant sum requirement (i.e. data must be expressed as proportions or percentages; compositional data fulfil this requirement, for the present case the input consists of normalized IRM acquisition curves) and mixing proportions should not be negative (Weltje 1997). The constant sum requirement ensures that the mixing plane (in the case of three end members, a two-dimensional equilateral plane triangle) contains the composition vector tips of all samples in the hyperplane or simplex. End-member modelling is an example of so-called bilinear unmixing. Bilinear implies here that both the end-member properties as well as their mixing proportions in each individual sample are determined from the same measured dataset. In our ternary diagram, all compositions throughout the diagram can be generated by combining the vertices of the triangle. However, any combination of three invariant points that span the data hull (the space that is occupied by the data) can be used to this end. The end members are therefore not unique in the mathematical sense and a geologically/palaeomagnetically reasonable set of end members should be sought that are subsequently used to calculate the mixing proportions for individual samples.

If the mixing was perfect this would yield (in matrix notation)

$$X = MB$$

where $X$ represents the measured data matrix, $M$ the mixing proportions for each individual sample and $B$ the invariant points, that is, the end members. Mixing proportions cannot be negative. In reality, mixing is always imperfect to some degree because of sampling and measurement errors. $X$ is thought to consist of $X = X' - E$ in the terminology of Weltje (1997) where $X'$ is the ideal mixing system and $E$ is the matrix of error terms. We therefore have:

$$X = MB + E$$

There exists a range of strategies to unmix which depend on the *a priori* information available. If the number of end members and their composition are known in advance, the mixing proportions $M$ can be calculated by solving a set of linear equations; this is referred to as linear unmixing. In the more general case of bilinear unmixing, both the end-member compositions $B$ and mixing proportions $M$ must be determined. In practice this implies determination of the number of end members $q$, their properties $B$, the mixing proportions $M$, the ideal mixing system $X'$ and the error matrix $E$ (Weltje 1997). As outlined by Weltje (1997), there exist various schemes to solve the bilinear unmixing problem and there is no unique solution since any end member would be appropriate (from a mathematical viewpoint) as long as it lies on the edge or outside the multivariate space spanned by the data points. Solving schemes that consider a convex hull approach are deemed to be the most realistic, but they are very sensitive to deviating

samples or outliers: a convex hull around the data must contain all data points. Weltje (1997) developed an elegant algorithm that alleviates the convex hull criterion: it is accepted that some data points remain outside the calculated data hull. In this way the end-member solution is (much) less sensitive to outlying samples. They are not considered in the solution, which makes the solution more robust to describe general patterns, groupings or trends in the data.

The unmixing problem is solved in a two-stage fashion which has also been adopted in several other schemes. First, the mixing space is partitioned into $X'$ and $E$ for each number of end members $q$ considered. For optimal $X'$ and $q$, $X'$ is then expressed as the product of $M$ and $B$. An exact solution does not exist and an 'optimal' solution is sought.

The idea is that we start from within the data hull by using the cluster centres obtained by probabilistic fuzzy $c$-means cluster analysis (Bezdek *et al.* 1984), guaranteed to lie within the data convex hull. Iteratively, the mixing polytope is enlarged by moving the end members in order to include an increasingly greater number of data points until either all data points are included or a stopping criterion for the algorithm is reached. End members as close as possible to the convex hull are calculated with this approach. Such end members have meaningful mixing proportions, easing the geological interpretation (it is 'easier' from a process viewpoint to explain silt by some fragmentation or winnowing of sand than to propose gross fragmentation of pebbles). Distant end members only yield very small mixing proportions to explain the observations, a situation that is undesirable from a geological interpretation point of view.

End-member modelling has been used rather extensively in sedimentology to unravel the contribution of various source areas of sediments (e.g. Weltje 1997; Prins *et al.* 2002). Other examples include colour spectra that have been unmixed into end members to provide an interpretation in terms of provenance area and diagenetic processes off the coast of Africa (Heslop *et al.* 2007). Magnetic properties, that is IRM acquisition curves, have also been unmixed (e.g. Heslop & Dillon 2007), again to shed light on provenance areas and diagenetic processes.

Typically, at least 30–50 input curves should be used to achieve a reasonable picture of the inherent data variability. The end-member modelling algorithm aims to 'fill' the mixing space (in the case of three end members, the equilateral plane triangle) as evenly as possible. As stated earlier, both the shape of the end members and their abundances are based on the variation in the measured dataset that serves as input of the end-member model effort. There are no criteria concerning the shape of the input curves, such as basis functions or type curves as in forward modelling techniques. The only constraints on the input data is that each input curve – here IRM acquisition curves – must be monotonic (i.e. each IRM value is greater than or equal to the foregoing measurement at a lower acquisition field) and must contain the same number of data points, each with the same set of $B$-values (or $H$-values; i.e. the IRM acquisition fields). Before the end-member model can be calculated, pre-processing of input IRM acquisition curves may be required (i.e. interpolation on to the same set of field values and forcing the IRM acquisition curves to be monotonic).

The end-member unmixing approach has a huge potential which we hope to outline in this contribution. However, it also has potential pitfalls. Since (almost) everything is possible, the user must have a certain idea of the significance of potential end members. The solution is mathematically non-unique and the most reasonable geological solution should be targeted (cf. Weltje 1997). The IRM acquisition curves are normalized to take out concentration effects (that can be visualized by back-transformation of the mixing proportions calculated). The input data are so-called closed data with their $q$ end members forming the apices of a $q - $ 1-dimensional hyperplane or simplex (Weltje 1997). In the case of three end members, these are the apices of an equilateral plane triangle. Magnetic interaction could bias the linear additivity of the mixing of IRM components. However, for the dilute 'magnetic suspensions' (i.e. magnetic particles dispersed at low concentration in a rock sample) of interest here, magnetic interaction has been shown to be insignificant (Heslop *et al.* 2006; Heslop & Dillon 2007). It is also important to realize that the end members need not be unimodal coercivity distributions. If magnetite (*sensu lato*) and hematite (*sensu lato*) co-vary throughout a sample collection, they could be captured as a single end member. The end members reflect processes rather than specific magnetic minerals. However, for an in-depth understanding of the processes involved, knowledge of the magnetic mineralogy that makes up the end-members is a distinct advantage. End-member modelling is therefore often used in conjunction with (independent) forward modelling techniques to unmix individual IRM acquisition curves.

## The optimal number of end members

Determining the number of meaningful end members which should be considered is at the heart of any end-member model interpretation. There exist various ways to calculate end members and related mixing proportions for a given dataset. By non-negative matrix factorization, the closest mixing polytope around the entire data hull is calculated iteratively. This makes the approach rather outlier-sensitive (one deviating input data curve can have an overly large influence). Iterative schemes that start from extreme observations also appear to be outlier-sensitive. If we start from within the convex hull of the data and iteratively expand the mixing polytope, the final solution is much less sensitive to outliers; this is especially true if we accept that not all data points must lie within the final mixing polytope (Weltje 1997). The effectiveness of the end-member model under consideration is evaluated from a mathematical viewpoint, primarily by the convexity error and the coefficient of determination.

The mismatch between the calculated end-member model and an ideal model that includes all data points is given by the so-called convexity error (Weltje 1997). This is determined by evaluating the location of the data points outside the mixing polytope. The convexity error conveys a measure of that mismatch by adding the logarithms of the proportion (Prop) of the samples outside the mixing polytope and of their mean-squared distance (Dist): $\log(\text{Prop}) + \log(\text{Dist})$. As an example, when 1% (0.01) of the observations are outside the mixing polytope and their average end-member contribution is $-10\%$ (0.1 thus squared 0.01), the convexity error would be $-4$. Remember that these data are outside the mixing polytope and therefore have a negative average end-member contribution. Logically 'good' models have a small convexity error: the program performs a maximum of 1000 iterations or stops after having reached a convexity error of $-6$ (equal to for example 1% of the data with an average negative end-member contribution of 1%). The algorithm accepts that some data points remain outside the mixing polytope in order to ensure a stable and interpretable solution. The algorithm converges because each mixing polytope encloses that of the previous iteration. The convexity error is therefore smaller for each successive iteration step.

Further, the coefficient of determination $(r^2)$ between the model and the input data is calculated for end-member models with an increasing number of end members. Obviously 'good' models have a high coefficient of determination. If $r^2$ is plotted against the corresponding end-member number, the break-in-slope provides a useful guideline for the optimum number of end members: the steeper slope at the low number of end members is considered 'information' and the almost-horizontal slope at the higher number of end members is 'interpretational noise', analogous to the scree-test regularly used in exploration geology. In this 'noise end', some end members differ only minutely from each other which often makes little sense in terms of interpretation: there is a serious risk that the dataset can be over-interpreted. In a way, this reflects the inherent natural variability of the end members; it is better to merge similar end members into a single end member with essentially the same meaning. Another case that bears the risk of over-interpretation is the following: if one end member would be represented by (almost) a single sample in the data without being present at least at some level in the others, there is a fair chance that that sample is anomalous in some way (it is so remarkable that it constitutes an end member). It should be discarded from the dataset and, after rerunning the analysis, both sets of solutions should be checked for consistency and interpretability.

Above all, end members must be geologically interpretable as emphasized by Weltje (1997). In the present case, end members must be understandable in terms of IRM acquisition curves of individual magnetic minerals or their combinations that subsequently can be tied to processes. Odd-shaped and noisy end-member IRM acquisition curves are unrealistic and the corresponding model should be discarded. Only meaningful end members should be linked to processes such as provenance area, lithological composition, remagnetization, etc.

*Input IRM acquisition curves*

Attention should be paid to the input IRM acquisition curves because the end-member model algorithm is sensitive to small and subtle effects. IRM acquisition curves can also be modelled by forward techniques. The cumulative log-Gaussian (CLG) approach (Robertson & France 1994; Kruiver *et al.* 2001) has been applied in a wide variety of settings. The CLG basis function has its limitations however because skewness and kurtosis are not incorporated. Skewness in particular can be inherent to some coercivity distributions (Egli 2003, 2004; Heslop *et al.* 2004) and a skewed generalized Gaussian (SGG) set of basis functions seem to be more appropriate (Egli 2003). However, the interpretation of datasets with both approaches shows large similarities (the effect of kurtosis is often insignificant and skewness can be modelled with an extra very low-coercivity component). It illustrates, however, that a non-parametric technique such as end-member modelling could be preferable as a starting point for the interpretation of IRM acquisition curves as it circumvents selection criteria pertaining to the optimal basis function (Heslop & Dillon 2007). If desired, the forward techniques could be implemented later to constrain the end-member interpretation. The input IRM acquisition curves are therefore determined in line with the criteria to be amenable to so-called IRM component analysis (Kruiver *et al.* 2001; Egli 2003). To minimize deviation from log-normality (Egli 2004; Heslop *et al.*

2004), IRM acquisition curves should be determined from the three-axis alternating field demagnetized starting state with the last demagnetization axis parallel to the IRM acquisition field. In this way, maximum discriminative power is achieved.

In addition, many marls and limestones may have surficially oxidized magnetite particles (Van Velzen & Zijderveld 1995), the magnetic effects of which have a bearing on the shape of IRM acquisition curves because the incipient surficial oxidation hardens the magnetic particles. Annealing at 150 °C undoes the effects of this oxidation (Van Velzen & Zijderveld 1995) and results in much better separated primary and secondary NRM components (e.g. Gong *et al.* 2008). Since remagnetization is often associated with magnetite, it is recommended to apply the annealing before the AF demagnetization and subsequent IRM acquisition (Gong *et al.* 2009*b*). However, when dealing with greigite-bearing remagnetized rocks such as those described by Rowan & Roberts (2006, 2008), 5–10 minutes of annealing at 150 °C might change the greigite properties and is not recommended. Greigite-bearing remagnetizations are not discussed in this contribution, however.

The end-member model algorithm is very sensitive to instrumental bias. The IRM acquisition curves should therefore be measured in a reproducible manner, avoiding large differences in time between field application and IRM measurement in a sample collection to be processed. In this way, short-term viscosity effects are minimized. Pulse magnetizers should also be checked for peak-field reproducibility and potential under-damping.

## Case studies

The IRM acquisition curves discussed here are measured with a robotized 2 G DC-SQUID (Direct Current Superconducting QUantum Interference Device) magnetometer (instrumental noise level $3 \times 10^{-12}$ Am$^2$) with 'in-line' set-up for IRM acquisition curves up to 700 mT at the Utrecht University (Netherlands) palaeomagnetic laboratory. All data were acquired with the so-called two-position measurement protocol. This allows the best correction for the (slightly varying) contribution of the sample tray. Despite the fact that this variation is very small, the end-member analysis may lead to flawed end members and mixing proportions. In these cases, the entire effort is understandably rather meaningless and may lead to incorrect conclusions and inferences. It must be realized that high-end mathematical algorithms (such as the end-member modelling algorithm) are sensitive to small biases in input data, so extra care must be executed to have really 'clean' input data. The MATLAB modules used for the calculation and an executable can be found online at: http://www.marum.de/ Unmixing_magnetic_remanence_curves_without_a_ priori_knowledge.html

The program calculates the solutions between 2 and 9 end members. The following criteria were used as guidelines to decide on the optimal number of end members: (1) $r^2$ between the input data and the end-member model (Heslop *et al.* 2007) must be $>0.8$; (2) inclusion of yet another end member provides little extra interpretational value (the equivalent of the scree-test); (3) virtual duplication of end members should be avoided: the limit of interpretational power has been reached; and (4) the program calculates the end-member model by iteration until a stopping convexity error has been reached (set at $-6$); if that convexity is not reached within 1000 iterations, the convexity error at termination can also be used to judge the quality of the model.

In addition, users must be certain that the entire data variability has been sampled by the input IRM acquisition curves; otherwise, the resulting end members are not representative. This is difficult to check but when end-member models from subsets of the data are essentially the same as those from the complete data array, it can be reasonably assumed that the complete data variability was indeed sampled.

The case studies described below were intended to explore whether end-member modelling could identify remagnetized and non-remagnetized strata in geological situations that were known beforehand. Importantly, this knowledge is not used in the end-member modelling effort, but provides a 'training set'. Importantly, this knowledge is not used in the end-member modelling effort. Instead it provides a 'training set' to investigate the performance of the end-member modelling algorithm in identifying remagnetized rocks. The goal in the longer term is to be able to use end-member modelling as an independent decision tool to diagnose whether or not a rock sequence is remagnetized. This is particularly useful when directional information may not be available (due to complex local tectonics) or is subject to rather large uncertainty (as is the case for many Precambrian rocks).

## Organyà Basin (Spain): remagnetized and non-remagnetized limestones

Across the Iberian Peninsula various Mesozoic sedimentary basins occur. Organyà Basin (Fig. 1) is the northernmost basin of the South Central Pyrenees thrust sheets, a piggy-bag series of nappes with the basin as the uppermost structural unit. Its sediments consist of mainly Lower Cretaceous limestones and marls. Marine sedimentation halted during the Cenomanian and nappe stacking began at the end of the Cretaceous and continued throughout the formation of the Pyrenees mountain chain. Triassic Keuper evaporates often acted as décollement.

Most Mesozoic basins in Iberia are remagnetized. Organyà Basin is atypical in the sense that the Cretaceous limestones up to the Barremian are remagnetized, but the Aptian and Albian marls with intercalated limestones contain a primary NRM (Dinarès-Turell & García-Senz 2000; Gong *et al.* 2008). The study of Dinarès-Turell & García-Senz (2000) indicated hysteresis parameters more or less in line with the remagnetized and non-remagnetized trend lines (Channell & McCabe 1994). However, the non-remagnetized samples plot close to the SD region where the diagnostic power of the trend lines is not large. Many samples studied by Gong *et al.* (2008) appeared to yield hysteresis loops subject to a fair amount of uncertainty because of low signals. Organyà Basin therefore constitutes an ideal setting to study the properties of remagnetized limestones and non-remagnetized limestones and marls with IRM acquisition curves and end-member modelling (Gong *et al.* 2009*b*). The rocks are close to each other in space and time and a fair number of samples must be discarded in hysteresis loop analysis.

In their study, Gong *et al.* (2009*b*) examined two sets of IRM acquisition curves of limestones and marls (288 IRM acquisition curves in total). Each set consists of three sites: one of remagnetized limestones, one of non-remagnetized limestones and one of non-remagnetized marls (unfortunately remagnetized marls do not occur in the study area). One set of samples (48 specimens for each of the 3 sites) was annealed at 150 °C before AF demagnetization and subsequent IRM acquisition and the other was not heated before this treatment (48 sister specimens from each of the 3 sites). As more or less anticipated, the annealed samples yielded better discriminative results in the end-member analysis and are discussed here. It appeared that in 16 remagnetized limestones (2 rows of 8 samples), the flux jump correction was incorrectly handled by the magnetometer software. This bug was discovered during measurement of this IRM acquisition series, which was the first to be measured with the automated system, and has since been corrected. After data selection, 32 specimens remained for the remagnetized limestones while all 48 specimens were retained for the two non-remagnetized groups.

At the initial stage the end members were calculated for each individual dataset, that is, the remagnetized (R) and non-remagnetized (NR) limestones, both annealed (Fig. 2). For both datasets, models with three end members were selected ($r^2$ c. 0.75–0.8 on the low side which is considered acceptable). The fourth end member almost duplicates one of the

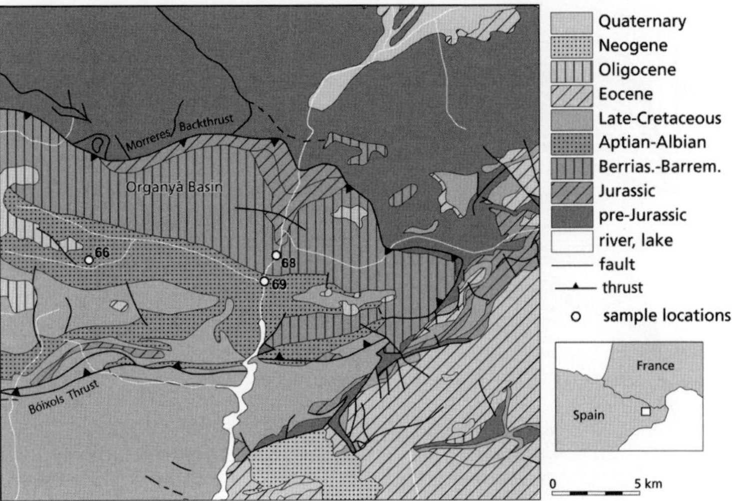


**Fig. 1.** Simplified geological map of the Organyá Basin with the sites investigated by Gong *et al.* (2009*b*, fig. 1) indicated.

existing end members, risking over-interpretation of the data; models with four or more end members are not considered further. Merging of the data (lowermost panels) shows essentially the same end members (Fig. 2); the entire data variability has therefore been collected and the end members are meaningful.

The shape of the end members indicate a high-coercivity end member labelled EM3 and interpreted to be hematite or goethite (goethite is less likely given its appreciable increase in fields from 200 mT upward already). The two low-coercivity end members are two grain-size distributions of magnetite. Importantly, one saturates at 300–400 mT which is compatible with fine-grained pseudo-single-domain magnetite, labelled EM1. The other low-coercivity end member (EM2) shows an appreciable increase in IRM up 100–200 mT acquisition fields, but does not saturate entirely in the maximum available field of 700 mT. This behaviour is in line with closely spaced very fine, nominally superparamagnetic (SP), magnetically interacting particles often associated with remagnetized limestones (e.g. Suk *et al.* 1993; Sun & Jackson 1994; Katz *et al.* 2000). It is envisaged that, by their non-stoichiometry, partially maghemitized surface layers raise the energy barrier for spin reorientation thus raising the coercivity. Industrial fly ashes show such behaviour in a much more extreme fashion: samples do not saturate at all even in 2 T fields (Dekkers & Pietersen 1992). It was shown via transmission electron microscopy that those samples contained such magnetically interacting fine SP particles in significant amounts. Non-remagnetized limestones should contain EM1 in fairly large

amounts while remagnetized limestones contain an appreciable fraction of EM2. Depicted in a ternary plot (Fig. 3) this is indeed the case: end-member modelling successfully discriminates remagnetized and non-remagnetized limestones solely on the basis of magnetic properties. As a rule, non-remagnetized limestones contain a minor contribution of the remagnetized end-member and vice versa so samples may indeed plot in between the two trend lines on the Day plot. EM3 is not relevant for the interpretation of remagnetized v. non-remagnetized limestone. Its contribution is therefore taken out by renormalizing end members 1 and 2 to 100% (Fig. 4). It shows that most samples indeed plot at their expected positions. When the averages (or more exactly, the measures of location) per group with the 95% confidence levels are calculated (including all samples), it appears that the remagnetized and non-remagnetized limestones are significantly different. Because the end-member proportions are closed data, the additive log-ratio transform was used to create a pseudo-open data space for this calculation (Aitchison 1982; Heslop 2009). For example, for a given EM1 this transform is $x = \log$ [EM1/(1– EM1)] and the plotted results are back-transformed. The marl samples appear to have very similar IRM acquisition curves which are perfectly compatible with their non-remagnetized nature: a low contribution of their EM2 and an EM1 that is saturating between 200 and 300 mT.

## Case studies in southern Turkey

The amalgamation of the Anatolian–Tauride Block (ATB) in present-day southern Turkey with Eurasia

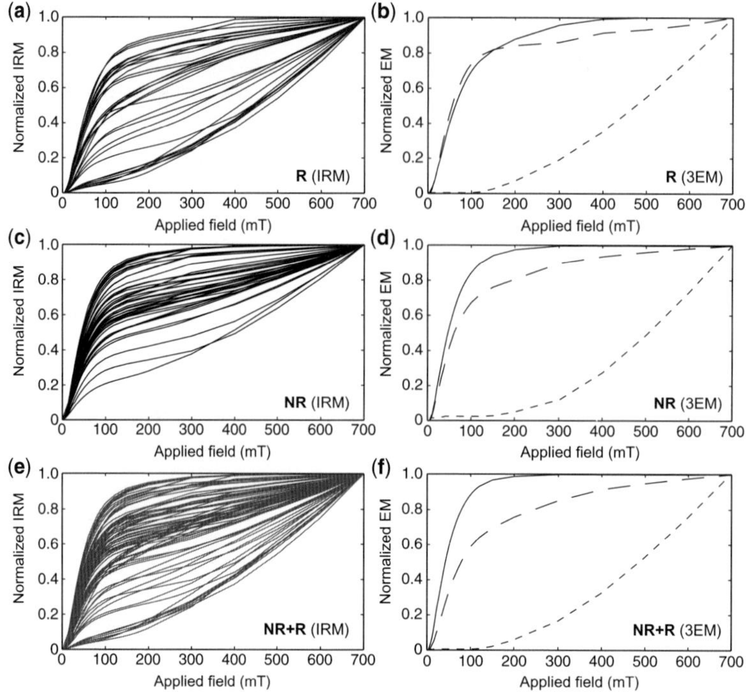

**Fig. 2.** End-member models for the IRM acquisition curves from the limestone groups that were pre-annealed at 150 °C. (**a**) Normalized IRM acquisition curves from the remagnetized (R) limestones. (**b**) Calculated end members for the remagnetized limestones in the three-end-member model. (**c**) Normalized IRM acquisition curves from the non-remagnetized (NR) limestones. (**d**) End members for the non-remagnetized limestones in the three-end-member model. (**e**) Normalized IRM acquisition curves from the merged dataset of the remagnetized and the non-remagnetized limestones. (**f**) End-members for the three-end-member model of merged pre-annealed limestones. Note the distinctly variable input IRM acquisition curves; the entire variability can be described by three end members. In the end-member diagrams (b, d and f), solid lines indicate the low-coercivity component (end member 1), long dashed lines indicate the low-coercivity component (end member 2) and short dashed lines indicate the high-coercivity component (end member 3) (Gong *et al.* 2009*b*, fig. 3).

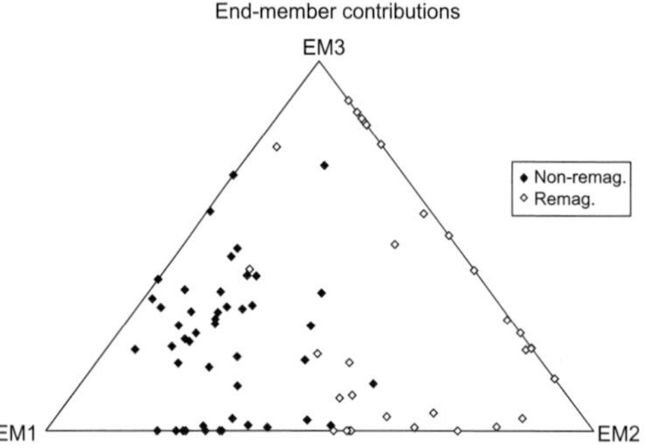


**Fig. 3.** Ternary plot with the percentages of the three end members in the end-member model from the pre-annealed limestones. End member 1 (EM1) is the down-left apex, end member 2 (EM2) is the down-right apex and end member 3 (EM3) is the top apex. Non-remagnetized (remagnetized) samples are indicated with full (open) diamonds according to their field-based site allocation (Gong *et al.* 2009*b*, fig. 4).

is a topic of intense geological research that has resulted in conflicting models to explain the geological history. The ATB with Cambrian–Tertiary sediments has a Gondwanan affinity and is considered to have rifted apart from Gondwana in the early Mesozoic (e.g. Şengör & Yilmaz 1981). The next two case studies deal with various aspects of this history and examine the remagnetized and non-remagnetized nature of the rocks under study with their direct geological implications. Unfortunately, extended databases of hysteresis properties or other magnetic property data for the study areas are not available. The outcome of the end-member analysis can therefore only be compared to palaeomagnetic directional information. In the Bey Dağlari study, non-remagnetized sediments are analysed while the Taurides study discusses remagnetized strata. The end-member model analysis appears to fully support the geological and palaeomagnetic information as detailed below.

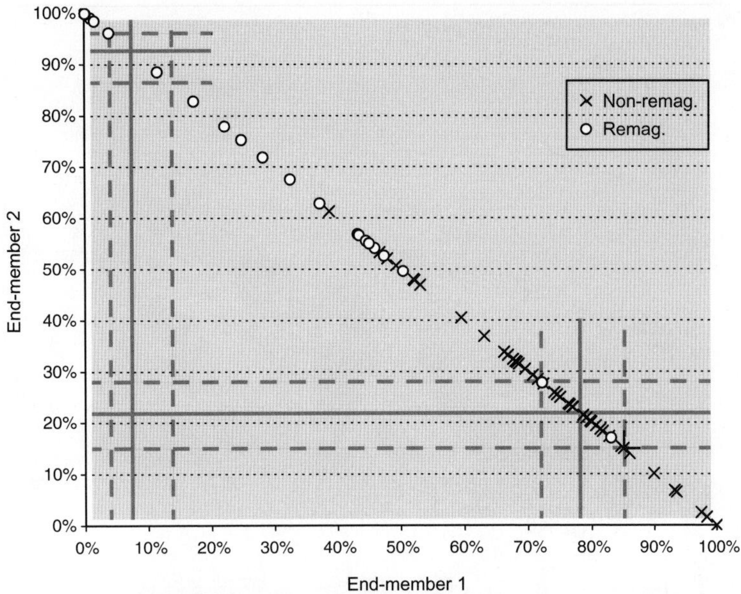


**Fig. 4.** The contribution of end member 1 v. end member 2 corrected for the (variable) contribution of the high-coercivity end member 3 for the pre-annealed limestone dataset. The open dots indicate the remagnetized data and the crosses denote the non-remagnetized data on the basis of site allocation. Note that these designations are independent of the end-member model calculation. The end-member averages (or measures of location) for the remagnetized and non-remagnetized datasets are indicated with the full lines. The lower and upper bounds of the 95% confidence areas are given with the dashed lines. The log-ratio transform was applied to account for the closed data (see text). Because the data are closed, the end-member 'averages' for each dataset added are 100%: for the non-remagnetized limestones, the average of end member 1 is 78% and thus that of end-member 2 is 22%. For the remagnetized limestones, the averages of end members 1 and 2 are 7% and 93%, respectively. While some individual samples would be classed incorrectly, the group averages are statistically significantly different (Gong *et al.* 2009*b*; fig. 5).

*Bey Dağlari: non-remagnetized strata.* The first case study involves the analysis of a set of non-remagnetized Miocene sediments from the Bey Dağlari region in southern Turkey (Fig. 5; Van Hinsbergen *et al.* 2010*a*). The results are included in an overall compilation of palaeomagnetic directions across southwestern Turkey to explain oroclinal bending of the eastern branch of the Aegean arc during Miocene times (Van Hinsbergen *et al.* 2010*b*) in relation to the formation of the Menderes Massif metamorphic core complex. This extension was accommodated by thrusting and strike-slip faulting in the so-called Isparta Angle in the Antalya region of southern Turkey. In the then prevailing geodynamic scenario, the Bey Dağlari region has rotated *c.* 25° anticlockwise post-dating a Middle Miocene remagnetization. To test this model, van Hinsbergen *et al.* (2010*a*) studied two Lower Miocene composite sections in the area spanning the period from the Aquitanian unconformity with the Bey Dağlari limestones to approximately the Burdigalian–Langhian boundary. Limestones occur dominantly in the Aquitanian while the

Burdigalian primarily consists of blue clays. A positive reversals test, correlatable magnetostratigraphy and the occurrence of inclination shallowing all point to a primary NRM. This was confirmed by the end-member analysis described below. The new palaeomagnetic results indicate that the Bey Dağlari region did not undergo rotation between the late Cretaceous and the late Burdigalian, and rotated *c.* 20° between 16 and 5 Ma. Hence, both limbs of the Aegean orocline rotated more or less coevally (Van Hinsbergen *et al.* 2010*a*).

Forty-nine IRM acquisition curves served as input to calculate the end-member model. Before IRM acquisition the samples were AF demagnetized, but not pre-annealed at 150 °C. The effects of pre-annealing on NRM demagnetization behaviour in the blue clay samples was shown to be insignificant in a pilot series. For this particular dataset, the optimal end-member solution was judged a three end-member model: $r^2$ is 0.84 and the final convexity level is −3.8. Models with a higher number of end members have only marginally higher $r^2$; in contrast their convexity level is

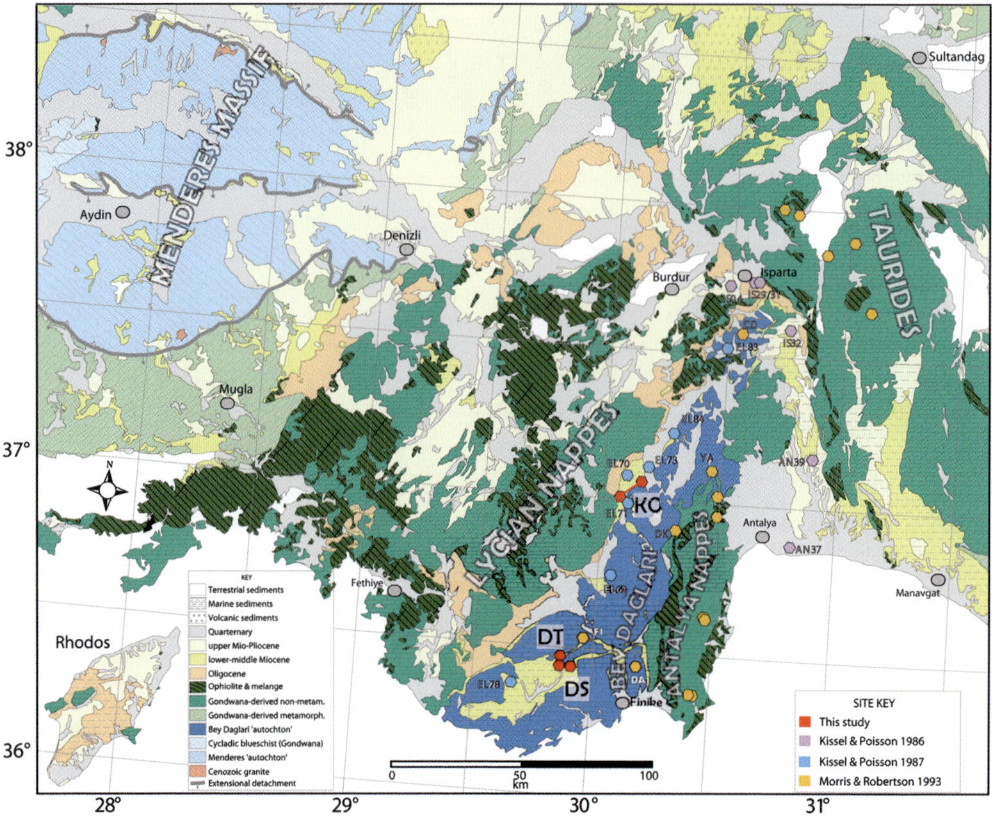

**Fig. 5.** Geological map of southwestern Turkey. Sample locations of Van Hinsbergen *et al.* (2010*a*) in red. The Lycian Nappes overthrust the Bey Dağlari platform from the NW and the platform is overthrust by the Antalya Nappes from the east. The central Tauride area in the east of this map is the topic of the Meijers *et al.* (2011) study (Van Hinsbergen *et al.* 2010*a*, fig. 2).

worse. The break-in-slope in the curve of $r^2$ v. the number of end members occurs at three end members. Further, from five end-member solutions onward, at least two of the end members essentially duplicate. The shape of the end members is shown in Figure 6. End member 1 is the comparatively hard SD-style magnetite typical of most limestones. End member 2 represents the softer PSD magnetite that occurs mainly in the blue clays. Note that both saturate at 300–400 mT, in line with their anticipated non-remagnetized nature and the results of Gong *et al.* (2009*b*). Lithological differences are tracked by subtle differences in magnetic properties. In the ternary plot (Fig. 6), samples with high contributions to end members 1 and 2 plot close to the baseline. Limestone samples are black and blue clay samples light grey.

The interpretation of the present end member 3, which has a prominent low-coercivity part and a high-coercivity tail up to 700 mT, is more

complicated. The shape of this end member resembles the remagnetized end member in the study of Gong *et al.* (2009*b*), which could imply that the samples with a high contribution of this end member would be remagnetized, as in the dataset of Gong *et al.* (2009*b*). In the present dataset, however, the hematite component is not identified as a separate end member as in the case study of Gong *et al.* (2009*b*). The hematite contribution up to 700 mT is often minute and only occasionally more prominent. The end-member algorithm (that imposes no constraints on the shape of the end members) therefore does not appear to resolve a separate hematite end member.

To constrain the end-member interpretation, Van Hinsbergen *et al.* (2010*a*) turned to a forward modelling approach as a supplementary tool. They utilized IRM acquisition curve coercivity component fits (cf. Kruiver *et al.* 2001). In samples where an appreciable amount of hematite was

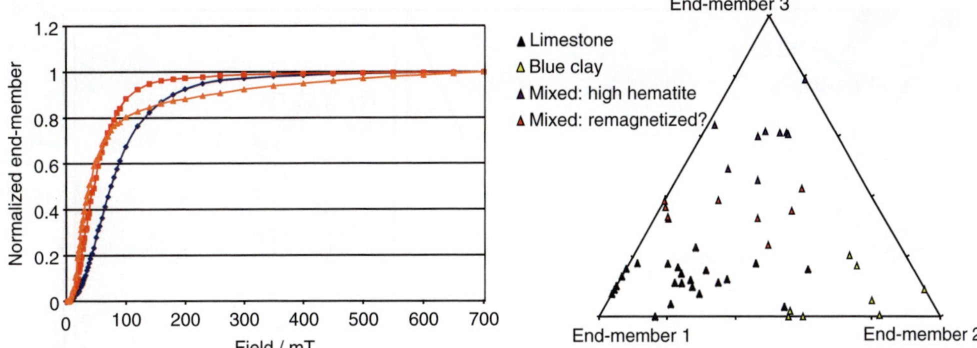


**Fig. 6.** End-member modelling results of Van Hinsbergen *et al.* (2010*a*). Left panel: the shape of the calculated end members in a three-end-member model. End member (EM) 1 (diamonds) is interpreted as SD magnetite and occurs dominantly in the limestones. EM2 (squares) is softer PSD magnetite and occurs mostly in the blue clays. Both saturate in *c.* 300 mT in line with their non-remagnetized nature. End member 3 (triangles) has a dual meaning (see also main text): it represents either a mixture of magnetite and hematite (in cases where IRM acquisition curve fitting yielded a fair amount of hematite) or the samples could be remagnetized (in cases where IRM acquisition curve fitting yielded a marginal amount of hematite). Right panel: ternary plot with the EM partitioning for each sample. Samples with a substantial hematite content are interpreted as a mixture of magnetite and hematite. In contrast, samples with *c.* 40% of end member 3 but virtually devoid of hematite are possibly remagnetized. These all occur close to the basal unconformity (Van Hinsbergen *et al.* 2010*a*, fig. 13).

required for the IRM component fit, end-member 3 was considered a mixture of soft magnetite and hematite. In samples where the amount of hematite required for the IRM fit is just a few percent, a mixture of magnetite and hematite cannot be the cause for the shape of end member 3. For these samples, the possibility of remagnetization is invoked. These samples almost all appeared to be located within *c.* 5 m from the basal unconformity. It is perfectly conceivable that those samples are actually remagnetized by the action of percolating groundwater using the unconformity as an aquifer. In this study, end member 3 therefore has a dual interpretation: in cases where appreciable hematite was fitted to the IRM acquisition curves, we interpret end member 3 as a mixture of magnetite and hematite; the samples probably carry a primary NRM as well. These samples have the highest end member 3 contribution. In those samples with raised end member 3 contributions but with hardly any hematite required in the IRM curve fits, the possibility of remagnetization cannot be excluded. Note that the latter group constitutes only a very small fraction of the total amount of samples. The vast majority of the samples (>95%) which determine the rotation and magnetic polarity of the sections were taken from blue clay, which provides no reason at all to suspect remagnetization.

*Taurides: remagnetized strata.* The Taurides sedimentary sequences of Cambrian–Tertiary age is mostly composed of shallow marine platform-type carbonates (e.g. Şengör & Yilmaz 1981; Mackintosh & Robertson 2009). The Taurides consist of several isopic units (Fig. 7) that are stacked as nappe sequences mainly during the late Cretaceous–Oligocene (Andrew & Robertson 2002; Özer *et al.* 2004). Palaeogeographically, these isopic zones include (from south to north) the Antalya, Alanya, Geyikdağı, Aladağ, Bolkardağı and Bozkır units. The Geyikdağı unit is structurally the lowest and relatively autochtonous unit (Fig. 7). Because of its similar structural position and stratigraphy, the unmetamorphosed and mildly deformed Bey Dağlari unit is generally considered as the lateral equivalent of the parautochthonous Geyikdağı unit. The tectonostratigraphic stacking was followed by the aforementioned middle–late Miocene anticlockwise vertical axis rotation in the western limb of the Isparta Angle (Kissel & Poisson 1987; Morris & Robertson 1993) determined for Lycian Nappes and Bey Dağlari region.

Meijers *et al.* (2011) conducted their study in the central Taurides, the eastern limb of the Isparta Angle. Here, late Cretaceous–Eocene compression formed the fold and thrust belt, structurally equivalent to the Lycian Nappes in the west. The stacking however took place at different times in each domain. The central Taurides fold and thrust belt is unconformably overlain by Miocene marine and terrestrial deposits.

This study of remagnetized rocks involved the end-member analysis of 174 IRM acquisition curves. The study was focused on the long Seydişehir

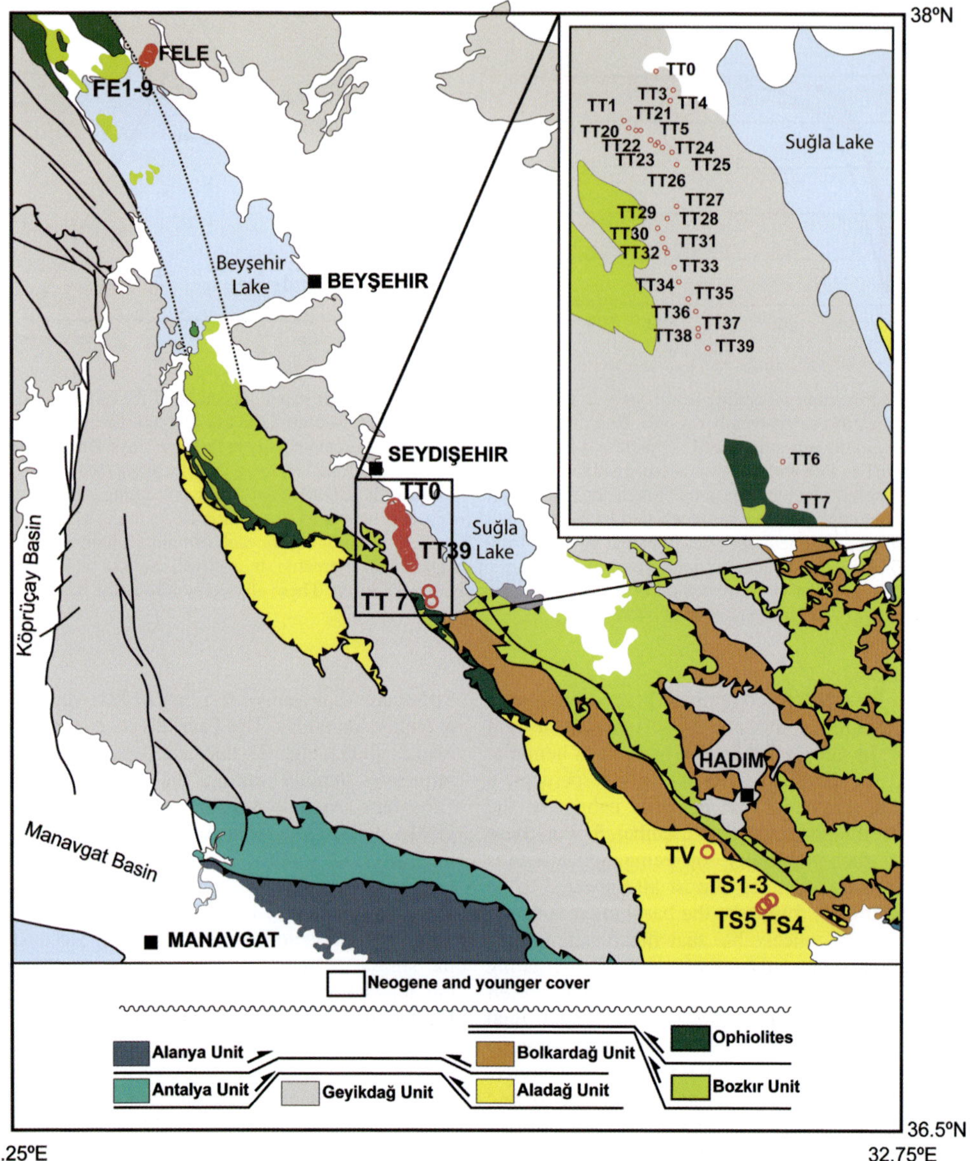



**Fig. 7.** Geological map of the study area of Meijers *et al.* 2011. In the lowermost part the schematized tectonostratigraphic build-up of the nappe sequence is provided. The discussion here is confined to the Seydişehir section (labelled TT; Meijers *et al.* 2011, fig. 3).

section (*c.* 2000 m stratigraphy) that consists of many rock types (limestones, sandy/shaly limestones, dolomitic limestones). Before IRM acquisition, the specimens were pre-heated to a temperature of 150 °C in a magnetically shielded oven and AF demagnetized according to the IRM acquisition curve protocol. The IRM was acquired in 57 steps up to 700 mT with the in-house developed robotized magnetometer.

First, data subsets were subjected to the end-member analysis. The merged data showed essentially the same end members so the entire data variability was included in the analysis and its outcome is not input-dependent. Meijers *et al.* (2011) propose either a two- or a three-end-member model. Coefficients of determination are 0.87 and 0.90 and the convexities at termination $-5.00$ and $-2.50$, respectively. From four end members onwards,

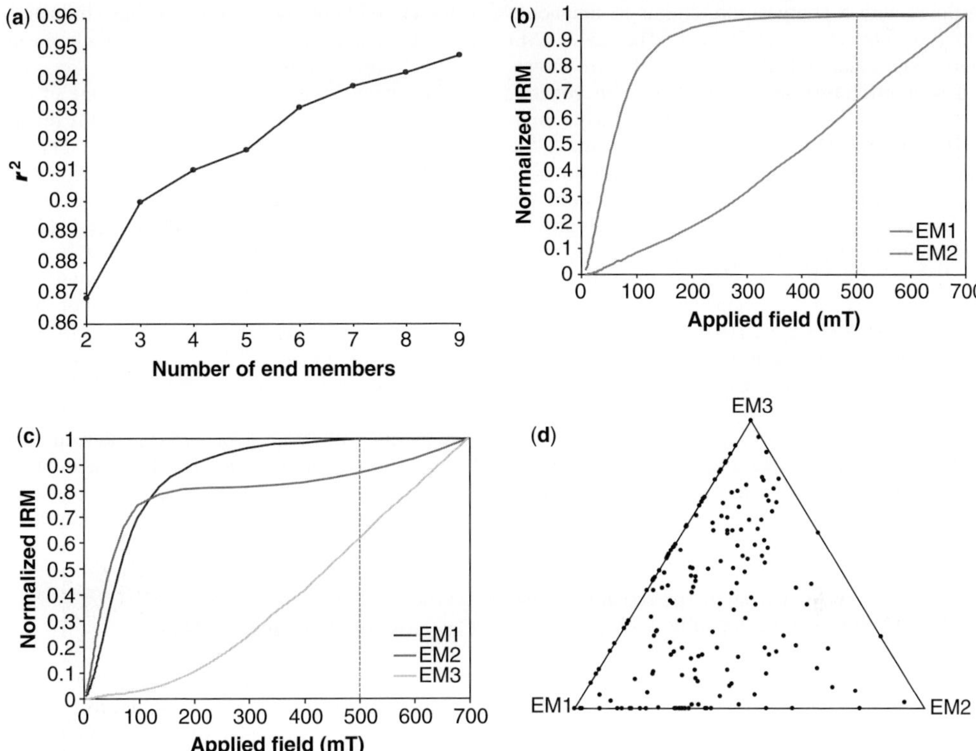

**Fig. 8.** (a) Plot of $r^2$ v. end-member (EM) number. A three-end-member model is preferable; in the five-end-member solution duplication of end members occurs. (**b, c**) End members for the normalized IRM acquisition curves for the (b) two and (c) three end-member models. (**d**) Ternary plot showing for each sample the percentages of the end members (EM) in the three-end-member model (Meijers *et al.* 2011, fig. 8).

duplication of end members occurs; evidently, the data become over-interpreted. A two-end-member model yields a discrimination in a magnetite (with a remagnetized nature) and a hematite end member (Fig. 8). Indeed, samples with a higher hematite end-member proportion also have a higher hematite contribution in (forward) IRM component fitting. The three-end-member model is preferred (with EM1, 2, 3) because it can be interpreted as reflecting lithological grouping and it represents the break-in-slope of the $r^2$ v. number-of-end-members curve (Fig. 8a). The two 'magnetite' end members in that model both have a remagnetized nature.

End-member shapes here reflect magnetic minerals. NRM directions (Meijers *et al.* 2011) are of dual polarity (without any correlation to the GPTS) and represent Early Tertiary directions for the ATB. The EM distribution and the polarity of the samples appear to be unrelated; the end members reflect magnetic processes and properties, not geomagnetic field behaviour. EM3 (light grey in Fig. 8) represents hematite and samples with a high

proportion of EM3 indeed show a considerable contribution of a high-coercivity IRM fraction in the forward log-Gaussian approach. EM1 (black in Fig. 8) is logically interpreted as representing magnetite. The SIRM of this end member is reached at 500 mT, a notably high field for 'classic' magnetite that typically reaches SIRM at 200–300 mT. Remagnetized magnetite particle suites can however reach saturation at much higher fields, supporting the remagnetized nature of the Tauride rocks under study.

The majority of the samples are mixtures of EM1 and EM3 in variable proportions (Fig. 8) with only minor contributions of EM2 (dark grey). As in the Van Hinsbergen *et al.* (2010a) study, the interpretation of end members can be complicated. This applies to EM2 in this case, which can be considered a mixture of (relatively soft) magnetite with a tail of hematite. This is supported by the slightly sandy nature of the limestones with an appreciable EM2 contribution. Alternatively, it could be associated with aggregates of ultrafine ferrimagnetic particles with a composition similar to magnetite which do not saturate at 700 mT. The EM2/EM3 ratio is substantially higher in dolomitic limestones and supports this latter interpretation. In general, dolomitic limestones are associated with high EM1 and low EM3 contributions, which make the existence of hematite tails in another end member unlikely (mathematically, its existence would not be likely either). Both EM1 and EM2 represent remagnetized magnetites.

Widespread Miocene remagnetization in the Bey Dağlari region and Lycian Nappes was negated by Van Hinsbergen *et al.* (2010*a*, *b*). For the Lycian Nappes this would concur with their high structural position. However, the evidence for remagnetization affecting the Antalya Nappes presented by Morris & Robertson (1993) remains firm.

In the Taurides, the oldest and structurally highest nappes are the 'ophiolites' (Bozkır, Antalya and Lycian) that are widespread across southern Turkey with metamorphic soles of 90–95 Ma (e.g. Çelik *et al.* 2006). This is older than the maximum age for remagnetization (palaeolatitude and reversed polarity are incompatible with the Cretaceous normal polarity superchron). After the emplacement of the ophiolites, folding and thrusting of the ATB eventually led to the development of the fold and thrust belt in the early Eocene (Özgül 1983). This implies that the remagnetization is more likely related to folding and thrusting during the Paleocene–early Eocene rather than to the ophiolite emplacement at <95–90 Ma. Based on the similar inclinations from all sites after tilt correction, remagnetization probably occurred before tilting or during the early stages of tilting, that is, in the early stages of folding and thrusting. This is supported by the small-circle-intersection method (Waldhör & Appel 2006) that shows best grouping at an early stage of folding (Meijers *et al.* 2011). Folding and thrusting could also be a candidate for remagnetization in the rocks sampled by Morris & Robertson (1993) in the Antalya Nappes in the Bey Dağlari region, which underwent nappe stacking during the same time interval (Juteau *et al.* 1977; Robertson & Woodcock 1981; Çelik *et al.* 2006).

Large amounts of external orogenic fluids are considered unlikely given the absence of related low-temperature Mississippi Valley Type ores in the region. Magnetite formation as a result of diagenetic reactions involving clay minerals is deemed more plausible. Influence of pressure solution, although equivocal in remagnetization processes (Evans *et al.* 2003; Elmore *et al.* 2006), cannot be excluded because stylolites are often observed. The Seydişehir section shows continuous sedimentation into the Palaeocene and lower Eocene foreland basin deposits. The timing of remagnetization is therefore likely Eocene, and may have influenced all Eocene and older fold and thrust rocks in the central Taurides axis.

Geochemical analysis and K–Ar dating of clay mineral suites may offer further clues to the remagnetization process in the Taurides. The approach described by Zwing *et al.* (2009) is particularly attractive. In their study of the NE Rhenish Massif in Germany, Zwing *et al.* (2009) were able to relate an older diagenetic illite suite (348 ± 7 Ma) to a regional mild thermal event that is interpreted to have caused remagnetization by action of fluids. A younger diagenetic illite suite (324 ± 3 Ma) was related to deformation. That younger event has overprinted earlier remagnetized NRM but not the illite that was already formed. If illite formation age trends with their geochemical signature can be established for the Taurides, this undoubtedly contributes to a robust unravelling of tectonostratigraphic, deformation and burial temperature arguments for that setting.

## Conclusions and prospects

It is clear that proper identification of remagnetized and non-remagnetized rocks is crucial to correctly understand the geological development of mountain belts, illustrated here with examples from the central Taurides area. End-member modelling is helpful to identify remagnetized and non-remagnetized limestones. End members reflect processes that have resulted in certain specific magnetic mineral signatures. When interpreting the end-member model options incorporation of forward techniques, such as log-Gaussian IRM acquisition curve fitting, is recommended in conjunction with the mathematical criteria to judge end-member solutions. A foremost criterion, however, is that any interpretation must make sense in the general geological context of the study area. Current end-member studies have concentrated on limestones. It is desirable that the 'training set', which is presently rather small, is enlarged by including more test case studies that also include other lithologies. In the longer term, this may result in the possibility of identifying chemical remagnetization for groups of samples and, in some cases, even on an individual sample basis (as for the samples close to the basal unconformity in the Van Hinsbergen *et al.* 2010*a* study) without having to consider palaeomagnetic directional behaviour.

The IRM acquisition data reviewed here were acquired with support of the Earth and Life Science Division of the Netherlands Science Foundation and the Netherlands Research Centre for Integrated Solid Earth Science (ISES).

## References

AITCHISON, J. 1982. The statistical analysis of compositional data. *Journal of the Royal Statistical Society Series B*, **44**, 139–177.

ANDREW, T. & ROBERTSON, A. H. F. 2002. The Beysehir–Hoyran–Hadim Nappes: genesis and emplacement of Mesozoic marginal and oceanic units of the northern Neotethys in southern Turkey. *Journal of the Geological Society, London*, **159**, 529–543.

BEZDEK, J. C., ERHLICH, R. & FULL, W. 1984. FCM: the fuzzy c-means clustering algorithm. *Computers & Geosciences*, **10**, 191–203.

ÇELIK, Ö. F., DELALOYLE, M. F. & FERAUD, G. 2006. Precise $^{40}$Ar–$^{39}$Ar ages from the metamorphic sole rocks of the Tauride Belt Ophiolites, southern Turkey: implications for the rapid cooling history. *Geological Magazine*, **143**, 213–227.

CHANNELL, J. E. T. & MCCABE, C. 1994. Comparison of magnetic hysteresis parameters of unremagnetized and remagnetized limestones. *Journal of Geophysical Research*, **99**, 4613–4623.

DAY, R., FULLER, M. D. & SCHMIDT, V. A. 1977. Hysteresis properties of titanomagnetites: grain size and composition dependence. *Physics of the Earth and Planetary Interiors*, **13**, 260–267.

DEKKERS, M. J. & PIETERSEN, H. S. 1992. Magnetic properties of low-Ca flyash: a rapid tool for Fe-assessment and a proxy for environmental hazard. *In*: GLASSER, F. P. *ET AL.* (eds) *Advanced Cementitious Systems: Mechanisms and Properties*. Materials Research Society Symposium Proceedings, **245**, 34–47, doi: http://dx.doi.org/10.1557/proc-245-37.

DINARÈS-TURELL, J. & GARCÍA-SENZ, J. 2000. Remagnetization of Lower Cretaceous limestones from the southern Pyrenees and relation to the Iberian plate geodynamic evolution. *Journal of Geophysical Research*, **105**, 19 405–19 418.

DUNLOP, D. J. 2002a. Theory and application of the Day plot ($M_{rs}/M_s$ v. $H_{cr}/H_c$) 1. Theoretical curves and tests using titanomagnetite data. *Journal of Geophysical Research*, **107**, doi: 10.1029/2001JB000486.

DUNLOP, D. J. 2002b. Theory and application of the Day plot ($M_{rs}/M_s$ versus $H_{cr}/H_c$): 2. Application to data for rocks, sediments, and soils. *Journal of Geophysical Research*, **107**, 2057, doi: 10.1029/2001JB000487.

EGLI, R. 2003. Analysis of the field dependence of remanent magnetization curves. *Journal of Geophysical Research*, **108**, 2081.

EGLI, R. 2004. Characterization of individual rock magnetic components by analysis of remanence curves. 2. Fundamental properties of coercivity distributions. *Physics and Chemistry of the Earth*, **29**, 851–867.

ELMORE, R. D., LEE-EGGER FOUCHER, J., EVANS, M., LEWCHUK, M. & COX, M. 2006. Remagnetization of the Tonoloway Formation and the Helderberg Group in the Central Appalachians: testing the origin of syntilting magnetizations. *Geophysical Journal International*, **166**, 1062–1076, doi: 10.1111/j.1365-246X.2006.02875.x.

EVANS, M. A. & ELMORE, R. D. 2006. Fluid control of localized mineral domains in limestone pressure solution structures. *Journal of Structural Geology*, **28**, 284–301.

EVANS, M. A., LEWCHUK, M. T. & ELMORE, R. D. 2003. Strain partitioning of deformation mechanisms in limestones: examining the relationship of strain and anisotropy of magnetic susceptibility (AMS). *Journal of Structural Geology*, **25**, 1525–1549.

FONT, E., TRINDADE, R. I. F. & NÉDÉLEC, A. 2006. Remagnetization in bituminous limestones of the Neoproterozoic Araras Group (Amazon craton): hydrocarbon maturation, burial diagenesis or both? *Journal of Geophysical Research*, **111**, B06204.

GONG, Z., LANGEREIS, C. G. & MULLENDER, T. A. T. 2008. The rotation of Iberia during the Aptian and the opening of the Bay of Biscay. *Earth and Planetary Science Letters*, **273**, 80–93.

GONG, Z., VAN HINSBERGEN, D. J. J. & DEKKERS, M. J. 2009a. Diachronous pervasive remagnetization in northern Iberian basins during Cretaceous rotation and extension. *Earth and Planetary Science Letters*, **284**, 292–301, doi: 10.1016/j.epsl.2009.04.039.

GONG, Z., DEKKERS, M. J., HESLOP, D. & MULLENDER, T. A. T. 2009b. End-member modelling of isothermal remanent magnetization (IRM) acquisition curves: a novel approach to diagnose remagnetization. *Geophysical Journal International*, **178**, 693–701.

HESLOP, D. 2009. Calculating descriptive statistics for the S-ratio. *Geophysical Journal International*, **178**, 159–161, doi: 10.1111/j.1365-246X.2009.04175.x.

HESLOP, D. & DILLON, M. 2007. Unmixing magnetic remanence curves without a priori knowledge. *Geophysical Journal International*, **170**, 556–566.

HESLOP, D., MCINTOSH, G. & DEKKERS, M. J. 2004. Using time and temperature dependant Preisach models to investigate the limitations of modelling isothermal remanent magnetisation curves with cumulative log Gaussian functions. *Geophysical Journal International*, **157**, 55–63.

HESLOP, D., WITT, A., KLEINER, T. & FABIAN, K. 2006. The role of magnetostatic interactions in sediment suspensions. *Geophysical Journal International*, **165**, 775–785.

HESLOP, D., VON DOBENECK, T. & HÖCKER, M. 2007. Using non-negative matrix factorization in the 'unmixing' of diffuse reflectance spectra. *Marine Geology*, **241**, 63–78.

JUTEAU, T., NICOLAS, A., DUBESSEY, J., FRUCHARD, J. C. & BOUCHEZ, J. L. 1977. Structural relationships in the Antalya Complex, Turkey: possible model for an oceanic ridge. *Geological Society of America*, **88**, 1740–1748.

KATZ, B., ELMORE, R. D., COGOINI, M., MICHAEL, H. E. & FERRY, S. 2000. Associations between burial diagenesis of smectite, chemical remagnetization, and magnetite authigenesis in the Vocontian trough, SE France. *Journal of Geophysical Research*, **105**, 851–868.

KENT, D. V. 1985. Thermoviscous remagnetization in some Appalachian limestones. *Geophysical Research Letters*, **12**, 805–808.

KENT, D. V. & OPDYKE, N. D. 1985. Multicomponent magnetizations from the Mississippian Mauch Chunk Formation of the central Appalachians and their tectonic implications. *Journal of Geophysical Research*, **90**, 5371–5383.

KISSEL, C. & POISSON, A. 1987. Etude paléomagnétique préliminaire des formations cénozoïques des BeyDağlari

(Taurides occidentales, Turquie). *Comptes Rendus de l'Académie des Sciences, Paris*, **304**, Serie II, 343–348.

KRUIVER, P. P., DEKKERS, M. J. & HESLOP, D. 2001. Quantification of magnetic coercivity components by the analysis of acquisition curves of isothermal remanent magnetization. *Earth and Planetary Science Letters*, **189**, 269–276.

LANCI, L. & KENT, D. V. 2003. Introduction of thermal activation in forward modeling of hysteresis loops for single-domain magnetic particles and implications for the interpretation of the Day diagram. *Journal of Geophysical Research*, **108**, 2148, doi: 10.1029/2001JB000944.

MACKINTOSH, P. W. & ROBERTSON, A. H. F. 2009. Structural and sedimentary evidence from the northernmargin of the Tauride platformin south central Turkey used to test alternative models of Tethys during Early Mesozoic time. *Tectonophysics*, **473**, 149–172.

MCCABE, C. & ELMORE, R. D. 1989. The occurrence of origin of late Paleozoic remagnetization in the sedimentary rocks of North America. *Reviews of Geophysics*, **27**, 471–494.

MCCABE, C., VAN DER VOO, R., PEACOR, D. R., SCOTESE, C. R. & FREEMAN, R. 1983. Diagenetic magnetite carries ancient yet secondary remanence in some Paleozoic sedimentary carbonates. *Geology*, **11**, 221–223.

MEIJERS, M. J. M., VAN HINSBERGEN, D. J. J., DEKKERS, M. J., ALTINER, D., KAYMAKCI, N. & LANGEREIS, C. G. 2011. Pervasive Palaeogene remagnetization of the central Taurides fold-and-thrust belt (southern Turkey) and implications for rotations in the Isparta Angle. *Geophysical Journal International*, **184**, 1090–1112, doi: 10.1111/j.1365-246X.2010.04919.x.

MOLINA-GARZA, R. S. & ZIJDERVELD, J. D. A. 1996. Paleomagnetism of Paleozoic strata, Brabant and Ardennes Massifs, Belgium: implications of prefolding and postfolding Late Carboniferous secondary magnetizations for European polar wander. *Journal of Geophysical Research*, **101**, 15 799–15 818.

MORRIS, A. & ROBERTSON, A. H. F. 1993. Miocene remagnetisation of carbonate platform and Antalya Complex units within the Isparta Angle, SW Turkey. *Tectonophysics*, **220**, 243–266.

OLIVA-URCIA, B., PUEYO, E. L. & LARRASOAÑA, J. C. 2008. Magnetic reorientation induced by pressure solution: a potential mechanism for orogenic-scale remagnetizations. *Earth and Planetary Science Letters*, **265**, 524–530.

ÖZER, E., KOÇ, H. & ÖZSAYAR, T. Y. 2004. Stratigraphical evidence for the depression of the northern margin of the Menderes–Tauride Block (Turkey) during the Late Cretaceous. *Journal of Asian Earth Sciences*, **22**, 401–412.

ÖZGÜL, N. 1983. Stratigraphy and tectonic evolution of the Central Taurides. *In*: TEKELI, O & GÖNCÜÖGLÜ, C. (eds) *Geology of the Taurus Belt*. International Symposium, Mineral Research and Exploration Inst., 1983 September 26–29, Ankara. Geological Society of Turkey and the Mineral Research and Exploration Institute, Ankara, 78–90.

PERROUD, H. & VAN DER VOO, R. 1984. Secondary magnetizations from the Clinton-type iron ores of the Silurian Red Mountain Formation, Alabama. *Earth and Planetary Science Letters*, **67**, 391–399.

PRINS, M. A., BOUWER, L. M. *ET AL.* 2002. Ocean circulation and iceberg discharge in the glacial North Atlantic: inferences from unmixing of sediment size distributions. *Geology*, **30**, 555–558.

ROBERTSON, D. J. & FRANCE, D. E. 1994. Discrimination of remanence carrying minerals in mixtures, using isothermal remanent magnetization acquisition curves. *Physics of the Earth and Planetary Interiors*, **82**, 223–234.

ROBERTSON, A. H. F. & WOODCOCK, N. H. 1981. Alakir Çay group, Antalya Complex, SW Turkey: a deformed Mesozoic carbonate margin. *Sedimentary Geology*, **30**, 95–131.

ROWAN, C. J. & ROBERTS, A. P. 2006. Magnetite dissolution, diachronous greigite formation, and secondary magnetizations from pyrite oxidation: unravelling complex magnetizations in Neogene marine sediments from New Zealand. *Earth and Planetary Science Letters*, **241**, 119–137.

ROWAN, C. J. & ROBERTS, A. P. 2008. Widespread remagnetizations and a new view of Neogene tectonic rotations within the Australia-Pacific plate boundary zone, New Zealand. *Journal of Geophysical Research*, **113**, B03103, doi: 10.1029/2006JB004594.

ROWAN, C. J., ROBERTS, A. P. & BROADBENT, T. 2009. Reductive diagenesis, magnetite dissolution, greigite growth and paleomagnetic smoothing in marine sediments: a new view. *Earth and Planetary Science Letters*, **227**, 223–235.

ŞENGÖR, A. M. C. & YILMAZ, Y. 1981. Tethyan evolution of Turkey: a plate tectonic approach. *Tectonophysics*, **75**, 181–241.

STEARNS, C. & VAN DER VOO, R. 1987. A paleomagnetic reinvestigation of the Upper Devonian Perry Formation: evidence for late Paleozoic remagnetization. *Earth and Planetary Science Letters*, **86**, 27–38.

SUK, D., VAN DER VOO, R. & PEACOR, D. R. 1993. Origin of magnetite responsible for remagnetization of early Paleozoic limestones of NewYork state. *Journal of Geophysical Research*, **98**, 419–434.

SUN, W. & JACKSON, M. 1994. Scanning electron microscopy and rock magnetic studies of magnetic carriers in remagnetized early Paleozoic carbonates from Missouri. *Journal of Geophysical Research*, **99**, 2935–2942.

VAN DER VOO, R. & TORSVIK, T. H. 2012. The history of remagnetization of sedimentary rocks: deception, developments, discoveries. *In*: ELMORE, R. D., MUXWORTHY, A. R., ALDANA, M. M. & MENA, M. (eds) *Remagnetization and Chemical Alteration of Sedimentary Rocks*. Geological Society, London, Special Publications, **371**, first published online 26 June 2012, http://dx.doi.org/10.1144/SP371.2

VAN DER VOO, R., STAMATAKOS, J. A. & PARÉS, J. M. 1997. Kinematic constraints on thrust-belt curvature from syndeformational magnetizations in the Lagos del Valle syncline in the Cantabrian Arc, Spain. *Journal of Geophysical Research*, **102**, 10 105–10 120.

VAN HINSBERGEN, D. J. J., DEKKERS, M. J. & KOÇ, A. 2010a. Testing Miocene Remagnetization of Bey Dağlari: timing and amount of Neogene Rotations in SW Turkey. *Turkish Journal of Earth Sciences*, **19**, 123–156, doi: 10.3906/yer-0904-1.

VAN HINSBERGEN, D. J. J., DEKKERS, M. J., BOZKURT, E. & KOOPMAN, M. 2010*b*. Exhumation with a twist: paleomagnetic constraints on the evolution of the Menderes metamorphic core complex (western Turkey). *Tectonics*, **29**(TC3009), doi: 10.1029/2009 TC002596.

VAN VELZEN, A. J. & ZIJDERVELD, J. D. A. 1995. Effects of weathering on single domain magnetite in early Pliocenemarls. *Geophysical Journal International*, **121**, 267–278.

WALDHÖR, M. & APPEL, E. 2006. Intersections of remanence small circles: new tools to improve data processing and interpretation in palaeomagnetism. *Geophysical Journal International*, **166**, 33–45.

WEIL, A. B. & VAN DER VOO, R. 2002. Insights into the mechanism for orogen related carbonate remagnetization from growth of authigenic Fe-oxide: a SEM and rock magnetic study of Devonian carbonates from northern Spain. *Journal of Geophysical Research*, **107**, 2063, doi: 10.1029/2001JB000200.

WEIL, A. B., VAN DER VOO, R., VAN DER PLUIJM, B. A. & PARÉS, J. M. 2000. Unraveling the timing and geometric characteristics of the Cantabria–Asturias Arc (northern Spain) through paleomagnetic analysis. *Journal of Structural Geology*, **22**, 735–756.

WELTJE, G. J. 1997. End-member modelling of compositional data: numerical–statistical algorithms for solving the explicit mixing problem. *Journal of Mathematical Geology*, **29**, 503–549.

ZEGERS, T. E., DEKKERS, M. J. & BAILLY, S. 2003. Late Carboniferous to Permian remagnetization of Devonian limestones in the Ardennes: role of temperature, fluids, and deformation. *Journal of Geophysical Research*, **108**, 2357–2376.

ZWING, A., BACHTADSE, V. & SOFFEL, H. C. 2002. Late Carboniferous remagnetisation of Palaeozoic rocks in the NE Rhenish Massif, Germany. *Physics and Chemistry of the Earth*, **27**, 1179–1188.

ZWING, A., CLAUER, N., LIEWIG, N. & BACHTADSE, V. 2009. Identification of remagnetization processes in Paleozoic sedimentary rocks of the northeast Rhenish Massif in Germany by K-Ar dating and REE tracing of authigenic illite and Fe oxides. *Journal of Geophysical Research*, **114**, 1, B06104, doi: 10.1029/ 2008JB006137.

# Rock magnetism and identification of remanence components in the Marcellus Shale, Pennsylvania


E. B. MANNING* & R. D. ELMORE

*School of Geology and Geophysics, University of Oklahoma, Norman, OK 73019, USA*

*Corresponding author (e-mail: earlbmanning@gmail.com)*



**Abstract:** With increasing interest in the middle Devonian Marcellus Shale as a gas play in the Appalachians, a study of the rock magnetic characteristics and remanence components was undertaken. Samples were collected from outcrops of the Union Springs Formation and the underlying Onondaga Formation in a syncline immediately east of the Broadtop synclinorium in the Valley and Ridge province in Pennsylvania. The rocks contain an intermediate-temperature (IT) component with south-southeasterly declinations and shallow up inclinations that was removed by 310–350 °C, interpreted as a chemical remanent magnetization (CRM) that resides in pyrrhotite. At higher temperatures (350–480 °C) a component with more southerly declinations and shallow down inclinations, interpreted as a CRM in magnetite, is removed. Low-temperature treatments resulted in more stable decay during thermal demagnetization which allowed the IT component to be more easily identified. Cumulative log-Gaussian analysis of an isothermal remanent magnetization (IRM) and triaxial thermal decay of the IRM indicate the presence of two dominant minerals: pyrrhotite and magnetite. Low-temperature saturation IRM experiments show the 32 K Besnus transition and saturation of magnetization between 200 and 250 K, indicative of pyrrhotite. Some specimens showed the Verwey transition at 120 K, indicating magnetite.


Studies focusing on shales are gaining attention in both academia and industry because of the current and future importance of shale gas plays. Although numerous studies have focused on aspects of shales, relatively little palaeomagnetic and rock magnetic work has been performed on them because of their fissile nature and general poor exposure. As the first stage of an integrated palaeomagnetic and diagenetic study of the Marcellus Subgroup, a prominent gas play, a palaeomagnetic and rock magnetic study was undertaken to identify the magnetic components and characterize the phases present. Insight gained from this study will lay the foundation for the subsequent detailed regional integrated palaeomagnetic and diagenetic study of the Marcellus Subgroup.

Standard palaeomagnetic techniques were used to identify the magnetic components present and provide information on the magnetic mineralogy. Rock magnetic experiments have identified the magnetic minerals present. The rock magnetic part of this study is crucial to set the stage for the integrated palaeomagnetic and diagenetic study which will investigate the origin of the magnetization and the diagenetic history of the Marcellus Subgroup in the Valley and Ridge province.

## Geological setting

The Marcellus Subgroup is the basal subgroup (previously a formation) of the Hamilton Group of middle Devonian-age (Eifelian–Givetian) rocks found throughout the Appalachians. Immediately above the Marcellus Subgroup lies the Devonian Mahantango Formation, and the units below it vary across the basin. Underlying the shale in eastern Pennsylvania is the Devonian (Eifelian) Onondaga Formation or Selinsgrove Member of the Needmore Formation. Across the study area in the Pennsylvania Salient, the Marcellus Subgroup has been defined into the basal Union Springs Formation, middle Purcell or Cherry Valley limestone and the upper Oatka Creek Formation (Ver Straeten 2010).

The Appalachian basin formed as a result of the flexure of the lithosphere due to thrusting to the east (Beaumont 1981; Quinlan & Beaumont 1984) during the Taconic, Acadian and Alleghanian orogenies. During the Acadian Orogeny, the Avalon terrane collided with the North American Craton in an oblique convergence (Ettensohn 1985) between the middle Devonian and early Mississippian (Ettensohn 1985). The Marcellus Subgroup was deposited at the beginning of the second tectophase of the Acadian orogeny and represents the distal part of the Catskill Delta (Osberg et al. 1989). The sediments were sourced from the Acadian Mountains to the SE (Sevon 1985) resulting in the 10–250 m thickness (Ettensohn 2008). A tectonoeustatic control of deposition has long been recognized in the basin; Ettensohn (2008) suggest that second–fourth-order cycles reflect tectonism while fifth–sixth-order cycles reflect eustatic controls.


*From*: ELMORE, R. D., MUXWORTHY, A. R., ALDANA, M. M. & MENA, M. (eds) 2012. *Remagnetization and Chemical Alteration of Sedimentary Rocks*. Geological Society, London, Special Publications, **371**, 271–282.
First published online September 03, 2012, http://dx.doi.org/10.1144/SP371.9

Deformation of the Marcellus Subgroup occurred during the Alleghanian Orogeny (early Pennsylvanian–late Permian; Hatcher *et al.* 1989). This final deformational event of the Appalachians resulted in the development of three distinct structural provinces: the Plateau, the Valley and Ridge, and the Blue Ridge (Hatcher *et al.* 1989). The Valley and Ridge province is represented by numerous blind thrusts and millimetre- to tens of kilometre-scale folds (Hatcher *et al.* 1989). The Palaeozoic rocks in the Appalachians have been the subject of many palaeomagnetic studies (see references in McCabe & Elmore 1989; Elmore *et al.* 2006), yet few have focused on shales. In a study of anisotropy of magnetic susceptibility, Hirt *et al.* (1995) has identified both pyrrhotite and magnetite in Devonian shales in the Plateau province.

## Methods

The fissile nature of the Marcellus Subgroup required careful examination of sites before sampling such that the best lithologic targets for sampling were determined to be concretions, followed by carbonate-rich beds and then shales. One-inch diameter cores were drilled using a modified chain-saw gas-powered drill and oriented using a compass-inclinometer. Nine sites were sampled from opposing limbs of a syncline just to the east of the Broadtop

synclinorium located in Newton–Hamilton, PA within the Valley and Ridge province (Fig. 1). Two sites were sampled from the SW-dipping limb where the Union Springs Formation conformably overlies and transitions into the Onondaga Formation. This transition zone consists of interbedded limestone and shale. From the NW-dipping limb, also within the transition, four sites were collected from the Onondaga Formation, two from the Union Springs Formation and one from the Purcell limestone. No acceptable sites were located for sampling in the Oatka Creek Formation. Six to eleven specimens were collected per site with an average of nine specimens per site.

Upon returning to the lab, each specimen was cut to 2.2 cm in length and the natural remanent magnetization (NRM) was measured for all specimens using a 2G Enterprises cryogenic magnetometer with DC superconducting quantum interference devices (SQUIDS) located in a magnetically shielded room. Prior to demagnetization, those specimens selected for thermal demagnetization were subjected to up to two low-temperature demagnetization (LTD) treatments in liquid nitrogen and warmed in a zero field to remove the unstable magnetization held in multi-domain (MD) magnetite (Dunlop & Argyle 1991). Stepwise thermal demagnetization in 16–21 steps from 100 °C up to 580 °C was performed in an ASC Scientific Thermal Specimen Demagnetizer. Thermal demagnetization

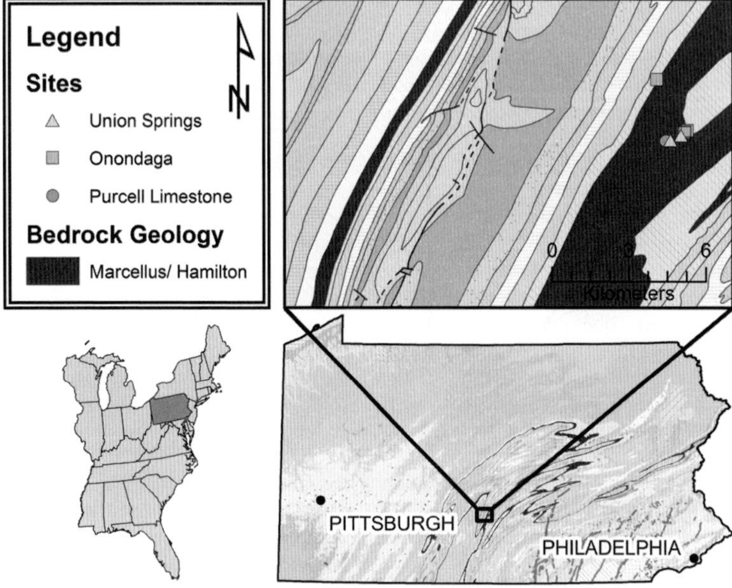



**Fig. 1.** Geological map of the study area with inset maps of Pennsylvania highlighting the mapped Marcellus and Hamilton units (Berg *et al.* 1980), and a map of eastern portion of the United States. Sites were collected from opposing limbs in the nose of a plunging syncline east of the Broadtop synclinorium.

continued until the magnetization was approximately zero or the growth of new magnetic minerals was indicated by an increase in magnetic intensity.

Thermal demagnetization data was analysed using the program Super-IAPD 2000. The data were displayed as vector orthogonal plots (Zijderveld 1967) and straight line segments were identified using principal component analysis (PCA) (Kirschvink 1980) to identify the characteristic remanent magnetizations (ChRMs). Only components with maximum angle of deviation (MAD) of less than 15° were accepted for further analysis. Mean directions for the sites were determined using Fisher (1953) statistics.

Up to two specimens from each site were selected for isothermal remanent magnetization (IRM) analysis. Specimens were first treated to a single step of alternating field (AF) demagnetization at 120 mT and the magnetization was measured. The specimens were subjected to an IRM at field strengths ranging from 10 to 2500 mT in 25 steps using an ASC Scientific Impulse Magnetizer. Immediately following each step, the IRM was recorded. Analysis of the IRM acquisition curves was then performed using the IRMUNMIX 2.2 software program (Heslop *et al.* 2002) followed by cumulative log-Gaussian (CLG) analysis with the IRM-CLG 1.0 spreadsheet (Kruiver *et al.* 2001) to model individual coercivity contributions (Heslop *et al.* 2004).

Thermal decay of a triaxial IRM was examined for at least one specimen per site. Specimens were treated to a single step of AF demagnetization at 120 mT and the magnetization was measured. To obtain and evaluate decay curves of the triaxial IRM for these specimens, a 120, 500 and 2500 mT magnetic field was imparted in the $X$, $Y$ and $Z$ axes (Lowrie 1990) respectively using an ASC Scientific Impulse Magnetizer. Each specimen was then subjected to stepwise thermal demagnetization up to 680 °C.

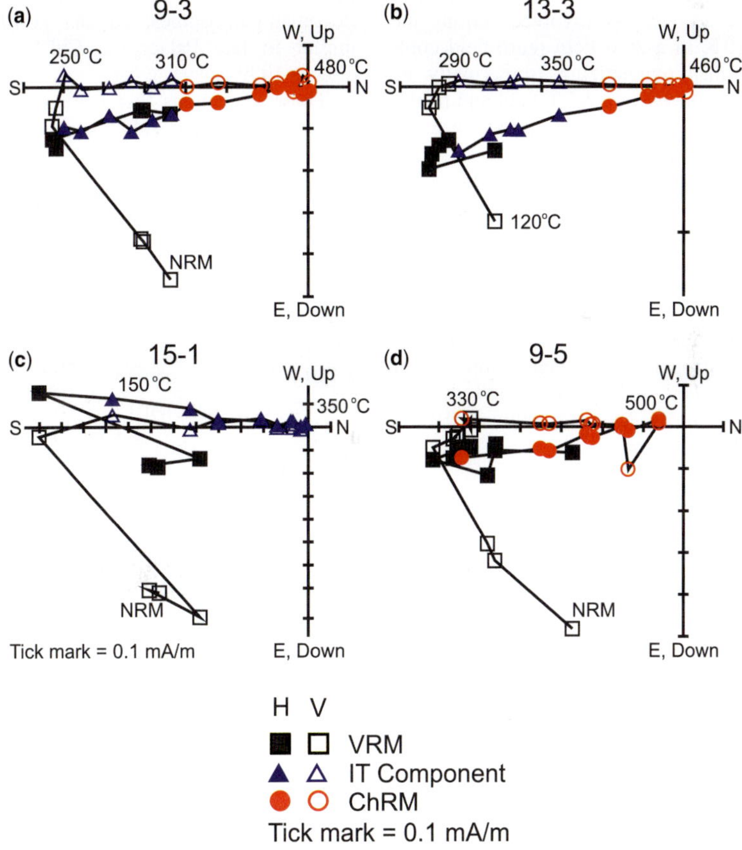


**Fig. 2.** Vector orthogonal plots (Zijderveld 1967) for representative specimens. (**a**) Three-component specimen with two low-temperature treatments. (**b**) Three-component specimen with two low-temperature treatments. The NRM has been removed to better illustrate the components. (**c**) Two-component specimen with two low-temperature treatments. (**d**) Two-component specimen with no low-temperature treatments. H, horizontal; v, vertical.

Low-temperature experiments were conducted for eight specimens using the Quantum Design Magnetic Property Measurement System (MPMS) cryogenic magnetometer at the Institute for Rock Magnetism, University of Minnesota. All specimens were imparted with a saturation IRM (SIRM) of 2.5 T at room temperature (RTSIRM). Specimens were then cooled to 10 K and warmed back to room temperature (300 K) with measurements of the magnetization taken every 5 K during the cooling and warming processes. These experiments resulted in the ability to establish the Morin, Verwey and Besnus transitions which indicate haematite, magnetite and pyrrhotite, respectively. Another set of low-temperature experiments was performed on three specimens to further test the presence of the Verwey transition. The first part of the experiment involved cooling the specimen to 10 K in an applied 2.5 T field (field cooling or FC), turning off the field and measuring the resulting remanence while warming the specimen to room temperature (300 K) in a zero field, with measurements taken every 5 K. The second part involved cooling the specimen to 10 K in a zero field (zero-field cooling or ZFC), then magnetizing isothermally by application and removal of a 2.5 T field and finally measuring the remanence while warming the specimen to 300 K still in a zero field.

Using a Princeton Measurements vibrating sample magnetometer at the Institute for Rock Magnetism, hysteresis loops were created. Samples were used from small ($<1$ g) chips from at least one specimen per site. The measurements were conducted using a 0.1–0.5 s averaging time and 5 mT steps up to 1 T. Corrections were made to the hysteresis loops to remove the paramagnetic component before further correction using both a hyperbolic and double-logistic functions filter (Jackson & Sølheid 2010) to determine the best-fit loop.

## Palaeomagnetic results and interpretations

The LTD resulted in a decrease of the NRM intensity for most specimens. In specimens with weak and/or unstable remanence where only one treatment was performed, there was a decrease in the NRM intensity ranging from 19–35% while in the stable remanence-carrying specimens a 3–24% decrease occurred. In weak and/or unstable remanence-carrying specimens where two treatments were performed, there was a decrease in the NRM intensity from the original NRM ranging from 9 to $-23\%$ while in the stable remanence-carrying specimens there was a 0–29% decrease in the NRM intensity. The decreases in remanence suggest that a significant portion (up to 35%) of the remanence is carried in MD magnetite (Dunlop & Argyle 1991).

In most samples, PCA of the stepwise thermal demagnetization data indicates a northerly and steep down component with maximum unblocking temperatures up to 275 °C that is interpreted as a viscous remanent magnetization (VRM) (Fig. 2a–d). Many of the weak and/or unstable remanence-carrying specimens do not contain this VRM. Those samples with a stable magnetization contain either one or two components at higher temperatures. Specimens identified with two additional components have an intermediate-temperature (IT) component with south-southeasterly declinations and shallow inclinations (Fig. 2a, b) that is removed between 250–290 °C and 310–350 °C. The other component is a higher-temperature ChRM that has more southerly declinations and shallow inclinations with a maximum unblocking temperature of 480 °C (Fig. 2a). Some specimens had either the ChRM or the IT component in addition to the VRM (Fig. 2c). Both higher-temperature components have directions that are inconsistent with Devonian-aged magnetizations (and corresponding Devonian palaeopoles; Van der Voo 1993) and are similar to late Palaeozoic ChRMs (McCabe & Elmore 1989).

Those specimens that were subjected to LTD also displayed more stable decay on the orthogonal

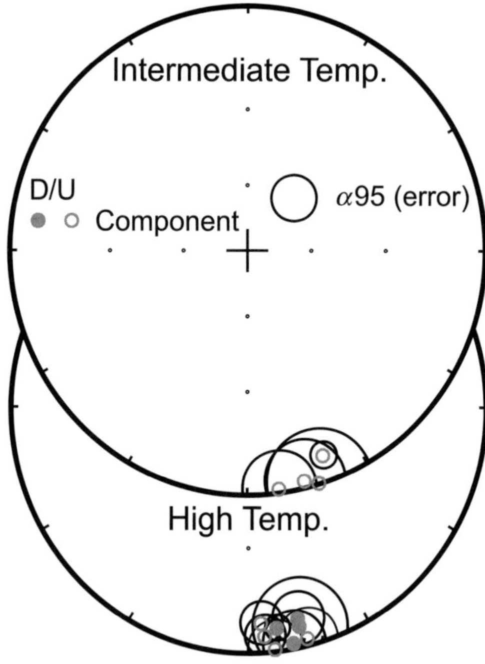


**Fig. 3.** Equal area projections showing site means with α95 cones of confidence for the intermediate- and high-temperature (ChRM) components in geographic coordinates.

**Table 1.** *Site statistics for components. $N/N_0$: number of specimens with direction v. number of specimens demagnetized; Dec, declination in degrees; Inc, inclination in degrees; k, precision parameter; $\alpha95$, cone of 95% confidence in degrees. All directions are in geographic coordinates*

| Comp | Lat/Long | Strike/Dip | Statistics | | | Geographic | |
|---|---|---|---|---|---|---|---|
| | | | $N/N_0$ | $k$ | $\alpha95$ | Dec | Inc |
| *IT* | | | | | | | |
| 9 | 40.38407/−77.84946 | 232/13NW | 4/10 | 53.85 | 12.6 | 162.2 | −0.3 |
| 10 | 40.38307/−77.85094 | 240/11NW | 8/10 | 31.97 | 9.9 | 165.5 | −1.8 |
| 11 | 40.38144/−77.85464 | 230/11NW | 6/11 | 359.6 | 3.5 | 159.0 | −6.5 |
| 13 | 40.38459/−77.8487 | 222/12NW | 7/9 | 45.07 | 9.1 | 172.2 | −1.5 |
| 15* | 40.38375/−77.84981 | 220/12NW | 2/8 | 23.22 | 19.5 | 173.9 | 4.4 |
| 16* | 40.40199/−77.85948 | 36/28SE | 2/7 | 29.76 | 47.5 | 159.4 | 11.1 |
| 17* | 40.40199/−77.85948 | 36/28SE | 1/6 | − | − | 177.5 | 15.6 |
| Mean | | | 4/7 | 167.6 | 7.1 | 164.7 | −2.5 |
| *ChRM* | | | | | | | |
| 9 | 40.38407/−77.84946 | 232/13NW | 9/10 | 97.73 | 5.2 | 175.6 | −3.7 |
| 10 | 40.38307/−77.85094 | 240/11NW | 3/10 | 472.4 | 5.7 | 176.4 | −7.1 |
| 11 | 40.38144/−77.85464 | 230/11NW | 4/11 | 58.05 | 12.2 | 166.3 | 4.8 |
| 12 | 40.38459/−77.84870 | 222/12NW | 3/5 | 255.1 | 7.7 | 168.6 | 1.0 |
| 13 | 40.38459/−77.84870 | 222/12NW | 7/9 | 63.84 | 7.6 | 173.2 | −0.3 |
| 15 | 40.38375/−77.84981 | 220/12NW | 4/8 | 215.2 | 4.6 | 165.0 | −1.4 |
| 16 | 40.40199/−77.85948 | 36/28SE | 7/7 | 339.6 | 3.3 | 172.2 | 5.4 |
| 17 | 40.40199/−77.85948 | 36/28SE | 6/6 | 93.1 | 7.0 | 166.3 | 6.9 |
| Mean | | | 8/9 | 152 | 4.5 | 170.4 | 0.7 |

*Site not used in analysis because $\alpha95 > 15°$ or low specimen numbers. No IT component was isolated from site 12.

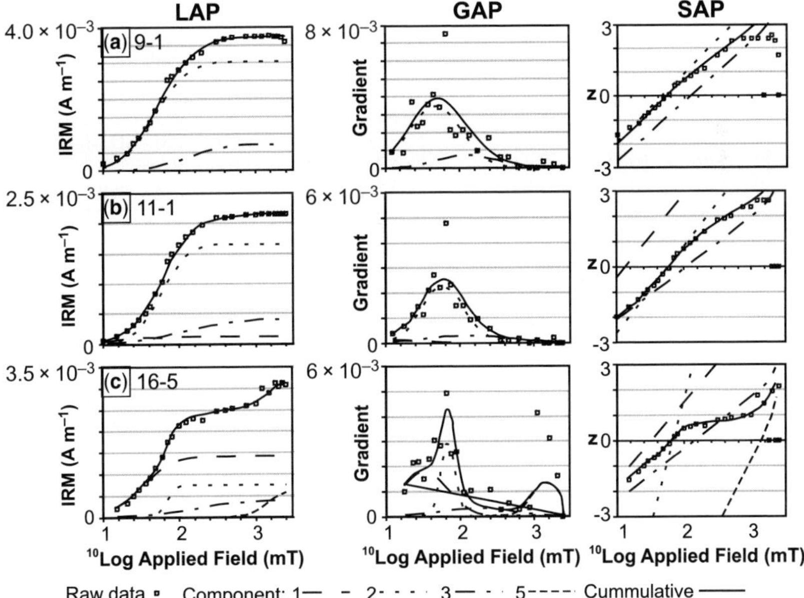


**Fig. 4.** Logarithmic acquisition plot (LAP), Gaussian acquisition plot (GAP), and standardized acquisition plot for representative specimens. Component number is consistent with text. (**a**) Specimen containing 2 components (SD/PSD magnetite and pyrrhotite). (**b**) Specimen containing 3 components (SD/PSD magnetite; MD magnetite or maghaematite; and pyrrhotite). (**c**) Specimen containing 4 components (SD/PSD magnetite; MD magnetite or maghaematite; pyrrhotite; and haematite or goethite).

plots than those that did not have the treatment. The generally higher percentage of the magnetization decay during LTD in specimens with weak NRM intensities suggests that these samples may have a higher percentage of MD magnetite. Without LTD it is difficult to identify the IT component (Fig. 2d) because the VRM partially overlaps the IT component. The ability to identify the IT component illustrates the value of the removal of the remanence carried in MD magnetite through LTD.

A subtle difference can be seen between the declinations (Figs 2a, b & 3; Table 1) and inclinations (Fig. 3; Table 1) of the IT and ChRM. The ChRM has southerly declinations and shallow down inclinations whereas the IT component has more southeasterly declinations and shallow up inclinations. The IT component is interpreted to reside in pyrrhotite ($Fe_7S_8$) because of the 310–350 °C maximum unblocking temperature, whereas the ChRM with a maximum unblocking temperature of 480 °C likely resides in magnetite ($Fe_3O_4$). The identification of the IT component in specimens is strengthened by results from vector-difference sums (e.g. Gee *et al.* 1993). A break in slope occurs in many samples between 300 °C and 350 °C which is consistent with an IT component residing in pyrrhotite.

## Rock magnetic results

CLG analysis (Kruiver *et al.* 2001) of the IRM was performed to identify the magnetic components in individual specimens. Specimens contained

**Table 2.** *Results for specimens selected for cumulative log-Gaussian analysis. Comp: the order the components are listed in text and displayed in Figure 4 (although component 4 is not shown in the figure); $B_{(1/2)}$, field where half of the SIRM is reached; DP, dispersion parameter is one standard deviation of the logarithmic distribution*

| Spec. | Comp. | SIRM ($A\ m^{-1}$) | $Log(B_{1/2})$ (mT) | $B_{1/2}$ (mT) | DP | Contr. (%) | Magnetic mineralogy |
|---|---|---|---|---|---|---|---|
| 9-1 | 2 | $3.03 \times 10^{-3}$ | 1.68 | 47.86 | 0.35 | 80.84 | PSD-SD Magnetite |
|  | 3 | $7.18 \times 10^{-4}$ | 2.10 | 125.89 | 0.40 | 19.16 | Pyrrhotite |
| 10-1 | 1 | $3.73 \times 10^{-4}$ | 1.23 | 16.98 | 0.20 | 11.76 | Maghaemite or MD Magnetite |
|  | 2 | $1.49 \times 10^{-3}$ | 1.74 | 54.95 | 0.18 | 46.96 | PSD-SD Magnetite |
|  | 3 | $1.31 \times 10^{-3}$ | 2.06 | 114.82 | 0.34 | 41.29 | Pyrrhotite |
| 11-1 | 1 | $1.07 \times 10^{-4}$ | 1.14 | 13.80 | 0.30 | 4.96 | Maghaemite or MD Magnetite |
|  | 2 | $1.65 \times 10^{-3}$ | 1.77 | 58.88 | 0.29 | 76.50 | PSD-SD Magnetite |
|  | 3 | $4.00 \times 10^{-4}$ | 2.05 | 112.20 | 0.50 | 18.54 | Pyrrhotite |
| 12-3 | 1 | $1.48 \times 10^{-4}$ | 0.98 | 9.55 | 0.08 | 6.42 | Maghaemite or MD Magnetite |
|  | 2 | $1.38 \times 10^{-3}$ | 1.70 | 50.12 | 0.30 | 59.90 | PSD-SD Magnetite |
|  | 3 | $7.76 \times 10^{-4}$ | 2.11 | 128.82 | 0.46 | 33.68 | Pyrrhotite |
| 13-1 | 1 | $1.41 \times 10^{-4}$ | 1.03 | 10.72 | 0.21 | 3.76 | Maghaemite or MD Magnetite |
|  | 2 | $2.37 \times 10^{-3}$ | 1.68 | 47.86 | 0.38 | 63.18 | PSD-SD Magnetite |
|  | 3 | $1.24 \times 10^{-3}$ | 2.17 | 147.91 | 0.45 | 33.06 | Pyrrhotite |
| 14-9 | 2 | $5.00 \times 10^{-4}$ | 1.71 | 51.29 | 0.32 | 70.32 | PSD-SD Magnetite |
|  | 3 | $1.70 \times 10^{-4}$ | 1.96 | 91.20 | 0.34 | 23.90 | Pyrrhotite |
|  | 4 | $4.11 \times 10^{-5}$ | 2.92 | 831.76 | 0.26 | 5.78 | Haematite |
| 15-2 | 2 | $2.66 \times 10^{-3}$ | 1.68 | 47.86 | 0.37 | 79.40 | PSD-SD Magnetite |
|  | 3 | $6.90 \times 10^{-4}$ | 2.14 | 138.04 | 0.42 | 20.60 | Pyrrhotite |
| 15-3 | 1 | $8.20 \times 10^{-5}$ | 1.09 | 12.30 | 0.25 | 3.19 | Maghaemite or MD Magnetite |
|  | 2 | $1.37 \times 10^{-3}$ | 1.67 | 46.77 | 0.30 | 53.27 | PSD-SD Magnetite |
|  | 3 | $1.12 \times 10^{-3}$ | 2.13 | 134.90 | 0.41 | 43.55 | Pyrrhotite |
| 16-1 | 1 | $5.20 \times 10^{-4}$ | 1.23 | 16.98 | 0.44 | 14.99 | Maghaemite or MD Magnetite |
|  | 2 | $1.00 \times 10^{-3}$ | 1.69 | 48.98 | 0.20 | 28.82 | PSD-SD Magnetite |
|  | 3 | $1.50 \times 10^{-4}$ | 2.15 | 141.25 | 0.30 | 4.32 | Pyrrhotite |
|  | 5 | $1.80 \times 10^{-3}$ | 3.35 | 2238.72 | 0.35 | 51.87 | Haematite or Goethite |
| 16-5 | 1 | $1.43 \times 10^{-3}$ | 1.52 | 33.11 | 0.32 | 44.30 | Maghaemite or MD Magnetite |
|  | 2 | $7.44 \times 10^{-4}$ | 1.84 | 69.18 | 0.10 | 23.05 | PSD-SD Magnetite |
|  | 3 | $3.90 \times 10^{-4}$ | 2.20 | 158.49 | 0.50 | 12.08 | Pyrrhotite |
|  | 5 | $6.64 \times 10^{-4}$ | 3.17 | 1479.11 | 0.20 | 20.57 | Haematite or Goethite |
| 17-4 | 1 | $8.54 \times 10^{-4}$ | 1.36 | 22.91 | 0.28 | 25.85 | Maghaemite or MD Magnetite |
|  | 2 | $1.03 \times 10^{-3}$ | 1.73 | 53.70 | 0.16 | 31.17 | PSD-SD Magnetite |
|  | 3 | $6.20 \times 10^{-4}$ | 2.02 | 104.71 | 0.27 | 18.77 | Pyrrhotite |
|  | 5 | $8.00 \times 10^{-4}$ | 3.43 | 2691.53 | 0.21 | 24.21 | Haematite or Goethite |

between two and four components, with most specimens containing three (Fig. 4a–c; Table 2). The lowest-coercivity component (9–33 mT) carries between 3% and 44% of the total IRM (Table 2). This component is interpreted to be maghaemite or MD magnetite based on the coercivity ranges identified by Peters & Dekkers (2003). MD magnetite was identified by LTD. An intermediate-coercivity component with values ranging over 46–69 mT contributes 28–80% of the total IRM (Fig. 4a–c; Table 2), and is interpreted to be

magnetite (Peters & Dekkers 2003). A separate intermediate-coercivity component (91–158 mT) contributes 4–41% of the total IRM in specimens (Fig. 4a–c) and is interpreted to be pyrrhotite (Peters & Dekkers 2003). One specimen contained a higher-coercivity component (832 mT) which contributes 6% of the total IRM, and is interpreted to be haematite. A few specimens contained a very high-coercivity (1479–2691 mT) component, which contributes 20–52% of the total IRM (Fig. 4c). According to the range established by Peters &

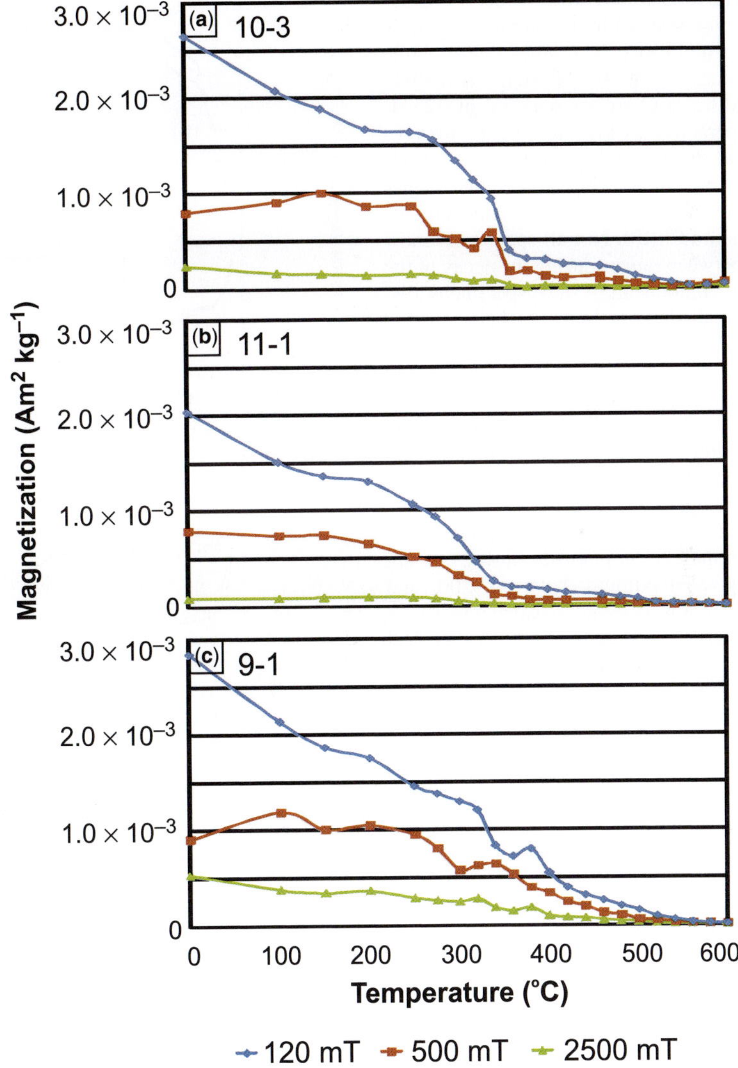


**Fig. 5.** Representative curves for thermal decay of a triaxial IRM. (**a**) Specimen 10-3 exhibits a decrease in intensity in the 120 mT component at *c.* 340 °C followed by continued decay. (**b**) Specimen 11-1 exhibits a decrease at *c.* 340 °C followed by continued decay. (**c**) Specimen 9-1 contains a minor loss of intensity at *c.* 340 °C followed by continued decay.

Dekkers (2003), this may be goethite. Most specimens examined using CLG analysis show evidence for components 2 (PSD-SD magnetite) and 3 (pyrrhotite) with many specimens containing component 1 (maghaemite or MD-magnetite).

Stepwise thermal demagnetization of the triaxial IRM was used to further differentiate the contribution of the minerals with different coercivities. The low-coercivity IRM component has the highest intensities in all specimens examined. Most specimens show steep decay of this component from the NRM to 340–360 °C followed by an inflection to a less steep decay that continues to 480–560 °C (Fig. 5a, b). The temperature at the inflection point (340 °C) is consistent with the unblocking temperature of pyrrhotite (Rochette *et al.* 1990) while the complete decay up to 560 °C suggests magnetite. Pyrrhotite has a Curie temperature of 325 °C although temperatures as high as 350 °C are possible (Rochette *et al.* 1990). Interestingly, specimens that have two unblocking-temperature ranges commonly come from sites that contain both the IT component and ChRM. Not all specimens show a clear break between ranges during the decay of the low-coercivity component. A few specimens display only a small drop in intensity between 320 and 340 °C with continued decay to 560 °C (Fig. 5c) which indicates the presence of pyrrhotite and magnetite, respectively (Rochette *et al.* 1990). When PCA is performed following stepwise thermal demagnetization on specimens from sites that display this behaviour, the IT component is rarely identifiable although the ChRM is.

Many specimens selected for RTSIRM experiments show an increase in intensity during cooling until reaching a maximum between 200 and 250 K. This is followed by decreasing remanence and a sharp decrease in intensity between 40 and 25 K (Fig. 6a, b). The decrease in remanence during cooling that occurs around 30 K (Fig. 6a) may indicate the Besnus transition which occurs at 32 K in pyrrhotite (Besnus & Meyer 1964; Dekkers *et al.* 1989; Rochette *et al.* 1990, 2011). This decrease in remanence has been noted in claystones and has been attributed to the presence of pyrrhotite (Aubourg & Pozzi 2010) as well as rhodocrosite and siderite (Kars *et al.* 2011). A maximum remanence between 200 and 250 K, along with the Besnus transition, is indicative of pyrrhotite (Rochette *et al.* 1990). However, this maximum remanence can also indicate the presence of oxidized magnetite (Özdemir & Dunlop 2010). Some specimens show a less well-defined Besnus transition but also contain a sharp increase in magnetic intensity from 10–25 K during cooling (Fig. 6b–d), which occurs due to a paramagnetic component induced by the imperfect zero field of the MPMS (Bowles *et al.* 2009). The nearly full

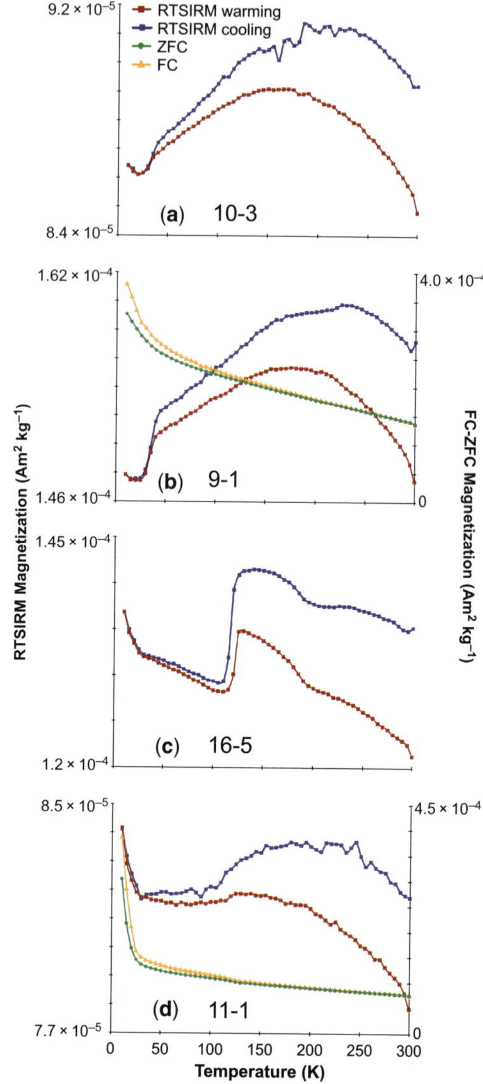



**Fig. 6.** Low-temperature SIRM results from representative specimens. (**a**) RTSIRM results during cooling indicate a maximum remanence between 200 and 250 K and a decrease around the 32 K (Besnus) transition followed by nearly full recovery of the warming curve at the Besnus transition. (**b**) RTSIRM results during cooling show a large decrease in intensity at the Besnus transition. FC-ZFC results do not indicate the Verwey transition at 120 K. (**c**) RTSIRM results during cooling show drop in intensity at *c.* 200 K (Morin transition) followed by a sharp Verwey transition and a small decrease near the Besnus transition. (**d**) FC-ZFC results show a large decrease in intensity between 10–40 K and a small decrease near the Verwey transition with continued decrease to 300 K. RTSIRM cooling results show the Verwey transition and a small decrease in intensity near the Besnus transition.

recovery of the magnetic intensity between the cooling and warming curves around the Besnus transition indicates a pyrrhotite grain size between 1 and 2 μm (Dekkers *et al.* 1989).

During cooling, a number of specimens also show a decrease in remanence between 100 and 150 K (Fig. 6b, d) with a few specimens showing a very sharp decrease at *c.* 120 K (Fig. 6c). This decrease in remanence, along with the increase while warming, indicates the Verwey transition which occurs at 120 K and is diagnostic of magnetite (Verwey 1939; Özdemir *et al.* 1993). The few specimens with the sharp Verwey transition also exhibit a decrease in intensity while cooling and an increase during warming at *c.* 200 K (Fig. 6c) which indicates the Morin transition in haematite (Morin 1950; De Boer *et al.* 2001). Furthermore, the increasing remanence during cooling and subsequent decreasing remanence during warming before and after these transitions in some specimens are likely due to a temperature-invariant phase such as goethite (pers. comm. Mike Jackson 2011), which has nearly complete reversibility of increasing remanence (Rochette & Fillion 1989).

Specimens that had a SIRM acquired at low temperatures show a considerable decrease in the intensity during warming between 10 and 300 K for both the ZFC and FC curves (Fig. 6b, d). Two specimens (e.g. Fig. 6d) show a nearly 80% loss of intensity between 10 and 40 K. The loss of intensity throughout the entire run may be partially attributed to a superparamagnetic (SP) magnetite component within the specimens while the steep drop at low temperatures (10–40 K) may be related to the oxidation of magnetite (Özdemir *et al.* 1993). Specimen 11–1 (Fig. 6d) also shows a suppressed Verwey transition (Özdemir *et al.* 1993) as noted by a decrease of intensity at *c.* 120 K. The suppression of Verwey transition in this and other specimens may be related to the oxidation of magnetite (Özdemir *et al.* 1993).

Generally, hysteresis loops did not show saturation due to the contribution of a paramagnetic component. The removal of this component results in fairly noisy loops that were subsequently filtered using a hyperbolic and double-logistic functions filter (Jackson & Sølheid 2010) and indicate saturation between 300 and 400 mT for all loops (Fig. 7). The curves for most samples are wasp-waisted, which indicates a mixture of minerals and/or coercivities (Roberts *et al.* 1995).

## Discussion

Examination of the vector orthogonal plots reveals that the sites contained up to two ancient components. An IT component with south-southeasterly

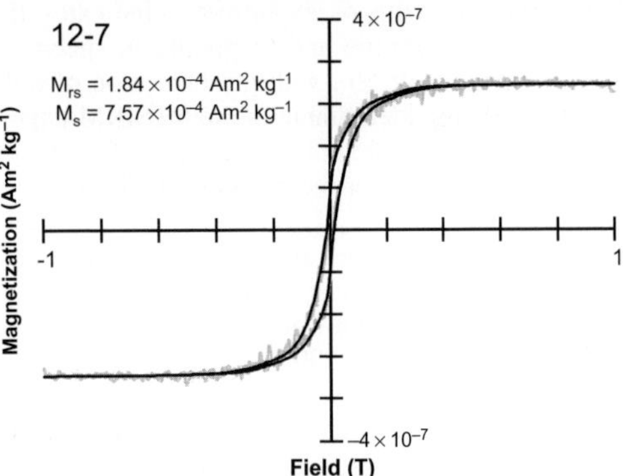



**Fig. 7.** Representative wasp-waisted hysteresis loop indicating a range of grain sizes and/or mixed magnetic mineralogy of magnetite and pyrrhotite (Roberts *et al.* 1995). Grey line represents the raw data and the black line is the filtered data.

declinations and shallow up inclinations has a maximum unblocking temperature of 350 °C. Based on interpreted burial temperatures (Epstein *et al.* 1977) from Conodont Alteration Index (CAI) and vitrinite reflectance data (Repetski *et al.* 2008), maximum unblocking temperatures and time v. temperature curves for pyrrhotite (Dunlop *et al.* 2000), this component is interpreted to be a chemical remanent magnetization (CRM) that resides in pyrrhotite. The prevalent ChRM resides in magnetite and has a shallow down inclination and more southerly declinations than the IT CRM. Based on the ChRM maximum unblocking temperature of 480 °C, the low burial temperatures and the time v. temperature curves for magnetite (Pullaiah *et al.* 1975; Dunlop *et al.* 2000), it is interpreted to be a CRM. The LTD removed up to 23% of the NRM in specimens with stable components, which suggests that the ChRM resides in pseudo-single-domain (PSD) or single-domain (SD) magnetite. It was not possible to identify the pyrrhotite CRM without LTD; this illustrates the value of low-temperature treatments in general, particularly in shales and other lithologies with pyrrhotite.

There is excellent agreement between the results in terms of the magnetic mineralogy from the different rock magnetic techniques. For example, specimen 11–1 contains pyrrhotite and magnetite as clearly indicated by CLG analysis (Fig. 4b), triaxial decay of the IRM (Fig. 5b) and RTSIRM and ZFC/FC experiments (Fig. 6d). In particular, the presence of pyrrhotite is indicated by the 32 K Besnus transition and a maximum remanence between 200 and 250 K in the RTSIRM experiments. Thermal demagnetization showing unblocking temperatures between 300 and 350 °C also suggests the presence of pyrrhotite and not siderite or rhodocrosite.

Similarly, the suite of experiments indicates the presence of pyrrhotite and magnetite in specimen 9-1 (Figs 4a, 5c & 6b), with the exception of only pyrrhotite being identifiable in the RTSIRM experiment. The nearly complete recovery of intensity in the RTSIRM experiments between cooling and warming suggest a *c.* 1 μm SD pyrrhotite grain size. The wasp-waisted hysteresis loops are also consistent with a mixture of magnetite and pyrrhotite, a variety of grain sizes or a combination of both (Roberts *et al.* 1995). The rock magnetic results indicate that pyrrhotite and magnetite are present and are consistent with the presence of components residing in both minerals, as identified in the demagnetization results.

A low-coercivity component (13–33 mT) identified in many specimens during CLG analysis is tentatively interpreted to be either maghaemite or MD magnetite. The loss of magnetic intensity in specimens subjected to LTD is consistent with MD magnetite. The suppressed Verwey transition and the significant loss of magnetization between 10 and 40 K documented in the ZFC curves may be due to oxidation of magnetite (Özdemir *et al.* 1993). This oxidation could also have produced maghaemite, which may be the low-coercivity component identified by CLG analysis. Some specimens also display the Morin transition during RTSIRM cooling experiments, which may also be due to the oxidation of magnetite. The haematite does not carry a component based on the demagnetization data and is interpreted to have formed during modern weathering. It is possible that maghaemite could carry the ChRM, but this is unlikely if it formed due to the modern weathering of magnetite.

Pyrrhotite is now recognized as an important mineral that can carry a secondary magnetization in a number of different rock types (e.g. Dekkers *et al.* 1989; Rochette *et al.* 1990; Jackson *et al.* 1993; Xu *et al.* 1998; Weaver *et al.* 2002; Crouzet *et al.* 2003; Gillett & Karlin 2004; Font *et al.* 2006; Preeden *et al.* 2008). Jackson *et al.* (1993) have suggested that the Cenozoic thermoviscous remanent magnetization (TVRM) widely reported throughout the Appalachian Fold and Thrust Belt may in fact be a CRM carried in pyrrhotite. Hirt *et al.* (1995) also report that pyrrhotite and magnetite are present in Devonian shales.

There are several chemical mechanisms that could explain the CRMs that reside in pyrrhotite and magnetite in the Marcellus Subgroup. These mechanisms include: pyrrhotite authigenesis due to pre-existing magnetite reacting with pyrite as a result of burial metamorphism under reducing conditions (Gillett 2003); oxidation of pre-existing pyrite (Salmon *et al.* 1988); thermochemical sulphate reduction (Pierce *et al.* 1998); hydrocarbon migration (Machel & Burton 1991); diagenesis associated with gas hydrates (Housen & Musgrave 1996; Larrasoaña *et al.* 2007); and release of trapped pore fluid (Urbat *et al.* 2000). CRMs residing in magnetite can be explained by burial diagenetic mechanisms (Banerjee *et al.* 1997; Katz *et al.* 2000; Blumstein *et al.* 2004) or by fluid migration events (Elmore *et al.* 1999, 2001). It is beyond the scope of this paper to determine the origin of the pyrrhotite and magnetite CRMs in the Marcellus Subgroup. This issue will be addressed in an integrated diagenetic and palaeomagnetic study of the Marcellus Subgroup.

## Conclusions

Palaeomagnetic analysis indicates that the Marcellus Subgroup contains two CRMs with similar late Palaeozoic directions that reside in pyrrhotite and magnetite. LTD was found to be a critical step which enables the identification of the CRM in pyrrhotite. Rock magnetic studies including CLG analysis, thermal decay of triaxial IRMs and low-temperature experiments confirm the presence of both pyrrhotite and magnetite. In particular, RTSIRM experiments show the 32 K Besnus transition and a maximum remanent magnetization between 200 and 250 K which are indicative of pyrrhotite. All of the experiments performed show excellent agreement in the interpreted magnetic mineralogy.

The authors would like to thank Devon Energy for providing the funding for this project. We would also like to thank the Institute for Rock Magnetism for providing funds and access to perform low-temperature and hysteresis experiments and M. Jackson for his assistance. Thank you to the reviewers, M. Jackson and one anonymous reviewer, for their excellent contributions that improved the manuscript. Finally, thank you to J. Pannalal, S. Anzaldua and V. Harvey for their help in collecting and processing samples in the lab.

## References

AUBOURG, C. & POZZI, J.-P. 2010. Toward a new 250 °C pyrrhotite-magnetite geothermometer for claystones. *Earth and Planetary Science Letters*, **294**, 47–57.

BANERJEE, S., ELMORE, R. D. & ENGEL, M. H. 1997. Chemical remagnetization and burial diagenesis; testing the hypothesis in the Pennsylvanian Belden Formation, Colorado. *Journal of Geophysical Research*, **102**, 24825–24842.

BEAUMONT, C. 1981. Foreland basins. *Geophysical Journal of the Royal Astronomical Society*, **65**, 291–329.

BERG, T. M., EDMUNDS, W. E., GEYER, A. R. *ET AL.* 1980. *Geologic map of Pennsylvania.* 2nd ed. Pennsylvania Geological Survey, 4th ser., Map 1, scale 1:250 000, 3 sheets [web release].

BESNUS, M. J. & MEYER, A. J 1964. Nouvelles données expérimentales sur le magnétisme de la pyrrhotine naturelle. *Proceedings of the International Conference on Magnetism*, Nottingham, 507–511.

BLUMSTEIN, A. M., ELMORE, R. D., ENGEL, M. H., ELLIOT, C. & BASU, A. 2004. Paleomagnetic dating of burial diagenesis in Mississippian carbonates, Utah. *Journal of Geophysical Research*, **109**, B04101, doi: 10.1029/2003JB002698.

BOWLES, J., JACKSON, M., CHEN, A. & SOLHEID, P. 2009. Interpretation of low-temperature data part 1: superparamagnetism and paramagnetism. *The IRM Quarterly*, **19**, 1,7–11.

CROUZET, C., GAUTAM, P., SCHILL, E. & APPEL, E. 2003. Multicomponent magnetization in western Dolpo (Tethyan Himalaya, Nepal); tectonic implications. *Tectonophysics*, **377**, 179–196.

DE BOER, C. B., MULLENDER, T. A. T. & DEKKERS, M. J. 2001. Low-temperature behaviour of haematite: susceptibility and magnetization increase on cycling through the Morin transition. *Geophysical Journal International*, **146**, 201–216.

DEKKERS, M. J., MATTEI, J. L., FILLION, G. & ROCHETTE, P. 1989. Grain-size dependence of the magnetic behavior of pyrrhotite during its low-temperature transition at 34 K. *Geophysical Research Letters*, **16**, 855–858.

DUNLOP, D. J. & ARGYLE, K. S. 1991. Separating multidomain and single-domain-like remanences in pseudo-single-domain magnetites (215–540 nm) by low-temperature demagnetization. *Journal of Geophysical Research*, **96**, 2007–2017.

DUNLOP, D. J., ÖZDEMIR, Ö., CLARK, D. A. & SCHMIDT, P. W. 2000. Time-temperature relations for the remagnetization of pyrrhotite (Fe7S8) and their use in estimating paleotemperatures. *Earth and Planetary Science Letters*, **176**, 107–116.

ELMORE, R. D., BANERJEE, S., CAMPBELL, T. & BIXLER, G. 1999. Paleomagnetic dating of ancient fluid-flow events and paleoplumbing in the Arbuckle Mountains, Southern Oklahoma. *In*: PARNELL, J. (ed.) *Dating and Duration of Fluid Flow Events and Rock-Fluid Interaction*. Geological Society, London, Special Publications, **144**, 9–25.

ELMORE, R. D., KELLEY, J., EVANS, M. & LEWCHUK, M. 2001. Remagnetization and orogenic fluids: testing the hypothesis in the central Appalachians. *Geological Journal International*, **144**, 568–576.

ELMORE, R. D., FOUCHER, J. L.-E., EVANS, M., LEWCHUK, M. & COX, E. 2006. Remagnetization of the Tonoloway Formation and the Helderberg Group in the Central Appalachians; testing the origin of syntilting magnetizations. *Geophysical Journal International*, **166**, 1062–1076.

EPSTEIN, A. G., EPSTEIN, J. B. & HARRIS, L. D. 1977. Conodont color alteration—an index to organic metamorphism. *Geological Survey Professional Paper*, **995**, 1–27.

ETTENSOHN, F. R. 1985. The Catskill Delta complex and the Acadian Orogeny: a model. *In*: WOODROW, D. L. & SEVON, W. D. (eds) *The Catskill Delta*. Geological Society of America, Boulder, CO, Special Paper, **201**, 39–49.

ETTENSOHN, F. R. 2008. Chapter 4 the appalachian foreland basin in Eastern United States. *In*: MIALL, A. D. (ed.) *Sedimentary Basins of the World*. Elsevier, Amsterdam, **5**, 105–179.

FISHER, R. A. 1953. Dispersion on a sphere. *Proceedings of the Royal Society of London, Series A*, **217**, 295–305.

FONT, E., TRINDADE, R. I. F. & NEDELEC, A. 2006. Remagnetization in bituminous limestones of the Neoproterozoic Araras Group (Amazon craton): hydrocarbon maturation, burial diagenesis, or both? *Journal of Geophysical Research*, **111**, 17, B06204, doi: 10.1029/2005JB004106.

GEE, J., STAUDIGEL, H., TAUXE, L. & PICK, T. 1993. Magnetization of the La Palma Seamount Series: implications for Seamount Paleopoles. *Journal of Geophysical Research*, **98**, 11742–11767.

GILLETT, S. L. 2003. Paleomagnetism of the Notch Peak contact-metamorphic aureole, revisited: pyrrhotite from magnetite + pyrite under submetamorphic conditions. *Journal of Geophysical Research*, **108**, 2446, doi: 10.1029/2002JB002386.

GILLETT, S. L. & KARLIN, R. E. 2004. Pervasive late Paleozoic–Triassic remagnetization of the miogeoclinal carbonate racks in the Basin and Range and vicinity, SW USA: regional results and possible tectonic implications. *Physics of the Earth and Planetary Interiors*, **141**, 95–120.

HATCHER, R. D. JR, THOMAS, W. A., GEISER, P. A., SNOKE, A. W., MOSHER, S. & WILTSCHKO, D. V. 1989. Alleghanian Orogen. *In*: HATCHER, R. D., JR, THOMAS, W. A. & VIELE, G. W. (eds) *The Appalachian-Ouachita Orogen in the United States*. Geological Society of America, Geology of North America, Boulder, CO, **F-2**, 233–318.

HESLOP, D., DEKKERS, M. J., KRUIVER, P. P. & VAN OORSCHOT, I. H. M. 2002. Analysis of isothermal remanent magnetisation urves using the expectation-maximisation algorithm. *Geophysical Journal International*, **148**, 58–64.

HESLOP, D., McINTOSH, G. & DEKKERS, M. J. 2004. Using time- and temperature-dependent Preisach models to investigate the limitations of modeling isothermal remanent magnetization acquisition curves with cumulative log Gaussian functions. *Geophysical Journal International*, **157**, 55–63.

HIRT, A. M., EVANS, K. F. & ENGELDER, T. 1995. Correlation between magnetic anisotropy and fabric for Devonian shales on the Appalachian Plateau. *Tectonophysics*, **247**, 121–132.

HOUSEN, B. A. & MUSGRAVE, R. J. 1996. Rock-magnetic signature of gas hydrates in accretionary prism sediments. *Earth and Planetary Science Letters*, **139**, 509–519.

JACKSON, M. & SØLHEID, P. 2010. On the quantitative analysis and evaluation of magnetic hysteresis data. *Geochemistry Geophysics Geosystems*, **11**, Q04Z15, doi: 10.1029/2009GC002932.

JACKSON, M., ROCHETTE, P., FILLION, G., BANERJEE, S. & MARVIN, J. 1993. Rock magnetism of remagnetized Paleozoic carbonates; low-temperature behavior and susceptibility characteristics. *Journal of Geophysical Research*, **98**, 6217–6225.

KARS, M., AUBOURG, C. & POZZI, J.-P. 2011. Low temperature magnetic behaviour near 35 K in unmetamorphosed claystones. *Geophysical Journal International*, **186**, 1029–1035, doi: 10.1111/j.1365-246X.2011.05121.x.

KATZ, B., ELMORE, R. D., COGOINI, M., ENGEL, M. H. & FERRY, S. 2000. Associations between burial diagenesis of smectite, chemical remagnetization, and magnetite authigenesis in the Vocontian Trough, SE France. *Journal of Geophysical Research*, **105**, 851–868.

KIRSCHVINK, J. L. 1980. The least-squares line and plane and the analysis of palaeomagnetic data. *Geophysical Journal of the Royal Astronomical Society*, **62**, 699–718.

KRUIVER, P. K., DEKKERS, M. J. & HESLOP, D. 2001. Quantification of magnetic coercivity components by the analysis of acquisition curves of isothermal remanent magnetization. *Earth and Planetary Science Letters*, **189**, 269–276.

LARRASOAÑA, J. C., ROBERTS, A. P., MUSGRAVE, R. J., GRÀCIA, E., PIÑERO, E., VEGA, M. & MARTÍNEZ-RUIZ, F. 2007. Diagenetic formation of greigite and pyrrhotite in has hydrate marine sedimentary systems. *Earth and Planetary Science Letters*, **261**, 350–366.

LOWRIE, W. 1990. Identification of ferromagnetic minerals in a rock by coercivity and unblocking temperature properties. *Geophysical Research Letters*, **17**, 159–162.

MACHEL, H. G. & BURTON, E. A. 1991. Causes and spatial distribution of anomalous magnetization in hydrocarbon seepage environments. *American Association of Petroleum Geologist Bulletin*, **75**, 1864–1876.

MCCABE, C. & ELMORE, R. D. 1989. The occurrence and origin of late Paleozoic remagnetization in the sedimentary rocks of North America. *Reviews of Geophysics*, **27**, 471–494.

MORIN, J. 1950. Magnetic susceptibility of $\alpha$Fe2O3 and $\alpha$Fe2O3 with added titanium. *Physical Review*, **78**, 819–820.

OSBERG, P. H., TULL, J. F., ROBINSON, P., HON, R. & BUTLER, J. R. 1989. The acadian orogen. *In*: HATCHER, R. D. JR, THOMAS, W. A. & VIELE, G. W. (eds) *The Appalachian-Ouachita Orogen in the United States*. Geological Society of America, Geology of North America, Boulder, CO, **F–2**, 153–164.

ÖZDEMIR, Ö. & DUNLOP, D. J. 2010. Hallmarks of maghemitization in low-temperature remanence cycling of partially oxidized magnetite nanoparticles. *Journal of Geophysical Research*, **115**, 02101, doi: 10.1029/2009JB006756.

ÖZDEMIR, Ö., DUNLOP, D. J. & MOSKOWITZ, B. M. 1993. The effect of oxidation on the Verwey transition in magnetite. *Geophysical Research Letters*, **20**, 1671–1674.

PETERS, C. & DEKKERS, M. J. 2003. Selected room temperature magnetic parameters as a function of mineralogy, concentration and grain size. *Physics and Chemistry of the Earth, Parts A/B/C*, **28**, 659–667.

PIERCE, J. W., GOUSSEV, S. A., CHARTERS, R. A., AMBERCROMBIE, H. J. & DEPAOLI, G. R. 1998. Intrasedimentary magnetization by vertical fluid flow and exotic geochemistry. *The Leading Edge*, **17**, 89–92.

PREEDEN, U., PLADO, J., MERTANEN, S. & PUURA, V. 2008. Multiply remagnetized Silurian carbonate sequence in Estonia. *Estonian Journal of Earth Sciences*, **57**, 170–180.

PULLAIAH, G., IRVING, E., BUCHAN, K. L. & DUNLOP, D. J. 1975. Magnetization changes caused by burial and uplift. *Earth and Planetary Science Letters*, **28**, 133–143.

QUINLAN, G. M. & BEAUMONT, C. 1984. Appalachian thrusting, lithospheric flexure, and the Paleozoic stratigraphy of the Eastern Interior of North America. *Canadian Journal of Earth Sciences*, **21**, 973–996.

REPETSKI, J. E., RYDER, R. T., WEARY, D. J., HARRIS, A. G. & TRIPPI, M. H 2008. Thermal maturity patterns (CAI and %Ro) in Upper Ordovician and Devonian rocks of the Appalachian basin: a major revision of USGS Map I-917-E using new subsurface collections. *United States Geological Survey Scientific Investigations Map 3006*, one CD-ROM.

ROBERTS, A. P., CUI, Y. & VEROSUB, K. L. 1995. Waspwaisted hysteresis loops: mineral magnetic characteristics and discrimination of components in mixed magnetic systems. *Journal of Geophysical Research*, **100**, 909–917.

ROCHETTE, P. & FILLION, G. 1989. Field and temperature behavior of remanence in synthetic goethite: paleomagnetic implications. *Geophysical Research Letters*, **16**, 851–854.

ROCHETTE, P., FILLION, G., MATTÉI, J.-L. & DEKKERS, M. J. 1990. Magnetic transition at 30–34 Kelvin in pyrrhotite: insight into a widespread occurrence of this mineral in rocks. *Earth and Planetary Science Letters*, **98**, 319–328.

ROCHETTE, P., FILLION, G. & DEKKERS, M. J. 2011. Interpretation of low-temperature data part 4: the low-temperature magnetic transision of monoclinic pyrrhotite. *The IRM Quarterly*, **21**, 1,7–10.

SALMON, E., EDEL, J. B., PIQUE, A. & WESTPHAL, M. 1988. Possible origins of Permian remagnetizations in Devonian and Carboniferous limestones from the Moroccan Anti-Atlas (Tafilalet) and Meseta. *Physics of the Earth and Planetary Interiors*, **52**, 339–351.

SEVON, W. D. 1985. Nonmarine facies of the Middle and Late Devonian Catskill coastal alluvial plain. *In*: WOODROW, D. L. & SEVON, W. D. (eds) *The Catskill Delta*. Geological Society of America, Boulder, CO, Special Paper, **201**, 79–89.

URBAT, M., DEKKERS, M. J. & KRUMSIEK, K. 2000. Discharge of hydrothermal fluids through sediment at the Escanaba Trough, Gorda Ridge (ODP Leg 169): assessing the effects on the rock magnetic signal. *Earth and Planetary Science Letters*, **176**, 481–494, doi: 10.1016/S0012-821X(00)00024-8.

VAN DER VOO, R. 1993. *Paleomagnetism of the Atlantic, Tethys and Iapetus Oceans*. Cambridge University Press, Cambridge, UK.

VER STRAETEN, C. A. 2010. Lessons from the foreland basin: Northern Appalachian basin perspectives on the Acadian orogeny. *In*: TOLLO, R. P., BARTHOLOMEW, M. J., HIBBARD, J. P. & KARABINOS, P. M. (eds) *From Rodinia to Pangea: the lithotectonic record of the Appalachian region*. Geological Society of America, Memoir, **206**, 251–282.

VERWEY, E. J. 1939. Electronic conduction of magnetite (Fe3O4) and its transition point at low temperature. *Nature*, **144**, 327–328.

WEAVER, R., ROBERTS, A. P. & BARKER, A. J. 2002. A late diagenetic (syn-folding) magnetization carried by pyrrhotite; implications for paleomagnetic studies from magnetic iron sulphide-bearing sediments. *Earth and Planetary Science Letters*, **200**, 371–386.

XU, W., VAN DER VOO, R. & PEACOR, D. R. 1998. Electron microscopic and rock magnetic study of remagnetized Leadville carbonates, central Colorado. *Tectonophysics*, **296**, 333–362.

ZIJDERVELD, J. D. A. 1967. A.C. demagnetization of rocks: analysis of results. *In*: COLLINSON, D. E., CREER, K. M. & RUNCORN, S. K. (eds) *Methods in Paleomagnetism*. Elsevier, Amsterdam, 254–286.

# Index

Page numbers in *italic* denote figures. Page numbers in **bold** denote tables.